Sweeteners

Nutritional Aspects, Applications, and Production Technology

Sweeteners

Nutritional Aspects, Applications, and Production Technology

Edited by
Theodoros Varzakas
Athanasios Labropoulos • Stylianos Anestis

CRC Press is an imprint of the
Taylor & Francis Group, an **informa** business

CRC Press
Taylor & Francis Group
6000 Broken Sound Parkway NW, Suite 300
Boca Raton, FL 33487-2742

First issued in paperback 2016

ISBN 13: 978-1-138-19962-0 (pbk)
ISBN 13: 978-1-4398-7672-5 (hbk)

Library of Congress Cataloging-in-Publication Data

Sweeteners : nutritional aspects, applications, and production technology / edited by Theodoros Varzakas, Athanasios Labropoulos, and Stylianos Anestis.
p. ; cm.
Includes bibliographical references and index.
ISBN 978-1-4398-7672-5 (hardcover : alk. paper)
I. Varzakas, Theodoros. II. Labropoulos, Athanasios. III. Anestis, Stylianos.
[DNLM: 1. Sweetening Agents. 2. Food Technology--methods. 3. Nutritive Value. WA 712]

664'.5--dc23 2012002702

Visit the Taylor & Francis Web site at
http://www.taylorandfrancis.com

and the CRC Press Web site at
http://www.crcpress.com

Contents

List of Figures

List of Tables

Preface

Sweeteners are of major interest today due to their extensive use in product development. From a functional standpoint, they offer a clean taste, and especially in combination, they can provide bulk and improve texture and mouth feel. Furthermore, they have fewer calories than sugars; foods developed with them offer potential benefits for diabetics since they do not cause increases in blood sugar levels and do not promote tooth decay. For example, chocolate may be formulated with polyols to improve flavor, digestibility, and texture.

Other functionality benefits include the enhancement of toffee and mint flavors, prolongation of the sweetness delivery in chewing gum, improvement of the taste and color of milk chocolate, and easy browning because of the Maillard reaction (which may or may not be a limitation).

In addition to their functionality, sweeteners also offer a variety of potential health benefits. They may be used as prebiotics, stimulating the beneficial bacteria in the digestive system; they could be suitable for individuals who are interested in a low-carbohydrate diet, including diabetics since they do not affect glucose levels in the blood; and, as already mentioned, they could be employed in product reformulation for consumers who are interested in cutting calories for weight management or other health reasons.

In attempting to place sweeteners into different discussion areas, the book has been structured as follows. Chapter 1 shows an overview of sweeteners describing general effects, safety, and nutrition. Chapter 2 talks about the chemistry and functional properties of monosaccharides, oligosaccharides, and polysaccharides. Chapter 3 addresses all issues related to sugar polyols including monosaccharide, disaccharide, and polysaccharide polyols. Chapter 4 describes all low-calorie nonnutritive sweeteners and the analytical methodologies for their determination. Chapter 5 addresses all issues related to honey and its products along with its applications in different products. Chapter 6 describes the use of different syrups such as corn syrup, maple syrup, sorghum, brown rice, glucose and fructose syrups, their production, physicochemical aspects, and their applications. Chapter 7 outlines the use of other sweeteners such as raisin juice concentrate, sykin, prune juice and apple juice concentrate, grape juice concentrate, petimezi, isomaltulose, and some syrups. Chapter 8, which discusses the application of sweeteners in food and drink processing, describes how multiple sweeteners could be used to replace sugar content in products without consumers detecting a change in taste. Application areas studied include beverages, confections, chewing gum, dairy products, table sweeteners, baked goods, and pharmaceuticals. Of course, new studies are ongoing delivering improved taste and reduced calorie consumption to consumers along with great economic advantages to food and beverage manufacturers. Blending different sweeteners in diet or light products and in full-sugared products takes advantage of distinctive qualitative synergies.

Some sweeteners, when used in combination, show even higher than expected levels of potency.

Bakery goods, confectionery products, candies, chewing gums, cookies, desserts, juices, and dairy products, as well as energy drinks and different phenomena occurring during the addition of sweeteners such as caramelization, browning, and gelatinization, etc., are well described in this chapter.

Chapter 9 deals with the very important aspect of quality control, production, handling, and storage. Different approved physicochemical methods are described along with sample preparation and specialized methodologies including sensory analysis and adulteration.

The latest reports on the result of an exhaustive evaluation of hundreds of articles related to sweeteners' safety that have been published in the scientific literature since 1988 are reported in Chapter 10, which discusses legislation covering not only aspects of European legislation but also Japanese, U.S., and Indian regulations.

Chapter 11 provides some background information on nutrition and health related to sugar replacement and indicates how sweeteners and sugar alternatives could have an impact on nutrition. However, the key is successful product formulation so that these products are consumed once they are tasted.

Consumer benefits including calorie reduction, dental health benefits, digestive health benefits, and improvements in long-term disease risk through strategies such as dietary glycemic control are mentioned here. Human performance applications are also taken into account.

Chapter 12 is unique in the whole area of sweeteners, giving us some examples of environmental and health concerns from the use of genetically modified (GM) herbicide-tolerant sugar beets or GM high fructose corn syrup, which is extensively used in processing.

Chapter 13 describes inulin and oligofructose as soluble dietary fibers derived from chicory root and promoted for their health benefits, including provision of fiber enrichment, improved health via prebiotics, and calcium absorption and use in the formulation of sugar-reduced foods and beverages. Maltodextrin and polydextrose are also described as bulking and fat-replacing agents.

Finally, Chapter 14 addresses all aspects of risk assessment of sweeteners as additives underlying the principles of safety evaluation along with a risk evaluation of sweeteners permitted for food use in Europe.

It is anticipated that the information contained in this book will have a significant impact not only on consumers but also on manufacturers who will try to produce healthier and safer products with better taste and appeal.

A concerted effort has been made to provide the audience with extensive, current information on a wide variety of bulk and nonnutritive, high-intensity sweeteners and substantial references accompanying it. It is essential to understand the alternative "sweetening" and "bulking agent" choices from a technical, functional, physiological, nutritional, application, safety, and regulatory perspective.

T. Varzakas

Editors

Theodoros Varzakas has a bachelor's degree (Honors) in microbiology and biochemistry (1992), a PhD in food biotechnology, and an MBA in food from Reading University, U.K. (1998). He has also worked as a postdoctoral research staff at the same university. He has worked in large pharmaceutical and multinational food companies in Greece for five years and has also obtained at least 11 years experience in the public sector. Since 2005, he has been serving as an assistant professor in the Department of Food Technology, Technological Educational Institute of Kalamata, Greece, specializing in issues of food technology, food processing, food quality, and food safety.

He is also a reviewer in many international journals such as the *International Journal of Food Science and Technology, Journal of Food Engineering, Waste Management, Critical Reviews in Food Science and Nutrition, Italian Journal of Food Science, Journal of Food Processing and Preservation, Journal of Culinary Science and Technology, Journal of Agricultural and Food Chemistry, Journal of Food Quality* and *Food Chemistry*. He has written more than 60 research papers and reviews and has presented more than 70 papers and posters in national and international conferences. He has written two books in Greek, one on genetically modified food and another on quality control in food.

He participated in many European and national research programs as a coordinator or a scientific member. He is a fellow of the Institute of Food Science and Technology (2007).

Athanasios Labropoulos received a bachelor of science (1964–1969) and pursued postgraduate study in agricultural science and business administration (1970–1972) at Aristotelian University and Industrial School of Business (Macedonian University), Greece, respectively. He received a master of science (1973–1975) in food science and nutrition and pursued post-master's study in food and resource economics at the University of Florida, Gainesville, Florida. Labropoulos received a PhD (1976–1980) in food technology and pursued postdoctoral (1980–1982) work in chemical engineering at Virginia Tech, Blacksburg, Virginia.

Labropoulos has served as a professor (1987–2005) of the Food Technology Department and the Oenology and Beverages Department at the Technological Educational Institution of Athens, Greece. He has contributed to more than 100 journals, magazines, books of abstracts, textbooks, and scientific reports and is referenced in more than 100 citation indices. Labropoulos is a member of various organizations, including Institute of Food Technologists (professional member for 30 years).

Stylianos Anestis received a bachelor's degree in mechanical engineering from the National Technical University of Athens (NTUA) in 1996 and a master's degree in computational fluid mechanics and business administration from NTUA in 2002. He has published many books on food processing, computational fluid mechanics, and computer software programs, all in Greek.

Contributors

Stylianos Anestis
Technological Educational Institute of Athens
Athens, Greece

Charikleia Chranioti
Laboratory of Food Chemistry and Technology
School of Chemical Engineering
National Technical University of Athens
Athens, Greece

Costas Chryssanthopoulos
Department of Physical Education and Sports Science
University of Athens
Athens, Greece

Gülsün Akdemir Evrendilek
Department of Food Engineering
Faculty of Engineering and Architecture
Abant Izzet Baysal Üniversity, Golkoy Campus
Bolu, Turkey

Virginia Giannou
Laboratory of Food Chemistry and Technology
School of Chemical Engineering
National Technical University of Athens
Athens, Greece

Athanasios Labropoulos
Technological Educational Institute of Athens
Athens, Greece

John Christian Larsen
Division of Toxicology and Risk Assessment
National Food Institute
Technical University of Denmark
Søborg, Denmark

Dimitra Lebesi
Laboratory of Food Chemistry and Technology
School of Chemical Engineering
National Technical University of Athens
Athens, Greece

Alicja Mortensen
Division of Toxicology and Risk Assessment
National Food Institute
Technical University of Denmark
Søborg, Denmark

Georgia-Paraskevi Nikoleli
Laboratory of Environmental Chemistry
Department of Chemistry
University of Athens
Athens, Greece

Dimitrios P. Nikolelis
Laboratory of Environmental Chemistry
Department of Chemistry
University of Athens
Athens, Greece

Barbaros Özer
Department of Food Engineering
Abant Izzet Baysal University
Bolu, Turkey

Vasiliki Pletsa
National Hellenic Research Foundation
Institute of Biological Research and Biotechnology
Athens, Greece

Constantina Tzia
Laboratory of Food Chemistry and Technology
School of Chemical Engineering
National Technical University of Athens
Athens, Greece

Theodoros Varzakas
Department of Food Technology
Technological Educational Institute of Kalamata
Kalamata, Greece

Todor Vasiljevic
School of Biomedical and Health Sciences
Victoria University
Melbourne, Victoria, Australia

List of Abbreviations

2-D ELPO	Two-dimensional electrophoresis
ADI	Acceptable daily intake
ADME	Absorption organ and tissue distribution metabolism and excretion
AG	Ammonium glycyrrhizin
AMA	American Medical Association
AMV	Apparent molar volume
AR	Adsorbent resin
ASV	Apparent specific volume
BMD	Benchmark dose
BMI	Body mass index
BP	Boiling point
BP-ANN	Back propagation artificial neural network
BRS	Brown rice syrup
BTC	Bis trichloromethyl carbonate
CBRS	Clarified brown rice syrup
CCC	Calorie control council
CCFA	Codex Committee on Food Additives
CDC	Center for Disease Control and Prevention
CE	Capillary electrophoresis
CITP	Capillary isotachophoresis
CL	Chemiluminescent
CMC	Carboxymethylcellulose
CP-ANN	Counter-propagation artificial neural networks
CWRS	Clarified white rice syrup
CWS	Cold water swelling starch
CZE	Capillary zone electrophoresis
DAUC	Area under the curve
DE	Dextrose equivalent
DIGE	Difference gel electrophoresis
DMF	Dimethylformamide
DP	Degree of polymerization
DP	Polymerization
DSC	Differential scanning calorimetry
DT	Decayed teeth
EC	European Commission
ECL	Electrochemiluminescence
EDTA	Ethylene diamine tetra acetic acid method
EFSA	European Food Safety Authority
ELISA	Enzyme-linked immunosorbent assays
ENGL	European Network of GMO Laboratories
ERF	European Ramazzini Foundation
ERH	Equilibrium relative humidity
EST	Expressed sequence tags
FAAS	Flame atomic absorption spectrometry
FAO	Food and Agriculture Organization
FCC	Food Chemicals Codex
FDA	Food and Drug Administration
FIA	Flow injection analysis

FID	Flame ionization detector
FOS	Fructoligosaccharide
FOSHU	Japanese Foods for Specified Health Use
FP	Freezing point
FSANZ	Food Standards Australia New Zealand
FSDU	Foods for special dietary uses
FT	Filled teeth
FTIR	Fourier transform infrared spectroscopy
GAC	Granular activated carbon
GADA	Grade A dark amber
GALA	Grade A light amber
GAMA	Grade A medium amber
GAPDH	Glyceraldehyde-3-phosphate dehydrogenase
GC	Gas chromatography
GI	Glycemic index
GL	Glycemic load
GMHT	Genetically modified herbicide tolerant
GMOs	Genetically modified organisms
GRAS	Generally recognized as safe
GSL	General state laboratory
H0	Dynamic indentation hardness
HATR	Horizontal attenuated reflectance
HF	High fructose
HFCS	High-fructose corn syrup
HFGS	High-fructose glucose syrup
HMF	Hydroxymethylfurfural
HPLC	High-performance liquid chromatography
HSH	Hydrogenated starch hydrolysates
HTST	High-temperature short time
IC	Immunochemical assays
IC	Ion chromatography
ICP	Induction coupled plasma spectroscopy
ICP OES	Inductively coupled plasma optical emission spectrometry
ICUMSA	International Commission for Uniform Methods of Sugar Analysis
IPA	Water isopropanol
IPRs	Intellectual property rights
JECFA	Joint Expert Committee on Food Additives
KNN	K-nearest neighbors
LC	Liquid chromatography
LCA	Life-cycle assessment
LC-ESI-MS	Liquid chromatography–electrospray mass spectrometry
LDA	Linear discriminant analysis
LLE	Liquid–liquid extraction
LOAEL	Lowest-observed-adverse effect level
LS-SVM	Least square support vector machine
MAG	Mono-ammonium glycyrrhizin
MC	Methylcellulose ethers
MEKC	Micellar electrokinetic capillary chromatography
MGC	Multigrain cookies
MGO	Methylglyoxal

MIR	Middle infrared
MRP	Maillard reaction products
MS	Mass spectroscopy
MS	Mixed sweetener
MUD	Maximum usable dose
MWV	Microwave vacuum
NDCs	Nondigestive carbohydrates
NDOs	Nondigestive oligosaccharides
NHDC	Neohesperidine dihydrochalcone
NIR	Near infrared
NMES	Nonmilk extrinsic sugars
NMP	N-methyl-pyrrolidone
NMR	Nuclear magnetic resonance
NOAEL	No-observed adverse effect level
NOS	Nopaline synthase
NoT	Number of teeth present
OF	Oligofructose
OP	Osmotic pressure
ORS	Organic rice syrups
PAD	Pulsed amperometric detection
PC	Phosphatidylcholine
PCA	Principal component analysis
PCR	Principal component regression
PD	Polydextrose
PKU	Phenylketonuria
PLS	Partial least squares
PNBCl	P-nitrobenzoyl chloride
qPCR	Polymerase chain reaction
RBF	River bank filtration
RDS	Refractometric dry solids
RFLP	Restriction fragment length polymorphism
RI	Refractive index
RJS	Raisin juice syrup
RO	Oligosaccharides
RP	Resistant polysaccharides
RSD%	Relative standard deviation
RT	Reverse transcription
RTE	Ready-to-eat cereals
RVA	Rapid viscoanalyzer
S.P.C.	Standard plate count
SAGE	Serial analysis of gene expression
s-BLMs	Stabilized bilayer lipid membranes
s-BLMs	Surface stabilized bilayer lipid membranes
SCF	Scientific Committee for Foods
SCFA	Short-chain fatty acids
SIRA	Stable isotope ratio analysis
SNIF	Site-specific natural isotope fractionation
SNP	Single nucleotide polymorphism
SPE	Solid-phase extraction
SRM	Single reaction monitoring

SSB Sugar-sweetened beverage
SSCP Single-strand conformation polymorphism
SSR Simple sequence repeat
SVM Support vector machine
T1DM Type 1 diabetes mellitus
TAIL Thermal asymmetric interlaced
TBHQ Tertiary butyl hydroquinone
TBR Tris (22-bipyridyl) ruthenium (II)
TG Triacylglycerol
TLC Thin layer chromatography
TMS Trimethylsilylation
tNOS *Agrobacterium tumefaciens* nopaline synthase terminator
TPA Tri-propylamine
UFW Unfrozen water content
UV Ultraviolet
WHO World Health Organization
WJM Whole juice molasses
WLM Weight loss maintainer group
WRS White rice syrup
WT Wavelet transformation
WTO World Trade Organization

CHAPTER 1

Sweeteners in General

Theodoros Varzakas, Athanasios Labropoulos, and Stylianos Anestis

CONTENTS

1.1 OVERVIEW

Sweeteners are defined as food additives that are used or intended to be used either to impart a sweet taste to food or as a tabletop sweetener. Tabletop sweeteners are products that consist of, or include, any permitted sweeteners and are intended for sale to the ultimate consumer, normally for use as an alternative to sugar. Foods with sweetening properties, such as sugar and honey, are not additives and are excluded from the scope of official regulations. Sweeteners are classified as either high intensity or bulk. High-intensity sweeteners possess a sweet taste, but are noncaloric, provide essentially no bulk to food, have greater sweetness than sugar, and are therefore used at very low levels. On the other hand, bulk sweeteners are generally carbohydrates, providing energy (calories) and bulk to food. These have similar sweetness to sugar and are used at comparable levels.

Sugar and other sweeteners are a major part of many diets. Sugars are carbohydrates. To clarify, carbohydrates are molecules of carbon, hydrogen, and oxygen produced by plants through photosynthesis. The term saccharide is a synonym for carbohydrate; a monosaccharide (mono = 1) is the fundamental unit of carbohydrates. Disaccharides (di = 2) are molecules containing 2 monosaccharide units. Disaccharides and monosaccharides are also known as sugars, simple sugars, or simple carbohydrates. Next are oligosaccharides and polysaccharides. Oligosaccharides are made of 3–9 monosaccharide links. Polysaccharides consist of 10 to thousands of monosaccharide links. A complex carbohydrate refers to many monosaccharide units linked together. Short carbohydrate chains are those under 10 sugar molecules, and long chains are those over 10 sugar molecules.

Diabetics and people looking to reduce calories are often looking for sugar substitutes. There were far more options than would have been imagined. There are still other sugars and sugar substitutes that are not included either because they are obscure and were not found or because they are not commonly found in our diets.

Although sugar (sucrose) is the most common sweetener in food and beverage industries, it is not suitable for all cases of applications, that is, some food and pharmaceutical products. Thus, alternative sweeteners are needed to fulfill that sugar gap by (1) providing choices to functional properties, for example, caloric control and sugar intake; (2) managing weight problems; (3) aiding diabetic concerns; (4) avoiding dental diseases; and (5) assisting in other sweetening aspects such as sugar shortages and sweetening costs.

Sugar alternatives should be colorless, odorless, and heat-stable with a clean and pleasant taste. Also, they should be water-soluble and stable to a wide range of acidic and basic food and beverage applications. Finally, they should be safe (nontoxic, normally metabolized, etc.) and easily produced, handled, and stored.

1.2 SUGARS

Sugar is a carbohydrate found in every fruit and vegetable. All green plants manufacture sugar through photosynthesis, but sugar cane and sugar beets have the highest natural concentrations. Beet sugar and cane sugar—identical products that may be used interchangeably—are the most common sources for the sugar used in the United States. Understanding the variety of sugars available and their functions in food will help consumers determine when sugar can be replaced or combined with nonnutritive sweeteners.

Sugar comes in many forms. The following includes many different sugars, mostly made from sugar cane or sugar beets, such as table sugar, fruit sugar, crystalline fructose, superfine or ultrafine sugar, confectioner's or powdered sugar, coarse sugar, sanding sugar, turbinado sugar, brown sugar, and liquid sugars.

1.3 SWEETENER CATEGORIES

Sweeteners can be categorized as low calories, reduced calories, intense calories, bulk calories, caloric alternative calories, natural sugar–based calories, sugar polyol calories, nonnutritive low calories, nonnutritive calories, nutritive calories, natural calories, syrups, intense sweeteners, and others.

Caloric alternative calories characterize the crystalline fructose, high fructose corn syrup, isomaltulose, and trehalose. *Intense sweeteners* characterize the acesulfame-K, alitame, aspartame, brazzeine, cyclamate, glycyrrhizin, neohesperidine, neotame, saccharin, stevioside, sucralose, and thaumati. *Bulk sweeteners* characterize the crystalline fructose, erythritol, isomalt, isomaltulose, lactitol, malitol, malitol syrup, mannitol, sorbitol, sorbitol syrup, trehalose, xylitol, and crystalline fructose. *Nonnutritive, high-intensity sweeteners* characterize the acesulfame-K, aspartame and neotame, saccharin and cyclamate, and sucralose. *Reduced calorie bulk sweeteners* characterize the erythritol, isomalt, lactitol, maltitol and maltitol syrups, sorbitol and mannitol, tagatose, and xylitol. *Low calorie sweeteners* include acesulfame-K, alitame, aspartame, cyclamate, neohesperidin dihydrochalcone, tagatose, neotame, saccharin, stevioside, sucralose, and less common high-potency sweeteners. *Reduced calorie sweeteners* include erythritol, hydrogenated starch hydrolysates and maltitol syrups, isomalt, maltitol, lactitol, sorbitol and mannitol, and xylitol. *Caloric alternatives* characterize the crystalline fructose, high fructose corn syrup, isomaltulose, and trehalose. *Other sweeteners* characterize the brazzeine, glycyrrhizin, thaumatin, polydextrose, sucrose, polyols, dextrose, fructose, galactose, lactose, maltose, stafidin, glucose, saccharoze, D-tagatose, thaumatin, glycerol, and glycerizim.

Sweeteners that contribute calories to the diet are called caloric or *nutritive* sweeteners. All common caloric sweeteners have the same composition: they contain fructose and glucose in essentially equal proportions (Hanover and White 1993). All caloric sweeteners require processing to produce a food-grade product. Common caloric sweeteners share the same general nutritional characteristics:

- Each has roughly the same composition—equal proportions of the simple sugars fructose and glucose.
- Each offers approximately the same sweetness on a per-gram basis; 1 g (dry basis) of each adds 4 calories to foods and beverages.
- Each is absorbed from the gut at about the same rate.
- Similar ratios of fructose and glucose arrive in the bloodstream after a meal, which are indistinguishable in the body.

Since caloric sweeteners are nutritionally equivalent, they are interchangeable in foods and beverages with no measurable change in metabolism (Widdowson and McCance 1935).

Replacing one caloric sweetener with another provides no change in nutritional value. Removing sweeteners entirely from their commonly used applications and replacing them with high-intensity sweeteners would drastically alter product flavor and sweetness, require the use of chemical preservatives to ensure product quality and freshness, result in a reduction in perceived food quality (bran cereal with caloric sweeteners removed would have the consistency of sawdust), and likely require the addition of bulking agents to provide the expected texture, mouth feel, or volume for most baked goods (White 1992, 2008).

High-intensity sweeteners (also called nonnutritive sweeteners) can offer consumers a way to enjoy the taste of sweetness with little or no energy intake or glycemic response, and they do not support growth of oral cavity microorganisms. Therefore, they are principally aimed at consumers in four areas of food and beverage markets: treatment of obesity, maintenance of body weight, management of diabetes, and prevention and reduction of dental caries. There are several different high-intensity sweeteners. Some of the sweeteners are naturally occurring, whereas others are synthetic (artificial) or semisynthetic. Most of the more commonly available high-intensity sweeteners and/or their metabolites are rapidly absorbed in the gastrointestinal tract. For example, acesulfame-K and saccharin are not metabolized and are excreted unchanged by the kidney. Sucralose, stevioside, and cyclamate undergo degrees of metabolism, and their metabolites are readily excreted. Acesulfame-K, aspartame, and saccharin are permitted as intense sweeteners for use in food virtually worldwide. In order to decrease cost and improve taste quality, high-intensity sweeteners are often used as mixtures of different, synergistically compatible sweeteners.

Bulk sweeteners, defined as those delivered, in solid or liquid form, for use in sweeteners per se or in foods in quantities greater than 22.5 kg, are disaccharides and monosaccharides of plant origin. Sucrose from sugarcane and sugar beet and starch-derived glucose and fructose from maize (corn), potato, wheat, and cassava are the major sweeteners sold in bulk to food and beverage manufacturing industry or packers of small containers for retail sale.

Unrefined sweeteners include all natural, unrefined, or low-processed sweeteners. Sweeteners are usually made with the fruit or sap of plants. But they can also be made from the whole plant or any part of it, and some are also made from starch with the use of enzymes. Sweeteners made by animals, especially insects, are put in their own section as they can come from more than one part of plants.

From sap. The sap of some species is concentrated to make sweeteners, usually through drying or boiling.

- Cane juice, syrup, molasses, and raw sugar, which has many regional and commercial names including demerara, jaggery, muscovado, panela, piloncillo, turbinado sugar, Florida Crystals, and Sucanat, are all made from sugarcane (*Saccharum* spp.).
- Sweet sorghum syrup is made from the sugary juice extracted from the stalks of *Sorghum* spp., especially *S. bicolor* (Nimbkar et al. 2006).
- Mexican or maize sugar can be made by boiling down the juice of green maize stalks.
- Agave syrup is made from the sap of *Agave* spp., including tequila agave (Agave tequilana; Beckley et al. 2007).
- Birch syrup is made from the sap of Birch trees (*Betula* spp.; Heikki et al. 2007).
- Maple syrup, taffy, and sugar are made from the sap of tapped maple trees (*Acer* spp.; Moerman 1998).
- Palm sugar is made by tapping of the flower stalk of various palms to collect the sap. The most important species for this is the Indian date palm (*Phoenix sylvestris*), but other species used include palmyra (*Borassus flabelliformis*), coconut (*Cocos nucifera*), toddy (*Caryota urens*), gomuti (*Arenga saccharifera*), and nipa (*Nypa fruticans*) palms.
- The sweet resin of the sugar pine (*Pinus lambertiana*) was considered by John Muir to be better than maple sugar (Saunders 1976).

From roots. The juice extracted from the tuberous roots of certain plants is, much like sap, concentrated to make sweeteners, usually through drying or boiling.

- Sugar beet syrup (*ZuckerrübenSirup* in German) is made from the tuberous roots of the sugar beet (*Beta vulgaris*; Emery 2003). Sugar beet molasses, a by-product of the processing to make refined sugar, also exists but is mainly used for animal feed (Draycott 2006).
- Yacón syrup is made from the tuberous roots of yacón (*Smallanthus sonchifolius*; Manrique et al. 2005).

From nectar and flowers. A "palatable" brown sugar can be made by boiling down the dew from flowers of the common milkweed (Asclepias syriaca).

From seeds. The starchy seeds of certain plants are transformed into sweeteners by using the enzymes formed during germination or from bacterian cultures. Some sweeteners made with starch are quite refined and made by degrading purified starch with enzymes, such as corn syrup.

- Barley malt syrup is made from germinated barley grains (Roehl 1996).
- Brown rice malt syrup is made from rice grains cooked and then cultured with malt enzymes (Belleme and Belleme 2007).
- Amazake is made from rice fermented with *Koji* (*Aspergillus oryzae*; Belleme and Belleme 2007).

From fruits. Many fresh fruits, dried fruits, and fruit juices are used as sweeteners. Some examples are as follows:

- Watermelon sugar, which is made by boiling the juice of ripe watermelons.
- Pumpkin sugar, which is made by grating the pumpkins, in the same manner as in making beet sugar (Hovey 1841).
- Dates, date paste, spread, syrup ("dibs"), or powder (date sugar) is made from the fruit of the date palm (*Phoenix dactylifera*).
- Jallab is made by combining dates, grape molasses, and rose water.
- Pekmez is made of grapes, fig (*Ficus carica*), and mulberry (*Morus* spp.) juices, condensed by boiling with coagulant agents.

A variety of molasses are made with fruit: carob molasses is made from the pulp of the Carob tree's fruit.

From leaves. In a few species of plants, the leaves are sweet and can be used as sweeteners.

- *Stevia* spp. can be used whole or dried and powdered to sweeten food or drink (Kinghorn 2002).
- Jiaogulan (*Gynostemma pentaphyllum*) has sweet leaves, although not as sweet as *Stevia*.

By animals:

- True honey, which is made by honey bees (*Apis* spp.) from gathered nectar.
- Sugarbag, the honey of stingless bees, which is more liquid than the honey from honey bees (Menzel and D'Aluisio 1998).

Artificial sweeteners, or sugar substitutes, are food additives that impart a sweet flavor but no nutritional value. Commonly used in sugar-free and reduced-calorie foods, artificial sugar substitutes vary in taste, level of sweetness, and stability when heated. For many diabetics and prediabetics, artificial sugars play an important role in blood sugar control and weight management.

1.4 ADDED SWEETENERS

Over the past 20 years or so, Americans have developed quite the sweet tooth, with an annual consumption of sugar at about 100 lbs per person. During these same years, many more Americans—particularly children—have become overweight and obese. Added caloric sweeteners may be one of the major reasons.

1.5 SUCROSE AND FRUCTOSE

Sucrose, or table sugar, has been the most common food sweetener. In the late 1960s, a new method was introduced that converts glucose in corn syrup to fructose. High-fructose corn syrup is as sweet as sucrose but less expensive, so soft-drink manufacturers switched over to using it in the mid-1980s. Now it has surpassed sucrose as the main added sweetener in the American diet.

Fructose once seemed like one of nutrition's good guys—it has a very low glycemic index. The glycemic index is a way of measuring how much of an effect a food or drink has on blood sugar levels; low glycemic index foods are generally better for consumers.

However, fructose, at least in large quantities, may have some serious drawbacks. Fructose is metabolized almost exclusively in the liver. It is more likely to result in the creation of fats, which increase the risk for heart disease. Moreover, recent work has shown that fructose may have an influence on the appetite hormones. High levels of fructose may blunt sensations of fullness and could lead to overeating.

1.6 FRUIT-JUICE CONCENTRATES: JUST EMPTY CALORIES

Fruit juices such as apple or white grape juice in concentrated form are widely used sweeteners. They are used to replace fats in low-fat products because they retain water and provide bulk, which improve the appearance and "mouth feel" of the food.

Although they may seem healthier and more natural than high-fructose corn syrup, fruit-juice concentrates also have high levels of fructose. Fruit-juice concentrates are another way that empty calories get into our diets.

1.7 SUGAR ALCOHOLS

In sweetening power, the sugar alcohols are closer to sucrose and fructose than to super-sweet artificial sweeteners. They do not affect blood-sugar levels as much as sucrose, a real advantage for people with diabetes, and they do not contribute to tooth decay. Sugar alcohols are used in candies, baked goods, ice creams, and fruit spreads. Reading the ingredients carefully, you will spot them in toothpaste, mouthwash, breath mints, cough syrup, and throat lozenges.

1.8 SWEETENED BEVERAGES

Sweeteners added to sports and juice drinks are particularly troubling because many people think those drinks are relatively healthful.

Researchers are beginning to document the adverse health outcomes. Harvard researchers reported that women who drank one or more sugar-sweetened soft drinks per day were 83% more likely to develop type 2 diabetes than women who drank less than one a month. Not surprisingly, the former were also more likely to gain weight.

When children regularly consume beverages that are sweetened, they are getting used to a level of sweetness that could affect their habits for a lifetime. A 2004 editorial in the *Journal of the American Medical Association* said that reducing the consumption of sugar-sweetened beverages "may be the best single opportunity to curb the obesity epidemic."

1.9 ARTIFICIAL SWEETENERS

Artificial sweeteners sing a siren song of calorie-free and, therefore, guilt-free sweetness. Those approved by the Food and Drug Administration (FDA) include acesulfame-K (Sunett), aspartame (NutraSweet, Equal), neotame, saccharin (Sweet 'N Low, others), and sucralose (Splenda). All are intensely sweet.

There is a cyberspace cottage industry dedicated to condemning the artificial sweeteners, especially aspartame. Some fears are based on animal experiments using doses many times greater than any person would consume. But even some mainstream experts remain wary of artificial sweeteners, partly because of the lack of long-term studies in humans.

Even if safety were not an issue, artificial sweeteners might still be a problem because they may set people (especially children) up for bad eating habits by encouraging a craving for sweetness that makes eating a balanced diet difficult (see http://www.health.harvard.edu/fhg/updates/Added-sweeteners.shtml).

1.10 NEW SWEETENERS STUDY SHOWS NO LINK WITH CANCER

A study of more than 16,000 patients has found no link between sweetener intake and the risk of cancer. This supports a previous ruling by the European Food Safety Authority (EFSA).

The safety of artificial sweeteners has been under scrutiny since the 1970s, when animal studies reported links with some forms of cancer. The studies were criticized because very high doses of sweeteners were used. More recent research in rats found that sweetener intakes similar to those consumed by humans could increase the risk of certain types of cancer. These findings were not replicated in studies of humans. After evaluating these and other studies in 2006, EFSA concluded that no further safety reviews of aspartame were needed and that the acceptable daily intake (ADI) of 40 mg/kg body weight should remain.

A new review, published in the *Annals of Oncology*, looked at the safety of a number of common sweeteners, particularly saccharin and aspartame. Italian case-control studies conducted over a 13-year period were brought together and checked for associations between sweetener consumption and the risk of developing cancer. Patients with various types of cancers formed the "test group," including those with colon, rectal, oral, and breast cancers. The "control group" comprised 7000 patients admitted to hospital for reasons other than cancer.

Dietary assessments were used to compare the intake of sweeteners in each group. No significant differences were found. When individual cancers were considered, it was found that women with breast or ovarian cancer tended to consume fewer sweeteners than controls. In the case of laryngeal cancer, a direct relationship was found between risk and total sweetener intake, although the sample size was relatively small.

The authors concluded that consumption of saccharin, aspartame, and other sweeteners did not appear to increase the risk of cancer. Average sweetener intake in Italy is lower than in other European countries, and little data were available for individual sweeteners, or the use of "Diet" drinks. Despite these shortcomings, the study nevertheless makes an important contribution to the debate. For more information, see Gallus et al. (2007) and http://www.eufic.org/page/el/show/latest-science-news/fftid/sweeteners-cancer/.

Historically, honey and maple syrup have been used to replace sugar.

Pure cornstarch is by far the biggest source of other carbohydrate sweeteners used by today's food manufacturers. Cornstarch is split into a variety of smaller fragments (called dextrins) with acid or enzymes. The smaller fragments are then converted into various cornstarch sweeteners used by today's food manufacturers.

Hydrolysis is the term used to describe the overall process where starch is converted into various sweeteners. Sweetener products made by cornstarch hydrolysis include dextrose, corn syrup, corn syrup solids, maltodextrin, high fructose corn syrup, and crystalline fructose.

A juice concentrate is the syrup produced after water, fiber, and nutrients are removed from the original fruit juice (see http://www.sugar.org/other-sweeteners/other-caloric-sweeteners.html).

1.11 SAFETY OF LOW-CALORIE SWEETENERS

All low-calorie sweeteners are subject to comprehensive safety evaluation by regulatory authorities; any unresolved issues at the time of application have to be investigated before they are approved for use in the human diet. Definitive independent information on the safety of sweeteners can be obtained from the Websites of the EFSA (http://www.efsa.europa.eu/EFSA/efsa_locale-1178620753812_home.htm), the European Scientific Committee on Food (http://ec.europa.eu/food/fs/sc/scf/reports_en.html), and Joint WHO/FAO Expert Committee on Food Additives (http://www.who.int/ipcs/food/jecfa/en/).

The safety testing of food additives involves *in vitro* investigations, to detect possible actions on DNA, and *in vivo* studies in animals, to determine what effects the compound is capable of producing when administered at high doses, or high dietary concentrations, every day.

The daily dose levels are increased until either some adverse effect is produced or 5% of the animal's diet has been replaced by the compound. The dose levels are usually very high because a primary purpose of animal studies is to find out what effects the compound can produce on the body irrespective of the dose level (hazard identification). The dose–response data are analyzed to determine the most sensitive effect (the so-called critical effect). The highest level of intake that does not produce the critical effect, that is, the no-observed adverse effect level (NOAEL), is used to establish a human intake with negligible risk, which is called the ADI. The NOAEL is normally derived from chronic (long-term) studies in rodents. The ADI is usually calculated as the NOAEL (in mg/kg body weight per day) divided by a 100-fold uncertainty factor, which allows for possible species differences and human variability (Renwick et al. 2003; Renwick 2009).

Low-calorie sweeteners are often added to foods as mixtures or blends because mixtures can provide an improved taste profile, and in some cases, the combination is sweeter than predicted from the amounts present. The only property that is common to all low-calorie sweeteners is their activity at the sweet-taste receptor. They do not share similar metabolic fates or high-dose effects.

Therefore, no interactions would arise if different low-calorie sweeteners are consumed together in a blend (Groten et al. 2000), and each sweetener would be as safe as if it were consumed alone.

REFERENCES

Beckley, J.H., J. Huang, E. Topp, M. Foley, and W. Prinyawiwatkul. 2007. *Accelerating New Food Product Design and Development*. Blackwell Publishing, San Francisco, USA, 36. ISBN 081380809X. Retrieved on May 13, 2011.

Belleme, J., and J. Belleme. 2007. *Japanese Foods That Heal*. Tuttle Publishing, Periplus Editions (HK) Ltd, California, USA, 55–58. ISBN 0804835942. Retrieved on May 13, 2011.

Draycott, P.A. 2006. *Sugar Beet*. Blackwell Publishing, San Francisco, USA, 451. ISBN 140511911X. Retrieved on May 13, 2011.

Emery, C. 2003. *The Encyclopedia of Country Living, An Old Fashioned Recipe Book*. Sasquatch Books, San Francisco, 313. ISBN 157061377X. Retrieved on May 13, 2011.

Gallus, S., L. Scotti, E. Negri, R. Talamini, S. Franceschi, M. Montella, A. Giacosa, L. Dal Maso, and C. La Vecchia. 2007. Artificial sweeteners and cancer risk in a network of case-control studies. *Ann. Oncol.*, 18:40–44.

Groten, J.P., W. Butler, V.J. Feron, G. Kozianowski, A.G. Renwick, and R. Walker. 2000. An analysis of the possibility for health implications of joint actions and interactions between food additives. *Regul. Toxicol. Pharmacol.*, 31:77–91.

Hanover, L.M., and J.S. White. 1993. Manufacturing, composition, and applications of fructose. *Am. J. Clin. Nutr.*, 58(suppl 5):724S–732S.

Heikki, K., T. Teerinen, S. Ahtonen, M. Suihko, and R.R. Linko. 2007. Composition and properties of birch syrup *(Betula pubescens)*. *J. Agric. Food Chem.*, 37:51–54.

Hovey, M.C. 1841. *The Magazine of Horticulture, Botany, and All Useful Discoveries*. Hovey and Co., Layton, Utah, USA. Retrieved on May 11, 2011.

Kinghorn, A.D. 2002. *Stevia: The Genus Stevia*. CRC Press. ISBN 0415268303.

Manrique, I., A. Párraga, and M. Hermann. 2005. Yacon syrup: Principles and processing (PDF). Series: Conservación y uso de la biodiversidad de raíces y tubérculos andinos: Una década de investigación para el desarrollo (1993–2003). 8B: 31p.

Menzel, P., and F. D'Aluisio. 1998. *Man Eating Bugs: The Art and Science of Eating Insects*. Ten Speed Press, Berkeley, California. ISBN 1580080227. Retrieved on June 2, 2011.

Moerman, D.E. 1998. *Native American Ethnobotany*. Timber Press, Portland, Oregon, 38–41. ISBN 0881924539. Retrieved on May 14, 2011.

Nimbkar, N., N.M. Kolekar, J.H. Akade, and A.K. Rajvanshi. 2006. Syrup production from sweet sorghum (PDF). Nimbkar Agricultural Research Institute (NARI), Phaltan, pp. 1–10.

Renwick, A.G. 2009. 10th Panhellenic Congress on Nutrition, Nov. 15, 2009, Athens, Greece. Sweetener risk roadshow booklet.

Renwick, A.G., S.M. Barlow, I. Hertz-Picciotto, A.R. Boobis, E. Dybing, L. Edler, G. Eisenbrand, J.B. Greig, J. Kleiner, J. Lambe, D.J.G. Muller, M.R. Smith, A. Tritscher, S. Tuijtelaars, P.A. van den Brandt, R. Walker, and R. Kroes. 2003. Risk characterization of chemicals in food and diet. *Food Chem. Toxicol.*, 41:1211–1271.

Roehl, E. 1996. *Whole Food Facts: The Complete Reference Guide*. Inner Traditions/Bear & Company, Rochester, Vermont, 134–135. ISBN 089281635X.

Saunders, C.F. 1976. *Edible and Useful Wild Plants of the United States and Canada*. Courier Dover Publications, New York, USA, 219. ISBN 0486233103.

White, J.S. 1992. Fructose syrup: Production, properties and applications, In *Starch Hydrolysis Products—Worldwide Technology, Production, and Applications*, F.W. Schenck and R.E. Hebeda (eds.), VCH Publishers, Inc., New York, USA. pp. 177–200.

White, J.S. 2008. Straight talk about high-fructose corn syrup: What it is and what it ain't. *Am. J. Clin. Nutr.*, 88(6):1716S–1721S.

Widdowson, E.M. and R.A. McCance. 1935. The available carbohydrate of fruits: Determination of glucose, fructose, sucrose and starch. *Biochem. J.*, 29(1):151–156.

INTERNET SOURCES

http://ec.europa.eu/food/fs/sc/scf/reports_ en.html, accessed July 2011.

http://www.efsa.europa.eu/EFSA/efsa_locale-1178620753812_home.htm, accessed August 2011.

http://www.who.int/ipcs/food/jecfa/en/, accessed July 2011.

htttp://www.health.harvard.edu/fhg/updates/Addedsweeteners.shtml, accessed July 2011.

http://www.eufic.org/page/el/show/latest-science-news/fftid/sweeteners-cancer/, accessed July 2011.

http://www.sugar.org/other-sweeteners/other-caloric-sweeteners.html, accessed July 2011.

CHAPTER 2

Chemistry and Functional Properties of Carbohydrates and Sugars (Monosaccharides, Disaccharides, and Polysaccharides)

Constantina Tzia, Virginia Giannou, Dimitra Lebesi, and Charikleia Chranioti

CONTENTS

2.1 INTRODUCTION

Carbohydrates are widely distributed in nature and are considered important food components, as they are a source of energy for the organism, structural materials (starch, glycogen, cellulose), and taste and flavor factors as well. Carbohydrates are formed in plants through photosynthesis (Equation 2.1) and can be used as a saving material in the form of starch. In animal tissues/organisms, carbohydrates are respectively found in the form of glycogen.

$$6CO_2 + 6H_2O \xrightarrow{h\gamma} C_6H_{12}O_6 + 6O_2 \tag{2.1}$$

The term "carbohydrate" was initially used, as the first known class of compounds from this category followed the general formula $C_n(H_2O)_n$; these compounds were derived from hydrated carbon atoms with hydrogen and oxygen contained in the same proportion as in water. The term "carbohydrate" is still used even though known compounds such as acetic acid ($C_2H_4O_2$) or lactic acid ($C_3H_6O_3$) follow the general formula of carbohydrates without having the same features.

Carbohydrates from the chemical point of view are aldehyde or ketone derivatives of polyvalent alcohols, namely, polyhydroxy aldehydes or polyhydroxy ketones, or condensation products thereof. The term "sugars" is commonly used for the simple members of the group, which are normally soluble in water and exhibit different properties from the senior members, for example, complex carbohydrates are relatively insoluble in water. Carbohydrates can be converted into sugars by hydrolysis by means of acids or enzymes (Southgate 1991; Robyt 1998; Izydorczyk 2005).

2.2 CHEMISTRY OF CARBOHYDRATES—CATEGORIES AND PROPERTIES

Carbohydrates can be divided into the following groups:

1. Monosaccharides or simple sugars that are the simplest members of carbohydrates and cannot be subjected to hydrolysis.
2. Oligosaccharides consisting of a small number of monosaccharides (usually two to 10 molecules), with disaccharides being the most interesting for foods.
3. Polysaccharides that consist of a large number of monosaccharides, for example, amylose contains 100–2000 units of the monosaccharide glucose in its molecule. The macromolecular polysaccharide compounds present variations in their physical and chemical properties compared with the monosaccharides from which they derived. The most important polysaccharides in nature are starch (energy storage in plants), glycogen (energy storage in animals), and cellulose (supporting material and structural component in plants) (BeMiller and Whistler 1996; BeMiller 2010).

2.3 MONOSACCHARIDES OR SUGARS

Monosaccharides or sugars are classified according either to the number of carbon atoms or the aldehyde or keto group they contain. Thus

1. Molecules containing 3, 4, 5, or 6 carbon atoms in their molecule can be distinguished into trioses, tetroses, pentoses (fouranozes), and hexoses (pyranoses), respectively, of which hexoses present the greatest interest in foods followed by pentoses.
2. Molecules containing an aldehyde or keto group are classified as aldoses or ketoses, respectively (Stick 2001; Belitz et al. 2009).

Numbering of carbon atoms in sugar molecules according to Emil Herman Fischer (1852 to 1919) starts from the closest to the reducing group carbon atom. Trioses are the most simple sugars, glyceraldehyde ($C_3H_6O_3$) is the simplest from the common aldoses, whereas dihydroxyacetone is the simplest from the ketoses.

Asymmetric carbon atoms (carbon atoms that are attached to four different types of atoms or four different groups of atoms) in sugar molecules lead to the presence of stereoisomer forms. Monosaccharides and disaccharides have the ability to rotate polarized light at different directions. Stereoisomer molecules are known as D or L isomers (enantiomers), whereas with regards to the direction of the rotation, they are symbolized as (+)(d) if it is clockwise or (–)(l) if it is counterclockwise.

The characterization of monosaccharide isomers is based on the glyceraldehyde molecule, which has an asymmetric carbon atom (C_2) with two isomers (D or L) that represent two enantiomers (R or S according to the priorities decrease in a clockwise or counterclockwise fashion), as shown in Figure 2.1.

Therefore, using glyceraldehyde as the point of reference, monosaccharides belong to the D- or L-series if the highest numbered asymmetric carbon atom has the same configuration with the middle carbon atom (chiral center) of D- or L-glyceraldehyde, respectively. Most sugars occurring in nature are in the D-form, and they belong to the R form. L-sugars occur rarely in nature; the most important of them is L-arabinose. Some L-sugars are of great interest as they can be used as alternative sweeteners.

The direction of polarized light rotation is characteristic for each sugar and is independent from the D or L enantiomerism (e.g., D-fructose is counterclockwise). The specific optical rotation $[\alpha]_D^{20}$ refers to 20°C and to monochromatic light D of sodium and can be determined with a polarimeter that corresponds to the angle of rotation caused by passing polarized light through a sugar solution of 1 g/mL with a path length of 1 dm. The specific optical rotation is equal to

$$[\alpha]_D^{20} = \frac{\alpha}{l \cdot c}$$

where l is the path length (in decimeters), and c is the concentration of the sugar solution (in grams per milliliter) used for identification and quantification (Robyt 1998; Coultate 2002; Izydorczyk 2005).

Sugars important for foods are considered: xylose from pentoses (nonfermentable sugars) and D-glucose, D-galactose, D-mannose (aldoses), and D-fructose (ketose) from hexoses (fermentable sugars). The most important of all sugars is glucose, which is found at the highest concentration in foods as free or bound (disaccharides, polysaccharides, and glycosides).

Since certain reactions could not be justified using the linear form of sugar molecules, it was concluded that the structure of simple sugars can also be cyclic through the formation of an intermediate hemiacetal form (Kuszmann 2006; Horton 2008). The circular forms of sugars are given by the Haworth projection or by Chair/Boat configuration, as shown for D-glucose (Figure 2.2).

The most important D-aldoses are presented in Figure 2.3.

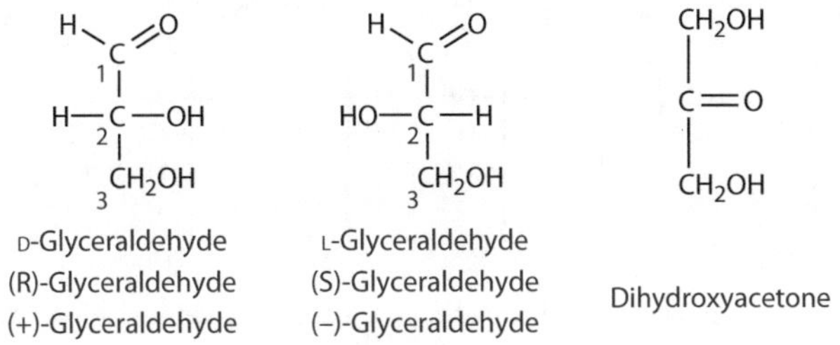

D-Glyceraldehyde
(R)-Glyceraldehyde
(+)-Glyceraldehyde

L-Glyceraldehyde
(S)-Glyceraldehyde
(–)-Glyceraldehyde

Dihydroxyacetone

Figure 2.1 Fischer projections of glyceraldehyde and dihydroxyacetone.

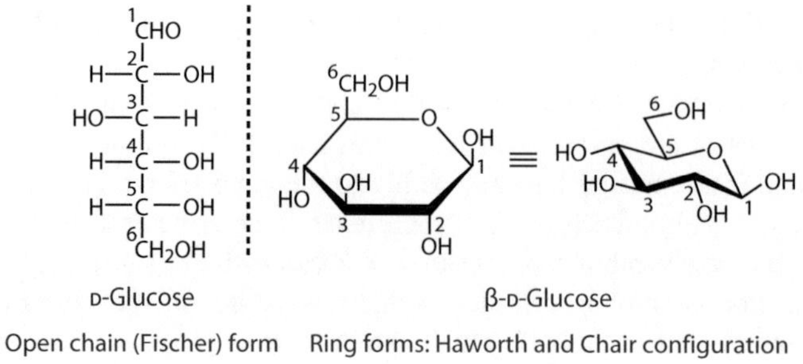

D-Glucose β-D-Glucose

Open chain (Fischer) form Ring forms: Haworth and Chair configuration

Figure 2.2 Open chain (Fischer projection) and ring forms (Haworth and Chair configuration) of D-glucose.

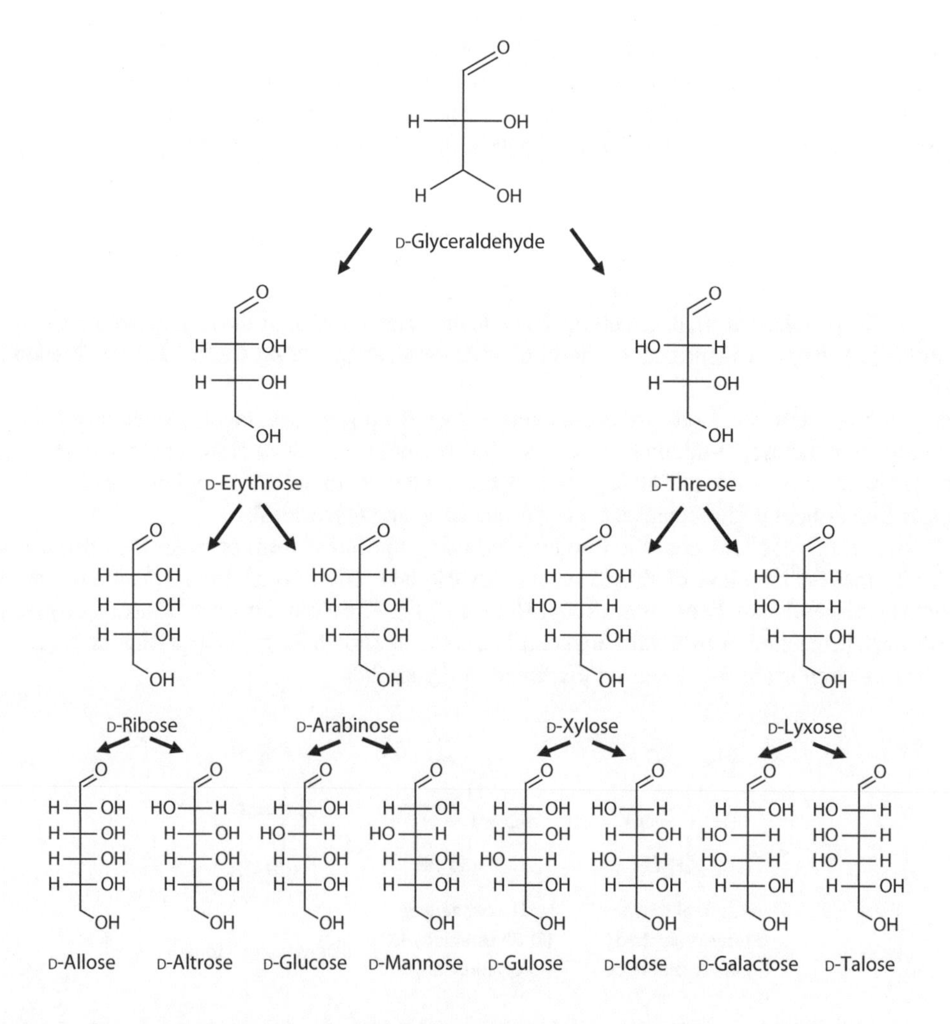

Figure 2.3 Structure and stereochemical relationships of D-aldoses having three to six carbons.

2.3.1 Properties of Monosaccharides or Sugars

Physical properties: Sugars are solid substances, sometimes difficult to crystallize, which often form supersaturated syrup solutions. Many of them have a sweet taste, whereas others are tasteless or even bitter. Simple sugars are soluble in water, less soluble in methanol, ethanol, and pyridine, and insoluble in lipophilic solvents such as ethylether, chloroform, benzene, and petroleum ether (Horton 2008; Belitz et al. 2009).

Mutarotation and transformations: In sugars, and particularly in aldohexoses, the open (linear) chain form cannot sufficiently outline the aldehyde character of these compounds, as previously mentioned. More specifically, the fact that certain reactions of aldehydes cannot proceed leads to the hypothesis that these molecules should appear in a cyclo-hemiacetal form. Using this assumption, C_1 in aldoses and C_2 in ketoses become asymmetric and thus may appear in two cyclic isomeric forms (anomers) α and β according to the position of the OH- in the anomeric carbon. The occurring forms for glucose and fructose are presented in Figure 2.4.

The cyclic forms of sugars have a ring with 5 (known as furanoses: with a ring of four carbon atoms and one oxygen atom) or 6 (known as pyranoses: with a ring of five carbon atoms and one oxygen atom) atoms. The pyranose structure is the one that occurs the most in nature due to its stability. The furanoses are often found in ketoses joined with other sugar molecules, that is, fructose (joined with glucose) in the sucrose molecule. The cyclic forms of sugars in aqueous solutions undergo ring opening to the acyclic aldehyde form, which, in turn, is converted to a mixture of cyclic forms, as described in Figure 2.5.

In the equilibrium mixture, the acyclic form is found in a small proportion, whereas the ratio of the cyclic forms varies from sugar to sugar. This phenomenon results in the transition of $[\alpha]_D^{20}$ with time during polarization of a sugar solution. The above are more clearly explained below in the case of glucose and fructose.

D-glucose is isolated in two forms that are mirror isomer to the C_1 atom. The forms that can be isolated are 1) α-D-(+)-glucose (melting point = 146°C and $[\alpha]_D^{20} = -113°$ in a fresh aqueous solution), which can be crystallized from an aqueous solution, and 2) β-D-(–)-glucose (melting point = 150°C and $[\alpha]_D^{20} = +18.7°$ in a fresh aqueous solution), which can be crystallized from a pyridine solution.

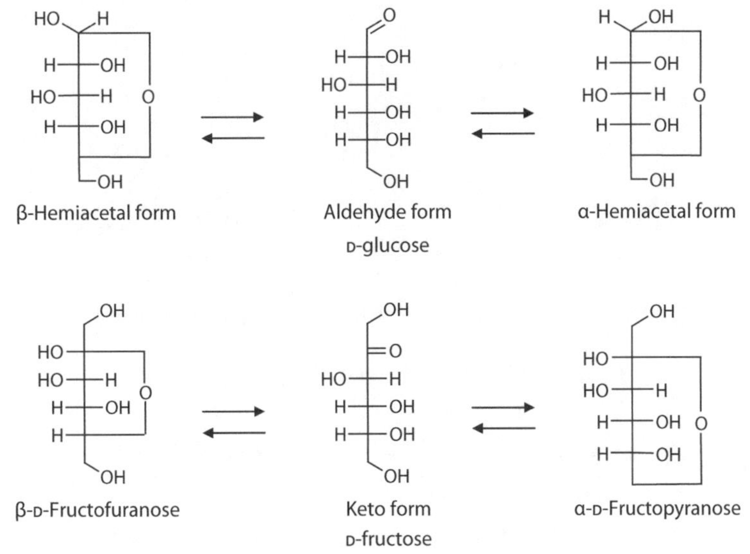

β-Hemiacetal form — Aldehyde form — α-Hemiacetal form

D-glucose

β-D-Fructofuranose — Keto form — α-D-Fructopyranose

D-fructose

Figure 2.4 Isomeric forms of D-glucose and D-fructose.

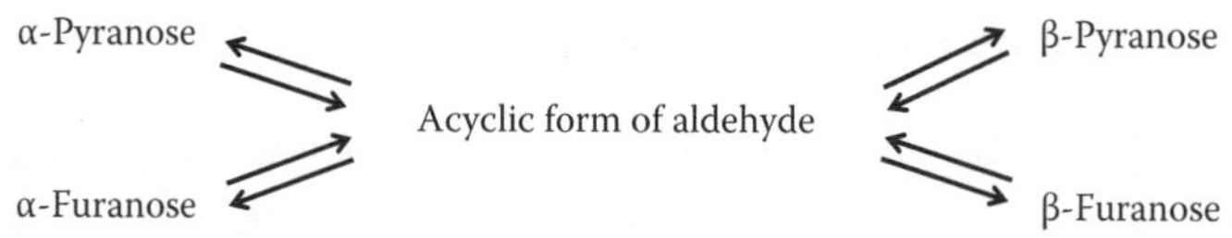

α-Pyranose ⇌ Acyclic form of aldehyde ⇌ β-Pyranose

α-Furanose ⇌ Acyclic form of aldehyde ⇌ β-Furanose

Figure 2.5 Conversion from acyclic to cyclic forms of sugars.

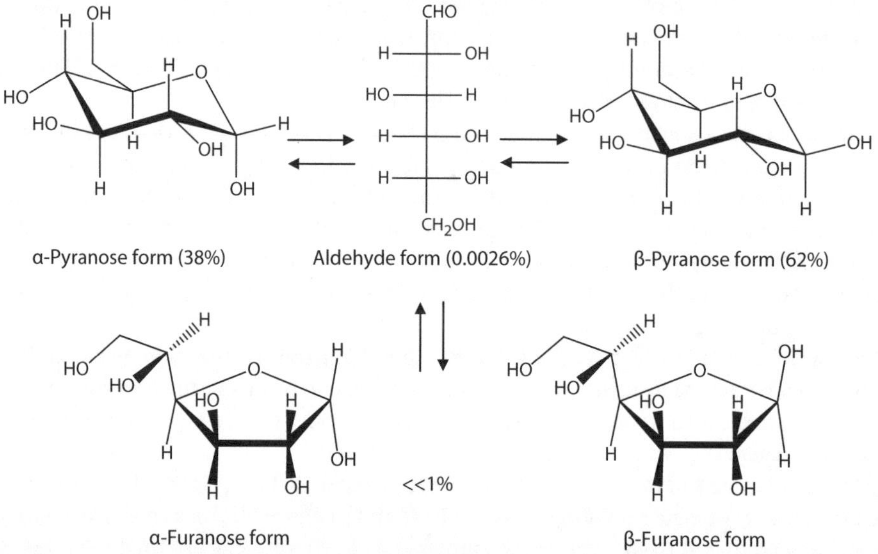

α-Pyranose form (38%) ⇌ Aldehyde form (0.0026%) ⇌ β-Pyranose form (62%)

α-Furanose form ⇌ Aldehyde form ⇌ β-Furanose form (<<1%)

Figure 2.6 Isomer forms of glucose in an aqueous solution at room temperature.

α-Pyranose form (traces) ⇌ Keto form ⇌ β-Pyranose form (75%)

α-Furanose form (4%) ⇌ Keto form ⇌ β-Furanose form (21%)

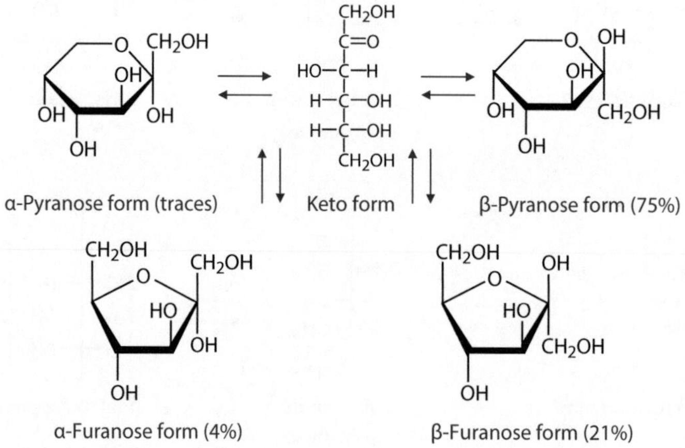

Figure 2.7 Isomer forms of fructose in an aqueous solution at room temperature.

Although recently prepared aqueous solutions of glucose present the above specific rotation ($[\alpha]_D^{20}$) values, after some time, the solutions of both forms tend to reach the same threshold $[\alpha]_D^{20} = +52.5°$. In the equilibrium state, the two forms of glucose (α and β) are encountered at rates of 38% and 62%, respectively. The existence of these two forms of glucose can explain the mutarotation of its aqueous solutions, as during sustention of the solution, the one form converts to the other until the equilibrium is attained (Pigman and Anet 1972; Shallenberger and Birch 1975; Coultate 2002; Izydorczyk 2005; Kuszmann 2006). The tautomer/isomer forms of glucose (α form has the OH- of the anomeric carbon in axial position and β form in equatorial position, respectively) in an aqueous solution at room temperature are presented in Figure 2.6.

In fructose, the equilibrium of the cyclic forms is reached at the threshold of $[\alpha]_D^{20} = -92°$, and the respective tautomers/isomers at the equilibrium are presented in Figure 2.7.

2.3.2 Monosaccharides' Reactions

Oxidation: By means of mild oxidizing agents (Br_2 in water, dilute nitric acid), aldoses can be converted to the corresponding acids (aldehyde group to carboxylic acid group forming aldonic acids), whereas ketoses under common conditions do not react. Based on this reactivity, it is easy to distinguish aldoses from ketones. By means of strong oxidizing agents (concentrated nitric acid), the two terminal groups of the molecule are oxidized, and dibasic acids (aldaric acids) are formed. By protecting the aldehyde group, the molecule is oxidized at the terminal group to uronic acids. By using very strong oxidizing agents as HIO4, sugar molecules are broken down, and several decomposition products (aldehydes/ketones and formic acid) are produced. The potential oxidation reactions of glucose, as shown below (Figure 2.8), lead to gluconic acid, glucaric acid (saccharic acid), and glucuronic acid. The oxidation reaction to saccharic acids is commonly used for quantitative determination of sugars; Fehling's solution is then used, and Cu(II) ions contained are reduced to Cu(I) forming a copper oxide red precipitate (Multon 1997; Stick 2001; Tomasik 2002; Stick and Williams 2009).

Reduction: Both aldoses and ketoses with reducing agents, that is, by $NaBH_4$ or by catalytic hydrogenation (H_2/Ni), can be reduced at the carbonyl group to polyvalent alcohols (alditols). These polyols generally have a sweet taste and are used as alternative sweeteners, such as pentitols from pentoses and hexitols from hexoses (BeMiller and Whistler 1996; Stick 2001; Izydorczyk 2005). Specifically D-glucose and the resulting D-sorbitol are presented as follows:

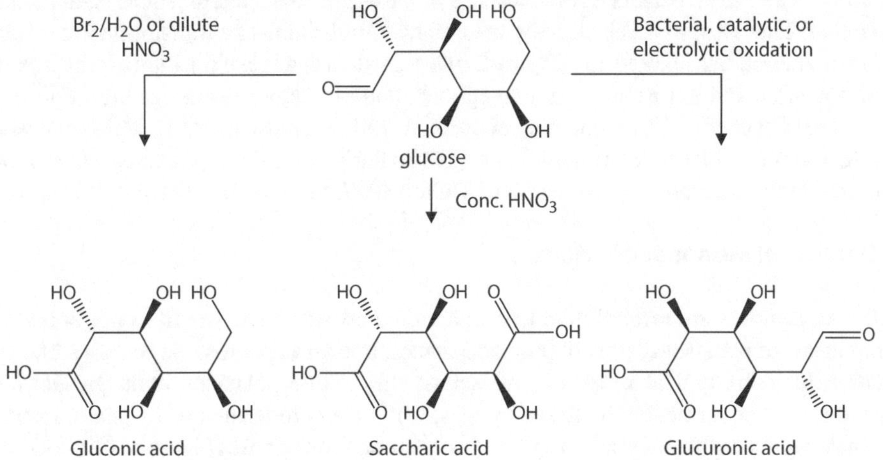

Br_2/H_2O or dilute HNO_3

HO OH HO O HO OH glucose

Bacterial, catalytic, or electrolytic oxidation

Conc. HNO_3

HO OH HO HO O HO OH Gluconic acid

HO OH O OH HO O HO OH Saccharic acid

HO OH O HO O HO OH Glucuronic acid

Figure 2.8 Oxidation products of glucose.

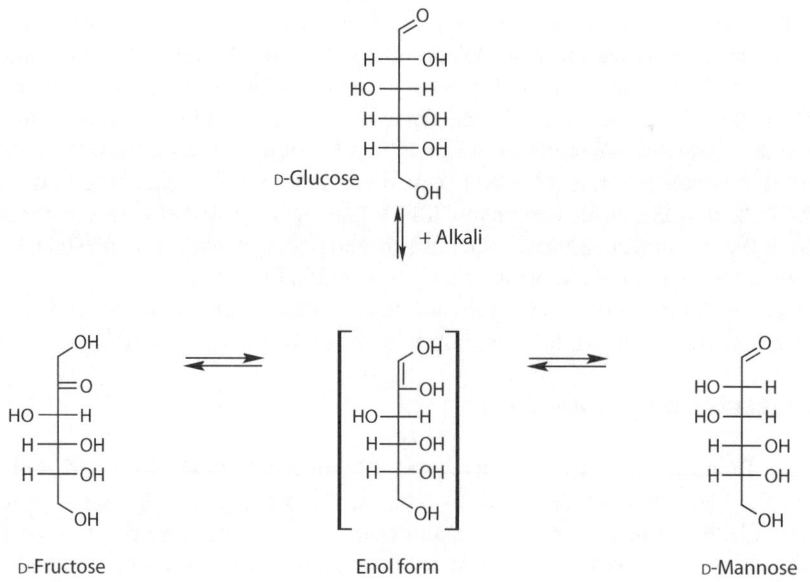

Figure 2.9 Enol forms of glucose.

$$\underset{\text{D-glucose}}{CH_2OH(CHOH)_4CHO} \rightarrow \underset{\text{D-sorbitol}}{CH_2OH(CHOH)_4CH_2OH} \tag{2.2}$$

Reactions with acids: Although monosaccharides do not react with weak acids even at mild temperatures, or react with strong inorganic acids sugars with slow rates, these reactions are accelerated by heat. Monosaccharides heated with hydrochloric or sulfuric acid form disaccharides and higher saccharides by intermolecular dehydration, whereas by further endomolecular dehydration, they form furfural and hydroxymethylfurfural from pentoses and hexoses, respectively. These compounds are condensed with α-naphthol (Molisch reaction) forming colored products characteristic of the sugars' presence (Multon 1997; Stick 2001; Stick and Williams 2009).

Reactions with bases: Sugars in the presence of bases are subjected to isomerization, known as epimerization. The reaction takes place by keto–enol equilibrium. The transformation involves the formation of intermediate enol forms (Figure 2.9) leading to a change of position of a hydroxyl group in the carbon atom and to formation of two epimers (aldose–ketose isomerization). Epimerization of glucose leads to enediol (the same may be formed from mannose as well), which can isomerize to fructose. On the influence of strong alkalies, particularly at high temperatures, sugar molecules are subjected to more severe transformations (Multon 1997; Stick 2001; Stick and Williams 2009).

2.3.3 Basics on Monosaccharides

Pentoses: Pentoses are rarely found free in nature; however, they usually occur bound in various components of plant sources as in polysaccharides, known as pentosans. Pentoses like hydroxy aldehydes exhibit the typical properties of sugars; they rotate polarized light, present reducing action, and cannot be fermented by baker's yeasts, while they are fermented by pentosanases.

The most significant pentoses are: D-xylose, L-arabinose, and D-ribose (Figure 2.10). D-xylose can be obtained by hydrolysis of plant/wood components, L-arabinose is a component of hemicellulose and pectin, while D-ribose is a basic constituent of nucleic acids (BeMiller and Whistler 1996; Izydorczyk 2005).

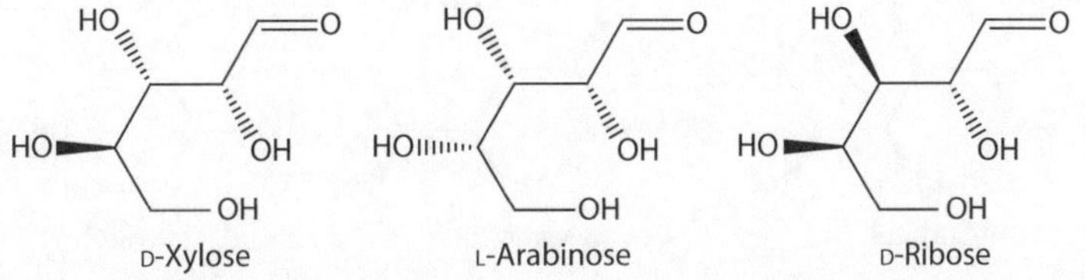

Figure 2.10 Important pentoses in food application.

Hexoses: Hexoses are the most interesting group of monosaccharides, as the most widespread in nature sugars (aldoses or ketoses) belongs to this category. They can be found in nature either free or as components of polysaccharides and glycosides. Hexoses have a sweet taste, are soluble in water and ethanol, and are insoluble in ethylether. They can reduce the Fehling's solution, and most of them are fermentable (BeMiller and Whistler 1996; Izydorczyk 2005).

The most important hexoses are from the aldohexoses (glucose, galactose, and mannose) and from the ketohexoses (fructose and sorbose; Figure 2.11).

D-*glucose* (grape sugar, dextrose): Along with fructose, D-glucose is found in fruit juices and in honey, and it is also a blood component. Glucose is a primary component of various oligosaccharides (such as sucrose, lactose, and maltose), polysaccharides (such as starch, glycogen, and cellulose), and glycosides. Glucose is industrially produced by acid or enzymatic hydrolysis of starch (potato or corn) and by acid hydrolysis of cellulose. It is a fermentable sugar both under aerobic and anaerobic (alcoholic) conditions (Stick 2001; Collins 2006; Brown 2008). Fermentation processes have many applications in foods, that is, production of ethanol in wine and beer.

D-*galactose*: It is a component of some oligosaccharides such as milk sugar (lactose), raffinose, and many complex polysaccharides known as galactans. It can be converted through oxidation to mucic acid, which is used for the detection and identification of galactose (BeMiller and Whistler 1996; Belitz et al. 2009).

D-*mannose*: It is rarely found in a free form and mostly bound in glycosides and polysaccharides. It can be produced by acid hydrolysis of polysaccharides known as mannanes (BeMiller and Whistler 1996; Belitz et al. 2009).

D-*fructose* (fruit sugar, levulose): It is the most important ketohexose. D-fructose is found in a free form in fruit juices, must, and honey along with glucose. It is a component of sucrose, raffinose, and inulin (fructan); fructose is produced only by hydrolysis of inulin, (deMan 1999; Collins 2006; Brown 2008).

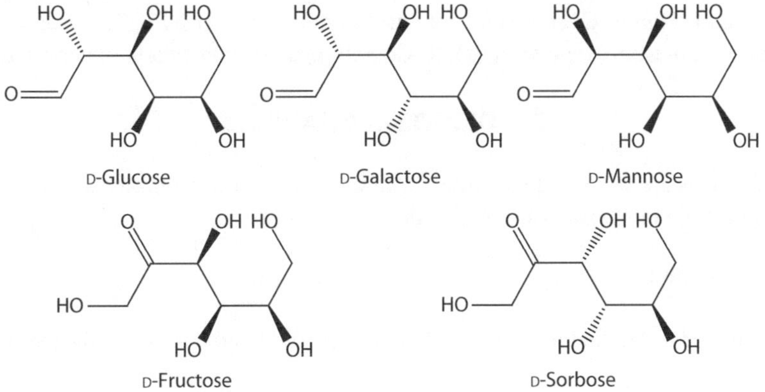

Figure 2.11 Important hexoses in food application.

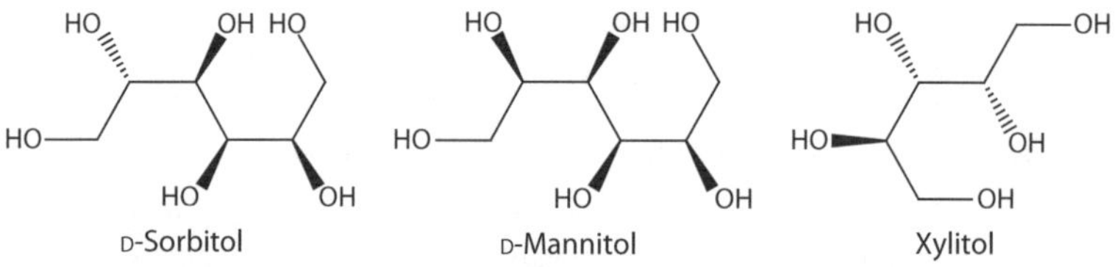

D-Sorbitol D-Mannitol Xylitol

Figure 2.12 Important sugar alcohols in food application.

2.3.4 Derivatives of Monosaccharides

Sugar alcohols: They are polyols (or polyalcohols) that are produced by reduction of sugars (deMan 1999; Duyff 2006). The most important polyols for foods are hexitols (D-sorbitol and D-mannitol) and pentitol (xylitol; Figure 2.12).

D-*sorbitol (glucitol)*: It is the most widespread in nature polyol. Fresh fruits contain 5%–10% sorbitol, whereas grapes contain little or none. Sorbitol cannot reduce Fehling's solution, and it is not fermented by yeasts. It is industrially produced by catalytic reduction of respective hexoses. Sorbitol is hygroscopic, and thus, it is used in confectionery products to contribute to freshness and soft texture. Because of its sweet taste, it is also used (in limited amounts) as an alternative to sucrose sweetener for diabetics. The well-known food emulsifiers TWEEN (polyoxyethylene sorbitan esters) and SPAN (sorbitan esters of fatty acids) are also produced from sorbitol (Coultate 2002; Collins 2006).

Xylitol: Xylitol is produced by catalytic hydrogenation of xylose. It is a pentitol that has sweet taste and caloric value equivalent to sucrose. Xylitol can also be used as an alternative to sucrose sweetener for diabetics because its metabolism in controlled amounts is independent of insulin (Collins 2006).

Uronic acids: They are derived from the corresponding aldoses by protecting the aldehyde group and oxidation of the terminal CH_2OH group to COOH as mentioned above. Uronic acids are not found in nature in a free form; however, they have an important role in the synthesis of some glycosides and polysaccharides such as pectins, alginic acids, and natural (plant) gums. The most common uronic acids are D-galacturonic and D-mannuronic acid (BeMiller and Whistler 1996).

Glycosides: In glycosides, a reducing sugar component (glycone) is bound to a noncarbohydrate moiety (aglycone) such as steroids and flavonoids by acetal hydroxyl (glycosidic linkage). Glucosides can be decomposed back to the primary components through acid hydrolysis. They do not show mutarotation and cannot reduce the Fehling's solution. The most common sugar in glycosides is glucose (from which their name derives), whereas the noncarbohydrate ingredient may be a substance or group containing N, S, and O resulting in N-glycosides, S-glycosides, and O-glycosides, respectively. Glycosides are widespread in nature, commonly in plants. Characteristic glycosides are anthocyanin (O-glycoside, a natural pigment), sinigrin (S-glycoside, responsible for the pungent taste of mustard), and gossypol (toxic polyphenolic constituent of cotton; deMan 1999).

2.4 OLIGOSACCHARIDES

Oligosaccharides are saccharide polymers containing two to 10 monosaccharides joined by glycosidic bonds through removing water molecules.

2.4.1 Disaccharides

Disaccharides are classified into two categories according to the way that the two simple sugars are joined:

1. Trehalose or sucrose type, in which the hemiacetal hydroxyl groups of the two monosaccharides are engaged in the glycosidic bond, so there is no free hemiacetal unit. Thus, these disaccharides cannot

reduce the Fehling's solution—they are nonreducing sugars and do not show mutarotation. Sucrose is the main representative of this class.

2. Maltose type, in which the hemiacetal hydroxyl of the one monosaccharide is joined through ether bond with one nonacetal hydroxyl of the other monosaccharide, allowing a free hemiacetal unit. Thus, these disaccharides can reduce the Fehling's solution—they are reducing sugars and present mutarotation. Maltose, lactose, and cellobiose belong to this category (Jackson 1999; deMan 1999).

Sucrose (saccharose): Sucrose (commonly known as sugar or table sugar) occurs in all plants, but it is commercially obtained from sugar cane (*Saccharum officinarum*) and sugar beet (*Beta vulgaris* ssp. *vulgaris* var. *altissima*). It is composed of an α-D-glucopyranosyl unit and a β-D-fructofuranosyl unit linked with an α-D-glycosidic bond; thus, it is a nonreducing sugar (Figure 2.13). It rotates the plane-polarized light to the right (+) and has a specific optical rotation $[\alpha]_D^{20} = +66.5°$. Sucrose is readily soluble in water, and its solubility increases with temperature. By cooling or evaporating a saturated sucrose solution, a supersaturated solution is obtained. Sucrose is lightly soluble in ethanol and insoluble in ethylether (Huberlant 1993; BeMiller and Whistler 1996; Izydorczyk 2005; Belitz et al. 2009).

Sucrose is not directly fermentable; it is hydrolyzed into D-glucose and D-fructose by the enzyme sucrase, which is present in the human intestinal tract, and therefore can be utilized by humans for energy. The monosaccharides D-glucose and D-fructose, being significant in our diets, do not need to undergo digestion before adsorption (BeMiller and Whistler 1996; Izydorczyk 2005).

Because of the unique carbonyl-to-carbonyl linkage, sucrose is highly labile in acid medium, and acid hydrolysis is more rapid than with other oligosaccharides. Sucrose can be enzymatic or acid hydrolyzed. The plant enzyme invertase is able to hydrolyze sucrose into its two constituent sugars in equimolar mixture of D-glucose and D-fructose (inversion reaction). Since the specific rotation is +66.5° for sucrose, +52.2° for D-glucose, and –93° for D-fructose, the mixture has a different value of specific rotation (–20.4°) and is termed invert sugar due to the inversion of the direction of rotation (deMan 1999; Izydorczyk 2005). Invert sugar is used in the production of jams, boiled sweets, and some other sugar confections. Sucrose is not a digestible carbohydrate, but it becomes one after hydrolysis.

Caramelization of sucrose requires a temperature of about 200°C. When sucrose is heated to 210°C, partial decomposition takes place and caramel is formed. Although caramelization occurs most readily in the absence of water, sugar solutions (syrups) will caramelize if heated strongly enough. Caramel, produced by heating sucrose solution in ammonium bisulfite, is used in colas and other soft drinks, baked goods, candies, and other food products as a colorant and flavor compound (Gaman and Sherrington 1990; Huberlant 1993; deMan 1999; Izydorczyk 2005; Owusu-Apenten 2005).

Sucrose is the traditional sweetening agent that is derived from sugar cane and sugar beet, and it has a purely sweet taste. The comparative sweetness of different sugars is commonly given using a point scale in which the sucrose sweetness is taken as 100 (Table 2.1). Sweetness is a sensorial characteristic, and its perception depends on the hydrogen bonding between glycol groups on the sugar molecules and receptor sites on the tongue. This property is difficult to measure reliably, being affected by the sugar concentration, the temperature of the solutions, and the viscosity of the carrier medium. Regardless of the conditions, fructose usually is ranked as the sweetest sugar and

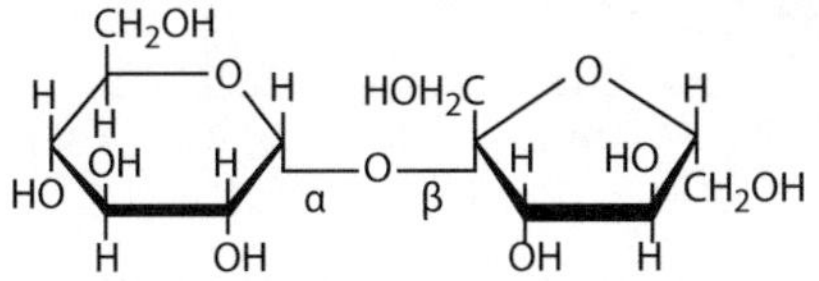

Figure 2.13 Structure of sucrose (Haworth type).

Table 2.1 Relative Sweetness of Various Sugars and Sugar Alcohols to Sucrose

Sugar/Sugar Alcohol	Relative Sweetness
Fructose	170
Invert sugar	130
Xylitol	85–120
Sucrose	100
Glucose	75
Maltitol	50–90
Sorbitol	50
Mannitol	40
Lactitol	30–40
Maltose	30
Galactose	30
Lactose	15

Source: Gaman, P. M. and Sherrington, K. B., *The Science of Food. An Introduction to Food Science, Nutrition and Microbiology*, Pergamon, Oxford, 1990; Emodi, A., *Polyols: Chemistry and Application*, AVI Publishing, Westport, CT, 1982.

lactose as the least sweet sugar. Figure 2.14 shows a comparison of common hexoses with sucrose. Only fructose is above the line while dulcin is a noncarbohydrate sweetener (Meyer 1987; Penfield and Campbell 1990).

Sucrose is used in many food product applications, either crystalline or as a refined aqueous solution known as liquid sugar. It can also be found as syrup in which sucrose is partly inverted to glucose and fructose. These syrups can be prepared with a higher concentration of solids since fructose has a very high solubility, and glucose does not readily crystallize. Figure 2.15 shows the improvement in solubility with increased inversion at various temperatures. It should be noted that after a certain critical concentration of invert sugar is reached, the solubility declines. The sweetness of these inverted sugars is comparable to sucrose (Meyer 1987; BeMiller and Whistler 1996).

Sucrose and most other low-molecular-weight carbohydrates (e.g., monosaccharides, alditols, disaccharides, and low-molecular-weight oligosaccharides), because of their great hydrophilicity

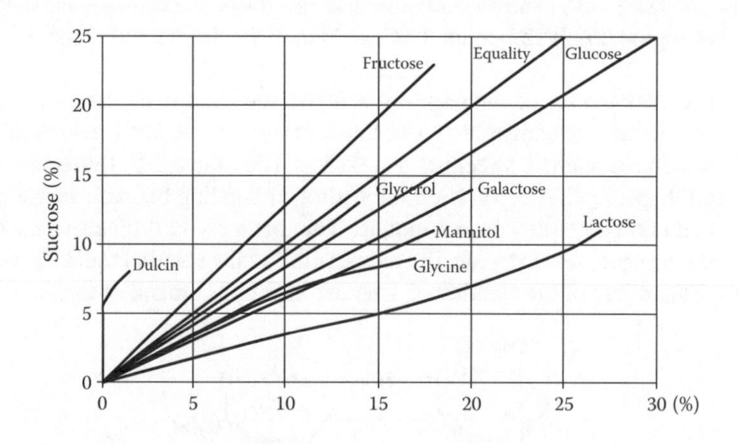

Figure 2.14 Relative sweetness of various compounds compared to sucrose from data of Cameron and Dahlberg. (Adapted from Cotton, R. H. et al., *The Role of Sugar in the Food Industry.* Copyright 1955, American Chemical Society.)

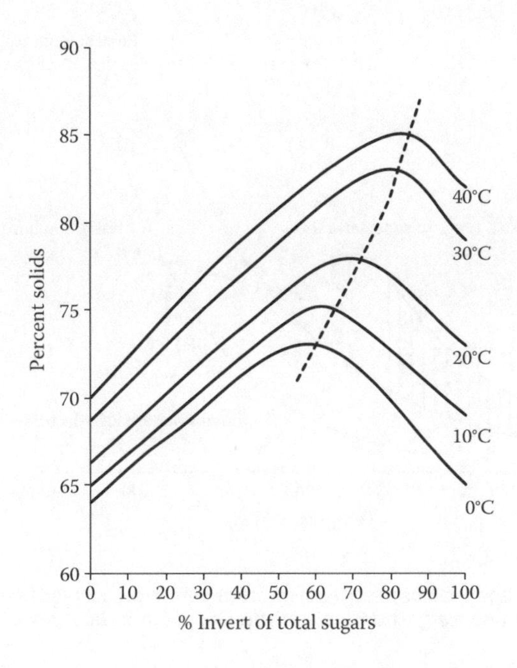

Figure 2.15 Solubility curves for sucrose–invert sugar mixtures at various temperatures. The dotted line shows peak solubility composition for various temperatures. (Adapted from Davis, P. R. and Prince, R. N., *Liquid Sugar in the Food Industry.* Copyright 1955, American Chemical Society.)

and solubility, can form highly concentrated solutions of high osmolality (BeMiller and Whistler 1996).

Lactose: It is a sugar that occurs almost only in the milk of mammals. Lactose is composed of two different sugar residues: D-glucose and D-galactose linked via β-(1→4)-linkage (Figure 2.16; Izydorczyk 2005). It can occur in two isomeric crystalline forms, the α-hydrate and the β-anhydrous, and can be found in an amorphous or glassy state; α-lactose is the most common form and can be obtained by crystallization from a supersaturated solution below 93.5°C, while β-lactose by crystallization above 93.5°C (Figure 2.17). Both forms exhibit mutarotation with final specific optical rotation $[\alpha]_D^{20} = +55.3$. Lactose solutions seek a state of equilibrium between the α and β forms. At room temperature, the equilibrium results in a ratio of about 40% α-lactose and 60% β-lactose. The fact that two forms of lactose exist with different molecular structure has profound effects on various properties of lactose, such as its solid state properties, crystal morphology, and solubility (Fox 2009). Lactose reduces Fehling's solution and is hydrolyzed by acid or enzymatically to glucose and galactose.

Lactose in mammals is enzymatically hydrolyzed in the small intestine by β-galactosidase. Lactose intolerance/malabsorption relates to the inadequate level of β-galactosidase to hydrolyze

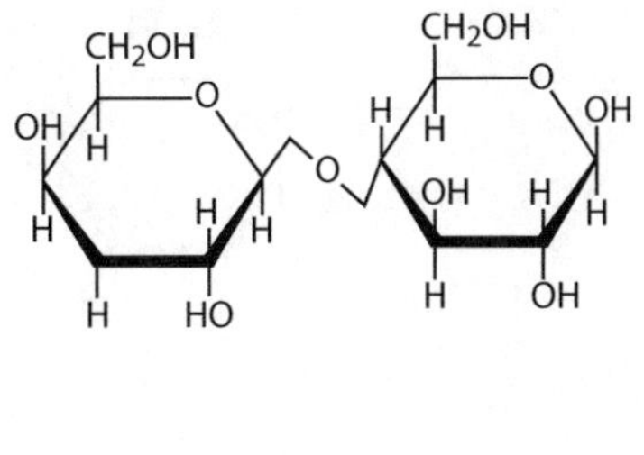

Figure 2.16 Structure of lactose.

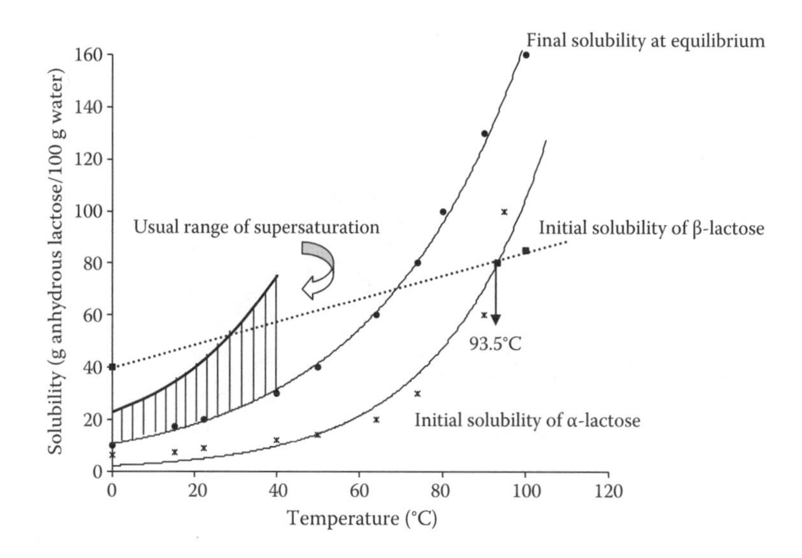

Figure 2.17 Solubility of α- and β-lactose as a function of temperature. (Adapted from Fox, P. F. and McSweeney, P. L. H., *Dairy Chemistry and Biochemistry*, Chapman & Hall, London, 1998.)

ingested lactose; it enters the large intestine, into which it draws water, causing diarrhea, and is metabolized by bacteria with the production of gas that causes cramps and flatulence. In humans, this may occur at eight to 10 years of age (Fox 2009).

Lactose is produced industrially from whey, a dairy by-product; by applying membrane technology, various protein products are separated, whereas lactose is recovered by crystallization or drying (Fox 2009; Paterson 2009). Lactose can serve as a substrate for fermentation resulting in the production of various useful products such as ethanol, lactic, acetic, and propionic acids. The *in situ* fermentation of lactose by lactic acid bacteria to lactic acid is widespread in the production of fermented dairy products.

During drying of milk or whey, lactose that has not been precrystallized forms an amorphous glass that is stable at low moisture content. Interestingly, crystalline lactose has very low hygroscopicity and is used in icing sugar blends. However, in certain foods, lactose crystallization is undesirable, such as in sweetened condensed milk or ice cream. Owing to its relatively low sweetness and low solubility, the applications of lactose are more limited than those of sucrose or glucose. The solubility of lactose is less than that of most other sugars, which may present problems in a number of foods containing lactose (Izydorczyk 2005). Finally, like all reducing sugars, lactose can undergo Maillard (nonenzymatic browning) reaction, resulting in (off-) flavor compounds and brown polymer production (Fox 2009).

Maltose: Maltose is a disaccharide formed from two units of glucose joined with an α-(1→4) glycosidic bond (Figure 2.18). It rotates the plane-polarized light to the right and has a specific optical rotation of $[\alpha]_D^{20} = +136°$. Maltose is the major end product of the enzymatic degradation of

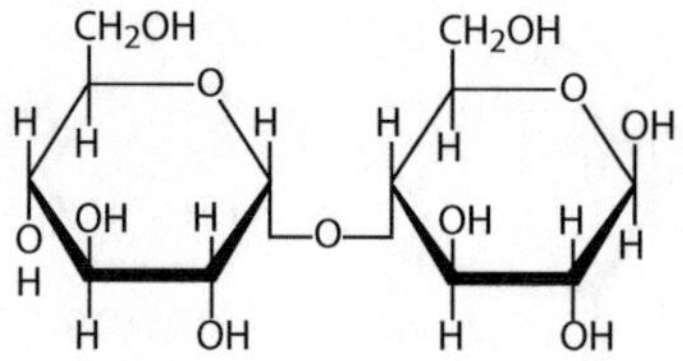

Figure 2.18 Structure of maltose.

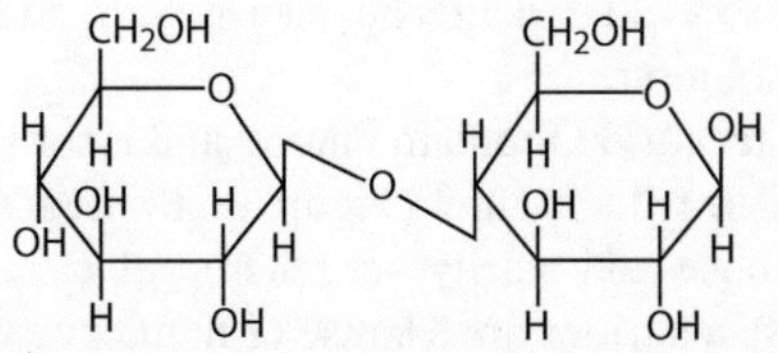

Figure 2.19 Structure of cellobiose.

starch by maltases, and it has a characteristic flavor of malt. It is a readily yeast-fermentable sugar, reduces Fehling's solution, is easily soluble in water and slightly soluble in ethanol, and presents mutarotation (deMan 1999).

Cellobiose: Cellobiose is a reducing disaccharide formed from two units of glucose joined with a β-(1→4) glycosidic bond (Figure 2.19). It can be obtained by enzymatic or acid hydrolysis of cellulose. Cellobiose is differentiated from maltose in the kind of glycosidic bond between the glucose units. It is enzymatically hydrolyzed by the enzyme emulsine (deMan 1999).

2.4.2 Trisaccharides and Tetrasaccharides

The most important trisaccharide is raffinose that is composed of galactose, glucose, and fructose (Figure 2.20). It can be found in seeds, roots, and underground stems (Meyer 1987). It is also found in sugar beets from where it is isolated. Raffinose negatively impacts sugar beet processing

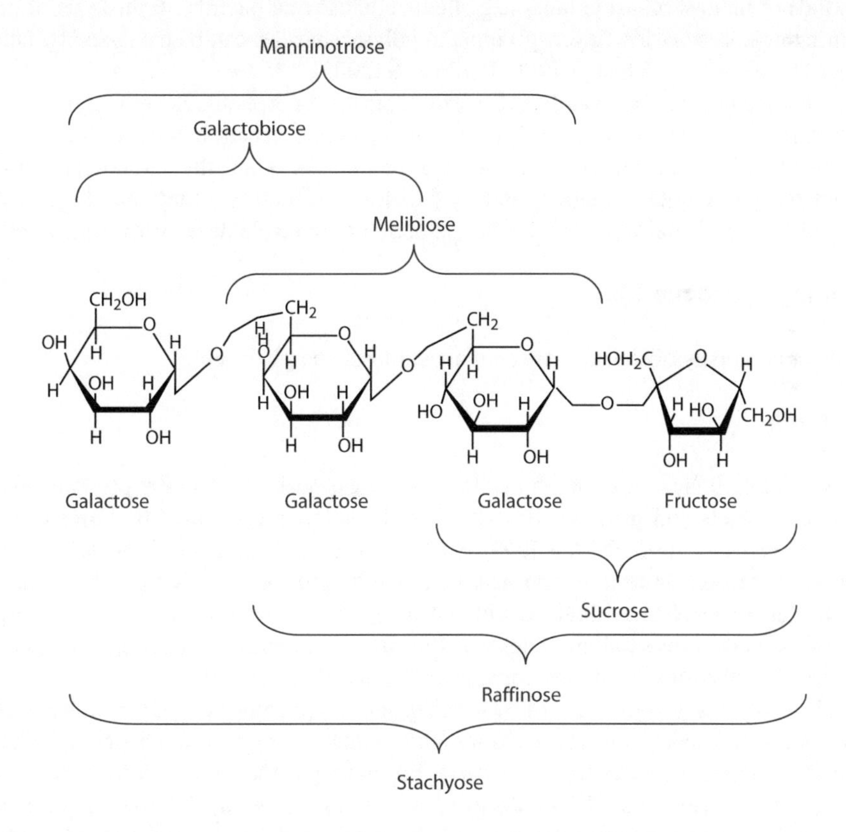

Figure 2.20 Composition of some major oligosaccharides occurring in foods. (Adapted from Shallenberger, R. S. and Birch, G. G., *Sugar Chemistry*, AVI Publishing, Co. Inc., London, 1975.)

by decreasing extractable sucrose yield and altering sucrose crystal morphology. It is water-soluble and does not reduce Fehling's solution.

Stachyose is the only tetrasaccharide found in nature, and it consists of two α-D-galactose units, one α-D-glucose unit, and one β-D-fructose unit (Figure 2.20). Legumes and pulses frequently contain valuable amounts of raffinose and stachyose, where galactose and glucose are linked with α-(1→6)-linkages. These oligosaccharides are a cause of flatulence when ingested because humans do not produce the α-galactosidase needed to hydrolyze the galactose–glucose bond (BeMiller and Whistler 1996).

2.5 POLYSACCHARIDES

Polysaccharides are very important for human nutrition and widely distributed in nature. It is estimated that more than 90% of the considerable carbohydrate mass in nature is in the form of polysaccharides. Examples include storage polysaccharides, such as starch and glycogen, and structural polysaccharides, such as cellulose and chitin. Polysaccharides (glycans) are high molecular weight polymers of monosaccharides and are named after their component monosaccharide. The degree of polymerization (DP), which is determined by the number of monosaccharide units in a chain, varies from a hundred to a few hundred thousand. Polysaccharides can be either linear or branched. Based on the number of different monomers present, polysaccharides can be divided into homopolysaccharides, consisting of only one kind of monosaccharide (e.g., cellulose and starch amylose, which are linear and starch amylopectin, which is branched), and heteropolysaccharides, consisting of two or more kinds of monosaccharide units (e.g., hemicellulose and pectins). Hydrolysis of glycosidic bonds joining monosaccharide (glycosyl) units in polysaccharides can be catalyzed by either acids or enzymes (BeMiller and Whistler 1996; Belitz et al. 2009).

Although it seems macroscopically and microscopically amorphous, X-ray analysis has revealed polysaccharides' microcrystalline structure. Depending on the structure, polysaccharides may have distinct properties from their monosaccharide building blocks. Thus, they do not have sweet taste, they do not reduce Fehling's solution, and they differ in solubility—they may be easily soluble (glycogen) or form colloidal solutions in water (starch) or are insoluble in warm water (cellulose).

2.5.1 Homopolysaccharides

The major homopolysaccharides are starch, cellulose, and glycogen.

2.5.1.1 Starch

Among food carbohydrates, starch occupies a unique position. It is the predominant storage carbohydrate in plants and provides 70%–80% of the calories consumed by humans worldwide (BeMiller and Whistler 1996; deMan 1999). Starch and starch hydrolysis products constitute most of the digestible carbohydrate in the human diet. Starch is formed by photosynthesis, and, unlike cellulose, its glucose can be metabolized with the aid of amylases in order to cover the energy needs of an organism. It occurs as small granules with the size range and appearance characteristic to each plant species, thus allowing the microscopic identification of the starch.

Starch is a polymer of D-glucose. Most starch granules are composed of a mixture of two polymer forms: a linear fraction (amylose) and a highly branched fraction (amylopectin). The ratio of amylose and amylopectin in starch sources varies depending on the starch origin, but typical values are 20%–30% for amylose and 70%–80% for amylopectin. Variations in structure and properties of starch may be associated with different species, growth conditions, environments, and genetic mutations of plants. Starch granules are birefringent, indicating a high degree of internal order.

Birefringence is the ability to refract light in two directions and is evidenced by distinctive polarization patterns in polarizing microscope (Penfield and Campbell 1990; Jane 2009). Pure starch is a white, tasteless, and odorless powder that is insoluble in cold water or ethanol. Starch is dispersed in cold water; the size, shape, and general integrity of the starch granules remain unchanged in cold water (Owusu-Apenten 2005). The viscosity-building (thickening) power of starch is realized only when a slurry of granules is thermally processed (starch paste).

Amylose is essentially a linear macromolecule consisting of α-D-glucose units linked together by (1→4) bonds (Figure 2.21). The DP is usually in the range of 250 to 350 (Meyer 1987), but can be many thousands. It has a molecular weight of 150,000–1,000,000, depending on its origin (Hood 1982). Amylose is found in the inner portion of a starch granule, it forms a colloidal dispersion in hot water without swelling, and its interaction with iodine results in formation of complexes with the characteristic blue color. Amylose constitutes 20%–30% of the total starch in the nonwaxy cereal starches and in potato starch. Waxy starches such as those from waxy maize have essentially no amylose. Some sources of relatively high-amylose starch are the wrinkled pea starch that contains about 66% amylose and some corn varieties (amylomaize) that contain up to 75% amylose (Penfield and Campbell 1990).

Amylopectin is a highly branched polysaccharide. Its structure consists of α-D-glucose units linked mainly by (1→4)-linkages (as in amylose) but with a greater proportion of nonrandom α-(1→6)-linkages, providing a highly branched structure (Figure 2.22). The DP ranges from 10^4

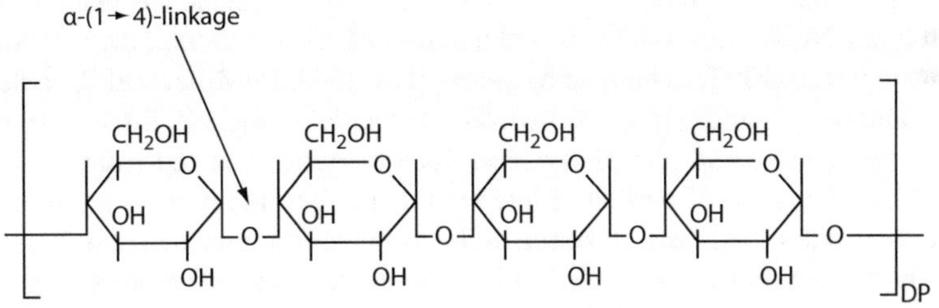

Figure 2.21 Glucose unit linkage in amylose.

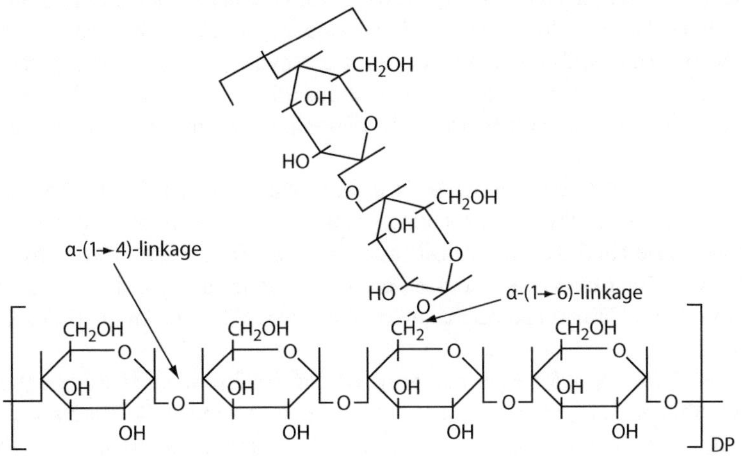

Figure 2.22 Glucose unit linkage in amylopectin.

to 10^5, making amylopectin one of the largest naturally occurring macromolecules. Its large size and highly branched structure are responsible for the high viscosity of amylopectin dispersions. On average, amylopectin has one branch point every 20 to 30 (typically 25) residues; the branches in amylopectin molecules are far shorter than those in amylose molecules and consist of 10–25 (typically 20) glucose units. The branch points are arranged in clusters and are not randomly located. The molecular weight (Mw) of amylopectin ranges from 10^6 to 10^9 g × mol^{-1}. Because of its very high Mw, large polydispersity, and susceptibility to shear degradation, amylopectin Mw determination is difficult, and results vary with the analytical technique and method used to disperse the sample. It is found in the outer portion of starch granule, it swells in water, and, when heated, it results in starch paste (Manners 1985; Penfield and Campbell 1990; Zobel and Stephen 1995; Thompson 2000; Wrolstad et al. 2005; Jane 2009).

Starch hydrolysis: Hydrolysis of starch may be brought about by the action of an acid or an enzyme. If starch is heated with an acid, it is broken down into successively smaller molecules, with the final product being glucose. There are various stages in this reaction. The large starch molecules are first broken down into shorter chains of glucose units known as dextrins. The hydrolysis for a short time produces amylodextrin (gives a blue color with iodine); further hydrolysis produces erythrodextrin (gives a red color with iodine) and, later, achrodextrin (not colorable by iodine). The dextrins produced from starch hydrolysis consist of a mixture of the above-mentioned types of dextrins, in proportions that vary according to the hydrolysis conditions. The dextrins are further broken down into maltose, and finally, maltose is broken down into glucose (Gaman and Sherrington 1990; Potter and Hotchkiss 1995). During enzymatic hydrolysis, similar products are formed. These polysaccharide fragments are called limit dextrins because the enzyme has reached the limit of its ability to hydrolyze. Dextrins are soluble in water and insoluble in ethanol, they do not have a sweet taste, and they are not readily fermentable by yeasts (Lee 1975; BeMiller and Whistler 1996).

The most common enzymes for starch hydrolysis include α-amylase, β-amylase, glucoamylase, oligosaccharide hydrolases, and phosphorylases. Such enzymes isolated from fungi, yeasts, bacteria, and plant kingdoms are of particular interest for the food industry because they are used to modify starch. α-Amylases can hydrolyze the α-(1→4) glucosidic bonds of starch in a random way at any (1→4)-linkage within the starch chain to rapidly reduce the molecular size of starch and the viscosity of the starch solution. α-Amylases result in a mixture of linear and branched oligosaccharides, and eventually maltotriose, maltose, glucose, and a range of branched α-limit dextrins.

β-Amylases also hydrolyze α-(1→4) bonds from the nonreducing ends of the outer chains of starch molecule and thus convert amylose almost completely to maltotriose, maltose, and glucose. Amylopectin is hydrolyzed in a similar way, starting from the nonreducing ends of the outer chains. β-Amylases free maltose from the branches of amylopectin until branch points are reached, resulting to a limit dextrin (in addition to maltose). Since β-amylases are unable to bypass or hydrolyze an α-D-(1→6) bond, the resulting limit dextrin contains all of the α-D-(1→6) bonds and has a high molecular weight. Because of the accumulation of maltose, β-amylases are called saccharifying enzymes.

A mixture of α-amylase and β-amylase is used in starch industry with the name "diastase." Diastase is also present in flour and germinating grain, and thus it is important in bread making and brewing. Glucoamylase catalyzes the hydrolysis of α-D-(1→4) and α-D-(1→6)-linkages to convert starch to D-glucose. Other enzymes also used are oligosaccharide hydrolases and phosphorylases (Lee 1975; Meyer 1987; Gaman and Sherrington 1990; Penfield and Campbell 1990; BeMiller and Whistler 1996).

Starch fractionation: Amylopectin and amylose can be fractionated using aqueous leaching, dispersion, and precipitation processes. Starch granules are completely dispersed in hot water or aqueous dimethyl sulfoxide, amylose is precipitated as a crystalline complex by the addition of hydrophilic organic solvents, while amylopectin is recovered from the supernatant by lyophilization (Liu 2005).

Moisture absorption by starch: When held at room temperature, starch equilibrates with the moisture in the atmosphere in which it is held and reversibly absorbs water. Under normal conditions, this amounts to about 10% to 17% moisture. The granules possess a limited amount of elasticity, which permits this to take place (Lee 1975). It has been suggested that water in starch may be held in three ways, namely, water of crystallization, absorbed water, or as interstitial water (Leach 1965). This is reversible.

Starch gelatinization: Native starch is insoluble in cold water, but when heated at a definite temperature (gelatinization temperature), which is unique for each kind of starch, the swelling begins and the viscosity of the mixture increases (Meyer 1987; Owusu-Apenten 2005). Starch gelatinization is the collapse (disruption) of molecular orderliness within the starch granule along with concomitant and irreversible changes in properties such as granular swelling, crystallite melting, loss of birefringence, viscosity development, and solubilization (Liu 2005). It is brought about in starch suspended in water, by heat at a critical temperature, and by certain chemicals at room temperature. Heat gelatinization does not occur all at once at a specific temperature, but over a range of about ~10°C. This temperature range varies with starches from different sources; for example, for potato starch, the gelatinization temperature range is 56°C–69°C, and for corn starch, it is 62°C–80°C (Pomeranz 1991). In general, the apparent temperature of initial gelatinization and the range over which gelatinization occurs depend on the starch/water ratio, granule type, and heterogeneities within the granule population under observation, which changes according to the method of measurement (Lee 1975; BeMiller and Whistler 1996).

Granule swelling is the most important phenomenon in the gelatinization of starch in aqueous medium. Swelling during gelatinization is favored in high amylopectin starches, and it depends on amylopectin structure and amylopectin density in the starch granules. Starch gelatinization begins during heating of starch, but it is finished on cooling when the starch molecules form a network with the water enclosed in its meshes, so producing a gel. The main gelatinization characteristics of starches are the swelling power that determines the critical concentration of starch to form a paste (g starch/100 g water at 95°C) and the exact temperature at which gelatinization begins (Lee 1975; Gaman and Sherrington 1990; Penfield and Campbell 1990).

Starch gelatinization process depends on a number of factors. The temperature at which gelatinization starts and the exact changes during the course of gelatinization are characteristics of, first, the starch variety. In addition, the pH at which gelatinization is attained, the temperature at which observations are made, the extent of heating, the presence of salts, sugars, and acids, and the size of the granules (the temperature of gelatinization decreases as granules decrease in size) are important for gelatinization (Meyer 1987; Gaman and Sherrington 1990).

Gelatinization of starch granules includes three stages: 1) In cold water, ~25%–30% of water is imbibed. This is apparently a reversible effect because the starch may be dried again with no observable change in structure. The viscosity of the starch–water mixture does not change during this phase. 2) It occurs at ~65°C for most starches, when the granules begin to swell rapidly and take up a large amount of water. The granules change in appearance during this second phase, and some of the more soluble starch molecules are leached out of the granules. This stage is not reversible. 3) More swelling is marked. The granule becomes enormous, often a void is formed, much more starch is leached out, and finally, the granule raptures, spilling more starch out into the surrounding fluid. The viscosity of the fluid increases markedly, and the starch granules stick together so that they can no longer be picked apart (Meyer 1987).

The swelling of starch, particularly amylose, which results in an increase in viscosity of a starch–water mixture and the formation, under proper conditions, of a gel is now believed to occur through the binding of water. In a starch granule, amylose and amylopectin molecules are loosely bound together by hydrogen bonds of the hydroxyls, forming micelles. As the temperature of the water–starch mixture rises, hydrogen bonding decreases, starch molecules are separated, the water molecules begin to freely penetrate between starch molecules, and they are imbibed by the starch molecules, resulting to starch granule swelling. The sticking together of granules is believed to be the result of molecules from adjacent granules becoming attracted and enmeshed in one another. Results of starch swelling can be monitored using a Brabender Visco/amylo/graph or a Rapid Viscoanalyzer, which records the viscosity continuously as the temperature is increased, held constant for a time, and then decreased (Meyer 1987; BeMiller and Whistler 1996; Liu 2005).

Gel formation occurs through the formation of a three-dimensional network of starch molecules, particularly the long straight chain amylose molecules. The degree of association may be affected by a number of factors, such as molecular weight, distribution of molecular weight and ratio of amylose to amylopectin. Other factors could be the length of outer branches in the amylopectin and its branching degree, as well as the presence of naturally occurring impurities of a noncarbohydrate nature. Since amylose aids gelling, high amylose starches are used where a rigid gel is needed and for instant gelatinization, whereas high amylopectin starches (i.e., waxy starches) do not crystallize readily and gel only at high concentrations (Lee 1975; Meyer 1987; Gaman and Sherrington 1990).

The subjection of an aqueous suspension of starch to heat above the critical temperature causes further swelling of the granules and loss of crystallinity, thus increasing the paste viscosity, paste clarity, and starch solubility. By stirring the starch paste, these highly swollen granules are easily broken and disintegrated, resulting in a rapid decrease in viscosity. On cooling, some starch molecules partially reassociate to form a precipitate or gel. This process is referred to as "setback" or retrogradation to indicate a partial return of a granule-like, ordered structure within the starch gel (Lee 1975; BeMiller and Whistler 1996; Smith 1982; Owusu-Apenten 2005).

High sugar concentration retards gelatinization, while lipids and surfactants also affect the pasting behavior of starch. The latter, by complexing with amylose, retard gelatinization and decrease the release of exudate during pasting (Penfield and Campbell 1990). The protein–starch interactions are also important and play a significant role in bakery products where gluten interacts with starch.

Starch gelatinization can occur also when food starch is dry heated (dextrinization). Thus, many foods containing starch also contain small amounts of dextrins. On heating, dextrins polymerize to form brown-colored compounds, called pyrodextrins. Pyrodextrins contribute to the brown color of many cooked foods including toast and bread crust (Gaman and Sherrington 1990).

Because gelatinization is of such great importance in food processing, a variety of analytical techniques have been employed to probe the phenomenon and to understand its mechanism. These include viscometry, optical microscopy, electron microscopy, differential scanning calorimetry (DSC), X-ray diffraction, nuclear magnetic resonance (NMR) spectroscopy, Fourier transform infrared (FTIR) spectroscopy, and simultaneous X-ray scattering (Liu 2005).

Starch retrogradation: Starch retrogradation is a process that occurs when the molecules comprising gelatinized starch begin to reassociate in an ordered structure. In its initial phases, two or more starch chains may form a simple juncture point, which then may develop into more extensively ordered regions. Ultimately, under favorable conditions, a crystalline order appears (Atwell et al. 1988).

Although amylose-containing starches gel best, they are less stable than high amylopectin starches. The amylose molecules tend to unwind, and the gel becomes opaque and like a pulpy sponge. Thus, starch gels are readily cut by shearing forces and reduced to liquid, which is often important in food preparation. On standing, the gel forms again. On aging, most starch gels show marked syneresis or weeping in which the water gradually passes out of the interstices of the gel. This phenomenon is made obvious as part of the starch aggregates and forms microcrystals that precipitate with time. Retrogradation occurs particularly when foods are frozen and then thawed (Meyer 1987; Gaman and Sherrington 1990).

Staling of bread is generally attributed to starch retrogradation. Retrogradation of amylose in the intergranular space is complete by the time the freshly baked loaf is cool, and further changes recognized as staling involve amylopectin chains that associate within the granules. Storage temperature affects the rate of staling. Staling is rapid at refrigerator temperatures but very slow at or below the freezing point (Penfield and Campbell 1990). Thus, lowering the storage temperature above the freezing point increases the staling rate. Finally, staling of bread is partly reversible when heated.

Several analytical techniques have been used to understand and control starch retrogradation, as well as starch gelatinization previously referred, during food processing and storage, for example, DSC, X-ray diffraction, NMR, rheological analysis, FTIR, and Raman spectroscopy and microscopy (Liu 2005).

Technology of starch: Starch is widely used in the food industry because of its unique properties. Corn is considered the main source for starch. The common procedure usually followed for starch production is as follows: corn is steeped for 35–45 h in water at 15°C with sufficient SO_2 (0.15%–0.20%) added to prevent growth of microorganisms and start the disintegration of protein. During steeping, the corn picks up water and is then readily ground to separate hulls and germ. The starch, protein, and water-soluble substances form a slurry that is separated from the germ. The slurry is then finely ground and separated from hulls and coarse particles. More SO_2 is then added, and the temperature increased to within the range of 29°C–33°C to make easier the separation of the starch granules from a proteinaceous material. The granules are separated by various methods such as centrifugation or filtering and are washed and dried (Meyer 1987).

Modified starches: Starch can be modified in order to increase its functionality and efficiency as a thickening agent and obtain properties necessary for specific uses in food systems. Natural starches can be modified by physical and chemical means (Potter and Hotchkiss 1995). Types of modifications that are most often made, sometimes singly, but often in combinations, are cross-linking of polymer chains, noncross-linking derivatization, depolymerization, and pregelatinization. Specific property improvements that can be obtained by proper combinations of modifications are reduction in the energy required to cook (improved gelatinization and pasting), modification of cooking characteristics, increased solubility, either increased or decreased paste viscosity, increased freeze–thaw stability of pastes, enhancement of paste clarity, increased paste sheen, inhibition of gel formation, enhancement of gel formation and gel strength, reduction of gel syneresis, improvement of interaction with other substances, improvement in stabilizing properties, enhancement of film formation, improvement in water resistance of films, reduction in paste cohesiveness, and improvement of stability to acid, heat, and shear (BeMiller and Whistler 1996).

The cold-water-soluble starch (pregelatinized) and the cold water swelling starch (CWS) are two types of instant starch that result from starch modification by physical means. Pregelatinized starch is widely used in the food industry because of its ability to act as a thickening agent, thus increasing the viscosity of the product and generating a pulpy texture. Pregelatinization is a relatively simple modification treatment that makes starches dispersible in cold water. In pregelatinized starch, gelatinization precedes and drying follows. It can be produced by spray cooking, drum drying, solvent-based processing, and extrusion (Xie et al. 2005). Both chemically modified and unmodified starches can be used to make pregelatinized starches. Pregelatinized starches are useful in the food industry (mainly in instant foods), as they gelatinize by swelling easily even in water at room temperature. Another type of instant starch is referred to as CWS. Granular starch that swells extensively in cold water is made by heating common corn starch in 75%–90% ethanol or by a special spray-drying process (BeMiller and Whistler 1996). It differs from pregelatinized starch in having intact granules and, therefore, greater stability and better texture (Luallen 1985). Mechanically modified starches and extruded starches are also starches modified by physical means (Belitz et al. 2009).

Many chemical treatments are used to modify starch. Chemically modified starches are the products of the treatment of any of several grain- or root-based native starches with small amounts of certain chemical agents that modify the physical characteristics of the native starches to produce desirable properties. Different treatments applied can be classified as mild oxidation (bleaching), moderate oxidation, acid depolymerization, monofunctional esterification, polyfunctional esterification (cross-linking), alkaline gelatinization, or certain combinations of these treatments. The two major groups of chemically modified starches are cross-linked and substituted (derivatized) starches. Modified food starches are usually produced as white or nearly white, tasteless, odorless powders, as intact granules, and, if pregelatinized, as flakes, amorphous powders, or coarse particles. They are insoluble in ethanol, and if not pregelatinized, they are practically insoluble in cold water (Lee 1975; Penfield and Campbell 1990; Wurzburg 1995).

Many modified starches are on the market, but those most commonly used in the food industry are the thin boiling types. They are the result of partial hydrolysis of the starch, usually using H_2SO_4 or other acids. Before drying, starch slurry is treated with 0.1N H_2SO_4 at 50°C for 6–24 h. Hydrolysis of some of the bonds in the starch results in a product that will disperse in boiling water to yield a dispersion with a viscosity not much greater than water. The more extensive the modification of the starch, the greater the effect on lowering viscosity (Meyer 1987).

Starch syrups: Starch is also used in sweet commodity products in a liquid form as syrup. Starch syrups are used to impart sweet taste in foods; they do not crystallize easily, they retard sucrose crystallization, and they act as softening agents.

They are prepared from starch, which undergoes partial hydrolysis (with hydrochloric acid and/or the enzyme amylase) in order to achieve the desired properties (Figure 2.23). Hydrolysis is not complete, and the syrups are a mixture of glucose, maltose, and longer chains of glucose units. For different applications, starch syrups subjected to extensive (soft syrups) and nonextensive hydrolysis (hard syrups) are available commercially. The extent of hydrolysis of a starch syrup is measured in terms of its dextrose equivalent (DE), which is an estimate of the percentage of reducing sugars (small molecular weight) present in the total starch product. The DE is defined as the reducing power, measured in a specific way and calculated as dextrose, expressed as a percentage of the dry substance. Products of particularly high DE are obtained by enzyme conversion. The wide range of starch syrups starts with those with a DE value of 10–20 (maltodextrins) and ends with those with a DE value of 96, while by definition, the DE of dextrose is 100. High DE syrups contain

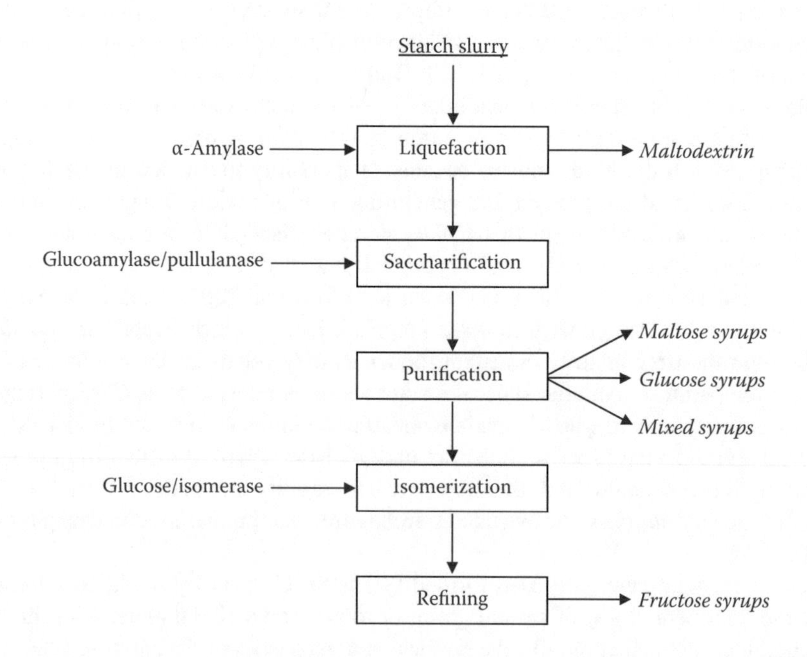

Figure 2.23 Major steps in enzymatic starch conversion. (Adapted from Olsen, H. S., *Enzymic Production of Glucose Syrups*, Chapman & Hall, London, 1995.)

more dextrose (glucose) and are therefore sweeter and have greater humectant properties than low DE syrups (Gaman and Sherrington 1990; Penfield and Campbell 1990; Blanchard and Katz 1995; Belitz et al. 2009).

2.5.1.2 Cellulose

Cellulose is the most abundant naturally occurring polysaccharide on earth. It is the major structural polysaccharide in the cell walls of higher plants, where it usually occurs together with hemicelluloses, pectin, and lignin (Belitz et al. 2009). It is also the major component of cotton boll (100%), flax (80%), jute (60% to 70%), and wood (40% to 50%). Cellulose can be found in the cell walls of green algae and membranes of fungi. *Acetobacter xylinum* and related species can synthesize cellulose. Cellulose can also be obtained from many agricultural by-products such as rye, barley, wheat, oat straw, corn stalks, and sugarcane (Izydorczyk et al. 2005).

Being present in fruits, vegetables, and other foods, cellulose is of great interest because it contributes to bulk in the diet. It is not a nutritious substance since it cannot be utilized by the human body, which lacks the necessary enzymes to digest it. However, when ingested, cellulose does contribute to the elimination process because of the bulk (Lee 1975).

Cellulose is a polymer composed of glucose units joined by β-D-(1→4)-glucosidic linkages (Figure 2.24). During its hydrolytic cleavage, the disaccharide of cellobiose is being formed. Cellobiose is a disaccharide of two glycose units, which differs from maltose in the configuration of the 1-carbon.

The DP depends on its nature and can range from 1000 to 14,000 (with corresponding molecular weights of 162 to 2268 kDa), while its mean molecular weight can be determined by viscosity measurement, ultracentrifugation, and end group analysis. Most determinations give ranges of values, indicating that the samples are composed of molecules of different molecular size (Meyer 1987).

Because of its high molecular weight and crystalline structure, cellulose is insoluble in water. Also, its swelling power or ability to absorb water, which depends partly on the cellulose source, is poor or negligible (Belitz et al. 2009).

Pure cellulose is a white, amorphous mass and insoluble in water, ethanol, ether, cold dilute alkali, and dilute acids. It is dissolved in the ammonium solution of copper oxide (AR Schweitzer). The parallel arrangement of its long molecules into fibers explains its mechanical properties.

The chemical structure of cellulose has been clarified by X-ray studies. There are two types of cellulose: 1) fiber, which is derived from the parallel arrangement of long molecules and is the rigid part of plant tissues, and 2) amorphous cellulose, which consists of disorganized parts and is the flexible part of plant tissues.

A purified cellulose powder is available as a food ingredient. The powdered cellulose used in foods has negligible flavor, color, and microbial contamination. Powdered cellulose is most often added to bread to provide noncaloric bulk. Reduced-calorie baked goods made with powdered cellulose not only have an increased content of dietary fiber but are also preserved moist and stay fresh longer.

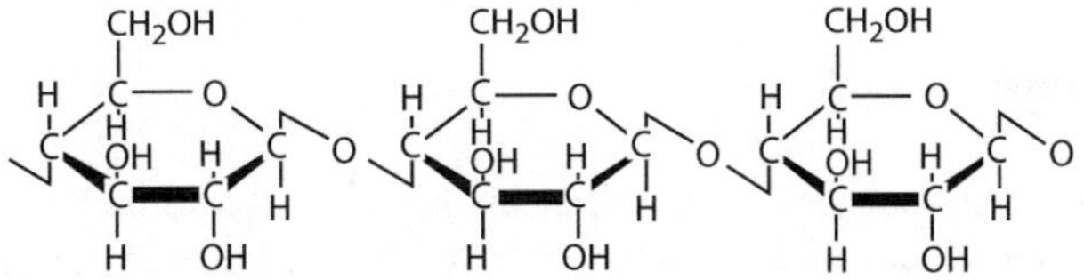

Figure 2.24 Structure of cellulose.

Cellulose forms esters and ethers, such as nitrocellulose, cellulose acetate, and viscose, which are products of great industrial importance. Only physically and chemically modified cellulose finds applications in various foodstuffs. For food products, methylcellulose ethers and carboxymethylcellulose present high interest since they can be applied as thickeners and binders (BeMiller and Whistler 1996).

2.5.1.3 Glycogen

Glycogen is a carbohydrate found only in animals. It can be thought of as the carbohydrate reserve in animals in the same way as starch is the carbohydrate reserve in plants (Gaman and Sherrington 1990). It is a glucose polymer and presents structural similarities with starch, but it includes more branches than that. It is stored more in the liver and less in the muscles. Glycogen is a white, amorphous, and tasteless powder soluble in water. It shows similar properties with starch because of the similarity in structure. Dilute acids hydrolyze glycogen to dextrin, maltose, and glucose, whereas enzymes hydrolyze it to maltose (Stick 2001).

2.5.1.4 Other Homopolysaccharides

Inulin: Inulin occurs as a reserve carbohydrate in many plant families such as scorzonera, topinambur, chicory, rye, onion, and dahlia bulb. It contains about 30 D-fructose units in a β-1,2-linkage. This linear polysaccharide has α-glucose residues in 2,1-bonding at its ends. Individual α-glucose residues in 1,3-bonding have also been detected in the interior of the polysaccharide. Inulin (molecular weight 5000–6000) is soluble in warm water and resistant to alkali. It is a white powder, soluble in hot water without being capable of forming gel, and it does not give color reaction with iodine solution. Inulin is nondigestible in the small intestine, but is degraded by the bacteria in the large intestine. It can be used in many foods as a sugar and fat substitute, for example, biscuits, yoghurt, desserts, and sweets. Inulin yields D-fructose on acid or enzymatic hydrolysis. Oligofructans have a slightly sweet taste due to the lower DP (Belitz et al. 2009).

Chitin: Chitin is a homopolysaccharide in which the macromolecular arrangement of the structural units of glucose is similar to that of cellulose, except that in position 2 of the glucose, a $NHCOCH_3$ group is attached as a ligand. It is the main component of the hard shells of crustaceans, thus being responsible for the rigidity of the shell tissue (Lee 1975).

Mannans and galactans: Mannans and galactans are homopolymers of mannose and galactose, respectively. They are usually found together in food sources as galactomannans. They occur in plant and animal tissues. They are used as thickeners and are produced commercially from seeds. Known galactomannans, of great interest for foods, are carrageenan and agar, which are found in plant gums (Meyer 1987).

2.5.2 Heteropolysaccharides

This group covers a substantial number of substances mainly of vegetable origin with complex composition, some of which have interesting properties and, subsequently, useful applications in foods. The most important heteropolysaccharides are hemicellulose and pectin components.

2.5.2.1 Hemicellulose

Hemicellulose is a polysaccharide consisting mainly of D-xylose molecules joined with α-D-(1→4) bonds. It occurs in plant cell walls along with other ingredients. After the removal of woody lignin components, an intimate mixture called "holocellulose" remains. Holocellulose appears to be a mixture of cellulose and other compounds that are soluble in aqueous alkali. These alkali-soluble compounds are called hemicellulose and are not well defined. The crude hemicellulose is composed by xylan chain and is separated into two fractions: a neutral fraction (hemicellulose A) that contains side chains of arabinose parts and an acidic one (hemicellulose B) that contains 4-O-methyl-D-glucuronic acid molecules.

It is generally agreed that the following differences exist between cellulose and hemicellulose molecules: (1) celluloses have a higher degree of polymerization (more monosaccharide units) per molecule than hemicelluloses; (2) celluloses are less soluble in alkali and less easily hydrolyzed by dilute acids than hemicelluloses; (3) celluloses are fibrous, whereas hemicelluloses are nonfibrous; (4) celluloses yield D-glucose on hydrolysis, whereas hemicelluloses yield predominantly D-xylose and other monosaccharides; and (5) celluloses have a higher ignition temperature than hemicelluloses (Meyer 1987).

However, there are some common properties between cellulose and hemicellulose molecules: (1) they are both abundant in the plant kingdom and act primarily as supporting structures in the plant tissues; (2) they are insoluble in cold and hot water; (3) they are not digested by humans and therefore do not yield energy to their nutrition; (4) they can be broken down to glucose units by certain enzymes and microorganisms; and (5) their long chains may be held together in bundles forming fibers (Potter 1995). These criteria are only general, qualitative means of differentiation.

2.5.2.2 Pectic Substances

The pectic substances are derivatives of polysaccharides and constitute an important group of carbohydrates for foods of plant origin. They are located between the cells, perhaps cell walls, serving as a binding and cell retention agent. Their structure and molecular weight are being investigated and have not been yet clearly defined. They comprise not only those substances in fruits and vegetables (pectins), which are capable of forming gels with sugar and acids, but a number of other compounds as well.

The pectic substances are polysaccharides of galacturonic acid or of its methyl ester (Figure 2.25). The main products of the hydrolysis of pectic components are galacturonic acid, a derivative of galactose in which the 6-carbon is oxidized to a carboxyl group, and methyl alcohol. Depending on their origin, they may contain D-glucose, L-arabinose, and L-rhamnose.

The pectic substances have been studied for many years, and concerning their nomenclature, the following definitions exist (Kertesz 1951):

1. *Pectic substances*: Complex colloidal carbohydrate derivatives that occur in, or are prepared from, plants and contain a large proportion of anhydrogalacturonic acid units that are thought to exist in a chainlike combination. The carboxyl groups of polygalacturonic acids may be partly esterified with methyl groups and partly or completely neutralized by one or more bases.
2. *Protopectin*: The term is applied to the water-insoluble parent pectic substance that occurs in plants, and that, upon restricted hydrolysis, yields pectinic acids.
3. *Pectinic acids*: The term "pectinic acids" is used for colloidal polygalacturonic acids containing more than a negligible proportion of methyl ester groups. Pectinic acids, under suitable conditions,

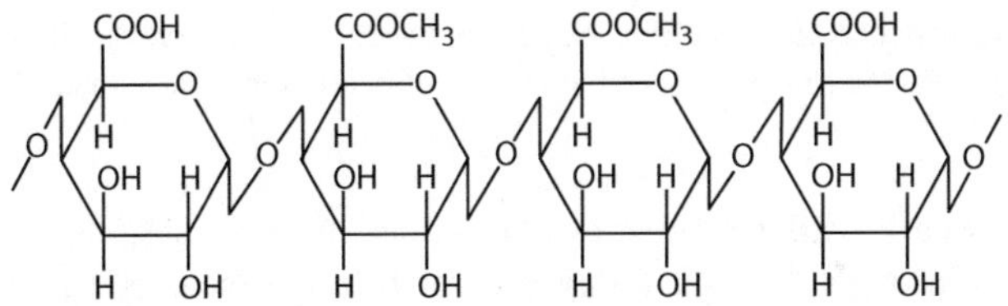

Figure 2.25 Structure of pectic substance.

are capable of forming gels (gellies) with sugar and acid or, if suitably low in methoxyl content, with certain metallic ions. The salts of pectinic acid are either normal or acid pectinates.
4. *Pectin*: The general term "pectin" (or pectins) designates those water-soluble pectinic acids of varying methyl ester content and degree of neutralization, which are capable of forming gels with sugar and acid under suitable conditions.
5. *Pectic acid*: The term "pectic acid" is applied to pectic substances mostly composed of polygalacturonic acids and essentially free from methyl ester groups. The salts of pectic acid are either normal or acid pectates.

Pectins: Pectin is a polysaccharide from terrestrial plants, which functions as an intercellular cement and helps to hold plant cells together. The texture of nonwoody vegetative plant material such as fruits and tubers is dependent on the pectin content. Commercial pectin is isolated from citrus fruit (apples, lemons, limes) or apple rind obtained from fruit juice processing (Owusu-Apenten 2005).

The main use of pectin is in gels' formations and high sugar content jams. These products are produced from pectins in which at least 50% of carboxyl groups are esterified with methyl alcohol (degree of methylation, DM = 50).

When a hot aqueous mixture of sugar, acid, and pectins is cooled, it sets into a gel; several theories have been proposed about the formation of this gel. The pectic substances that produce firm gels are those that have large molecular weight with a high DM (few free-radical -COOH). With increasing DM, the temperature in which a gel can be formed gets higher. Pectins with low DM (many free-radical -COOH) form gels by junction of their molecules (through carboxyl groups) with divalent cations, especially calcium; they do not require sugars to form gels and can give firm gels on a wide range of pH (Meyer 1987).

Gel formation: The gel formation depends on many factors:

1. The percentage of pectin (%): As the percentage of pectin increases in mixtures, the firmness of the gel produced in cooling increases the stability of the gel; usually 1% percentage of pectin is used.
2. The molecular weight of pectic substances: with increasing molecular weight, the gelling power increases.
3. The methyl ester content (%): the more the methyl esters, the firmer the gel (in practice, at least 8% of methyl ester content is used, which represents esterification of half of the carboxyl groups).
4. The most pH-firm gels are formed as the pH decreases approximately at a value of 3.2. With very low pHs, the amount of pectin can be decreased and a satisfactory gel is formed (usually there is an optimum pH value; and if the pH is lowered below this point, the firmness of the gel diminishes and excessive syneresis develops).
5. Sugars (mainly sucrose) are necessary for the formation of the pectin gels and must be present in a minimum concentration, usually 65%. The high concentration of sucrose is thought to strengthen hydrophobic interactions between methylated pectin chains (Owusu-Apenten 2005).

Gel characteristics: The gel that gives a pectin is characterized by the following:

1. The degree of gelation, which evaluates the ability of a pectic substance to produce gel. As it is difficult to measure the degree of gelation, it is determined by the amount of sugar needed to form "a satisfactory gel."
2. The setting time, which is the time that elapses between the addition of all components and the formation of a gel. Setting time evaluates the quality of pectic substance since rapid setting does not imply a firm gel.

Commercial pectins are available either as a "rapid set" or "slow set," depending on the application. A hot mixture forms a gel during cooling, and the setting time of the gel is influenced by several factors such as the pot shape and size and the cooling rate. A "rapid set" begins at about 88°C, whereas a "slow set" forms a gel below 54°C. The pH of the mixture has an influence on setting time—the setting is more rapid at lower pH values. Salts tend to delay setting, and some such as disodium phosphate are regularly added to "slow-set" pectins (Meyer 1987).

2.5.2.3 Crude Fibers

The term "crude fibers" is used to define the sum of all those organic components of plant cell membranes and supporting structures that, during chemical analysis of plant foodstuffs, remain after removal of crude protein, crude fat, and nitrogen-free extractives. Thus, the fibers are mainly composed of cellulose, hemicellulose, and some of the materials that encrust the cell walls such as pectin substances and lignins. The fibers constitute the nondigestible carbohydrate food ingredients.

The determination of fibers is difficult and depends on the method used. The methods generally include removal of crude fat by ether extraction while crude protein is dissolved by hydrolyzing with dilute acid. Crude fiber values for a given food show the same variation with several factors such as climate, soil conditions, and degree of maturity, as do other values (BeMiller and Whistler 1996).

Interest in crude fiber determinations centers either in the detection of adulteration or from the nutritional standpoint in the bulkiness of the diet. It is known that the compounds in crude fiber make up most of the bulk or residue of the diet and are not hydrolyzed by the digestive fluids, and thus they are discarded. Nevertheless, it has been shown that fibers serve a very real function in the diet as they remain long enough in the stomach, covering the sense of hunger, while they assist in the removal of other waste products from the body. Their action is important because some of the fibers absorb large amounts of water, which contributes to the cure of hyperglycemia and other diseases.

2.5.2.4 Gelling Agents: Plant Gums and Seaweed Extract

Plant gums and seaweed extracts contain substances that belong in carbohydrate category and have many applications in food; for example, they act as gelling agents, adhesives, or thickeners. The term "gum" was originally used for gummy and sticky plant secretions, while later it was used to describe the characteristic of being insoluble in water products or/and water-soluble substances. Nowadays, gum is used for water-soluble thickening and gelling agents, while those insoluble in water substances are classified as resins. Various plant gums (guar gum, carrageenan, alginate, agar) are used in foods as gelling agents or stabilizers.

Gums are polymer ingredients that, in a solution or when dispersed in water, can give a thickening or gelling effect. As these substances are of colloidal nature, they can still be referred as hydrocolloids. Most gums that find application in foods are polysaccharides.

Seed gums: This category includes guar or guar flour and carob seed flour. Both these gums contain mainly galactomannans; guar flour content in galactomannans is 80%–85%, and carob flour content is 75%–80%, but it contains galactomannans with fewer branches. They are substances that are stable to heat, while they get readily hydrated forming viscous solution.

The guar gum can be easily dispersed in cold water compared with carob seed flour, which needs warm water. Galactomannans are capable of forming highly viscous solutions at relatively low concentrations and can interact with many polysaccharides, improving their functionality, thus having many applications in foods.

Plant exudates: They are gummy substances derived from secretions of trees (acacia, cherry, plum), which, on contact with air, get dried in chambers with gummy texture. The most famous gums of this class are arabic gum, tragacanth gum, and karaya gum. These gums are heteropolysaccharide mixtures containing D-glucuronic or D-galacturonic acid and three or more monosaccharides. Arabic gum, which has many branches in its molecule, presents higher solubility regardless of the fact that it has higher molecular weight. On the contrary, karaya gum is less water soluble (Izydorczyk et al. 2005).

Seaweed extracts: The algae gums are a group of hydrocolloids that are received by algae through extraction and purification and are widely used in food industry. The most important of these are alginate or alginate salts, carrageenan, and agar.

Alginate: Alginate is a linear polysaccharide consisting of mannuronic acid and L-guluronic acid in different proportions. Alginate salts, which are obtained by replacing alginate cations with sodium or potassium, are commonly used. The alginates have the ability to form gels that are stable at high temperatures. In various food applications, the gel formation with calcium ions is of great interest. The alginates may also act as stabilizers in oil in water emulsions, like ice cream, sauces, and fruit pulp, inhibiting aggregation and phase separation.

Carrageenan: Carrageenan is nearly a linear molecule consisting of a polysaccharide mixture (galactan derivatives) mainly of sulfate or disulfate galactose esters. There are various types of carrageenan characterized by the Greek letters κ, λ, μ, and ι, which differ with respect to the molecule size and the extent and location of esterification with sulfuric acid, thus having different gelling and thickening properties that find applications in foods (Coultate 2002).

All carrageenans are soluble in hot water, whereas λ-carrageenan and sodium salts of κ- and ι-carrageenan are soluble in cold water. λ-Carrageenan gives viscous solutions, which are used as thickeners especially in dairy products in order to offer a creamy texture. Hot solution of κ- and ι-carrageenan can form a range of different texture gels when cooled to temperatures between 40°C and 60°C, depending on the existing cations. Carrageenan presents synergistic interaction with proteins and other stabilizers.

Agar: Agar consists of partially esterified galactans with sulfuric acid and can be divided into two fractions: the neutral agarose and the acid agaropectin. Agar, as a carrageenan, is soluble in hot water and, even in small concentrations, can give consistent gels (Lee 1975).

Other polysaccharides used as thickeners or gelling agents in foods are as follows.

Xanthan: Xanthan is the first of a new generation of polysaccharides produced biotechnologically. The structure of xanthan consists of a β-1,4 linked D-glucose central chain supporting a trisaccharide chain (3,1,-α-D-mannopyranose-2,1-β-D-mannopyranose) at alternating glycose units.

Some characteristics of xanthan that are important for industrial applications include its high stability with respect to pH, heat, and reaction to enzymes. The xanthan gels maintain their viscosity at a certain temperature and are not greatly affected by pH. Xanthan is a permitted thickening agent in many foods. It functions as a stabilizer and imparts smooth texture and creamy consistency. It is added to pourable dressings, dairy products (such as milk shakes and ice cream), whipped desserts, sauces, gravies, and dry mixes. Owing to its low digestibility, xanthan functions as a dietary fiber (Owusu-Apenten 2005).

2.6 FUNCTIONAL PROPERTIES OF CARBOHYDRATES

Carbohydrates possess functional properties due to which they find useful applications in food. The main effects on foods concern the following characteristics (Tomasik 2002).

Taste: Carbohydrates are associated with sweet taste, although some of them do not possess this feature. Commonly used carbohydrates as sweeteners are glucose, fructose, lactose, sucrose, maltose, and amylosyrup. A small degree of sugar caramelization contributes to the flavor and aroma of thermally processed foods.

Color: Carbohydrates help in producing a dark brown color when sugar caramelization and dextrinization occur, which are favored at high temperatures and in the presence of various additives (Coultate 2002).

Flavor: During heating of carbohydrates, substances that contribute to food flavor are produced (carbohydrate derivatives such as furo-2-aldehyde).

Texture: The dense solutions of carbohydrates that present high viscosities while interacting with other food components contribute to texture enhancement (increasing viscosity, thickening, adhesion). Also, carbohydrates interact with other food ingredients such as minerals, proteins, and fats that contribute to texture, offering a cryoprotectant role in frozen products. Some carbohydrates have found application as fat substitutes because of the relative texture that they offer to foods to which they are added (BeMiller and Whistler 1996).

2.7 QUANTIFICATION OF SUGARS AND CARBOHYDRATES

The identification and quantification of monosaccharides (reducing sugars) are based on their reducing properties. The Fehling solution, which contains $CuSO_4$ in alkaline solution, is reduced by sugars to Cu_2O through which the amount of sugars is estimated. Because proteins can cause interferences during analysis, they should therefore be removed before analysis. For disaccharide or polysaccharide quantification, hydrolysis must precede, and then identification of the resulting simple sugars follows (Belitz et al. 2009).

The detection of starch is based on the color reaction between starch and iodine. The quantification of starch is accomplished after enzymatic hydrolysis via glycoamylase or acid hydrolysis and conversion to glucose, which is determined as reducing sugar. For the determination of dietary fibers the food samples are defatted, then heated to gelatinize the starch and then subjected to enzymatic digestion by protease, amylase, and amyloglucosidase to remove the digestible components of the food. At this point the total dietary fibers can be distinguished into those being soluble in water (pectin) and insoluble (cellulose, hemicelluloses) via precipitation with ethanol. The respective residues are quantitated and adjusted for protein and ash.

The quantification of sugars is also accomplished with the use of enzymes, with photometric methods (the most common photometric method involves the use of dinitrosalicylic, DNS, acid), whereas the analysis of individual sugars is performed by high-performance liquid chromatography (Mort and Pierce 1995).

2.8 CHANGES OF CARBOHYDRATES DURING FOOD PROCESSING—BROWNING OR MAILLARD REACTIONS

The reducing sugars during thermal processing of foods react with proteins to form brown-colored polymers (nonenzymatic browning). With heating in dry state, carbohydrates are being caramelized. Browning or nonenzymatic reactions are complex reactions that occur during food processing or long storage.

Depending on the food type, the reactions are sometimes desirable, such as the dark color and the characteristic odor ("baked" odor), and others are not desirable, such as the loss of nutrients, the reduction of food protein value, and/or the formation of toxic products. The browning reactions exhibit antioxidant activity against linoleic acid only in some cases (BeMiller and Whistler 1996).

The nonenzymatic browning reactions include a series of consecutive reactions that have been investigated for many food products. Three general types of browning reactions have been identified:

1. The reaction of aldehydes and ketones (among them, the reducing sugars glucose, galactose, maltose, lactose, and fructose) with amino compounds such as amino acids, peptides, and proteins, reactions known as Maillard reactions. These reactions are independent of the presence of oxygen.
2. Caramelization, the change that occurs in polyhydroxycarbonyl compounds such as reducing sugars and sugar acids when they are heated to high temperatures and which is also independent of oxygen.
3. The oxidative change of polyphenols to dicarbonyl or polycarbonyl compounds and possibly the oxidation of ascorbic acid. This may be partially or wholly enzymatic (Levy and Fugedi 2006).

During Maillard reactions, the amino group reacts with sugar to form condensation products, which act as autocatalysts for further reactions such as enolization reaction, Amadori rearrangement, and dehydration. The presence of carbonyl components is necessary in the first two types and at the first stage of the third type. The final products of these reactions are brown pigments, which are called melanins or melanoidins. The rate of the Maillard reactions is influenced by the temperature, pH, the type and concentration of reducing sugar and amino compound, the water activity, and the presence of metal ions. The Maillard reactions are favored by intermediate humidity (aw = 0.6–0.8). The acids inhibit or prevent the Maillard reactions because they inactivate the free amino groups of amino acids, peptides, and proteins by forming salts with them. The bases increase the reaction because they release amino groups that are probably inactivated in a salt form. Blocking of the reaction can be caused by chemical means, for example, with $NaHSO_3$, which reacts with the carbonyl group of the reducing sugars. The study of parameters affecting the reaction is very useful for the design of many food products (especially dehydrated and intermediate moisture ones) and the determination of their preservation (Coultate 2002).

2.9 CARBOHYDRATE METABOLISM—HUMAN NEEDS IN CARBOHYDRATES

The carbohydrates during metabolism are hydrolyzed by enzymes or acids to glucose, which is needed for the physiological operations of the body. During chewing, food is mixed with salivary amylase, so starch is broken down in small extent to maltose. After swallowing, food reaches the stomach, where it is being acidified, and the degradation of starch continues by the pancreatic amylases. After the completion of digestion, the macromolecular components, having been broken down into smaller components, can be absorbed by the body through the intestinal walls, so monosaccharides pass through the blood vessels of the intestinal wall and lead to the liver, while the nondigestible components (fiber) are removed from the intestine (Gaman and Sherrington 1990).

The nondigestive carbohydrates consist mainly of nonstarch polysaccharides, resistant starch, and nondigestive oligosaccharides. Carbohydrates provide the body with energy (4 kcal/g). Along with fat, they cover most of the energy needs of the human body. A normal diet should contain 10% of carbohydrates. The minimum daily required amount of carbohydrate is 40 g. The calories obtained from carbohydrates should be 55%–65%, and the indicative daily requirement is 300–400 g. Both lack and excessive consumption of carbohydrates can cause adverse effects on the body, such as lack of glucose or obesity (BeMiller 2007).

2.10 ALTERNATIVE SUGARS—SWEETENERS

Alternative sugars are food constituents or additives used to substitute sugar and duplicate its effect in taste, usually with less energy intake. They are derived from the requirements of reducing sucrose in the diet of diabetics, avoiding risks connected with obesity or overweight caused by the excessive caloric intake and reactive hypoglycemia, and preventing dental caries. These substances can either be carbohydrates or carbohydrate derivatives [glucose, fructose, L-sugars, invert sugar syrup, polyols (lactitol, maltitol, mannitol, sorbitol, xylitol)] or noncarbohydrate substances (aspartame, acesulfame-K, cyclamate, saccharin, stevioside, thaumatin, etc.). They can also be categorized as nutritive and nonnutritive sweeteners. Nonnutritive sweeteners are usually not carbohydrate-based and therefore have different chemical and physical properties. They also often have flavor characteristics that differ from those of carbohydrate sweeteners and are intensely sweet compared with carbohydrate sweeteners (Salminen and Hallikainen 2002).

The main alternative sugars are the following.

L-*sugars*: L-sugars are produced through synthesis and are noncaloric substances. They have the same taste as the respective D-sugars. They are stable, noncarcinogenic, and suitable for diabetics.

Invert sugar syrup: It is a glucose–fructose syrup made through the enzymatic isomerization of glucose. A solution of D-glucose is passed through a column containing bound (immobilized) glucose isomerase that catalyzes the isomerization of D-glucose to D-fructose. Since an isomerization of only 42% is achieved, the production of higher concentrations (e.g., 55% D-fructose syrup is used as a soft-drink sweetener) requires the addition of fructose. The fructose is obtained from the syrup by chromatographic enrichment (BeMiller and Whistler 1996; Belitz et al. 2009).

Polyols: The polyols most commonly used as alternative sugars are lactitol, maltitol, mannitol, sorbitol, xylitol, and sorbitol.

Lactitol is a colorless and odorless sugar alcohol and has a pleasant mild sweetness. Its relative sweetness is about 50% that of glucose. It also promotes the amount of bifidobacteria and lactic acid bacteria in the human colon and therefore can be used as a prebiotic substance as well.

Maltitol is produced by enzymatic hydrolysis of starch (potato or corn) to obtain a high maltose syrup from which crystalline maltitol is obtained. Both liquid and crystalline maltitol are used in food products. They are soluble in water and are very stable both at different pH and temperature conditions. The sweetening power for crystalline maltitol is 0.9 and is 0.6 for liquid (sucrose = 1).

Mannitol is commonly formed via the hydrogenation of fructose and has 50%–70% of the relative sweetness of sugar. However, it lingers in the intestines for a long time and therefore often causes bloating and diarrhea.

Sorbitol is found naturally in fruits and vegetables. It is manufactured from corn syrup and has only 50% of the relative sweetness of sugar. It has less of a tendency to cause diarrhea compared to mannitol. It is often an ingredient in sugar-free gums and candies.

Xylitol is a sweetener with the same caloric content as sucrose. It is shown to be a noncariogenic sweetener, and it is also suitable for diabetic and dietetic foods (Salminen and Hallikainen 2002).

REFERENCES

Atwell, W. A., L. F. Hood, D. R. Lineback, E. Varriano-Marston and H. F. Zobel. 1988. The terminology and methodology associated with basic starch phenomena. *Cereal Foods World* 33:306–311.

Belitz, H.-D., W. Grosch and P. Schieberle. 2009. *Food Chemistry*, pp. 248–270. Heidelberg: Springer-Verlag.

BeMiller, J. N. 2007. *Carbohydrate Chemistry for Food Scientists*. St. Paul, Minnesota: AACC International.

BeMiller, J. N. 2010. Carbohydrate analysis (ch. 10). In *Food Analysis*, ed. S. S. Nielsen. New York: Springer Science & Business Media, LLC.

BeMiller, J. N. and R. L. Whistler. 1996. Carbohydrates (ch. 4). In *Food Chemistry*, ed. O. R. Fennema. New York: Marcel Dekker, Inc.

Blanchard, P. H. and F. R. Katz. 1995. Starch hydrolysates. In *Food Polysaccharides and Their Applications*, ed. A. M. Stephen. New York: Marcel Dekker, Inc.

Brown, A. M. 2008. *Understanding Food: Principles and Preparation*. Belmont: Thomson Wadworth.

Collins, P. M. 2006. *Dictionary of Carbohydrates*. Boca Raton: Taylor & Francis Group, LLC.

Cotton, R. H., P. A. Rebers, J. E. Maudru and G. Rorobaugh. 1955. The role of sugar in the food industry. In *Use of Sugars and Other Carbohydrates in the Food Industry*. Washington: American Chemical Society.

Coultate, T. P. 2002. *Food—The Chemistry of Its Components*. Cambridge: The Royal Society of Chemistry.

Davis, P. R. and R. N. Prince. 1955. Liquid sugar in the food industry. In *Use of Sugars and Other Carbohydrates in the Food Industry*. Washington: American Chemical Society.

deMan, J. M. 1999. *Principles of Food Chemistry*. Maryland: Aspen Publishers, Inc.

Duyff, R. L. 2006. *American Dietetic Association Complete Food and Nutrition Guide*. New Jersey: American Dietetic Association.

Emodi, A. 1982. Polyols: chemistry and application. In *Food Carbohydrates*, eds. D. R. Lineback and G. E. Inglett. Westport, CT: AVI Publishing, Co.

Fox, P. F. 2009. Lactose: chemistry and properties. In *Advanced Dairy Chemistry, Volume 3: Lactose, Water, Salts and Minor Constituents*, eds. P. L. H. McSweeney and P. F. Fox, 3rd ed., New York: Springer.

Fox, P. F. and P. L. H. McSweeney. 1998. *Dairy Chemistry and Biochemistry*. London: Chapman & Hall.

Gaman, P. M. and K. B. Sherrington. 1990. *The Science of Food. An Introduction to Food Science, Nutrition and Microbiology*, 3rd ed. Oxford: Pergamon Press.

Hood, L. F. 1982. Current concepts of starch structure. In *Food Carbohydrates*, eds. D. R. Lineback and G. E. Inglett. Westport Connecticut: AVI Publishing, Co.

Horton, D. 2008. The development of carbohydrate chemistry and biology (ch. 4). In *Carbohydrate Chemistry, Biology and Medical Applications*, eds. H. G. Garg, M. K. Cowman and C. A. Hales. London: Elsevier Ltd.

Huberlant, J. 1993. Sucrose: properties and determination. In *Encyclopedia of Food Science, Food Technology and Nutrition*, eds. R. Macrae, R. K. Robinson and M. J. Sadler. London: Academic Press.

Izydorczyk, M. 2005. Understanding the chemistry of food carbohydrates (ch. 1). In *Food Carbohydrates: Chemistry, Physical Properties, and Applications*, ed. S. W. Cui. New York: CRC Press.

Izydorczyk, M., S. W. Cui and Q. Wang. 2005. Polysaccharide gums: Structures, functional properties, and applications. In *Food Carbohydrates, Chemistry, Physical Properties, and Applications*. Boca Raton: Taylor & Francis Group, CRC Press.

Jackson, E. B. 1999. *Sugar Confectionary Manufacture*. Maryland: Aspen Publishers, Inc.

Jane, J.-L. 2009. Structural features of starch granules II. In *Starch: Chemistry and Technology*, eds. J. BeMiller and R. Whistler, 3rd ed. Orlando: Academic Press.

Kertesz, Z. I. 1951. *The Pectic Substances*. New York: Interscience Publishers, Inc.

Kuszmann, J. 2006. Introduction to carbohydrates (ch. 2). In *The Organic Chemistry of Sugars*, eds. D. E. Levy and P. Fugedi. Boca Raton: Taylor & Francis Group, LLC.

Leach, H. W. 1965. Gelatinization of starch. In *Starch: Chemistry and Technology*, eds. R. L. Whistler and E. F. Paschall, Vol. 1. New York: Academic Press.

Lee, F. A. 1975. *Basic Food Chemistry*. Westport, Connecticut: AVI Publishing Co. Inc.

Levy, D. E. and P. Fugedi. 2006. *The Organic Chemistry of Sugars*. Boca Raton: Taylor & Francis Group, CRC Press.

Luallen, T. E. 1985. Starch as a functional ingredient. *Food Technol.* 39(1):59–63.

Manners, D. J. 1985. Some aspects of the structure of starch. *Cereal Foods World* 30:461–467

Meyer, L. H. 1987. *Food Chemistry*. Indian edition. Delhi: CBS Publishers & Distributors.

Mort, A. Z. and M. L. Pierce. 1995. Preparation of carbohydrates for analysis by HPLC and HPCE (ch 1). In *Carbohydrate Analysis, High Performance Liquid Chromatography and Capillary Electrophoresis*, ed. Z. E. Rassi. Amsterdam: Elsevier Science B.V.

Multon, J.-L. 1997. *Analysis of Food Constituents*. New York: Wiley-VCH, Inc.

Olsen, H. S. 1995. Enzymic production of glucose syrups. In *Handbook of Starch Hydrolysis Products and Their Derivatives,* eds. M. W. Kearsley and S. Z. Dziedzic. London: Chapman & Hall.

Owusu-Apenten, R. 2005. *Introduction to Food Chemistry*. Boca Raton: CRC Press.

Paterson, A. H. J. 2009. Production and uses of lactose. In *Advanced Dairy Chemistry, Volume 3: Lactose, Water, Salts and Minor Constituents,* eds. P. L. H. McSweeney and P. F. Fox, 3rd ed. New York: Springer.

Penfield, M. P. and A. M. Campbell. 1990. *Experimental Food Science*. San Diego: Academic Press, Inc.

Pigman, W. W. and E. F. L. J. Anet. 1972. Mutarotations and actions of acids and bases (ch. 4). In *The Carbohydrates: Chemistry and Biochemistry*, eds. W. W. Pigman and D. Horton, pp. 165–194. San Diego: Academic Press.

Pomeranz, Y. 1991. *Functional Properties of Food Components*, 2nd ed., San Diego: Academic Press, Inc.

Potter, N. N. and J. H. Hotchkiss. 1995. *Food Science*, 5th ed. New York: Chapman & Hall.

Robyt, J. F. 1998. *Essentials of Carbohydrate Chemistry*. New York: Springer-Verlag Inc.

Salminen, S. and A. Hallikainen. 2002. Sweeteners (ch. 15). In *Food Additives*, eds. A. L. Branen, P. M. Davidson, S. Salminen and J. H. Thorngate III. New York: Marcel Dekker, Inc.

Shallenberger, R. S. and G. G. Birch. 1975. *Sugar Chemistry*. London: AVI Publishing Co. Inc.

Smith, P. S. 1982. Starch derivatives and their use in foods. In *Food Carbohydrates*, eds. D. R. Lineback and G. E. Inglett. Westport Connecticut: AVI Publishing, Co.

Southgate, D. A. T. 1991. *Determination of Food Carbohydrates*. London: Elsevier Applied Science.

Stick, R. V. 2001. *Carbohydrates. The Sweet Molecules of Life*. San Diego: Academic Press.

Stick, R. V. and S. J. Williams. 2009. *Carbohydrates: The Essential Molecules of Life*. Amsterdam: Elsevier Ltd.

Thompson, D. B. 2000. On the non-random nature of amylopectin branching. *Carbohydrate Polymers* 43:223–239.

Tomasik, P. 2002. Saccharides (ch. 5). In *Chemical and Functional Properties of Food Components*, ed. Z. E. Sikorski. Boca Raton: CRC Press, LLC.

Wrolstad, R. E., E. A. Decker, S. J. Schwartz and P. Sporns. 2005. *Handbook of Food Analytical Chemistry. Lipids, Proteins, Enzymes, Lipids, and Carbohydrates*. New Jersey: John Wiley & Sons, Inc.

Wurzburg, O. B. 1995. Modified starches. In *Food Polysaccharides and Their Applications*, ed. A. M. Stephen. New York: Marcel Dekker, Inc.

Xie, S. X., Q. Liu and S. W. Qui. 2005. Starch modification and applications (ch. 8). In *Food Carbohydrates: Chemistry, Physical Properties, and Applications*, ed. S. W. Cui. Boca Raton: Taylor & Francis Group, LLC.

Zobel, H. F. and A. M. Stephen. 1995. Starch: structure, analysis and application. In *Food Polysaccharides and Their Applications*, ed. A. M. Stephen. New York: Marcel Dekker, Inc.

CHAPTER 3

Sugar Alcohols (Polyols)

Gülsün Akdemir Evrendilek

CONTENTS

3.1 INTRODUCTION

Sugar alcohols, also known as polyols, polyalcohols, or polyhydric alcohols, are naturally present in many fruits and vegetables, contributing to the sweetness, and are a hydrogenated form of carbohydrates of which the carbonyl group (aldehyde or ketone, reducing sugar) has been reduced to a primary or secondary hydroxyl group (Wang 2003). Sugar alcohols have the general formula of $H(HCHO)_{n+1}H$, whereas sugars have the general formula of $H(HCHO)_nHCO$. The defining characteristic is the occurrence of an alcohol group (>CH-OH) in place of the carbonyl group (>C=O) in the aldose and ketose moieties of monosaccharides, disaccharides, oligosaccharides, and polysaccharides; hence, polyols are not sugars and generally carry the suffix "-itol" in place of the suffix "-ose" according to modern carbohydrate nomenclature (McNaught 1996). The name "polyol" is an abridgement of "polyalcohol" or "polyhydric alcohol." Preferred names are "polyol" or "hydrogenated carbohydrate"; the latter makes explicit that these substances are carbohydrate (Livesey 2003).

Production of sugar alcohols started in the 1920s with a health concern related to consumption of too much sugar with foods. After their first production, they have started to become one of the most consumed food and pharmaceutical ingredients; hence, their digestion does not require insulin synthesis, they are noncariogenic, they stimulate salivation when consumed, and they can be used in the production of foods for diabetics (Livesey 2003; Marie and Piggott 1991).

3.2 IMPORTANCE OF SUGAR ALCOHOLS FOR FOOD INDUSTRY AND HUMAN HEALTH

Sugar alcohols can be obtained from monosaccharides, disaccharides, and mixture of monosaccharides, disaccharides, and oligosaccharides (Tables 3.1 and 3.2). Depending on the saccharide type, they can have different physical, chemical, and sensory properties (Table 3.3). With regard to the sweetness of sugar alcohols, the relative sweetness level varies from 1 for xylitol to about 0.4 for lactitol and mannitol. Generally, compared to other carbohydrate sweeteners, the relative sweetness level of polyols is more concentration dependent. Due to the fact that the enthalpy of the polyol solutions is different from that of sugar, the sweetness levels of the polyols are different (Table 3.4; Counsel 1987; O'Brien and Gelardi 1991). Cooling sensation (refreshment) in the oral cavity is caused by the polyols with a higher positive enthalpy of solution (Schiweck et al. 2011).

The hygroscopic abilities of individual polyols differ greatly. For example, sorbitol (crystalline and aqueous solutions) and maltitol syrups are very hygroscopic, but mannitol and isomalt are practically nonhygroscopic with water activities lower than those of sugar (Schiffman and Gatlin 1993; Schiweck et al. 2011). The melting (sintering) points are changed among the polyols. They range from 92°C for xylitol and sorbitol to 165°C for mannitol. Polyols with a lower melting point (mp) give products with a smooth surface when used in tablets (Schiweck et al. 2011; Table 3.5).

The solubility of polyols in water changes depending on the temperature (Figure 3.1); the solubility of the polyols sorbitol, xylitol, and maltitol at 20°C is of the same order of magnitude as that of sucrose. However, the solubility of isomalt and mannitol is significantly lower. The temperature dependence of solubility of all polyols is greater than that of sucrose. The most important physical and chemical parameters such as boiling point, elevation, density, and viscosities of aqueous solutions and melts are changed among polyols (O'Brien and Gelardi 1991; Schiweck et al. 2011).

Sugar alcohols affect the water activity (a_w) of solutions differently. Compared to sucrose, sorbitol and erythritol have bigger impact on lowering the water activity of the solutions (Figure 3.2). The ability of sugar alcohols to lower water activity is dependent on the concentration.

Because sugar alcohols lack a carbonyl group, their sweetness differs considerably from that of carbohydrates (Table 3.3). Therefore, Maillard reactions and the Strecker degradation are inhibited. Moreover, monosaccharide and disaccharide alcohols are much more resistant to enzyme systems,

Table 3.1 Common Sugar Alcohols

2 Carbon Polyols	3 Carbon Polyols	4 Carbon Polyols	5 Carbon Polyols	6 Carbon Polyols	12 Carbon Polyols	Others
Glycol	Glycerol	Erythritol	Arabitol	Mannitol	Isomalt	Polyglycitol
		Threitol	Xylitol	Sorbitol	Maltitol	
			Ribitol	Dulcitol	Lactitol	
				Iditol		

Table 3.2 Specifications of Common Sugar Alcohols

Polyol	Formula	Type of Sugar	Generic Form	Synonyms	Molecular Weight
Erythritol	$C_4H_{10}O_4$	Mono-	Tetritol	Hydrogenated erythrose Meso-erythritol Tetra-hydroxybutane 1,2,3,4-butanetetrol Erythrol Physitol	122:12
Xylitol	$C_5H_{12}O_5$	Mono-	Pentitol	Hydrogenated xylose Xylite	152:15
Mannitol	$C_6H_{14}O_6$	Mono-	Hexitol	Hydrogenated mannose D-mannitol Mannite	182:17
Sorbitol	$C_6H_{14}O_6$	Mono-	Hexitol	Hydrogenated glucose D-sorbitol Glucitol Sorbol Sorbit	182:17
Sorbitol syrup	Mixed mono- and smaller amounts of other hydrogenated saccharides			Hydrogenated glucose syrup D-Glucitol syrup	
Lactitol	$C_{12}H_{24}O_{11}$	Di-	Hexopyranosyl-hexitol	Hydrogenated lactose β-D-Galactopyranosyl-1-4-D-sorbitol β-D-Galactopyranosyl-1-4-D-glucitol Lactositol Lactitol Lactobiosit	344.3
Isomalt	$C_{12}H_{24}O_{11}$	Mixed di-	Hexopyranosyl-hexitol	Hydrogenated isomaltulose Hydrogenated palatinose Mixture of α-D-glucopyranosyl-1-6-D-sorbitol and α-D-glucopyranosyl-1-1-D-mannitol	344.3
Maltitol	$C_{12}H_{24}O_{11}$	Di-	Hexopyranosyl-hexitol	Hydrogenated maltose α-D-Glucopyranosyl-1-4-D-sorbitol α-D-Glucopyranosyl-1-4-D-glucitol	344.3
Maltitol syrups	Mixed, ≥50% di- and lesser amounts of mono- and higher saccharides			Hydrogenated high-maltose glucose syrup Hydrogenated starch hydrolysate Dried maltitol syrup Maltitol syrup powder Several forms available, regular, intermediate, high, high polymer	
Polyglycitol	Mixed, ≥50% di- and of other especially oligo- and polysaccharides			Polyglucitol Hydrogenated starch hydrolysate	

acids, alkalis, and the action of heat than sugar or glucose syrups (Counsel 1987; O'Brien and Gelardi 1991).

The noncariogenicity (nonacidogenicity) of the polyols also differs from each other, and this noncariogenicity value decreases as follows: xylitol ≤ isomalt < sorbitol, maltitol < mannitol ≈ lactitol (Schiweck et al. 2011). In addition, monosaccharide and disaccharide alcohols are absorbed and metabolized differently. The monosaccharide alcohols such as sorbitol, xylitol, mannitol, and erythritol are directly absorbed from the small intestine. However, the disaccharide alcohols must be cleaved first by intestinal carbohydrases into their constituents (i.e., monosaccharides and hexitols/polyols). For this reason, the disaccharide alcohols isomalt, lactitol, and maltitol are also called second-generation sugar alcohols. The rate of cleavage of individual disaccharide alcohols by the digestive enzyme systems of the human mucosa varies (Ziesenitz and Siebert 1987).

Table 3.3 Comparison of Sucrose with Sugar Alcohols for Sweetness, Glycemic Index, and Calorie

Ingredient	Sweetness (%)	GI	Cal/g
Sucrose (sugar)	100	60	4
Xylitol	100	13	2.5
Maltitol syrup	75	53	3
Maltitol	75	36	2.7
Erythritol	70	0	0.2
Sorbitol	60	9	2.5
Mannitol	60	0	1.5
Isomalt	55	9	2.1
Lactitol	35	6	2
Polyglycitol (hydrogenated starch hydrolysate)	33	39	2.8

Note: GI—glycemic index.

Table 3.4 Saccharides and Sugar Alcohol Solutions Enthalpy Values

	Enthalpy Values		
Compound	kJ/kg	kJ/mol	Source
Sucrose	18.2	6.21	Schiweck et al. 2011
Erythritol	180.3	22.0	Muller 2007
Xylitol	153.1	23.27	Schiweck et al. 2011
Mannitol	120.9	22.0	Counsel 1987
Sorbitol	111.0	20.2	Counsel 1987
Lactitol monohydrate	65.3	23.7	Kruger 2007
Isomaltulose	60.2	21.66	Schiweck et al. 2011
Isomalt (palatinit)	39.4	14.6	Schiweck et al. 2011
Maltitol	23.0	6.7	Schiweck et al. 2011

Table 3.5 Properties of Sugar Alcohols

Sugar	Heat of Solution (kJ/kg)	Cooling Effect	Calorie (kJ/g)	Hygroscopicity
Sucrose	−18	None	16.74	Medium
Xylitol	−153	Very cool	10.05	High
Maltitol	−79	None	8.79	Medium
Sorbitol	−111	Cool	10.88	Medium
Erythritol	−180	Cool	1.67	Very low
Mannitol	−121	Cool	6.70	Low
Isomalt	−39	None	8.37	Low
Lactitol	−53	Slightly cool	8.37	Medium

Sugar alcohols are suitable for the production of noncariogenic and reduced-calorie (“light”) products and products with a defined nutritional purpose (e.g., diabetics) because of their noncariogenic properties and their reduced calorific value compared to carbohydrate sweeteners. The sensory and food manufacturing properties of individual sugar alcohols vary substantially. Therefore, proper selection of raw materials or the combination of several sugar alcohols enables sugar and glucose syrup to be replaced as bulking agents and sweeteners in all food categories in such a manner that the resulting products are acceptable to consumers (Schiweck et al. 2011; Ziesenitz and Siebert 1987).

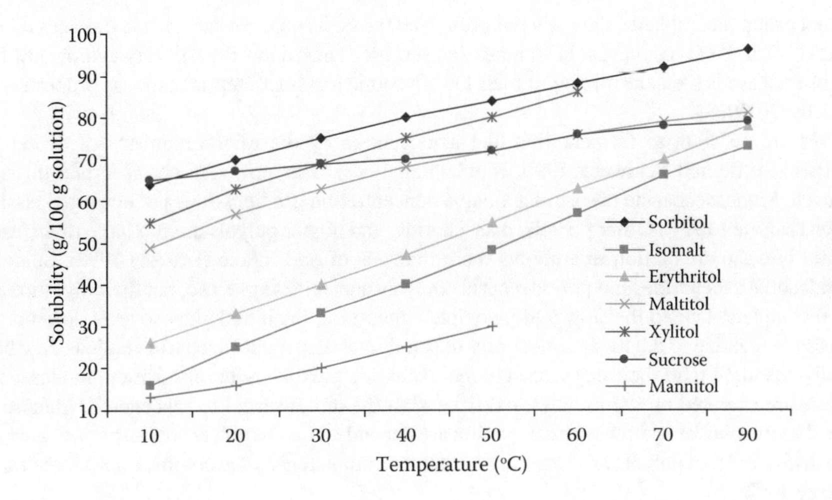

Figure 3.1 Solubility of sugar alcohols and sucrose.

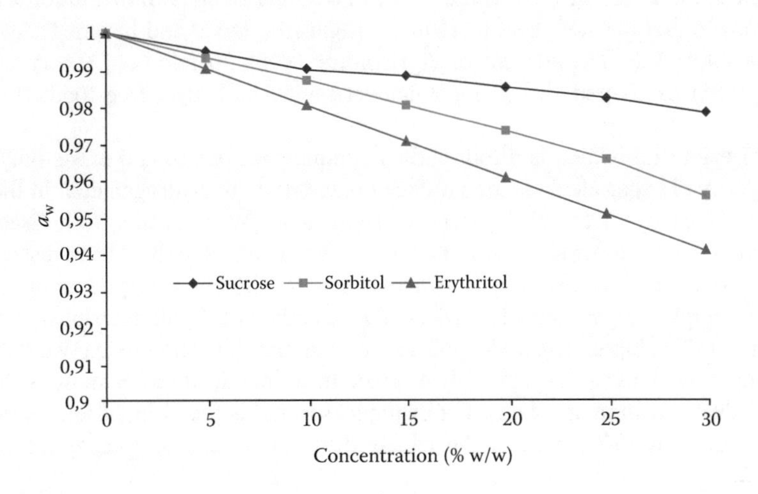

Figure 3.2 Impact of sweeteners on water activity (a_w).

Consumption of calorie-reduced foods is in demand because of increased awareness of consumers with reduced-calorie diet, tooth decays with sweet foods, and health concerns with regard to being overweight and cardiovascular diseases. Regarding some important properties such as reduced-calorie intake, sugar alcohols–polyols gain importance due to their refreshing taste and reduced-calorie content. Generally, sugar alcohols are used instead of table sugar in food industry, and often they are combined with high-intensity artificial sweeteners to counter the low sweetness (Table 3.3; McNutt and Sentko 1996; Alper et al. 2002).

The benefits of polyols can be summarized as follows:

- Polyols taste like sugar.
- Polyols have low energy value (Table 3.3). They are either metabolized and provide calories or are not metabolized and thus are noncaloric.
- Polyols have a low cariogenicity, low glycemia, and low insulinemia value. They ferment relatively easily in the colon. This property results from the hindrance to digestion and absorption by the

alcohol group that replaces the carbonyl group and the occurrence of saccharide linkages other than the α (1–4) and α (1–6) present in starches and sucrose. Thus, a low digestibility and/or slow hepatic glucose release is the determinant of their low glycemic and insulinemic response properties (Table 3.3; Livesey 2003).

- Polyols are resistant to fermentation and acidogenesis by the microorganisms of dental plaque (Willibald-Ettle and Schiweck 1996; Kandelman 1997). They are also not absorbed through the stomach. Monosaccharide absorption along a concentration gradient does not occur by passive diffusion (Herman 1974). Other polyols, disaccharide, and higher polyols are too large to diffuse from the gut into the circulation in amounts more than 2% of oral intake (Livesey 1992). Some disaccharide, oligosaccharide, and polysaccharide polyols may release glucose, but their digestion is slow and incomplete. Due to the slow and incomplete digestion, the blood glucose level does not rise. It is possible that due to the less permeability of small intestine, monosaccharide polyols may be more readily absorbed than their coreleased monosaccharide polyols. After absorption, monosaccharide polyols are excreted by the kidneys, oxidized directly, or converted to glycogen or glucose in the liver. The metabolism pathway and excretion are dependent on the polyol structure. Any unabsorbed carbohydrates from polyols are generally fermented completely by the colonic microflora (Table 3.6; Livesey 1992).
- Many low-calorie and sugar-free foods are sweetened with polyols.

As a result of these findings, the United States Food and Drug Administration has approved a "does not promote tooth decay" health claim for sugar-free foods and beverages sweetened with polyols (Deis 2005; U.S. Department of Agriculture 1995). In addition, the American Dental Association (1998) has issued an official statement saying that sugar-free foods do not promote dental caries.

Except for erythritol, which is a four-carbon symmetrical polyol and exists only in the meso form, the majority of sugar alcohols are produced industrially by hydrogenation in the presence of Raney-nickel as a catalyst from their parent-reducing sugar (Wang 2003). Both disaccharides and monosaccharides can form sugar alcohols; however, sugar alcohols derived from disaccharides (e.g., maltitol and lactitol) are not entirely hydrogenated because only one aldehyde group is available for reduction. The simplest sugar alcohol, ethylene glycol, is the sweet, but notoriously toxic, chemical used in antifreeze. The higher sugar alcohols are, for the most part, nontoxic (Wang 2003).

All the sugar alcohols are acyclic polyols since they contain three or more hydroxyl groups. Polyols naturally occur in many plants. For example, sorbitol is found in various berries and some higher plants (Lohmar 1962). One of the widely distributed polyols, D-mannitol, is being most

Table 3.6 Digestibility of Common Sugar Alcohols

Polyol	Fermentation (g/100 g)	Absorption (g/100 g)	Urinary Secretion (g/100 g)
Lactitol	98	2	<2
Isomalt	90	10	<2
Mannitol	75	25	25
Sorbitol	75	25	<2
Maltitol	60	40	<2
Xylitol	50	50	<2
Erythrithol	10	90	90
Polyglycitol	60[a]	40[a]	<2
Maltitol syrup			
Regular, intermediate, high	*ca.* 50[b]	*ca.* 50[b]	<2
High polymer	*ca.* 50[b]	*ca.* 50[b]	<2

[a] Based on *in vitro* digestion.
[b] Data based solely on glycemic and insulinemic responses.

frequently present in plant exudates (Lohmar 1962). Erythritol exists in both fruits (Shindou et al. 1989) and mushrooms (Yoshida et al. 1986), but the amounts in these products are extremely low. Sorbitol (Winegrad et al. 1972), D-mannitol (Laker and Gunn 1979), and xylitol are also endogenous metabolites in mammals.

Polyols are important sugar substitutes and are utilized in different food formulations for their different sensory, special dietary, and functional properties, which make them feasible. Thus, polyols are used in low-calorie food formulations in different countries including the United States and in most European countries. Since they are absorbed more slowly in the digestive tract than sucrose, they are useful in certain special diets. However, if consumed in large quantities, some of them can have a laxative effect. Polyols offer the same preservative benefit and a similar bodying effect to food as sucrose. They are more resistant to either thermal breakdown or hydrolysis than sugar. Moreover, most polyols are resistant to fermentation by oral bacteria and are therefore prime ingredients for tooth-friendly confectionaries such as sugarless or sugar-free chewing gums (Wang 2003; Ziesenitz and Siebert 1987).

Exogenous polyols are absorbed slowly, and metabolism mechanism in the liver mainly includes conversion to the corresponding 2-keto sugars by the action of nonspecific NAD-dependent polyol dehydrogenase (McCorkindale and Edson 1954). Unlike xylitol, sorbitol, and D-mannitol, erythritol is not a precursor of liver glycogen (Mäkinen 1994). Xylitol and sorbitol are metabolized completely after moderate administration, whereas D-mannitol is poorly utilized due to its low affinity for L-iditol dehydrogenase, causing an increased D-mannitol concentration in urine (Dills 1989). Exogenous erythritol is very poorly metabolized, being excreted almost completely in urine without degradation (Noda et al. 1994).

Polyols are also used in the diets of diabetic subjects and in infusion therapy solutions as ingredients. Compared to xylitol and sucrose, which are approximately of equal sweetness (Moskowitz 1971), D-mannitol is 45%–57% (Moskowitz 1974), erythritol is 75%–80% (Kawanabe et al. 1992), and sorbitol is 35%–60% (Wright 1974) as sweet at equal weight. Like xylitol, sorbitol has an energy value similar to that of sucrose. D-Mannitol, when consumed as part of a mixed diet, has a reduced energy value (Dills 1989). Erythritol is a very low-energy sweetener, with the available energy value being under 10% of that of sucrose (Noda and Oku 1992; Table 3.3). Other polyols share some properties with xylitol, also with regard their association with calcium metabolism. Sorbitol and D-mannitol increase calcium absorption and urinary calcium excretion in rats (Hämäläinen and Mäkinen 1986; Knuuttila et al. 1989; Vaughan and Filer 1960). Dietary sorbitol also increases the concentration of bone calcium, although less than xylitol (Knuuttila et al. 1989). Absorption, fermentation, and urinary secretion of polyols are different from each other (Table 3.4). Therefore, depending upon these properties, they are used in different formulas.

3.3 MONOSACCHARIDE POLYOLS

3.3.1 Erythritol

Erythritol ((2R,3S)-butane-1,2,3,4-tetraol) is an acyclic carbohydrate consisting of four carbon atoms, each carrying a hydroxyl group (Figure 3.3). Since it exhibits a meso structure, the molecule is achiral, although exhibiting two asymmetric carbon atoms (2R and 3S). Therefore, the molecule does not have optical rotation (Schiweck et al. 2011).

Erythritol has been approved for use as a food additive in the United States (Kawanabe et al. 1992) and throughout much of the world. It was discovered in 1848 by British chemist John Stenhouse and was commercialized during the 1990s in the Japanese market. It was the first polyol to be industrially produced by a fermentation process (Schiweck et al. 2011). It has a clean sweet taste that is similar to sucrose, and it is an anhydrous, nonhygroscopic white crystalline powder

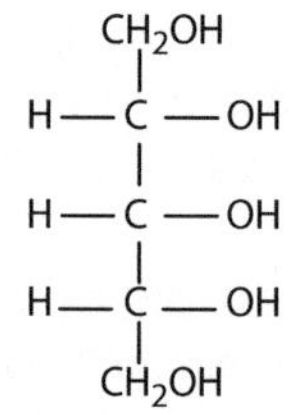

CH_2OH
H—C—OH
H—C—OH
H—C—OH
CH_2OH

Figure 3.3 Erythritol structure.

that is odorless (Arrigoni et al. 2005). Erythritol has a glass transition temperature of about –42°C and mp of 121°C. The molecule has a high thermal stability up to 180°C. No decomposition takes place in acidic and basic environment (pH 2–10). It shows a very high stability toward microbiology. When heating erythritol above 121°C, it forms a colorless nonviscous melt. Solubility of erythritol at 20°C is modest, but increases significantly at higher temperatures. At 80°C, the solubility is comparable to that of maltitol (Schiweck et al. 2011). Unlike other polyols, it combines the uniqueness of both being noncalorific and possessing a high digestive tolerance.

Actually, erythritol is a natural occurring polyol, which is present in different fruits, vegetables, and fermented fruits (Schiweck et al. 2011; Table 3.7). It can also be found in the human body (Schiweck et al. 2011). When erythritol dissolves in water, it gives a significant cooling effect that is larger than that of other polyols. This cooling effect is important in some products such as chewing gums and mint candies since freshness is desirable (Perko and DeCock 2006).

Similar to other polyols, erythritol does not promote tooth decay and is safe for people with diabetes. It is approximately 70% as sweet as sucrose and flows easily due to its nonhygroscopic character. However, erythritol has the caloric value of 0.2 calorie per gram and high digestive tolerance that makes it distinguishable from some other polyols. It has approximately 7% to 13% the calories of other polyols and 5% the calories of sucrose. Because erythritol is rapidly absorbed in the small intestine and rapidly eliminated by the body within 24 hours, laxative side effects sometimes associated with excessive polyol consumption are unlikely when consuming erythritol-containing foods (Arrigoni et al. 2005; Muller 2007).

The industrial production of erythritol uses a fermentation process with an osmophilic yeast (*Moniliella* sp. and *Trichosporonoides* sp.) or a fungus (*Aureobasidium* sp.), whereas other polyols are obtained by the hydrogenation process (Kasumi 1995). Osmotolerant microorganisms ferment the D-glucose resulting from hydrolyzed starch. As a result, a mixture of erythritol and minor amounts of glycerol and ribitol is formed (Sasaki 1989; Ishizuka et al. 1989; Aoki et al. 1993). During the fermentation process, the microorganisms have the characteristics of tolerating high sugar concentrations resulting in high erythritol yields. Approximately conversion yields of 40% to 50% have been achieved with the fermentation process, and erythritol is crystallized at over 99% purity from the filtered and concentrated fermentation broth (Wang 2003). This fermentation process for erythritol production is exothermic, and thus the reaction temperature and pH have to be controlled (Wang 2003).

Table 3.7 Erythritol Content in Different Food Materials

Food Material	Erythritol Concentration
Sake	1550 mg/L
Soy sauce	910 mg/L
Wine	130–300 mg/L
Melons	22–47 mg/kg
Pears	0–40 mg/kg
Grapes	0–42 mg/kg

Erythritol has been certified as one of the tooth-friendly polyols (Kawanabe et al. 1992). This sugar alcohol cannot be metabolized by oral bacteria and therefore does not contribute to tooth decay. Erythritol carries some of the properties similar to xylitol to starve harmful bacteria. However, it is actually absorbed into blood stream after consumption but before excretion. It has been explained that this small molecule is absorbed readily by diffusion, with approximately 10% escaping to the large intestines in humans (Bornet et al. 1996; Noda et al. 1994; Oku and Nada 1990). Absorbed erythritol is distributed in human tissue, but its metabolism is poorly reabsorbed via the kidneys. It is essentially excreted unused in urine (Bernt et al. 1996).

Erythritol has a strong cooling effect (endothermic in nature; Wohlfarth 2006) when it dissolves in water. Therefore, it is often combined with the cooling effect of mint flavors, but proves distracting with more subtle flavors and textures. The cooling effect is only present when erythritol is not already dissolved in water. This property can be seen in an erythritol-sweetened frosting, chocolate bar, chewing gum, or hard candy. The cooling effect of erythritol is very similar to that of xylitol (Jasra and Ahluwalia 1982).

Erythritol can be used in calorie-reduced beverages in combination with intense sweeteners like aspartame or acesulfame-K (sweeteners). Using erythritol at levels of 1%–3.5%, body and mouthfeel are provided while masking certain off-flavors. These combinations provide better sweetness quality than high-potency sweeteners of equal sweetness intensity because of better mimicking of the sensorial profile of sucrose (Muller 2007). Erythritol can be used in calorie-reduced dairy products and ice cream, substituting sucrose and puddings. In ice cream, erythritol is used in combination with maltitol to replace sugar. Erythritol increases the baking stability and shelf life with added amounts of 7%–19%. For calorie-reduced baking products, sucrose is largely substituted by erythritol. Using erythritol in fat creams has the advantage that the fatty mouthfeel is reduced by the cooling effect of erythritol. Erythritol can also be used in cereal bars (Schiweck et al. 2011).

Erythritol is used in tooth-friendly chewing gums due to its excellent properties. It can be used in the gum base and also for coating. Erythritol creates a taste of freshness in combination with its clean, sweet taste and high cooling effect. With the addition of erythritol to the formula, calorie-free, crystalline hard-boiled candies can be produced (Schiweck et al. 2011). In chocolate, sucrose can be substituted by erythritol up to a level of 50% without changing the formulation. In order to steer the crystallinity, erythritol can be used for the preparation of fondants in combination with maltitol syrup. Amounts of erythritol up to 99% are possible for the production of lozenges since erythritol shows very good application suitability. Because of the high crystallization tendency, erythritol is well suited for use in dragées. No addition of seeding crystals is needed. Due to its noncariogenicity and other oral health benefits, it is used in oral care products like toothpaste and mouth washes (Muller 2007).

Moreover, erythritol is used to (partially) mask unwanted off-tastes caused by some ingredients like chlorhexidine. It can also be applied for tabletting by using grades that were granulated with isomalt or sorbitol as a binder enabling a high compressibility. Erythritol can be used as a carrier for high-potency sweeteners to formulate table-top sweeteners (Muller 2007; Schiweck et al. 2011).

3.3.2 Xylitol

Xylitol, (2*R*,3*r*,4*S*)-pentane-1,2,3,4,5-pentol, is a five-carbon polyalcohol pentitol (Figure 3.4), which is widely distributed in nature. It forms orthorhombic crystals when crystallized from ethanol or methanol. A metastable, monoclinic form, mp 61°C–61.5°C, has also been described (Froesch and Jakob 1974). Commercial, food-grade xylitol has an mp of 92°C–96°C (Kim and Jeffrey 1969).

Xylitol has a high solubility in water (122.0 g/100 g H_2O at 4°C, 168.8 g/100 g at 20°C, 291.3 g/100 g at 40°C, and 571.1 g/100 g at 60°C) and sparingly soluble in absolute ethanol (1.2 g/100 g solution at 25°C) and absolute methanol (6 g/100 g solution). Xylitol is optically inactive. When it is a crystal form, it has a relatively low hygroscopicity. It is stable at temperatures up

$$
\begin{array}{c}
CH_2OH \\
| \\
H-C-OH \\
| \\
HO-C-H \\
| \\
H-C-OH \\
| \\
CH_2OH
\end{array}
$$

Figure 3.4 Xylitol structure.

to its boiling point (216°C) and does not caramelize. Xylitol is the sweetest of all polyols and has the highest positive enthalpy of solution (+153.1 kJ/kg; Schiweck et al. 2011).

Xylitol was discovered in the late nineteenth century almost simultaneously by German and French chemists (Wang and Van Eys 1981). It was first obtained by E. Fischer in 1891 as syrup by reduction of D-xylose with sodium amalgamate. After that, it was first popularized in Europe as a safe sweetener for people with diabetes mellitus that would not impact insulin levels. In the early 1970s, its dental significance was researched in Finland (Wang and Van Eys 1981). Xylitol has been produced on a commercial scale by catalytic reduction of D-xylose with hydrogen in the presence of Raney nickel since the early 1970s. D-Xylose is obtained by acid-catalyzed hydrolysis of xylan-containing plant materials such as birch wood, corn cobs, and straw (Aminoff et al. 1978).

One of the common methods in obtaining xylitol described in Figure 3.5 starts with hydrolysis of xylitol-containing material. This raw material after addition of pentose sugar material is exposed

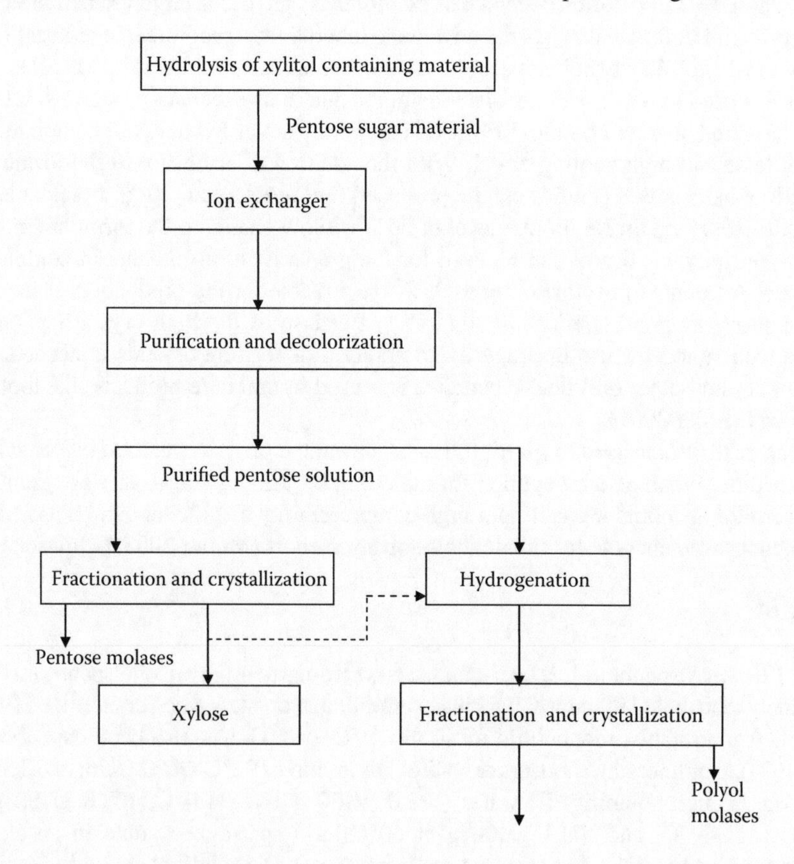

Figure 3.5 Xylose and xylitol production.

to an ion exchanger. After purification–decolorization and following fractionation–crystallization, xylose is obtained. In addition, xylitol is obtained by hydrogenation and following the fractionation–crystallization process (Artik et al. 1993).

Production of xylose from xylan is a common process from which D-xylulose is an intermediate product (Figure 3.6). This two-step process is realized by hydrolysis in the presence of H_2O + acid in the first step and H_2O + catalyst in the second step. Quality and purity of the raw product are important to obtain a high amount of xylose (Artik et al. 1993).

Xylitol is produced by hydrogenation of xylose, which converts the sugar (an aldehyde) into a primary alcohol (Gare 2003). Endogenous xylitol is produced in the liver from L-xylulose by an nicotinamide adenine dinucleotide phosphate (NADP)-linked dehydrogenase, as a metabolite of the glucuronate–xylulose pathway (Touster et al. 1956). The function of this cycle is obscure, but production of glucuronic acid for synthetic processes and detoxification reactions has been assumed (Touster 1974; Sochor et al. 1979).

Ingested xylitol is absorbed by passive or facilitated diffusion from the intestine (Bässler 1969; Lang 1969). The absorption rate is quite slow, which means that high oral doses may induce transient osmotic diarrhea. Unadapted persons can consume 30–60 g oral xylitol per day without side effects. After adaptation, up to 400 g of xylitol have been taken daily without side effects (Mäkinen and Scheinin 1975). Proposed adaption mechanisms involve induction of polyol dehydrogenase activity in the liver (Bässler 1969) and selection of intestinal microflora (Krishnan et al. 1980).

Xylitol is apparently excreted by simple glomerular filtration (Wyngaarden et al. 1957). Although there is no reabsorptive mechanism for xylitol (Lang 1969), very little is excreted in the urine, probably due to the fast diffusion from the blood to the tissues (Demetrakopoulos and Amos 1978). The net xylitol utilization in humans is over 90% after moderate xylitol administration (Lang 1969).

Most of the exogenous xylitol is metabolized in the liver (Jakob et al. 1971; Wang and van Eys 1981), although other tissues like the kidney, testes, adipose tissue, adrenal cortex, muscles, and erythrocytes are also able to metabolize it (Lang 1969; Wang and Meng 1971). Xylitol is oxidized mainly to D-xylulose by a nonspecific nicotinamide adenine dinucleotide (NAD)-linked polyol dehydrogenase (Smith 1962; Froesch and Jakob 1974), which then enters the pentose phosphate shunt via D-xylulose-5-phosphatase. Another possible pathway of xylitol metabolism is oxidation to L-xylulose by a specific NADP-linked polyol dehydrogenase. In both these reactions, a reduced redox state is produced (the ratios of reduced nicotinamide adenine dinucleotide (NADH)/NAD and reduced nicotinamide adenine dinucleotide phosphate (NADPH)/NADP increased), which has been regarded as a primary metabolic effect of xylitol. The final metabolic products of xylitol in the liver are glucose and glycogen (Froesch and Jakob 1974).

Most fruits including berries and plants contain xylitol. It is found in the fibers of many fruits and vegetables, including different berries, corn husks, oats, and mushrooms (Gare 2003). The richest natural sources of xylitol are plums, strawberries, raspberries, cauliflower, and endives (Washüttl et al. 1973). Xylitol is also an intermediate of mammalian carbohydrate metabolism. In the human body, 5–15 g/day of xylitol is formed daily. It is roughly as sweet as sucrose with only two-thirds the food energy (Hollman and Touster 1964).

Xylitol is used as a source of energy in intravenous nutrition because tissues can use xylitol under postoperative and posttraumatic conditions, when considerable insulin resistance prevents the effective utilization of glucose (Georgieff et al. 1985). Xylitol is also used as a sugar substitute due

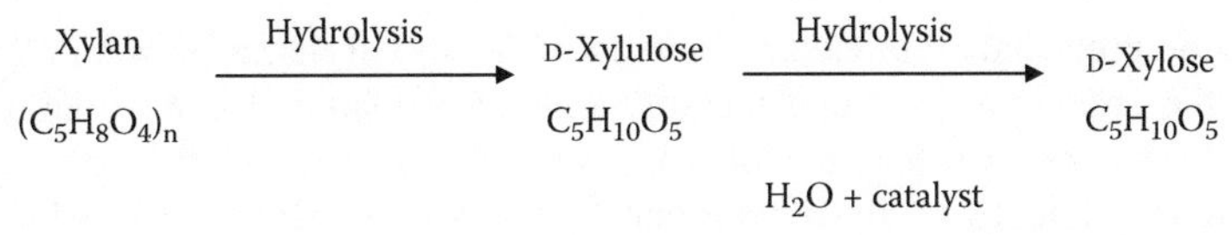

Xylan $(C_5H_8O_4)_n$ —Hydrolysis→ D-Xylulose $C_5H_{10}O_5$ —Hydrolysis (H_2O + catalyst)→ D-Xylose $C_5H_{10}O_5$

Figure 3.6 Production of xylitol from xylan.

to its anticariogenic properties (Mäkinen 1994). Moreover, it is used in the diet of diabetic subjects because it is slowly absorbed, its initial metabolic steps are independent of insulin, and it does not cause rapid changes in blood glucose concentration (Lang 1969; Förster 1974). According to Uhari et al. (1996), xylitol-containing chewing gum has been shown to reduce the occurrence of acute otitis media in day-care children.

3.3.3 Sorbitol

Sorbitol, (2S,3R,4R,5R)-hexane-1,2,3,4,5,6-hexol, belongs to the group of naturally occurring hexitols. It (D-glycitol) was discovered initially in the fresh juice of mountain ash berries *Sorbus aucuparia L.* in 1872. Sorbitol is a reduced form of dextrose and the most available polyol in nature. The fruits of certain rosaceae are especially rich in sorbitol: plums 1.7–4.5 wt%, pears 1.2–2.8 wt%, peaches 0.5–1.3 wt%, and apples 0.2–1 wt%. In fruits and leaves, sorbitol is formed as a biochemical intermediate in the synthesis of starch, cellulose, sorbose, or vitamin C. Small amounts are found in the plane tree, the African snowdrop tree, and in various algae. Because sorbitol occurs to a very small extent in grapes, assay of the sorbitol content of wine (qv) has been used to detect its adulteration with other fruit wines or apple cider. An anhydride of sorbitol, polygallitol (1,5-sorbitan), is found in the *Polygala* shrub (Takiura et al. 1974).

In animals, sorbitol can be detected as an intermediate in the absorption of glucose or the formation of fructose via glucose (Boussingault 1872; Steuart 1955). It forms γ-polymorphous, finely crystalline white crystals that are odorless, freely flowable, and slightly hygroscopic. Sorbitol is also available in the form of water-clear syrup solutions. Sorbitol has a sweet, cooling taste and has approximately half the sweetness of sucrose. Its physiological calorific value corresponds to that of other sugar alcohols (Schiweck et al. 2011).

Sorbitol is readily soluble in water, and 70% solutions of sorbitol in water are available commercially, but tend to crystallize when cooled to <10°C for long periods. Sorbitol is also soluble in dilute acetic acid, methanol, and warm ethanol; it is practically insoluble in organic solvents. The melting range of sorbitol is 92°C–96°C. The optical rotation of aqueous solutions is increased greatly by the addition of complexing salts (borax, ammonium molybdate). The enthalpy of solution is +111 kJ/kg. The bulk density of instant grades is 40–50 g/100 mL, and that of crystallized grades is 60–70 g/100 mL (Burt 2006; Schiweck et al. 2011).

Depending on conditions (crystallization from solvents or from the melt), sorbitol crystallizes in several modifications (Steuart 1955), with the most stable being the γ-form (mp 101°C). In neutral media, sorbitol is temperature-resistant up to 150°C and, consequently, is bake and boil proof. As a sugar alcohol, sorbitol does not undergo Maillard reactions with amino acids or proteins at high temperature. Like other hexitols, sorbitol forms chelate complexes with some metal ions (e.g., Fe, Cu, Co, and Ni) in aqueous solutions that are stable in the alkaline range. Thus, sorbitol inhibits the prooxidative effect exerted by trace metals in autoxidation (e.g., of fats; Levin et al. 1995; Schiweck et al. 2011; Suzuki et al. 1985).

Sorbitol cannot be fermented by yeast. It prevents the growth of bacteria by osmosis in highly concentrated aqueous solutions (sorbitol content of about 50%). On the contrary, at low concentration, sorbitol solutions serve as a nutrient medium for bacteria. *Acetobacter xylinium* oxidizes sorbitol enzymatically to give L-sorbose (an intermediate in the synthesis of ascorbic acid; Schiweck et al. 2011).

Sorbitol can be produced from electrochemical reduction of dextrose in an alkaline medium, which leads to the production of a considerable amount of mannitol due to the alkali-catalyzed epimerization of dextrose to fructose, which is more readily available than dextrose. Sorbitol is industrially produced either by catalytic hydrogenation of dextrose (Figure 3.7), which produces less than 2% mannitol, or by catalytic hydrogenation of sucrose as a mixture with mannitol (Schiweck et al. 2011; Wang 2003).

```
     CH                        CH2O
     |                          |
 H—C—O                     H — C — O
     |          H2              |
 H—C—H        ———→         H — C — H
     |                          |
 H—C—O                     H — C — O
     |                          |
 H—C—O                     H — C — O
     |                          |
     CH2                       CH2O

  Glucose                    Sorbitol
```

Figure 3.7 Sorbitol formation from glucose.

It is synthesized by sorbitol-6-phosphate dehydrogenase and converted to fructose by succinate dehydrogenase and sorbitol dehydrogenase (Petrash 2004; Figure 3.8). Sorbitol is synthesized commercially by high-pressure hydrogenation of glucose, usually using a nickel catalyst. Catalyst promoters include magnesium salts, nickel phosphate, and iron. Other heterogeneous catalysts used for glucose hydrogenation include cobalt, platinum, palladium, and ruthenium (Boyers 1959). Reduction of glucose to sorbitol can also be effected using ruthenium dichlorotriphenyl phosphine as a homogenous hydrogenation catalyst, preferably in the presence of a strong acid such as HCl (Kruse 1976). In order to form sorbitol, glucose is usually hydrogenated in the pH range of 4–8. Under alkaline conditions, glucose isomerizes to fructose and mannose; hydrogenation of the fructose and mannose yields mannitol as well as sorbitol. In addition, under alkaline conditions, the Cannizzaro reaction occurs, and sorbitol and gluconic acid are formed. Gluconic acid formation during hydrogenation can be minimized if anion exchange resins in the basic form are the source of alkalinity (Jacot-Guillarmod et al. 1963). Although aqueous solutions are customarily used, the monomethyl ethers of ethylene glycol or diethylene glycol are satisfactory solvents (Jacot-Guillarmod et al. 1963). Electrolytic reduction of glucose was used formerly for the manufacture of sorbitol (Creighton 1939). Both the γ- and δ-lactones of D-gluconic acid may be reduced to sorbitol by sodium borohydride (Hough et al. 1962). Sorbitol occurs from simultaneous hydrolysis and hydrogenation of starch (Kool et al. 1952), cotton cellulose (Vasyunina et al. 1964), or sucrose (Montgomery and Wiggins 1947; Figure 3.9).

The use of sorbitol is approved by more than 40 countries in foods, cosmetics, and pharmaceuticals. It has been acknowledged as a generally recognized as safe (GRAS) substance in the United States for its use in foods (LSRO for FDA 1986). The powdered grades are employed as a sugar

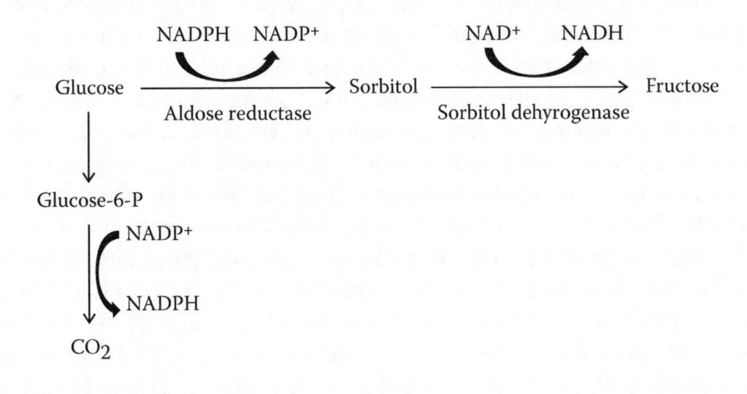

Figure 3.8 Sorbitol synthesis pathway.

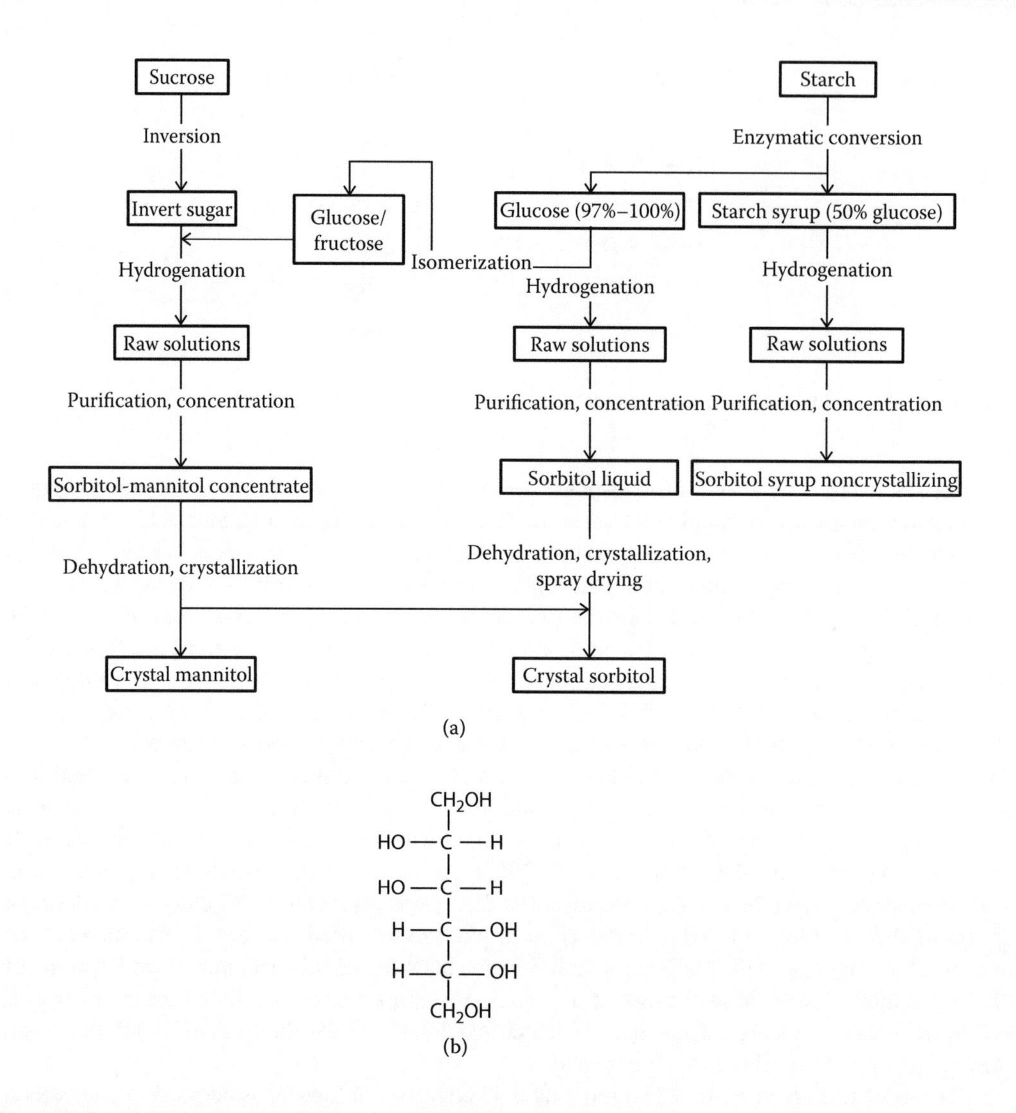

Figure 3.9 (a) Production of sorbitol and mannitol. (b) Mannitol structure.

substitute for diabetic foods; for sugar-free items such as chocolates, dragées, tablets, chewing gum, and filled hard candy; for reduced-calorie sweetening agents; and for infusion solutions used in parenteral nutrition. Sorbitol liquid and sorbitol syrup serve as softeners and moisture stabilizers for confectionery of all types: marzipan; cakes and pastries; aroma substances; delicacies; mayonnaise and dressings; products made of fish, shrimp, and roe; baking emulsifiers; candied fruits; sauerkraut; natural and artificial skins; ice cream; enzyme preparations; tonics; and tobacco products. The concentration of sorbitol powder used is usually 10%–100%, depending on the food. Sorbitol liquid and sorbitol syrup are employed in concentrations of 5%–10% (Burt 2006; Petersen and Razanamihaja 1999). Sorbitol syrup is preferred over glycerol in the production of water-containing cosmetics such as creams, ointments, emulsions, lotions, gels, and especially toothpastes (Petersen and Razanamihaja 1999; Schiweck et al. 2011). Depending on the desired effect, the concentration required is 3%–5% for water stabilization and 10%–20% for softening effects. Sorbitol can also be used in pharmaceutical preparations (Grenby et al. 1989). Sorbitol powder has gained acceptance as a tablet excipient (Gwilt et al. 1963). As a result of its surface structure, spray-dried sorbitol is especially suitable for the uniform binding of solid agents. It can be molded directly into tablets without granulation (Schmidt and Benke 1985). This type of sorbitol is also suited to the production of stable dry suspensions (e.g., with antibiotics) of vitamin tablets or antacid granules (Nikolakakis and Newton 1989). Sorbitol syrup is important primarily as a moisture stabilizer and softener. About 25% of the worldwide production of sorbitol is used for this purpose (Schiweck et al. 2011).

3.3.4 Mannitol

Mannitol, (2*R*,3*R*,4*R*,5*R*)-hexane-1,2,3,4,5,6-hexol, is the alcohol form of mannose with a molecular weight of 182, prepared commercially by the reduction of dextrose (Figure 3.9a). It is sparingly soluble in many organic solvents such as ethanol (1.2 g/100 mL) and glycerol (5.5 g/100 mL), and practically insoluble in ether, ketones, and hydrocarbons. Mannitol has an mp of 165°C–168°C and relative density of $d^{20} = 1.52$; the particle-size distribution of crystalline grades is 20–250 μm and that of granulated grades is 100–850 mm; its bulk density is 550–650 g/L. The viscosities of mannitol and sorbitol solutions are comparable. Crystalline mannitol is only slightly hygroscopic; appreciable water absorption occurs at >90% relative humidity. Mannitol, like other hexitols, forms chelate complexes with metal ions (e.g., iron, copper, and nickel) in aqueous solution. The iron complexes are especially stable. Even at 150°C–170°C, mannitol does not react with proteins according to the Maillard reaction. Nitration of mannitol with nitrating acid yields mannitol hexanitrate (hexanitromannitol), which is used as an igniter for detonators and electrical fuses. Although it has an important place in current therapy, few physicians fully understand its chemistry, mode of action, or clinical applicability. Of the commercially used sugar alcohols, mannitol is the least soluble in water. Because of its positive heat of solution in water, a cooling sensation occurs when mannitol dissolves in the mouth. This effect is, however, less pronounced than that observed with sorbitol or xylitol (Grenby 1989; Schiweck et al. 2011; Ziesenitz and Siebert 1987).

D-Mannitol is widespread in nature. It is found to be a significant extent in the exudates of trees and shrubs such as the plane tree (80%–90%), manna ash (30%–50%), and olive tree. Mannitol occurs in the fruits, leaves, and other parts of various plants. This hexitol is present in pumpkins, hedge parsley, onions, celery, strawberries, the genus *Euonymus*, the genus *Hebe*, the cocoa bean, grasses, lilac, *Digitalis purpurea*, mistletoe, and lichens. Mannitol also occurs in marine algae, especially brown seaweed, with a seasonal variation in mannitol content that can reach over 20% in the summer and autumn (Kearsley and Deis 2006; Song and Vieille 2009). It is extracted from seaweed for use in food manufacturing and is sometimes used as a sweetener in dietetic products. Mannitol is found in the mycelia of many fungi and is present in the fresh mushroom to the extent of about 1.0% (Kearsley and Deis 2006). It was reported that there is a direct relation between mushroom yield and mannitol content (Kulkarni 1990).

Mannitol is commonly used as a nutritive sweetener, stabilizer, humectant, and bulking agent in foods and supplements. This sugar alcohol is approximately 0.7 times as sweet as sucrose. It is used as a bulking agent in powdered foods and as a dusting agent for chewing gum. It can be derived from the Manna plant or from seaweed; however, for commercial use, it is manufactured via a catalytic hydrogenation process (Schiweck et al. 2011).

The chemical structure of mannitol allows it to be absorbed more slowly by the body than regular sugars. Therefore, it has a smaller impact on blood insulin levels, making it and other sugar alcohols useful for diabetic foods. Mannitol is not absorbed in the GI tract; moreover, it does not cross an intact blood–brain barrier (BBB). It does not enter cells and is cleared by glomerular filtration. The glomeruli filter mannitol, which does not get reabsorbed from renal tubules; therefore, mannitol increases the osmolarity of renal tubular fluid and prevents reabsorption of water. Sodium, the renal tubular fluid, is diluted and less is reabsorbed. Mannitol works osmotically to create diuresis. It works by creating an osmotic gradient. Side effects result from its osmotic effect. It increases intravascular volume, which can precipitate pulmonary edema and/or congestive heart failure, especially in patients with poor cardiac function. Hypovolemia can occur along with hypotension.

Nephrotoxins and prolonged renal ischemia can damage renal tubules so that they become permeable to mannitol and its osmotic effect is lost. Finally, venous thrombosis is a rare event, and tissue necrosis is unlikely with extravasation of mannitol (the opposite is true for urea). Mannitol also increases plasma osmolarity and draws fluid from intracellular to extracellular spaces relieving increased intracranial pressure (ICP) and increased ocular pressure. If the BBB is damaged, worsened elevations in ICP could occur as mannitol enters the brain tissue and causes an osmotic gradient into the tissue (Cosolo et al. 1989).

On the other hand, because it is slowly absorbed, excessive consumption may have a laxative effect, similar to certain high-fiber foods. Because of this, products containing mannitol must include a laxative warning on the label if the mannitol content in a serving exceeds 20 g. FDA allows the use of a caloric value of 1.6 cal/g. It is permitted for use in many countries, including the United States (Calorie Control Council 2011).

Mannitol does not promote tooth decay. It is approximately 72% as sweet as sugar (sucrose) and is reported to have a cool, sweet taste (Sicard and Leroy 1983). Mannitol is widely used in the food and pharmaceutical industries because of its unique functional properties. It is about 50% as sweet as sucrose and has a desirable cooling effect often used to mask bitter tastes. Mannitol is noncariogenic and has a low caloric content. It is suitable for ingestion and has been used safely around the world for over 60 years (Sicard and Leroy 1983; Schiweck et al. 2011).

Production of mannitol by extraction of plant raw materials is no longer economically important. Only in China mannitol is still obtained by acid extraction of marine algae. Mannitol can be obtained from different sources; microbial formation of D-mannitol occurs with fungi (Vega and Tourneau 1971) or bacteria (Foster 1949), starting with glucose, fructose, sucrose, or the tubers of Jerusalem artichokes (Schlubach 1953). The precursor of mannitol in its biosynthesis in the mushroom *Agaricus bisporus* is fructose. Mannitol is produced from glucose in 44% yield after six days by aerobic fermentation with *Aspergillus candidus* (Smiley et al. 1967) and in 30% yield after 10 days with *Torulopsis mannitofaciens* (Onishi 1971). It is formed by submerged culture fermentation of fructose with *Penicillium chrysogenumin* 7.3% conversion (Abdel-Akher et al. 1967). Small quantities of mannitol are found in wine (Barker et al. 1958). Today, mannitol is usually produced industrially (Figure 3.9) from sucrose, after inversion to glucose–fructose syrup (ratio 1:1), or from a glucose solution, which is also converted to high-fructose corn syrup. Theoretically, 25% mannitol and 75% sorbitol are formed in the neutral pH range from a 1:1 D-glucose–D-fructose syrup. Hydrogenation occurs under high pressure in the presence of Raney nickel catalysts. As a result of its lower solubility compared to sorbitol, mannitol is isolated from the concentrated hydrogenation solution by crystallization. The filtrate still contains considerable amounts of mannitol, which can be separated chromatographically (Oy 1973). If sucrose is used as starting material and hydrogenation is performed in alkaline solution, mannitol yields of up to 31% can be obtained. In a pure glucose solution, moreover, D-glucose can be epimerized to D-mannose before hydrogenation by the addition of an acidic solution of ammonium molybdate. Mannose is then hydrogenated directly to D-mannitol (Schiweck et al. 2011).

Pure D-mannose is not yet commercially available, but it can be obtained by acid hydrolysis of the mannan of ivory nutmeal in 35% yield and from spent sulfite liquor or prehydrolysis extracts from conifers through the sodium bisulfite mannose adduct or methyl α-D-mannoside. Reduction of fructose leads to sorbitol and D-mannitol in equal parts. Sucrose, on reduction under hydrolyzing conditions, also yields the same products in the ratio of three parts of sorbitol to one of D-mannitol. Commercially, D-mannitol is obtained by the reduction of invert sugar. In alkaline media, glucose, fructose, and mannose are interconverted (Lobry de Bruyn and Alberda van Ekenstein 1895; Wolfrom and Lewis 1928).

L-Mannitol does not occur naturally; however, it is obtained by the reduction of L-mannose or L-mannonic acid lactone (Baer and Fischer 1939). It can be synthesized from the relatively abundant L-arabinose through the L-mannose and L-glucose cyanohydrins, conversion to the phenylhydrazines that are separated, liberation of L-mannose, and reduction with sodium borohydride. Another synthesis is from L-inositol (obtained from its monomethyl ether, quebrachitol) through the diacetonate, periodate oxidation to the blocked dialdehyde, reduction, and removal of the acetone blocking groups (Kuhn and Klesse 1958).

3.3.5 D-Tagatose

Tagatose, (*3S,4S,5R*)-1,3,4,5,6-pentahydroxy-hexan-2-one, is a ketohexose C-4 fructose epimer potentially obtainable by the oxidation of the corresponding hexitol D-galactitol (Figure 3.10). It is obtained from D-galactose by isomerization under alkaline conditions in the presence of calcium (Manzoni et al. 2001). It has an excellent sucrose-like taste, has no cooling effect, has no aftertaste or potentiation of off-flavor, and can thus be considered a possible alternative bulking sweetener. Although tagatose is present in some foods and is obtainable from natural sources, its availability appears limited and its recovery expensive, a major impediment to its use in the food industry (Levin et al. 1995). However, the initial FDA proposal states the use of fungal enzymes in the preparation of this product, in its manufacture from lactose.

Compared to other polyols, tagatose is not common in nature. It is a sugar naturally found in small amounts in milk. It can be produced commercially from lactose, which is first hydrolyzed to glucose and galactose. The galactose is isomerized under alkaline conditions to D-tagatose by calcium hydroxide. The resulting mixture can then be purified and solid tagatose produced by crystallization (Figure 3.11).

It was reported that tagatose does not support fat deposition and appears to deliver less usable energy than sucrose (Livesey and Brown 1996). Similar metabolic studies using [14C] tagatose have demonstrated that it is metabolized in both the oral and intravenous routes of administration, with an effective energy value of zero (Levin et al. 1995; Livesey and Brown 1995). Recent studies

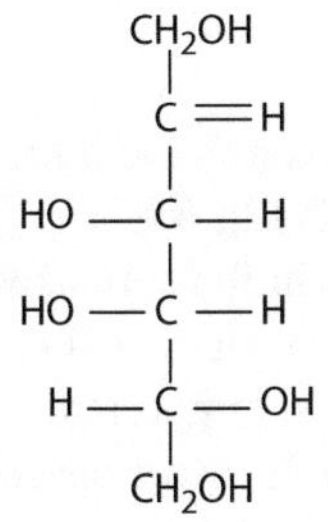

$$\begin{array}{c} CH_2OH \\ | \\ C{=}H \\ | \\ HO-C-H \\ | \\ HO-C-H \\ | \\ H-C-OH \\ | \\ CH_2OH \end{array}$$

Figure 3.10 Structure of tagatose.

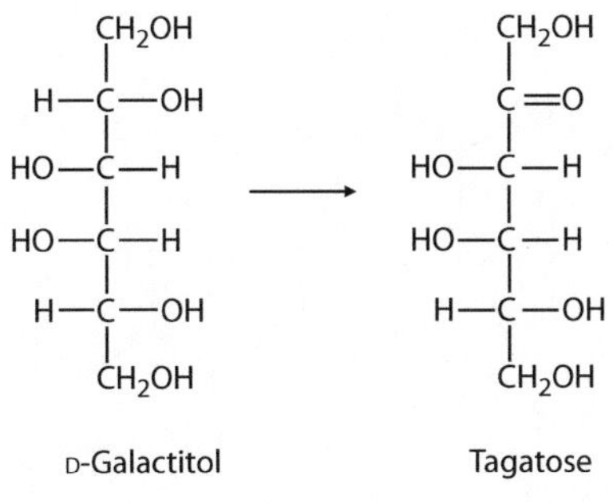

$$\begin{array}{c} CH_2OH \\ | \\ H-C-OH \\ | \\ HO-C-H \\ | \\ HO-C-H \\ | \\ H-C-OH \\ | \\ CH_2OH \\ \text{D-Galactitol} \end{array} \longrightarrow \begin{array}{c} CH_2OH \\ | \\ C{=}O \\ | \\ HO-C-H \\ | \\ HO-C-H \\ | \\ H-C-OH \\ | \\ CH_2OH \\ \text{Tagatose} \end{array}$$

Figure 3.11 Formation of tagatose from D-galactitol.

have demonstrated that the intake of tagatose produces a significant increase in lactic acid bacteria, important inhabitants of the intestinal tract of man and animals where they provide functional benefits (Livesey 2003).

Tagatose is noncariogenic and also exhibits a probiotic effect, being a not-digestible food ingredient able to benefit humans by selectively stimulating the growth and/or activity of a limited number of bacteria, thus improving host health (Bertelsen et al. 1999; Johnson 1995). Tagatose induces butyrate production by a limited number of microorganisms, especially enterococci, lactobacilli, and a few other butyrate-producing bacteria, with a significant decrease in coliforms in the colon. Many studies highlight the importance of butyrate in the colon, where it is the major and preferred fuel for the epithelium (Bertelsen et al. 1999; Roediger 1980), and where it plays an important role in controlling the proliferation and differentiation of the epithelial cells (Johnson 1995). Based on the documented selection of bacteria and butyrate effects in the colon, tagatose represents a promising food bulk sweetener with a prebiotic action (Manzoni et al. 2001).

Originally developed as a reduced-calorie sugar substitute, D-tagatose is poorly absorbed in the small intestine, thus preventing the stimulation of insulin secretion and lowering blood glucose levels. D-Tagatose passes directly to the lower intestine, where it is fermented by bacteria to produce short-chain fatty acids and carbon dioxide. It is considered a prebiotic, promoting more favorable microbial flora in the colon (O'Brien and Gelardi 1991).

Tagatose gained GRAS status, with the subsequent possibility of commercializing it as a bulk sweetener (GAIO, Arla Foods) in April 2001. Although tagatose is obtainable from natural sources, its availability appears limited and its recovery expensive, a major impediment to its use in the food industry (Levin et al. 1995).

3.4 DISACCHARIDE POLYOLS

3.4.1 Maltitol

Maltitol, 4-O-α-D-glucopyranosyl-D-glucitol, is a disaccharide consisting of a glucose unit linked to a sorbitol one via an a-1,4 bond (Figure 3.12). It has an mp of 146°C–147°C. Maltitol forms orthorhombic crystals. It is weakly hygroscopic in the solid state. The viscosity of aqueous maltitol solutions corresponds to that of sucrose solutions. In the pH range 5–7, maltitol and maltitol solutions are stable up to about 150°C. Lower pH values, temperature >50°C, and longer contact times result in cleavage of the disaccharide bond. Maltitol is also cleaved by enzyme systems (e.g., of the mucosa). The relative sweetness level of maltitol is generally given as 0.65, although values up to 0.9 are encountered in chocolates (Dziedzic and Kearsley 1984; Rugg-Gunn 1989; Schiweck et al. 2011).

Because of the nonreducing character of its molecular structure, maltitol has a high stability on both the thermal and chemical levels (Schouten et al. 1999). Although several of its properties are

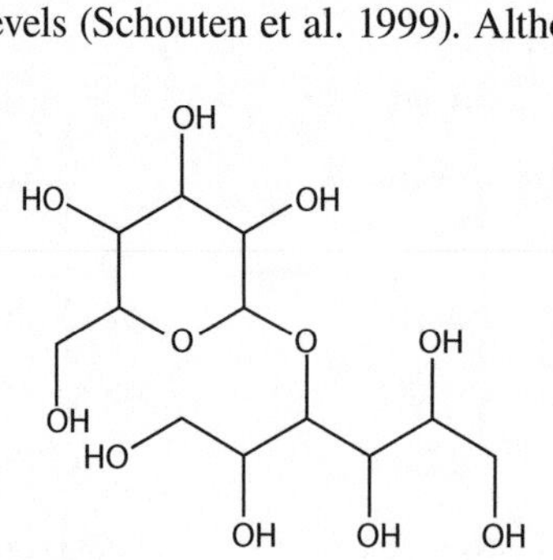

Figure 3.12 Maltitol structure.

comparable to those of sucrose such as solubility in water, viscosity of aqueous solutions, hygroscopicity, and sweetening potency, maltitol has a lower energy value and higher heat of dissolution than that of sucrose, which prepares this polyol for specific applications like food and pharmaceutical products with low calorie. Maltitol finds more and more applications as a low calorie sweetener and a cariogenic additive in toothpastes, mouthwash, and tablets (Maguire et al. 2000).

High-maltose glucose corn syrups are the starting material from which pure maltose is obtained by chromatographic separation on ion-exchange resins for maltitol production. Depending on the separation technique, maltose syrups with different purities (85%–98%) are obtained. The syrups are evaporated to a solids content of 50% and then hydrogenated. Initially, hydrogenated syrups were spray-dried together with a carrier (alginates, methyl celluloses). The products have a maltitol content of >85% and are very hygroscopic. Crystallization of maltitol from water was achieved in the early 1980s. Continuous crystallization processes are also being used now (Kristott and Jones 1992; Ohno et al. 1982).

Crystallization is the most difficult step to control during all stages of maltitol preparation, and to be able to control the size distribution and the purity of the produced crystals is the important task. Crystallization is an important unit of operation for the manufacture of many sugars, polyols, salts, etc. It represents a separation technique that is also used for product purification. The product obtained by crystallization has to show well-defined chemical and physical properties, as well as a certain crystal size distribution. This process involves many factors not always easy to control and largely affects the final product quality and particularly crystal purity. The presence of impurities in industrial sugar syrups leads to important modifications in the crystal shape. Indeed, the presence of maltotriitol in the crystallization medium has been shown to control the formation of bipyramidal or prismatic maltitol crystals (Capet et al. 2004; Leleu et al. 2008). On the other hand, if the crystallization process is not technically optimized, amorphous structures may be obtained, and their evolution to more stable structures during storage leads generally to bad shelf-life stability (Gharsallaou et al. 2010).

Digestion of maltitol requires hydrolysis before absorption. Absorption in human subjects is reported to range from 5% to 80% (Beaugerie et al. 1990); the wide range is partly due to the use of invasive methods and partly due to the incorrect evaluation of results from noninvasive methods. Previous studies revealed a lower limit to absorption of 35% for maltitol (10 g) in the solution. Second, glycemia and insulinemia indicate a lower limit to absorption of 35% to 27%, respectively, for maltitol (25–50 g) in the solution. Third, based on indirect calorimetry following the ingestion of a high-polymer maltitol syrup containing 50% maltitol and 50% polymer and separate study of the polymer fraction (Sinaud et al. 2002), the energy value of maltitol can be estimated. This estimated energy value corresponds to maltitol absorption of approximately 32% when consumed in three mixed solid meals interspersed by three maltitol drinks (totaling 50 g maltitol in 50 g polymer daily). On the basis of energy values for maltitol proposed by several authorities, absorbability by consensus is 45% (American Diabetes Association 2001; Australia New Zealand Food Authority 2001; Bar 1990; Bernier and Pascal 1990; Brooks 1995; Dutch Nutrition Council 1987; Life Sciences Research Office 1994, 1999). The products of hydrolysis by intestinal brush-border disaccharidases are glucose and sorbitol (Livesey 2003).

The enhancing effects of maltitol on intestinal calcium absorption were first reported by Goda et al. (1992), who demonstrated that the consumption of a 10% maltitol diet by rats resulted in elevated calcium absorption. *In vitro* experiments using everted ileal segments of rats suggested that maltitol accelerated passive diffusion of calcium in the lower part of the small intestine (Goda et al. 1993; Kishi et al. 1996). Another disaccharide sugar alcohol, lactitol (Ammann et al. 1988), and the monosaccharide sugar alcohols sorbitol (Vaughan and Filer 1960; Suzuki et al. 1985), mannitol (Armbrecht and Wasserman 1976), and xylitol (Hämäläinen et al. 1985) were reported to increase intestinal calcium absorption. Although it has been assumed that the enhancement of intestinal calcium absorption by maltitol and other sugar alcohols results from enhancement of passive diffusion

(Bronner 1987; Goda et al. 1993; Kishi et al. 1996), which might be triggered by interactions of the sugar alcohols with the brush-border membranes (Suzuki et al. 1985; Kishi et al. 1996), the exact mode of action is unknown. Although intake of large quantities (approximately 20–70 g/day) of nondigestible saccharides and sugar alcohols causes diarrhea (Ellis and Krantz 1941; Patil et al. 1987), the maximum noneffective dose of maltitol is relatively high, approximately twice that of sorbitol (Koizumi et al. 1983a,b). Maltitol is thus a potentially useful agent as an enhancer of the intestinal calcium absorption (Fukahori et al. 1998).

Because maltitol's hygroscopicity and capacity for browning or flavor development differ, the applications of maltitol correspond largely to those of sugar. Maltitol can be used in the production of chocolates, hard and soft caramel, toffees, gummy bears, chewing gum, jams, ice cream, etc. (Fabry and Grenby 1987).

3.4.2 Isomalt

Isomalt, (2R,3R,4R,5R)-6-[[(2S,3R,4S,5S,6R)-3,4,5-trihydroxy-6-(hydroxymethyl)-2-tetrahydropyranyl]oxy]hexane-1,2,3,4,5-pentol, is a mixture of the diastereomers 6-O-a-D-glucopyranosyl-D-sorbitol (isomaltitol, 1,6-GPS) and 1-O-a-D-glucopyranosyl-D-mannitol (GPM; the 1-O and 6-O-bound mannitol derivatives are identical because of the symmetry of the mannitol part of the molecule) obtained by hydrogenating isomaltulose, which is enzymatically derived from sucrose (Figure 3.13). The two isomers are formed in different ratios, depending on reaction conditions for the hydrogenation of isomaltulose (Wohlfarth 2006). Indeed, 1,6-GPS is also formed in small amounts in the catalytic hydrogenation of starch hydrolysates. These isomers belong to the group of disaccharide alcohols (Schiweck et al. 2011).

With a comparable water content of about 2%, isomalt, in the melt, has a lower viscosity, higher specific heat capacity, and higher boiling-point elevation than sucrose (Mende and Schiweck 1991). The components of isomalt are hydrolyzed by acid to give 1 mol each of glucose and sorbitol or mannitol. Compared to sucrose and similar disaccharides, the hydrolysis is clearly slower. For this reason, isomalt can be regarded as acid stable in most food applications. Like those of sucrose, the hydroxyl groups of the components of isomalt can be etherified and esterified (Grillo-Werke 1992).

Isomalt is manufactured in a two-stage process in which sugar is first transformed into isomaltulose, a reducing disaccharide (6-O-α-D-glucopyranosido-D-fructose). The isomaltulose is then hydrogenated using a Raney nickel catalytic converter. The final product—isomalt—is an equimolar composition of 6-O-α-D-glucopyranosido-D-sorbitol (1,6-GPS) and 1-O-α-D-glucopyranosido-D-mannitol-dihydrate (1,1-GPM-dihydrate; Figure 3.14). Now, isomalt has been approved for use in the United States since 1990. It is also permitted for use in Australia, New Zealand, Norway, Iran, and the Netherlands (McNutt and Sentko 1996).

Because of its pure sweet taste (sweetening power = 0.45 relative to sucrose in about 10% solution), its noncariogenicity, and its low calorific value (0.5–0.6 relative to sucrose), isomalt is almost the ideal sugar substitute for most applications in confectionery. In the production of foods, isomalt

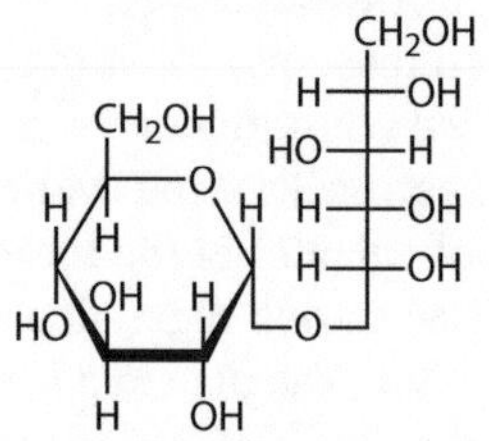

Figure 3.13 Isomalt structure.

1. Step sucrose (nonreducing) $\xrightarrow{\text{Biocatalyst}}$ Isomaltulose (reducing)

2. Isomaltulose $\xrightarrow{\text{Hydrogenation}}$ 6-O-α-D-glucopyranosyl-D-sorbitol (1,6-GPS) + 1-O-α-D-glucopyranosyl-D-mannitoldihydrate(1,1-GMP)

Figure 3.14 Production of isomalt from sucrose.

generally does not react with other components of the formulation. It can usually be processed in plants designed for sucrose as the raw material. Isomalt is used primarily for its sugar-like physical properties. It has only a small impact on blood sugar levels and does not promote tooth decay. It has 2 kcal/g, half the calories of sugars (Wohlfarth 2006; Ndindayino et al. 2002).

Due to its properties such as taste, mouthfeel, low calorie content, carcinogenicity, suitability to diabetics, and low hygroscopicity, it offers several advantages over most polyols when formulated in pharmaceutical dosage forms, particularly when used as a tablet excipient. Isomalt is typically blended with a high-intensity sweetener such as sucralose, so that the mixture has approximately the sweetness of sugar (Strater 1989; Fritzsching 1993).

Although isomalt-containing tablets can be manufactured by direct compression, the drug concentration of these tablets was limited to 30% as higher drug loads yielded tablets of unacceptable quality. However, literature reports indicated that the compression properties of polyols were successfully enhanced when these products were melted prior to compression (Kanig 1964; Serpelloni 1990). The melting range of isomalt is between 145°C and 150°C, and no decomposition is detected when the product is melted (Cammenga and Zielasko 1996); this material can be thermally treated using the hot stage extrusion technique (melt extrusion). Furthermore, neither browning reactions nor caramel tasting was observed after melting of isomalt (Strater 1989; Fritzsching 1993).

However, like most sugar alcohols (with the exception of erythritol), it carries a risk of gastric distress, including flatulence and diarrhea, when consumed in large quantities. Therefore, isomalt is advised not to be consumed in quantities larger than about 50 g per day for adults and 25 g per day for children. Isomalt may prove upsetting to the stomach because the body treats it as a dietary fiber instead of as a simple carbohydrate. Therefore, like most fibers, it can increase bowel movements, passing through the bowel in virtually undigested form. As with other dietary fibers, regular consumption of isomalt might eventually lead one to become desensitized to it, decreasing the risk of stomach upset (Wohlfarth 2006).

Several studies including noninvasive methods in human subjects and methods in animals (Livesey 1990a,b, 2000) on glycemia and insulinemia suggest that 0% to 14% of isomalt is available as carbohydrate in man. On the basis of the energy values of isomalt, it is concluded that approximately 90% isomalt is fermented in the colon, with a stoichiometry *in vivo* and *in vitro* indicating relatively little H_2 gas production (Brooks 1995; Dutch Nutrition Council 1987; Life Sciences Research Office 1994; Livesey 1992; Livesey et al. 1993).

3.4.3 Lactitol

Lactitol (4-O-α-D-galactopyranosyl-D-glucitol; Figure 3.15) is a synthetic disaccharide sugar alcohol produced by catalytic hydrogenation of the glucose moiety of lactose (Abrahamse et al. 1999; Anonymous 1988; Soontornchai et al. 1999). It has a good solubility in water of 57.1 g/100 g solution at 20°C, which is roughly in the range of that of sucrose (66.7 g/100 g solution). With increasing temperature, the solubility of lactitol rises considerably, exceeding the solubility of sucrose at temperatures above about 50°C. In contrast to that, the solubility in ethanol (0.75 g/100 g

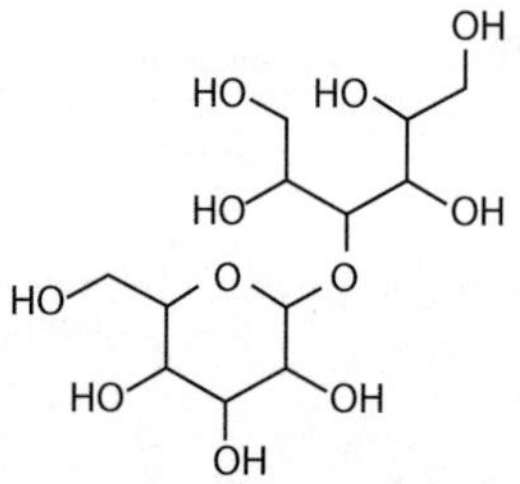

Figure 3.15 Lactitol structure.

solution of the monohydrate at 20°C) is low. Anhydrous lactitol has a very low hygroscopicity. It has the lowest hygroscopicity with the exception of mannitol. This effect can be used for application as food ingredient if moisture pick-up should be avoided, that is, as dusting agent in chewing gum or hard-boiled candies (Schiweck et al. 2011).

The sweetness of lactitol is 30%–40% that of sucrose. The taste is clean and sweet. Lactitol exhibits the lowest sweetness among the polyols. This might be beneficial for applications where a significant sweetness is not desired, for example, in products (Schiweck et al. 2011) like marzipan or surimi. On the other hand, sweetness can be slightly increased by adding high-intensity sweeteners (aspartame, acesulfame-K, and sucralose). With addition of 0.03% aspartame or acesulfame-K to a 10% aqueous solution of lactitol, the sweetness of a 10% sucrose solution is achieved (Smith and Hong-Shun 2003).

The preparation of lactitol by hydrogenation of lactose is realized with a nickel catalyst (van Velthuijsen 1979; Wolfrom et al. 1938). Lactitol can be prepared by reduction of lactose with sodium borohydride (Scholnick et al. 1975; van Velthuijsen 1979), but technically lactitol is prepared by hydrogenation of a lactose solution at about 100°C with a Raney nickel catalyst, as lactitol is produced essentially in the same way from lactose as sorbitol is produced from glucose. The reaction is carried out in an autoclave under a pressure of 40 atm or more. Due to the lower solubility of lactose, compared to glucose, the lactose concentration at the beginning of the reaction is only 30–40 wt%. When the hydrogenation reaction is completed, the catalyst is sedimented and filtrated, and the lactitol solution is purified by ion-exchange resins and activated carbon. The purified lactitol solution is then concentrated by evaporating the water, the obtained syrup is crystallized, and the crystals are separated with a centrifuge and finally dried. The lactitol mother liquor is concentrated again, giving more crops of the lactitol monohydrate. The final mother liquor can be used as a 64% solution or in mixtures with sorbitol to prepare noncrystallizing lactitol syrup at a concentration of 70% (Figure 3.16; van Velthuijsen 1979).

On hydrolysis, the molecule is split into D-galactose and D-sorbitol, both occurring widely in nature. Due to the absence of a carbonyl group, lactitol is chemically more stable than the related

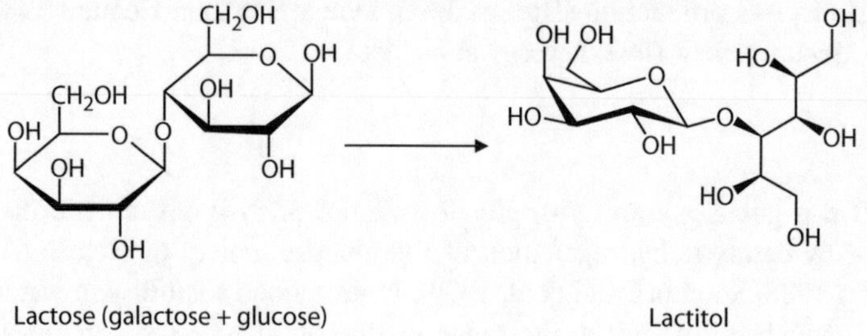

Lactose (galactose + glucose)

Lactitol

Figure 3.16 Lactitol formation from lactose.

disaccharides like lactose. The stability of lactitol in the presence of alkali is markedly higher than that of lactose. The stability of lactitol in the presence of acids is comparable to that of lactose (Wolfrom et al. 1938). When heated at 100°C for 4 h at pH 1 and 2 (adjusted with HCl), the rates of hydrolysis of 10% lactitol solutions are 5.6% and 1.4%, respectively. The solutions remain colorless under these conditions. The comparable hydrolysis of lactose are 5.4% and 1.3%, respectively. Like other sugar alcohols, lactitol shows only a very slight discoloration, as a result of caramelization reactions, when heated at temperatures of 150°C–200°C (van Velthuijsen 1979).

It is currently used as a bulk sweetener in calorie-controlled foods. Discovered in 1920, it was first used in foods in the 1980s (van Velthuijsen and Blankers 1991). The relative sweetness level of lactitol is about 35% as compared to sucrose and 110%–140% to lactose (Uyl 1987; van Velthuijsen 1979; Siijonmaa et al. 1978). Lactitol has a clean sweet taste that closely resembles the taste profile of sucrose. Its mild sweetness makes it an ideal bulk sweetener to partner with low-calorie sweeteners such as acesulfame-K, aspartame, neotame, saccharin, and sucralose (Van Es et al. 1986).

Lactitol offers many interesting applications for the food industry, especially for the fields of dietetic and low-caloric foods. In many products, lactitol can replace sucrose without drastic changes in formulations or manufacturing processes. Depending on the specific application, lactitol can be used as a solution (e.g., for jams, beverages, hard candies) or as the crystalline, nonhygroscopic monohydrate (e.g., for chocolate, chewing gum, bakery products; van Velthuijsen 1979).

As a sweetening ingredient, lactitol has a low glycemic index, does not induce an increase in blood glucose or insulin levels, and contributes half the calories of most other carbohydrates (2 calories per gram). Foods using lactitol to replace sugar can be used by people with diabetes, giving them a wider variety of low-calorie and sugar-free choices. Lactitol is not metabolized by oral bacteria, which break down sugars and starches to release acids that may lead to cavities or erode tooth enamel (Grenby and Desai 1988).

In a patent application (Hayashibara 1974), it was demonstrated with rats that lactitol inhibits the absorption of sucrose and also the formation of cholesterol. The increase in the blood glucose content after consumption of a 1:1 mixture of sucrose and lactitol was about half of the increase after consumption of sucrose only (the amount of sucrose intake was the same), whereas the formation of liver glycogen with the 1:1 mixture was only one-fifth of that with only sucrose. Diets with added cholesterol showed that the inclusion of lactitol in the diets of rats resulted in a reduction to about 50% of the contents of liver cholesterol and of the total serum cholesterol (van Velthuijsen 1979).

Unlike the metabolism of lactose, lactitol is not hydrolyzed by lactase. It is neither hydrolyzed nor absorbed in the small intestine. Lactitol is metabolized by bacteria in the large intestine, where it is converted into biomass, organic acids, carbon dioxide, and a small amount of hydrogen. The organic acids are further metabolized resulting in a caloric contribution of 2 calories per gram (carbohydrates generally have about 4 calories per gram; van Velthuijsen and Blankers 1991).

Very little of this disaccharide polyol is absorbed, perhaps 2% as lactitol and its hydrolysis products galactose and sorbitol. This is due to a very low activity of β-galactosidase in the human intestine (Grimble et al. 1988; Nilsson and Jagerstad 1987). The liver readily uses absorbed galactose and sorbitol in either hepatic glycogen storage or hepatic glucose production. Unabsorbed lactitol is completely fermented with a stoichiometry giving a generous yield of H_2 gas *in vivo* and *in vitro* (Livesey et al. 1993) and butyric acid *in vitro* (Clausen et al. 1998).

Lactitol is fermented in the colon and consequently has beneficial effects on the colonic microflora. A reduction in the pH of the colon, along with an increase in probiotic bacteria and a significant reduction in potential pathogens, emphasizes the beneficial effects of lactitol. In essence, lactitol functions as a prebiotic (van Velthuijsen and Blankers 1991).

Lactitol has been shown to be as effective as lactulose for the treatment of chronic stable hepatic encephalopathy, but it was more acceptable as a medication (Lanthier and Mosgan 1985; Patil et al. 1987; Scevola et al. 1989). Its laxative effect and other side effects (i.e., osmotic diarrhea, flatulence) diminish when administered regularly. The metabolism of lactitol is different from that of carbohydrates. In certain respects, lactitol is similar to dietary fiber. It is also largely metabolized by the bacterial gut flora (Dills 1989). The clinical benefits of administrating lactitol have been investigated in adults suffering from chronic functional constipation (Ravelli et al. 1995; Vanderdonckt et al. 1990) and for the management of episodic and acute hepatic encephalopathy in cirrhotic patients (Lanthier and Mosgan 1985; Scevola et al. 1993; Tarao et al. 1995).

3.4.4 Trehalose

Trehalose (α,α-trehalose; α-D-glucopyranosyl-(1→1)-α-D-glucopyranoside) is a disaccharide composed of two glucose molecules bound by an alpha, alpha-1, 1 linkage (Figure 3.17). The bonding makes trehalose very resistant to acid hydrolysis and therefore is stable in solution at high temperatures, even under acidic conditions. The bonding also keeps nonreducing sugars in closed-ring form, such that the aldehyde or ketone end groups do not bind to the lysine or arginine residues of proteins (a process called glycation). Trehalose is broken down by the enzyme trehalase into glucose and has about 45% the sweetness of sucrose. It is less soluble than sucrose, except at high temperatures (>80°C). Trehalose forms a rhomboid crystal as the dihydrate and has 90% of the calorific content of sucrose in that form. Anhydrous forms of trehalose readily regain moisture to form the dihydrate and can show interesting physical properties when heat-treated (Higashiyama 2002).

Trehalose was previously being manufactured through an extraction process from cultured yeast, but, since production costs were prohibitive, use was limited to only certain cosmetics and chemicals. In 1994, Hayashibara, a saccharified starch maker in Okayama prefecture, Japan, discovered a method of inexpensively mass-producing trehalose from starch (Hayashibara 1994). The following year, Hayashibara started manufacturing trehalose by activating two enzymes, the glucosyltrehalose-producing enzyme that changes the reducing terminal of starch into a trehalose structure and the trehalose free enzyme that detaches this trehalose structure. As a result, a high-purity trehalose from starch can be mass-produced for a very low price (Higashiyama 2002).

Since the reducing end of a glucosyl residue is connected with the other, trehalose has no reducing power. Trehalose is widely distributed in nature. It is known to be one of the sources of energy in most living organisms and can be found in many organisms including bacteria, fungi, insects, plants, and invertebrates. Mushrooms contain up to 10%–25% trehalose by dry weight. Furthermore, trehalose protects organisms against various stresses such as dryness, freezing, and osmopressure. In the case of resurrection plants, which can live in a dry state, when the water dries up, the plants also dry up. However, they can successfully revive when placed in water. The anhydrobiotic organisms are able to tolerate the lack of water owing to their ability to synthesize large quantities of trehalose, and the trehalose plays a key role in stabilizing membranes and other macromolecular assemblies under extreme environmental conditions (Higashiyama 2002).

Trehalose's relative sweetness level is 45% of sucrose. Trehalose has high thermostability and a wide pH-stability range. Therefore, it is one of the most stable saccharides. When 4% trehalose solutions with pH 3.5–10 were heated at 100°C for 24 h, no degradation of trehalose was observed in any

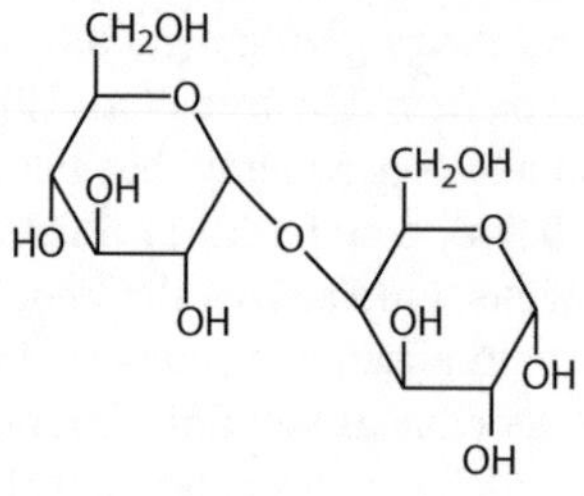

Figure 3.17 Structure of trehalose.

case. Because of nonreducing sugar, this saccharide does not show Maillard reaction with amino compounds such as amino acids or proteins. Its particular physical features make it an extremely attractive substance for industrial applications. Furthermore, this saccharide shows good sweetness like sucrose, and in the food industry, this saccharide is used as a sweetener (Yuan-Tseng et al. 2008).

Trehalose can play a number of different roles in biological systems, including serving as a reservoir of glucose for energy and/or carbon (Susman and Lingappa 1959); functioning as a stabilizer or protectant of proteins and membranes during times of stress (Arguillis 2000); acting as a regulatory molecule in the control of glucose metabolism (Thevelein and Hohmann 1995); serving as a transcriptional regulator (Burklen et al. 1998); and playing a structural and functional role as a component of various cell wall glycolipids in mycobacteria and related organisms (Brennan and Nikaido 1995).

Trehalose has good stabilizing functions, namely, preventing starch retrogradation, protein denaturation, and lipid degradation. For example, trehalose shows poor reactivity against amino compounds in food because it has high thermostability, wide pH-stability range, and no reducing power. It also masks unpleasant tastes and odors in food. Therefore, it is superb for the maintenance of food quality. It was also reported that trehalose is capable of suppressing degradation of fatty acid. It is indicated that trehalose has a suppressive effect on the auto-oxidation of unsaturated fatty acids (Higashiyama 2002).

It was suggested that trehalose might have a kind of suppressive effect on the development of osteoporosis (Nishizaki et al. 2000). The results further imply that the daily ingestion of trehalose-containing foods could be useful both for bone metabolism and prevention of osteoporosis. Furthermore, it has been shown that trehalose could protect corneal epithelial cells in culture from death by desiccation and suppress tissue denaturalization (Hirata et al. 1994; Matsuo 2001). Trehalose has been considered as a potential new drop for dry eyes syndrome and for effective preservation of organs (Higashiyama 2002).

3.5 POLYSACCHARIDE POLYOLS

3.5.1 Hydrogenated Starch Hydrolysates

Hydrogenated starch hydrolysates (HSH) are a mixture of several sugar alcohols (a type of sugar substitute). They were developed by a Swedish company in the 1960s. HSH are produced by the partial hydrolysis of starch. Most often, corn starch is used; however, potato starch or wheat starch is also used. This creates dextrins (glucose and short glucose chains). The hydrolyzed starch (dextrin) then undergoes hydrogenation to convert the dextrins to sugar alcohols (Livesey 2003; Wheeler et al. 1990).

HSH are produced by the partial hydrolysis of corn, wheat, or potato starch and subsequent hydrogenation of the hydrolysate at high temperature under pressure. The end product is an ingredient composed of sorbitol, maltitol, and higher hydrogenated saccharides (maltitriitol and others). By varying the conditions and extent of hydrolysis, the relative occurrence of various monomeric, dimeric, oligomeric, and polymeric hydrogenated saccharides in the resulting product can be obtained. A wide range of polyols (also known as sugar alcohols) that can satisfy varied requirements with respect to different levels of sweetness, viscosity, and humectancy can, therefore, be produced (McFetridge et al. 2004; Modderman 1993).

Hydrogenated starch hydrolysate is similar to sorbitol. If the starch is completely hydrolyzed, there are only single glucose molecules, and then after hydrogenation, the result is sorbitol. Because in HSH the starch is not completely hydrolyzed, a mixture of sorbitol, maltitol, and longer chain hydrogenated saccharides (such as maltotriitol) is produced. When there is no single dominant polyol in the mix, the generic name "hydrogenated starch hydrosylate" is used. However, if 50% or more of the polyols in the mixture are of one type, it can be labeled as "sorbitol syrup," "maltitol syrup," etc. (McFetridge et al. 2004; Schiffman and Gatlin 1993).

HSH or polyglycitols, including hydrogenated glucose syrups, maltitol syrups, and sorbitol syrups, are a family of products found in a wide variety of foods. They serve a number of functional roles, including use as bulk sweeteners, viscosity or bodying agents, humectants, crystallization modifiers, cryoprotectants, and rehydration aids. They can also serve as sugar-free carriers for flavors, colors, and enzymes (Modderman 1993).

HSH have the following properties:

- Pleasant-tasting bulk sweeteners that blend well with other sweeteners and are synergistic with low-calorie sweeteners (e.g., acesulfame-K, aspartame, neotame, saccharin, and sucralose)
- Blend well with flavors and can mask unpleasant off-flavors
- Reduced-calorie alternatives to sugar, having not more than 3 calories per gram
- Used in a variety of products; exceptionally well suited for sugar-free candies because they do not crystallize
- Do not contribute to the formation of dental caries
- May be useful as alternatives to sugar for people with diabetes on the advice of their physician

Polyols, including HSH, are resistant to metabolism by oral bacteria that break down sugars and starches to release acids that may lead to cavities or erode tooth enamel. They are, therefore, non-cariogenic. HSH absorption is slow and incomplete. Therefore, the rise in blood glucose and insulin response associated with the ingestion of glucose is significantly reduced when HSH are used as alternative sweeteners. The reduced caloric value (75%, or less, that of sugar) of HSH is consistent with the objective of weight control. Products in which HSH replace sugar may, therefore, be of use providing a wider variety of reduced-calorie and sugar-free choices to people with diabetes (Modderman 1993).

Absorption of HSH by the body is slow, allowing a portion of HSH to reach the large intestine where metabolism yields fewer calories. Therefore, unlike sugar that contributes 4 calories per gram, the caloric contribution of HSH is not more than 3 calories per gram. For a product to qualify as "reduced calorie" in the United States, it must have at least a 25% reduction in calories. HSH may, therefore, be of use in formulating reduced-calorie food products (Livesey 2003).

3.6 CONCLUSIONS

Due to their importance in the chemistry, food and pharmaceutical industry extensive studies were conducted on sugar alcohols. They provide different taste, flavor, and sensory benefits to consumers because of their properties, which make them unique in food and pharmaceutical formulations. However, safety issues of the sugar alcohols need to be studied extensively as their use dramatically increases in food and pharmaceutical formulations. The polyols and similar substances used as bulk sugar substitutes in the United States are also safe, but consumers need to be aware of their presence in food products so that they can limit their intake sufficiently to avoid gastrointestinal discomfort.

REFERENCES

Abdel-Akher, M., Foda, I. O., and El-Nawawy, A. S. 1967. *Penicillium chrysogenum* 4176 to convert D-fructose to D-mannitol h submerged conditions. *Journal of Chemistry* U.A.R.10, 355.

Abrahamse, S. L., Pool-Zobel, B. L., and Rechkemmer, G. 1999. Potential of short chain fatty acids to modulate the induction of DNA damage and changes in the intracellular calcium concentration by oxidative stress in isolated rat colon cells. *Carcinogenesis* 20: 629–634.

Alper, H., Garcia, J. L., and Jang, E. J. 2002. New heterogeneous catalysis for the synthesis of poly(ether polyol)s. *Journal of Applied Polymer Science* 86(7): 1553–1557.

American Dental Association. 1998. *Position Statement on the Role of Sugar-Free Foods and Medications in Maintaining Good Oral Health*. ADA Publishing, Chicago, IL, USA.

American Diabetes Association. 2001. Postprandial blood glucose (a consensus statement). *Diabetes Care* 24: 775–778.

Aminoff, C., Vanninen, E., and Doty, T. E. 1978. The occurrence, manufacture and properties of xylitol. In *Xylitol*. Counsell, J. N. (ed.). Applied Science Publ., London, pp. 1–9.

Ammann, P., Rizzoli, R., and Fleisch, H. 1988. Influence of the disaccharide lactitol on intestinal absorption and body retention of calcium in rats. *Journal of Nutrition* 118: 793–795.

Anonymous. 1988. *Association of the British Pharmaceutical Industry (ABPI). Patient Information and Consent for Clinical Trials*. ABPI, London.

Aoki, M. A. Y., Pastore, G. M., and Park, Y. K. 1993. Microbial transformation of sucrose and glucose to erythritol. *Biotechnology Letters* 15: 338–383.

Arguillis, J. C. 2000. Physiological roles of trehalose in bacteria and yeast: a comparative analysis. *Archives of Microbiology* 174: 217–224.

Armbrecht, H. J., and Wasserman, R. H. 1976. Enhancement of Ca++ uptake by lactose in the rat small intestine. *Journal of Nutrition* 106: 1265–1271.

Arrigoni, E., Brouns, F., and Amadò, R. 2005. Human gut microbiota does not ferment erythritol. *British Journal of Nutrition* 94(5): 643–646.

Artık, N., Velioğlu, S., and Kavalci, B. 1993. Xylitol as a sugar alcohol: properties, production, and its uses in foods. *Gıda* 18(2): 101–109.

Australia New Zealand Food Authority. 2001. *Inquiry Report: Derivation of Energy Factors*. ANZFA, Canberra, Australia.

Baer, E., and Fischer, H. O. L. 1939. Studies on acetone-glyceraldehyde. VII. Preparation of l-glyceraldehyde and l(-)acetone glycerol. *Journal of American Chemical Society* 61: 761–765.

Bar, A. 1990. Factorial calculation model for the estimation of the physiological caloric value of polyols. In *Caloric Evaluation of Carbohydrates*. Hosoya, N. (ed.). Research Foundation for Sugar Metabolism, Tokyo, pp. 209–257.

Barker, S. A., Gomez-Sanchez, A., and Stacey, M. 1958. Studies of *Aspergillus niger*. X. Polyol and disaccharide production from acetate. *Journal of Chemical Society* 1958: 2583–2586.

Bässler, K. H. 1969. Adaptive processes concerned with absorption and metabolism of xylitol. In *International Symposium on Metabolism, Physiology and Clinical Use of Pentoses and Pentitols*. Horecker, B. L., Lang, K., and Takagi, Y. (eds.). Springer, Berlin, pp. 190–196.

Beaugerie, L., Fourie, B., Marteau, P., Pellier, P., Franchisseur, C., and Rambaud, J.-C. 1990. Digestion and absorption in the human intestine of three sugar alcohols. *Gastroenterology* 99: 717–723.

Bernier, J. J., and Pascal, G. 1990. The energy value of polyols (sugar alcohols). *Medicine et Nutrition* 26: 221–238.

Bernt, W. O., Borzelleca, J. F., Lamm, G., and Munro, I. C. 1996. Erythritol: a review of biological and toxicological studies. *Regulatory Toxicology and Pharmacology* 24: 191–197.

Bertelsen, H., Jensen, B. B., and Buemann, B. 1999. D-Tagatose—A novel low calorie bulk sweetener with prebiotic properties. In *Low-Calorie Sweeteners: Present and Future*. Corti, A. (ed.). S Karger Medical and Scientific Publisher, Basel, Switzerland, pp. 98–109.

Bornet, F. R. J., Blayo, A., Dauchy, F., and Slama, G. 1996. Plasma and urine kinetics of erythritol after oral ingestion by healthy humans. *Regulatory Toxicology and Pharmacology* 24: 220–285.

Boussingault, I. 1872. *Pharmaceutical Chemistry* 40: 36.

Boyers, G. G. 1959. U.S. Patent no. 2,868,847. Engelhard Industries, Inc.

Brennan, P. J., and Nikaido, H. 1995. The envelope of mycobacteria. *Annual Review of Biochemistry* 64: 29–63.

Bronner, F. 1987. Intestinal calcium absorption: mechanisms and applications. *Journal of Nutrition* 117: 1347–1352.

Brooks, S. P. J. 1995. *Report on the Energy Value of Sugar Alcohols*. Ministry of Health, Ottawa, Canada.

Burklen, L., Schock, F., and Dahl, M. K. 1998. Molecular analysis of the interaction between the Bacillus trehalose repressor TreR and the tre operator. *Molecular Genetics* 260: 48–55.

Burt, B. A. 2006. The use of sorbitol- and xylitol-sweetened chewing gum in caries control. *Journal of American Dental Association* 137(2): 190–196.

Calorie Control Council. 2011. Sorbitol. Available at http://www.caloriecontrol.org/sweeteners-and-lite/polyols/sorbitol. Accessed July 1, 2011.

Cammenga, H. K., and Zielasko, B. 1996. Glasses of sugars and sugar substitutes. *Physical Chemistry* 100: 1607–1609.

Capet, F., Comini, S., Odou, G., Louten, P., and Descamps, M. 2004. Orientated growth of crystalline anhydrous maltitol (4-o-alpha-d-glucopyranosyl-d-glucitol). *Carbohydrate Research* 339: 1225–1231.

Clausen, M. R., Jorgensen, J., and Mortensen, P. B. 1998. Comparison of diarrhea induced by ingestion of the fructo-oligosaccharide idolax and the disaccharide lactulose. *Digestive Diseases and Sciences* 43: 2696–2707.

Cosolo, W. C., Martinello, P., Louis, W. J., and Christophidis, N. 1989. Blood–brain barrier disruption using mannitol: time course and electron microscopy studies. *American Journal of Physiology* 256(2): 443–447.

Counsel, J. N. 1987. *Xylitol*. Applied Science Publishers, London, p. 7.

Creighton, H. J. 1939. The electrochemical reduction of sugars. *Transactions of the Electrochemical Society* 75: 301.

Deis, R. C. 2005. Low-Glycemic Foods—Ready for Prime Time? *Food Product Design*. http://www.foodproductdesign.com/articles/2005/04/food-product-design-health-nutrition--april-2005.aspx.

Demetrakopoulos, G. E., and Amos, H. 1978. Xylose and xylitol. *World Review of Nutrition and Dietetics* 32: 96–122.

Dills, W. L. 1989. Sugar alcohols as bulk sweeteners. *Annual Reviews of Nutrition* 9: 161–186.

Dutch Nutrition Council. 1987. *The Energy Values of Polyols. Recommendations of the Committee on Polyols*. Nutrition Council, The Hague.

Dziedzic, S. Z., and Kearsley, M. W. 1984. Physicochemical properties of glucose syrups. In *Glucose Syrups Science and Technology*. Dziedzic, S. Z., and Kearsley, M. W. (eds), Elsevier Appl. Sci. Publ., London.

Ellis, F. W., and Krantz, J. C. Jr. 1941. Sugar alcohols XXII. Metabolism and toxicity studies with mannitol and sorbitol in man and animals. *Journal of Biological Chemistry* 141: 147–154.

Fabry, I., and Grenby, T. H. 1987. *Developments in Sweeteners—3*. Elsevier Applied Science, London, pp. 83–108.

Förster, H. 1974. Comparative metabolism of xylitol, sorbitol and fructose. In *Sugars in Nutrition*. Sipple, H. L., and McNutt, K. W. (eds.). Academic Press, New York, pp. 259–280.

Foster, J. W. 1949. *Chemical Activities of Fungi*. Academic Press, Inc., New York, pp. 470–472.

Fritzsching, B. 1993. Isomalt, a sugar substitute ideal for the manufacture of sugar-free and calorie-reduced confectionery. In *Conference Proceedings—Food Ingredients Europe*. Maarssen, The Netherlands, pp. 371–377.

Froesch, E. R., and Jakob, A. 1974. The metabolism of xylitol. In *Sugars in Nutrition*. Sipple, H. L., and McNutt, K. W. (eds.). Academic Press, New York, pp. 241.

Fukahori, M., Sakurai, H., Akatsu, S., Negishi, M., Sato, H., Goda, T., and Takase, S. 1998. Enhanced absorption of calcium after oral administration of maltitol in the rat intestine. *Journal of Pharmacy and Pharmacology* 50: 1227–1232.

Gare, F. 2003. *The Sweet Miracle of Xylitol*. Basic Health Publications, Inc., North Bergan, NJ.

Georgieff, M., Moldawer, L. L., Bistrian, B. R., and Blackburn, G. L. 1985. Xylitol, an energy source for intravenous nutrition after trauma. *Journal of Parenteral and Enteral Nutrition* 9: 199–209.

Gharsallaoui, A., Roge, B., and Mathlouthi, M. 2010. Study of batch maltitol (4-O-a-D-glucopyranosyl-D-glucitol) crystallization by cooling and water evaporation. *Journal of Crystal Growth* 312: 3183–3190.

Goda, T., Yamada, M., Takase, S., and Hosoya, N. 1992. Effect of maltitol intake on intestinal calcium absorption in the rat. *Journal of Nutrition Science and Vitaminology* 38: 277–286.

Goda, T., Takase, S., and Hosoya, N. 1993. Maltitol-induced transepithelial transport of calcium in rat small intestine. *Journal of Nutrition Science and Vitaminology* 39: 589–595.

Grenby, T. H. 1989. *Progress in Sweeteners*. Elsevier Science Publ. Ltd., London.

Grenby, T. H., and Desai, T. 1988. A trial of lactitol in sweets and its effects on human dental plaque. *British Dental Journal* 164: 383–387.

Grenby, T. H., Phillips, A., and Mistry, M. 1989. Studies of the dental properties of lactitol compared with five other bulk sweeteners in vitro. *Caries Research* 23: 315–319.

Grillo-Werke, S. 1992. WO 92/13 866.

Grimble, G. K., Patil, D. H., and Silk, D. B. A. 1988. Assimilation of lactitol, an unabsorbed disaccharide, in the normal human colon. *Gut* 29: 1666–1671.

Gwilt, J. R., Robertson, A., Goldman, L., and Blanchard, A. W. 1963. The absorption characteristics of paracetamol tablets in man. *Journal of Pharmacy and Pharmacology* 15(1): 445–453.

Hämäläinen, M. M., and Mäkinen, K. K. 1986. Alterations in electrolyte and iron metabolism in the rat in relation to peroral administration of galactitol, mannitol and xylitol. *Journal of Nutrition* 116: 599–609.

Hämäläinen, M. M., Miikinen, K. K., Parviainen, M. T., and Koskinen, T. 1985. Peroral xylitol increases intestinal calcium absorption in the rat independently of vitamin D action. *Mineral and Electrolyte Metabolism* 11: 178–181.

Hayashibara Co., 1994. Netherlands Patent Application. 73.13151.

Hayashibara, K. K. 1974. Method for the preparation of sucrose containing sweeteners and sucrose containing groceries. Dutch patent application No. 7313151, submitted by CCA to WHO.

Herman, R. H. 1974. Hydrolysis and absorption of carbohydrates and adaptive response of the jejunum. In *Sugars in Nutrition*. Sipple, H. L., and McNutt, K. W. (eds.). Academic Press Inc., New York, pp. 145–172.

Higashiyama, T. 2002. Novel functions and applications of trehalose. *Pure Applied Chemistry* 74(7): 1263–1269.

Hirata, T., Fukuse, T., Liu, C. J., Muro, K., Yokomise, H., Yagi, K., Inui, K., Hitomi, S., and Wada, H. 1994. Effects of trehalose in canine lung preservation. *Surgery* 115: 102–107.

Hollman, S., and Touster, O. 1964. *Non-glycolytic Pathways of Metabolism of Glucose*. Academic Press, New York, pp. 276.

Hough, L., Priddle, J. E., and Theobald, R. S. 1962. Carbohydrate carbonates. Part II. Their preparation by ester-exchange methods. *Journal of the Chemical Society* 1934–1938.

Ishizuka, H., Wako, K., Kasumia, T., and Sasaki, T. 1989. Breeding of a mutant of *Aureobasidium* sp. with high erythritol production. *Journal of Fermentation and Bioengineering* 68(5): 310–314.

Jacot-Guillarmod, A., von Bézard, A., and Haselbach, C. H. 1963. Étude de l'hydrogénation du glucose en présence d'un échangeur d'anions. *Helvetica Chimica Acta* 46(1): 45–48.

Jakob, A., Williamson, J. R., and Asakura, T. 1971. Xylitol metabolism in perfused rat liver. *Journal of Biological Chemistry* 246: 7623–7631.

Jasra, R. V., and Ahluwalia, J. C. 1982. Enthalpies of solution, partial molal heat capacities and apparent molal volumes of sugars and polyols in water. *Journal of Solution Chemistry* 11(5): 325–338.

Johnson, I. T. 1995. Butyrate and markers of neoplastic change in the colon. *European Journal of Cancer Prevention* 4: 365–371.

Kandelman, D. 1997. Sugar, alternative sweeteners and meal frequency in relation to caries prevention: new perspectives. *British Journal of Nutrition* 77(Suppl. 1): 121–128.

Kanig, J. L. 1964. Properties of fused mannitol in compressed tablets. *Journal of Pharmacological Science* 53: 188–192.

Kasumi, T. 1995. Fermentative production of polyols and utilization for food and other products in Japan. *Japan Agricultural Research Quarterly* 29: 49–55.

Kawanabe, J., Hirasawa, M., Takeuchi, T., Oda, T., and Ikeda, T. 1992. Noncariogenicity of erythritol as a substrate. *Caries Research* 26: 358–362.

Kearsley, M. W., and Deis, R. C. 2006. Sorbitol and mannitol. In *Sweeteners and Sugar Alternatives in Food Technology*. Oxford, Ames, pp. 249–261.

Kim, H. S., and Jeffrey, G. A. 1969. The crystal structure of xylitol. *Acta Crystallica* B25: 2607–2613.

Kishi, K., Goda, T., and Takase, S. 1996. Maltitol-increased transepithelial diffusional transfer of calcium in rat ileum. *Life Science* 59: 1133–1140.

Knuuttila, M., Svanberg, M., and Hämäläinen, M. 1989. Alterations in bone composition related to polyol supplementation of the diet. *Bone and Mineral* 6: 25–31.

Koizumi, N., Fujii, M., Ninomiya, R., Inoue, Y., Kagawa, T., and Tsukamoto, T. 1983a. Study on transitory laxative effects of sorbitol and maltitol I: estimation of 50% effective dose and maximum non-effective dose. *Chemosphere* 12: 45–53.

Koizumi, N., Fujii, M., Ninomiya, R., Inoue, Y., Kagawa, T., and Tsukamoto, T. 1983b. Study on transitory laxative effects of sorbitol and maltitol II: differences in laxative effects among various foods containing the sweetening agents. *Chemosphere* 12: 105–116.

Kool, C. M. H., Westen, H. A. V., and Hartstra, L. 1952. United States Patent 2,609,3 NVWA Scholten's Chemische Fabriken.

Krishnan, R., Wilkinson, I., Joyce, L., Rofe, A. M., Bais, R., Conyers, R. A. J., and Edwards, J. B. 1980. The effect of dietary xylitol on the ability of caecal flora to metabolise xylitol. *Australian Journal of Experimental Biology and Medical Science* 58: 639–652.

Kristott J. U., and Jones, S. A. 1992. Physical Properties of Polyols, Polydextrose and Their Solutions. Leatherhead Food RA, Scientific and Technical Surveys, no. 173, Surrey, June 1992.

Kruse, W. M. 1976. U.S. Patent no. 3,935,284. United States Inc.

Kuhn, R., and Klesse, P. 1958. Darstellung von L-glucose und L-mannose. *Chemische Berichte* 91: 1989–1991.

Kulkarni, R. K. 1990. Mannitol metabolism in *Lentinus edodes*, the shiitake mushroom. *Applied and Environmental Microbiology* 56(1): 250–253.

Laker, M. F., and Gunn, W. G. 1979. Natural occurrence disqualifies mannitol as an internal standard when urinary monosaccharides are determined by gas–liquid chromatography. *Clinica Chimica Acta* 96: 265–267.

Lang, K. 1969. Utilization of xylitol in animals and man. In *International Symposium on Metabolism, Physiology and Clinical Use of Pentoses and Pentitols*. Horecker, B. L., Lang, K., and Takagi, Y. (eds.). Springer, Berlin, pp. 151–157.

Lanthier, P. L., and Morgan, M. Y. 1985. Lactitol in the treatment of chronic hepatic encephalopathy; an open comparison with lactulose. *Gut* 26: 415–420.

Leleu, J. B., Haon, P., Duflot, P., and Looten, P. 2008. EP0905138B2, 23.07.2008.

Levin, G. V., Zehner, L. R., Saunders, J. P., and Beadle, J. R. 1995. Sugar substitutes: their energy values, bulk characteristics and potential health benefits. *American Journal of Clinical Nutrition* 62: 1161–1168.

Life Sciences Research Office for FDA. 1986. *Early Childhood Dental Caries: Risk Assessment for Dental Caries in Young Children. No. 223-83-2020*. Federation of American Societies for Experimental Biology (FASEB), Bethesda, MD, p. 85.

Life Sciences Research Office. 1994. *The Evaluation of the Energy of Certain Sugar Alcohols Used as Food Ingredients*. Life Sciences Research Office, Federation of American Societies for Experimental Biology, Bethesda, MD.

Life Sciences Research Office. 1999. *Evaluation of the Net Energy Value of Maltitol*. Life Sciences Research Office, Federation of American Societies for Experimental Biology, Bethesda, MD.

Livesey, G. 1990a. The impact of the concentration and dose of Palatinit in foods and diets on energy value. *Food Sciences and Nutrition* 42F: 223–243.

Livesey, G. 1990b. On the energy value of sugar alcohols with the example of isomalt. In *International Symposium on Caloric Evaluation of Carbohydrates*. The Japan Association of Dietetic and Enriched Foods, Kyoto, Japan, pp. 141–164.

Livesey, G. 1992. Energy values of dietary fibers and sugar alcohols for man. *Nutrition Research Reviews* 5: 61–84.

Livesey, G. 2000. *Studies on Isomalt—Published and Unpublished*. Independent Nutrition Logic, Wymondham, UK.

Livesey, G. 2003. Health potential of polyols as sugar replacers, with emphasis on low glycaemic properties. *Nutrition Research Reviews* 16(2): 163–191.

Livesey, G., and Brown, J. 1995. Whole-body metabolism is not restricted to D-sugars as energy metabolism of L-sugars fits a computational model in rats. *Journal of Nutrition* 125: 3020–3029.

Livesey, G., and Brown, J. C. 1996. D-Tagatose is a bulk sweetener with zero energy determined in rats. *Journal of Nutrition* 126: 1601–1609.

Livesey, G., Johnson, I. T., Gee, J. M., Smith, T., Lee, W. A., Hillan, K. A., Meyer, J., and Turner, S. C. 1993. Determination of sugar alcohol and polydextrose absorption in humans by the breath hydrogen (H2O) technique: the stoichiometry of hydrogen production and the interaction between carbohydrates assessed in vivo and in vitro. *European Journal of Clinical Nutrition* 47: 419–430.

Lobry de Bruyn, C. A., and Alberda van Ekenstein, W. 1895. Action des alcalis sur les sucres, II. Transformation réciproque des uns dans les autres des sucres glucose, fructose et mannose. *Recueil des Travaux Chimiques des Pays-Bas* 14(7): 203–207.

Lohmar, R. L. 1962. The polyols. In *The Carbohydrates, Chemistry, Biochemistry, Physiology*. Pigman, W. (ed.). Academic Press, New York, pp. 241–298.

Maguire, A., Rugg-Gunn, J., and Wright, G. 2000. Adaptation of dental plaque to metabolise maltitol compared with other sweeteners. *Journal of Dentistry* 28(1): 51–59.

Mäkinen, K. K., and Scheinin, A. 1975. Turku sugar studies. VI. The administration of the trial and the control of the dietary regimen. *Acta Odontologica Scandinavica* 33(Suppl 70): 105–127.

Mäkinen, K. K. 1994. Sugar alcohols. In *Functional Foods; Designer Foods, Pharmafoods, Nutraceuticals*. Goldberg, I. (ed.). Chapman and Hall, New York, pp. 219–241.

Manzoni, M., Rollini, N., and Bergomi, B. 2001. Biotransformation of D-galactitol to tagatose by acetic acid bacteria. *Process Biochemistry* 36(10): 971–977.

Marie, S., and Piggott, J. R. 1991. *Handbook of Sweeteners*. Blackie and Son Ltd., Glasgow.

Matsuo, T. 2001. Trehalose protects corneal epithelial cells from death by drying. *British Journal of Ophthalmology* 85: 610–612.

McCorkindale, J., and Edson, N. L. 1954. Polyol dehydrogenases. The specificity of rat-liver polyol dehydrogenase. *Biochemistry Journal* 57: 518–523.

McFetridge, J., Rades, T., and Lim, M. 2004. Influence of hydrogenated starch hydrolysates on the glass transition and crystallisation of sugar alcohols. *Research International* 37(5): 409–415.

McNaught, A. D. 1996. Nomenclature of carbohydrates (JCBN). *Pure and Applied Chemistry* 868: 1919–2008.

McNutt, K., and Sentko, A. 1996. Sugar replacers: a growing group of sweeteners in the United States. *Nutrition Today* 31: 255–61.

Mende, K., and Schiweck, H. 1991. Physical principles of continuous hard caramel production. Part 2. Rheological studies on saccharide solutions and melts. *Alimenta* 30(6): 115–122.

Modderman, J. P. 1993. Safety assessment of hydrogenated starch hydrolysates. *Regulatory Toxicology and Pharmacology* 18(1): 80–114.

Montgomery, R., and Wiggins, L. F. 1947. Anhydrides of polyhydric alcohols. Part VI. 1: 4-3: 6-Dianhydro mannitol and 1:4-3:6-dianhydro sorbitol from sucrose. *Journal of the Chemical Society* 433–436.

Moskowitz, H. R. 1971. The sweetness and pleasantness of sugars. *American Journal of Psychology* 84: 387–405.

Moskowitz, H. R. 1974. The psychology of sweetness. In *Sugars in Nutrition*. Sipple, H. L., and McNutt, K. W. (eds.). Academic Press, New York, pp. 37–64.

Muller, T. 2007. Erythritol (E968). In *Handbuch Sußungsmittel*. Rosenplenter, K., and Nohle, U. (eds.). 2nd ed. Behr's Verlag, Hamburg, pp. 323–339.

Ndindayino, F., Henrist, D., Kiekens, F., Van den Mooter, G., Vervaet, C., and Remon, J. P. 2002. Direct compression properties of melt-extruded isomalt. *International Journal of Pharmaceutics* 235: 149–157.

Nikolakakis, I., and Newton, J. M. 1989. Solid state adsorption of antibiotics onto sorbitol. *Journal of Pharmacy and Pharmacology* 41: 145–148.

Nilsson, U., and Jagerstad, M. 1987. Hydrolysis of lactitol, maltitol and palatinint by human intestinal biopsies. *British Journal of Nutrition* 58: 199–206.

Nishizaki, Y., Yoshizane, C., Toshimori, Y., Arai, N., Akamatsu, S., Hanaya, T., Arai, S., Ikeda, M., and Kurimoto, M. 2000. Disaccharide-trehalose inhibits bone resorption in ovariectomized mice. *Nutrition Research* 20: 653–664.

Noda, K., and Oku, T. 1992. Metabolism and disposition of erythritol after oral administration to rats. *Journal of Nutrition* 122: 1266–1272.

Noda, K., Nakayama, K., and Oku, T. 1994. Serum and insulin levels and erythritol balance after oral administration of erythritol in healthy subjects. *European Journal of Clinical Nutrition* 48: 286–292.

Nordin, B. E. C., and Morris, H. A. 1989. The calcium deficiency model for osteoporosis. *Nutrition Review* 47: 65–72.

O'Brien, L. N., and Gelardi, R. C. 1991. *Alternative Sweeteners*, 2nd ed. Marcel Dekker, Inc., New York.

Ohno, S., Hirao, M., and Kido, M. 1982. X-ray crystal structure of maltitol (4-O-α-D-glucopyranosyl-D-glucitol. *Carbohydrate Research* 108: 163–171.

Oku, T., and Noda, K. 1990. Erythritol balance study and estimation of metabolisable energy of erythritol. In *Caloric Evaluation of Carbohydrates*. Hosoya, N. (ed.). Research, Tokyo, pp. 65–75.

Onishi, H. 1971. U.S. Patent. 3,622,456. Noda Institute for Scientific Research.

Oy, S. S. 1973. U.S. Patent 2 418 801.

Patil, D. H., Grimble, G. K., and Silk, D. B. A. 1987. Lactitol, a new hydrogenated lactose derivative: intestinal absorption and laxative threshold in normal human subjects. *British Journal of Nutrition* 57: 195–199.

Patil, D. H., Westaby, D., Mahida, Y. R., Palmer, K. R., Rees, R., Clark, M. L., Dawson, A. M., and Silk, D. B. A. 1987. Comparative modes of action of lactitol and lactulose in the treatment of hepatic encephalopathy. *Gut* 28: 255–259.

Perko, R., and DeCock, P. 2006. Erythritol. In *Sweeteners and Sugar Alternatives in Food Technology*. Mitchell, H. (ed.). Blackwell, Oxford, pp. 151–176.

Petersen, P. E., and Razanamihaja, N. 1999. Carbamide-containing polyol chewing gum and prevention of dental caries in schoolchildren in Madagascar. *International Dentistry Journal* 49: 226–230.

Petrash, J. M. 2004. All in the family: aldose reductase and closely related aldo-keto reductases. *Cellular and Molecular Life Science* 61(7–8): 737–49.

Ravelli, G. P., Whyte, A., Spencer, R., Hotten, P., Harbron, C., and Kenan, R. 1995. Effect of lactitol intake upon stool parameters and the faecal bacterial flora in chronically constipated women. *Acta Therapeutica* 21: 243–255.

Roediger, W. E. W. 1980. Role of anaerobic bacteria in metabolic welfare of the colonic mucosa in man. *Gut* 21: 21793–21798.

Rugg-Gunn, A. J. 1989. Lycasin B and the prevention of dental caries. In *Progress in Sweeteners*. Grenby, T. H. (ed.). Elsevier Applied Science, London, pp. 311–329.

Sasaki, T. 1989. Production of erythritol. *Journal of the Agricultural Chemical Society of Japan* 63(6): 1130–1132.

Scevola, D., Bottari, A., Franchini, A., Guanziroli, A., Faggi, A., Monzillo, V., Pervesi, L., and Oberto, L. 1993. The role of lactitol in the regulation of the intestinal microflora in liver disease. *Giorn Ital Malatt Infett Parassit* 45: 906–918.

Scevola, D., Zambelli, A., Concia, E., Perversi, L., and Candiani, C. 1989. Lactitol and neomycin; monotherapy or combined therapy in the prevention and treatment of hepatic encephalopathy? *Clinical Therapy* 129: 105–111.

Schiffman, S. S., and Gatlin, C. A. 1993. Sweeteners: state of knowledge review. *Neuroscience and Biobehavioral Reviews* 17(3): 313–345.

Schiweck, H., Bar, A., Vogel, R., Schwarz, E., Kunz, M., Lüssem, B., Moser, M., and Peters, S. 2011. Sugar alcohols. *Ullmann's Encyclopedia of Industrial Chemistry*: 1–28.

Schlubach, H. H. 1953. German Patent. 871,736 (Mar. 26, 1953).

Schmidt, P. C., and Benke, K. 1985. Supersaturated ordered mixtures on the basis of sorbitol. *Drugs Made in Germany* 28: 49–55.

Scholnick, F., Ben-Et, G., Sucharski, M. K., Maurer, E. W., and Linfield, W. M. 1975. Lactose-derived surfactants: II. Fatty esters of lactitol. *Journal of American Oil Chemistry Society* 52: 256–258.

Schouten, A., Kanters, J. A., Kroon, J., Looten, P., Duflot, P., and Mathlouthi, M. 1999. A redetermination of the crystal and molecular structure of maltitol (4-O-α-D-glucopyranosyl-D-glucitol). *Carbohydrate Research* 322(3–4): 298–302.

Serpelloni, M. 1990. Directly compressible maltitol powder and its preparation procedure. EP No 0 220 103 B1.

Shindou, T., Sasaki, Y., and Miki, H. 1989. Identification of erythritol by HPLC and GC-MS and quantitative measurement in pulps of various fruits. *Journal of Agriculture and Food Chemistry* 37: 1474–1476.

Sicard, P. J., Leroy, P. 1983. Mannitol, sorbitol and lycasin: properties and food applications. In *Developments in Sweeteners—2*. Grenby, T. H., Parker, K. J., and Lindley, M. G. (eds.). Applied Science Publishers Ltd., London.

Siijonmaa, T., Heilonen, M., Kreula, M., and Linko, P. 1978. Preparation and characterization of milk sugar alcohol, lactitol. *Milchwissenschaft* 33: 733–736.

Sinaud, S., Montaurier, C., Wils, D., Vernet, J., Brandolini, M., Boutloup-Demange, C., and Vermorel, M. 2002. Net energy value of two low digestible carbohydrates, Lycasin HBC and the hydrogenated polysaccharide constituent of Lycasin HBC in healthy human subjects and their impact on nutrient digestive utilisation. *British Journal of Nutrition* 87: 131–139.

Smiley, K. L., Cadmus, M. C., and Liepins, P. 1967. Biosynthesis of D-mannitol from D-glucose by *Aspergillus candidus*. *Biotechnology Bioengineering* 9: 365.

Smith, M. G. 1962. Polyol dehydrogenases. 4. Crystallization of the L-iditol dehydrogenase of sheep liver. *Biochemistry Journal* 83: 135–144.

Smith, J., and Hong-Shun, L. 2003. *Sweeteners*. Food Additive Data Book, Blackwell Publishing, pp. 940–942.

Sochor, M., Baquer, N. Z., and McLean, P. 1979. Regulation of pathways of glucose metabolism in kidney. The effect of experimental diabetes on the activity of the pentose phosphate pathway and the glucuronate-xylulose pathway. *Archives of Biochemistry and Biophysics* 198: 632–646.

Song, S. H., and Vieille, C. 2009. Recent advances in the biological production of mannitol. *Applied Microbiology and Biotechnology* 84: 55–62.

Soontornchai, S., Sirichakwal, P., Puwastien, P., Tontisirin, K., Krüger, K., and Grossklaus, R. 1999. Lactitol tolerance in healthy Thai adults. *European Journal of Nutrition* 38: 218–226.

Steuart, D. W. 1955. The sorbitol test. d-Sorbitol: a new source, method of isolation, properties. *Journal of the Science of Food and Agriculture* 6(7): 387–390.

Strater, P. J. 1989. Palatinit®, the ideal ingredient for confectionery. In Conference Proceedings—Food Ingredients Europe. Maarssen, The Netherlands, pp. 260–266.

Susman, A. S., and Lingappa, B. T. 1959. Role of trehalose in ascospores of *Neurospora tetrasperma*. *Science* 130: 1343–1344.

Suzuki, K., Endo, Y., Uehara, M., Yamada, H., Goto, S., Imamura, M., and Shiozu, S. 1985. Effect of lactose, lactulose and sorbitol on mineral utilization and intestinal flora. *Journal of Japanese Society Nutrition and Food Science* 38: 3942.

Takiura, K., Yamamoto, M., Murata, H., Takai, H., Honda, S., and Yuki, H. 1974. Studies on oligo saccharides part 13. Oligo saccharides in Polygala SenegavarLati folia and structures of glycosyl 15 anhydro D-glucitols. *Yakugaku Zasshi* 94: 998–1003.

Tarao, K., Tamai, S., Ito, Y., Okawa, S., and Hayashi, M. 1995. Effects of lactitol on faecal bacterial flora in patients with liver cirrhosis and hepatic encephalopathy. *Japanese Journal of Gastroenterology* 92: 1037–1050.

Thevelein, J. M., and Hohmann, S. 1995. Trehalose synthase: guard to the gate of glycolysis in yeast? *Trends in Biochemical Science* 20: 3–10.

Touster, O. 1974. The metabolism of polyols. In *Sugars in Nutrition*. Sipple, H. L., and McNutt, K. W. (eds.). Academic Press, New York, pp. 229–240.

Touster, O., Reynolds, V. H., and Hutcheson, R. M. 1956. The reduction of L-xylulose to xylitol by guinea pig liver mitochondria. *Journal of Biological Chemistry* 221: 697–702.

Uhari, M., Kontiokari, T., Koskela, M., and Niemelä, M. 1996. Xylitol chewing gum in prevention of acute otitis media: double blind randomized trial. *British Medical Journal* 313: 1180–1183.

U.S. Department of Agriculture/U.S. Department of Health and Human Services. 1995. *Nutrition and Your Health: Dietary Guidelines for Americans*, 4th ed.

Uyl, C. H. 1987. Technical and commercial aspects of the use of lactitol in foods as a reduced-calorie bulk sweetener. In *Developments in Sweetener*. Grenby, T. H. (ed.). Elsevier Applied Science, London, pp. 65–81.

Van Es, A. J. H., de Groot, L., and Vogt, J. E. 1986. Energy balances of eight volunteers fed on diets supplemented with either lactitol or saccharose. *British Journal of Nutrition* 56: 545–554.

van Velthuijsen, J. A. 1979. Food additives derived from lactose: lactitol and lactitol palmitate. *Journal of Agriculture and Food Chemistry* 27: 680–686.

van Velthuijsen, J. A., and Blankers, I. H. 1991. Lactitol: a new reduced-calorie sweetener. In *Alternative Sweeteners*, 2nd ed. Nabors, L. O., and Gelardi, R. C. (eds.). Marcel Dekker, Inc., New York.

Vanderdonckt, J., Coulon, J., Denys, W., and Ravelli, G. P. 1990. Study of the laxative effect of lactitol (Importal) in the elderly institutionalized, but not bedridden, population suffering from chronic constipation. *Journal of Clinical and Experiment Gerontology* 12: 171–189.

Vasyunina, N. A., Barysheva, G. S., Balandin, A. A., Chepigos, S. V., and Pogosov, Y. L. 1964. Hydrolytic hydrogenation of cotton cellulose. *Zhurnal Prikladnoi Khimii* 12: 2725–2729.

Vaughan, O. W., and Filer, L. J. Jr. 1960. The enhancing action of certain carbohydrates on the intestinal absorption of calcium in the rat. *Journal of Nutrition* 71: 10–14.

Vega, R. R., and Le Tourneau, D. 1971. Trehalose and polyols as carbon sources for *Yerticillium* spp. *Phytopathology* 61: 339–340.

Wang, M. C., and Meng, H. C. 1971. Xylitol metabolism in extrahepatic tissues. *Zeitschrift für Ernährungswissenschaft* 11(suppl): 8–16.

Wang, Y. M., and van Eys, J. 1981. Nutritional significance of fructose and sugar alcohols. *Annual Review of Nutrition* 1: 437–75.

Wang, Y.-J. 2003. Saccharides: modifications and applications. In *Chemical and Functional Properties of Food Saccharides*. Piotry, T. (ed.). CRC Press, Boca Raton, FL, USA, pp. 35–47.

Wang, Y.-M., and van Eys, J. 1981. Nutritional significance of fructose and sugar alcohols. *Annual Review of Nutrition* 1: 437–475.

Washüttl, J., Reiderer, P., and Bancher, E. 1973. A qualitative and quantitative study of sugar-alcohols in several foods. *Journal of Food Science* 38: 1262–1263.

Wheeler, M. L., Fineberg, S. E., Gibson, R., and Fineberg, N. 1990. Metabolic response to oral challenge of hydrogenated starch hydrolysate versus glucose in diabetes. *Diabetes Care* 13(7): 733–740.

Willibald-Ettle, I., and Schiweck, H. 1996. Properties and applications of isomalt and other bulk sweeteners. In *Advances in Sweeteners*. Grenby, T. H. (ed.). Chapman and Hall, New York, pp. 134–149.
Winegrad, A. I., Clements, R. S., and Morrison, A. D. 1972. Insulin-independent pathways of carbohydrate. In *Handbook of Physiology*. Field, J. (ed.). Williams & Williams, Baltimore, Maryland, pp. 457–471.
Wohlfarth, C. 2006. CRC Handbook of Enthalpy Data of Polymer-Solvent Systems. CRC/Taylor & Francis, Boca Raton, FL, USA.
Wolfrom, M. L., and Lewis, W. L. 1928. The reactivity of methylated sugars II. The action of dilute alkali on tetramethyl glucose. *Journal of American Chemical Society* 50: 837.
Wolfrom, M. L., Burke, W. J., Brown, K. R., and Rose, R. S. 1938. Crystalline lactitol monohydrate and a process for the preparation thereof, use thereof, and sweetening agent. *Journal of American Chemical Society* 60: 571–573.
Wright, L. W. 1974. Sorbitol and mannitol. *Chemtechnology* 4: 42–46.
Wyngaarden, J. B., Segal, S., and Foley, J. B. 1957. Physiological disposition and metabolic fate of infused pentoses in man. *Journal of Clinical Investigation* 36: 1395–1407.
Yoshida, H., Sugahara, T., and Hayashi, J. 1986. Studies on free sugars and free sugar alcohols of mushrooms. *Nippon Shokuhin Kogyo Gakkaishi* 33: 426–433.
Yuan-Tseng, P., Carroll, J. D., Asano, N., Pastuszak, I., Edavana, V. K., and Elbein, A. D. 2008. Trehalose synthase converts glycogen to trehalose. *FEBS Journal* 275: 3408–3420.
Ziesenitz, S. C., and Siebert, G. 1987. The metabolism and utilization of polyols and other bulk sweeteners compared with sucrose. In *Developments in Sweeteners*. Grenby, T. H. (ed.) Elsevier Applied Science, London, pp. 109–149.

CHAPTER 4

Low Calorie Nonnutritive Sweeteners

Georgia-Paraskevi Nikoleli and Dimitrios P. Nikolelis

CONTENTS

4.1 INTRODUCTION

Sweeteners are defined as food additives that are used to provide a sweet taste to food or as a tabletop sweetener. Tabletop sweeteners are products that include any permitted sweeteners and are intended for use as an alternative to sugar. Foods with sweetening properties, such as sugar and honey, are not additives and are excluded from the scope of official regulations. Sweeteners are classified as either high intensity or bulk. High-intensity sweeteners possess a sweet taste, but are noncaloric, provide essentially no bulk to food, have greater sweetness than sugar, and are therefore used at very low levels. On the other hand, bulk sweeteners are generally carbohydrates, providing energy (calories) and bulk to food. These have a similar sweetness to sugar and are used at comparable levels. High-intensity sweeteners are classified as synthetic artificial [i.e., acesulfame, alitame, aspartame (ASP), cyclamate, neotame (NTM), saccharin, and sucralose], semisynthetic [neohesperidin dihydrochalcone (NHDC)], and natural (stevioside, traumatin, and glycyrrhizin).

4.1.1 High-Intensity Sweeteners

High-intensity sweeteners (also called nonnutritive sweeteners) can offer consumers a way to enjoy the taste of sweetness with very low or no energy intake, and they have the advantage that they do not provide growth of oral cavity microorganisms. Therefore, they are principally aimed at consumers in four areas of food and beverage markets: treatment of obesity, maintenance of body weight, management of diabetes, and prevention and reduction of dental caries. Most available high-intensity sweeteners and/or their metabolites are rapidly absorbed in the gastrointestinal tract. For example, acesulfame-K (ACS-K) and saccharin are not metabolized and are excreted unchanged by the kidney. Sucralose, stevioside, and cyclamate undergo degrees of metabolism, and their metabolites are readily excreted.

Acesulfame-K (ACS-K), ASP, and saccharin are permitted as intense sweeteners for use in food virtually worldwide. In order to decrease cost and improve taste quality, high-intensity sweeteners are often used as mixtures of different, synergistically compatible sweeteners. Sweetness characteristics of high-intensity sweeteners are shown in Table 4.1.

The approvals of new-generation sweeteners like ACS-K are too recent to establish any epidemiological evidence about possible carcinogenic risks (Weihrauch and Diehl 2004). Following

Table 4.1 Sweetness Characteristics of High-Intensity Sweeteners (Scale Uses Sucrose as Sweetness of 1 and Compares Sweetness of Other Sweeteners to Sucrose)

Sweetener	Relative Sweetness (Sucrose = 1)	Aftertaste
Acesulfame-K	150–200	Very slight bitter
	Alitame	Not unpleasant
Aspartame	160–220	Prolonged sweetness
Aspartame–acesulfame salt	350	–
Cyclamate	30–40	Prolonged sweetness. At high concentrations, a distinct sweet–sour lingering
Glycyrrhizin	50–100	Prolonged sweetness (licorice)
Neohesperidin dihydrochalcone	1000–2000	Lingering menthol–licorice
Neotame	7000–13,000	Not unpleasant
Saccharin	300–600	Bitter metallic
Stevioside	250–300	Bitter and unpleasant
Sucralose	400–800	Not unpleasant
Thaumatin	2000	Licorice

Source: Yebra-Biurrun, M. C., *Food Add. Contam.*, 17, 9, 733–738, 2000.

the adoption of the EC Sweeteners Directive in 1994 (Commission of the European Communities 1994) and its implementation into the national laws, member states are required to establish a system of consumer surveys to monitor additive intake (European Commission 1994). For this reason, intake data are required through developed and validated methods for the determination of these two artificial sweeteners.

The acceptable daily intake (ADI) has been defined by the World Health Organization (WHO) as "an estimate by JECFA (the Joint FAO/WHO Expert Committee on Food Additives) of the amount of a food additive, expressed on a body weight basis, that can be ingested daily over a lifetime without appreciable health risk" and is based on an evaluation of available toxicological data. For example, in Europe, the ADI is set at 9 mg/kg of body weight/day for ACS-K (Wilson et al. 1999). For ASP, there is a safety margin, even in high consuming diabetics (Ilbäck et al. 2003). The U.S. Food and Drug Administration (FDA) have set the ADI for ASP at 50 mg/kg of body weight/day. An ADI of 40 mg/kg body weight/day set by the committee of experts of the Food and Agriculture Organization (FAO) and the WHO is not likely to be exceeded, even by children and diabetics. A European Commission (EC) report gives a theoretical maximum estimate for adults' consumption of 21.3 mg/kg body weight/day of ASP. However, the actual consumption is likely to be lower, even for high consumers of ASP. The report also gives refined estimates for children, which show that they consume 1% to 40% of the ADI. People with diabetes are high consumers of foods containing ASP; their highest reported intake varies between 7.8 and 10.1 mg/kg body weight/day. Health at international level (JECFA) and of United States (FDA) have set the ADI for ACS-K at 9 mg/kg body weight. At the European level on March 13, 2000, the ADI has been set at 9 mg/kg body weight [Scientific Committee for Foods (SCF)].

Generally speaking, in the European Union (EU), sweeteners are thoroughly assessed for safety by the European Food Safety Authority (EFSA) before they are authorized for use. EU Directives 94/35/EC (European Commission 1994), 96/83/EC (European Commission 1996), 2003/115/EC (European Commission 2003), and 2006/52/EC (European Commission 2006) define which sweetener has been approved to be added to food products and beverages.

Today, ACS-K and ASP are used in foods including baked goods (dry bases for mixes), beverages (dairy beverages, instant tea, instant coffee, fruit-based beverages), soft drinks (colas, citrus-flavored drinks, fruit-based soft drinks), sugar preserves and confectionery (calorie-free dustings, frostings, icings, toppings, fillings, syrups), alcoholic drinks (beer), vinegar, pickles, and sauces (sandwich spreads, salad dressings), dairy products (yogurt and yogurt-type products, puddings, desserts and dairy analogues, sugar-free ice cream), fruit, vegetables, and nut products, sugar-free jams and marmalades, low-calorie preserves, and other food products (i.e., chewing gums, liquid concentrates, frozen and refrigerated desserts). Hard-boiled candies can be manufactured using ACS-K as the intense sweetener. ACS-K rounds the sweetness and brings the taste close to standard, sugar-containing products. In chocolate and related products, ACS-K can be added at the beginning of the production process (e.g., before rolling). It withstands all treatments including conching without detectable decomposition (Baron and Hanger 1998). In reduced-calorie baked goods, bulking agents like polydextrose substitute for sugar and flour may help in reducing the level of fats. ACS-K combines well with suitable bulking ingredients and bulk sweeteners and therefore allows production of sweet-tasting baked goods having fewer calories. In diabetic products, combinations of ACS-K and sugar alcohols like isomalt, lactitol, maltitol, or sorbitol can provide volume and sweetness. Texture and sweetness intensity can be similar to sucrose-containing products.

The more important properties of the high-intensity sweeteners that are permitted for use in food and drink applications are shown in Table 4.2.

ACS-K, ASP, and saccharin are the most used between the artificial sweeteners and have been approved for uses in the United States and EU. Therefore, their examination and description below are more extensive than other artificial sweeteners.

Table 4.2 Characteristics and Selected Physical Properties of High-Intensity Sweeteners

Sweetener	Molecular Formula	Molecular Weight	m.p. (°C)	Solubility in H_2O at 20°C (%)	ADI (mg/kg body weight)	
					JECFA	SCF
Acesulfame-K	$C_4H_4NO_4SK$	201.2	200	27	15	9
Alitame	$C_{14}H_{25}N_3O_4S$	331.43	136–147	14.3 (pH 7)	1	0.3
Aspartame	$C_{14}H_{18}N_2O_5$	294.31	246	1	40	40
A–A salt	$C_{18}H_{23}O_9N_3S$	457.46	–	–	Covered by the ADI values previously established for aspartame and acesulfame-K	
Cyclamate	$C_6H_{13}NO_3S$	179.24	169–170	7.7 Na salt: 19.5	11	7
Glycyrrhizin	$C_{42}H_{62}O_{16}$	822.93	–	–	Not specified	Not evaluated (100 provisional)
NHDC	$C_{28}H_{36}O_{15}$	612.6	156–158	0.05	Not evaluated	5
Neotame	$C_{20}H_{30}N_2O_5$	378.46	80.9–83.4	1.3	2	Not evaluated
Saccharin	$C_7H_5NO_3S$	183.18	228.8–229.7	0.3 Na salt: 83	5	5
Stevioside	$C_{38}H_{60}O_{18}$	804.9	198	0.125	Not evaluated	Not acceptable
Sucralose	$C_{12}H_{19}O_8Cl_3$	397.63	125	25.7	15	15
Thaumatin	–	~22000	–	60	Not specified	Unlimited

Source: Yebra-Biurrun, M. C., *Food Add. Contam.*, 17, 9, 733–738, 2000.
Note: A–A salt, aspartame–acesulfame; NHDC, neohesperidin dihydrochalcone; m.p., melting point; ADI, acceptable daily intake; JECFA, Join Expert Committee on Food Additives of the Agriculture Organization/World Health Organization; SCF, Scientific Food Committee of the European Community.

4.2 ACESULFAME-K

ACS-K (6-methyl-1,2,3-oxathiazin-4(3H)-one-2,2-dioxide, MW 201.24) was accidentally discovered in 1967 from studies on novel ring compounds at Hoechst Corporation in West Germany (Sardesai and Waldshan 1991). Its full name is potassium acesulfame and consists of a 1,2,3-oxathiazine ring, a six-heterocyclic system in which oxygen, sulfur, and nitrogen atoms are adjacent to each another (Figure 4.1). ACS-K is about 200 times sweeter than sucrose but has a slightly bitter aftertaste (especially at high concentrations); it is soluble in water and has an extremely long storage life. Unlike ASP, it is stable at high temperatures, which makes it ideal for use in baking (Zygler et al. 2009). ACS-K is not metabolized by the body and is excreted unchanged by the kidneys. A large number of pharmacological and toxicological studies have been conducted, and the sweetener has been found to be safe (Sardesai and Waldshan 1991). ACS-K can be used as a sweetening agent in a wide range of products (low-calorie products, diabetic foods, sugarless products). It is suitable for low-calorie beverages because it has a pronounced stability in aqueous solutions and is even suitable for diet soft

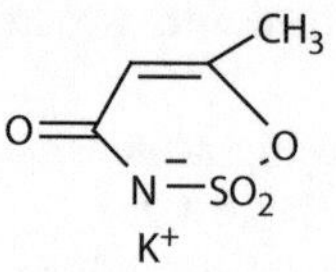

Figure 4.1 Chemical structure of ACS-K.

drinks that maintain a low pH. ACS-K [I] has excellent stability under high temperatures and good solubility that makes it suitable for numerous products. It is approved for use in food and beverage products in about 90 countries including the United States, Switzerland, Norway, United Kingdom, Canada, Australia, and the EU.

4.3 ALITAME

Alitame [L-α-aspartyl-*N*-(2,2,4,4-tetramethyl-3-thioethanyl)-D-alaninamide] is an amino acid-based sweetener (Figure 4.2) developed by Pfizer Central Research from L-aspartic acid, D-alanine, and 2,2,4,4-tetraethylthioethanyl amine. A terminal amide group instead of the methyl ester constituent of ASP was used to improve the hydrolytic stability. The incorporation of D-alanine as a second amino acid in place of L-phenylalanine has resulted in optimum sweetness. The increased steric and lipophilic bulk on a small ring with a sulfur derivative has provided a very sweet product and good taste qualities.

The formula of alitame is $C_{14}H_{25}O_4N_3S$, with a molecular weight of 331.06. It is produced under the brand name Aclame. It is a crystalline, odorless, and nonhygroscopic powder, with a good solubility in most polar solvents such as water (130 g/L at pH 5.6) and alcohol. Alitame is 2000 times sweeter than sucrose, 12 times sweeter than ASP, and 6 times sweeter than saccharin. It has a clean sweet taste with no unpleasant aftertaste. It is blended with other sweeteners such as saccharin, cyclamate, and ACS-K to maximize the quality of sweetness.

Alitame offers several benefits such as stability at high temperatures and a broader pH range. For instance, it is stable for over a year at pH 6–8 and room temperature and withstands pasteurization. However, prolonged storage of acidic solutions at high temperatures or in combination with certain ingredients (hydrogen peroxide or sodium bisulfite) may produce off-flavors. In the presence of high levels of reducing sugars, alitame can undergo Maillard reactions.

Alitame is noncariogenic. From an oral intake, 7%–22% is unabsorbed and excreted in the feces. The remainder is hydrolyzed to aspartic acid and alanine amide. The aspartic acid is normally metabolized, and the alanine amide is excreted in the urine as a sulfoxide isomer, sulfone, or conjugated with glucuronic acid. The incomplete absorption and metabolism result in a core value of 1.4 kcal g^{-1}.

The Joint Expert Committee on Food Additives (JECFA) concluded that alitame was not carcinogenic and did not show reproductive toxicity. In 1996, an ADI of 0–1 mg/kg of body weight was allocated. It is approved for use in Australia, New Zealand, Mexico, and China. A food additive petition was submitted to the FDA in 1986, and approval is awaited. In the petition, the estimated daily intake is 0.34 mg/kg of body weight, which represents the amount if alitame is the only sweetener in the diet. The level at which no observed adverse effects occur in animals is 100 mg/kg. Potential uses include baked goods, baking mixes, hot and cold beverages, dry beverage mixes, tabletop sweeteners, chewing gum, candies, frozen desserts, and pharmaceuticals. Alitame has been approved for use in some countries such as Australia, Mexico, New Zealand, and China, but not in the United States or the EU.

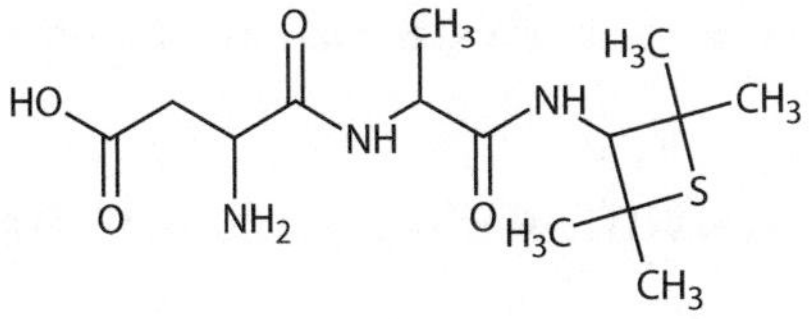

Figure 4.2 Chemical structure of alitame.

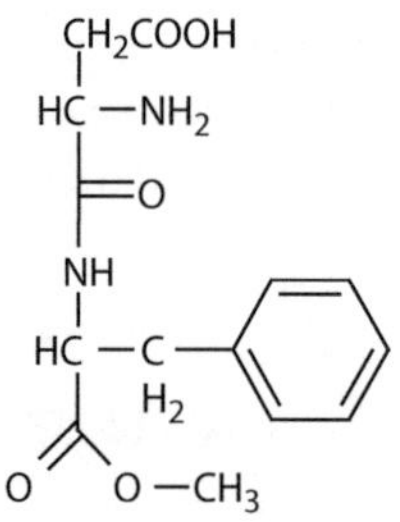

Figure 4.3 Chemical structure of ASP.

4.4 ASPARTAME

ASP (N-L-a-aspartyl-L-phenylalanine-1-methylester) is a dipeptide (Figure 4.3). It is 200 times sweeter than sugar, and its sweet taste is almost identical to that of sucrose but lacks in aftertaste. This sweetener was discovered accidentally in 1965 by James M. Schlatter at G.D. Searle Co. This nutritive sweetener provides 4 calories per gram, but the amount required to give the same sweetness as sugar is only 0.5% of the calories (Sardesai and Waldshan 1991). It is a white crystalline powder with a molar mass of 294.31 and a calorie value of 17 kJ/g. Because of its high level of sweetness, the amounts that can be used are small. Although it is relatively stable in a dry form, the compound can undergo pH and temperature-dependent degradation; for this reason, ASP is undesirable as a baking sweetening agent. Below pH 3, ASP is unstable and hydrolyzes to produce aspartylphenylalanine, and above pH 6, it changes to form 5-benzyl-3,6-dioxo-2-piperazineacetic acid (Zygler et al. 2009). ASP was approved in 1981 and produced for the market as "Nutra-Sweet." For the first time, dairy products (ice cream, yogurt, etc.) were calorie-reduced and could be sold with the prefixes "diet" or "light" (Weihrauch and Diehl 2004).

The metabolism of ASP was extensively studied. The major point that was concluded from metabolism studies is that ASP is broken down in the gastrointestinal tract to its constituents (i.e., aspartic acid, phenylalanine, and methanol; Sardesai and Waldshan 1991).

ASP is probably the most controversial artificial high-intensity sweetener on the market. It has been reported that ASP can cause medical effects to consumers, such as multiple sclerosis, systemic lupus, brain tumors, and methanol toxicity (Zygler et al. 2009). Recent reports suggest that ASP can be related with cancer, lymphomas, and leukemias after investigations using rats. In 2006 and 2007, these findings were published by the European Ramazzini Foundation (ERF) of Oncology and Environmental Sciences (Soffritti et al. 2007). These allegations regarding ASP and its effects in humans have been carefully evaluated by scientists at regulatory agencies around the world including the EU and the United States. The conclusion was that ASP does not cause anything of the effects mentioned. In March 2009, EFSA investigated the results of these studies and found no indications of any genotoxic or carcinogenic potential of ASP (EFSA 2009). The only disadvantage of ASP is that it cannot be used by individuals suffering from phenylketonuria (PKU). For this reason, foodstuff in which ASP is a constituent must be labeled that it cannot be used by phenylketonurics as it contains phenylalanine. The intake of ASP must be limited in these individuals (MacKinnon 2003). ASP is approved in more than 90 countries (the EU, the United States, Canada, South America, Australia, Japan, etc.) for use in numerous foodstuffs.[III]

4.5 ASPARTAME–ACESULFAME SALT

ACS-K and ASP are characterized as artificial high-intensity sweeteners. They are also called nonnutritive sweeteners. These two are commonly used in foods, beverages, and confectionery products. They are exclusively used for low-calorie intake that helps obese consumers to maintain their weight. It is also medically suggested to diabetics to use foods containing these two artificial sweeteners instead of sugar.

A substantial number of low-calorie beverages, however, are sweetened with mixtures of ACS-K and ASP. The combination of ACS-K and ASP has led to an improvement of the quality of sweetened products. In soft drinks, a combination of these two has found a broad application. The production and use of combination of these two sweeteners to create a mixture in which each molecule contains both sweeteners are constantly rising. The compounds that contain combined sweeteners are called "twinsweets." The reasons for marketing these compounds are obvious. These two sweeteners together offer two very important advantages for the food industries. "Twinsweets" offer a greater sweetness stability and longer stability as compared with the individual use of ASP or ACS-K. ASP and ACS-K exhibit quantitative synergy. When these two artificial sweeteners are used together, they provide a more potent sweetener than when they are used independently. These beverages benefit from a synergism and an improved taste that is provided by such blends. For example, a sweetness level equivalent to approximately 10% of sucrose in beverage is replaced by concentrations in the range of 500–600 mg/L of ACS-K or ASP. If a blend of the two sweeteners is used, the same sweetness level can be achieved by using only 160 mg/L of each of these sweeteners. The blend to achieve this is about 60:40 (%) of ASP–ACS-K, which in reality is a unimolar ratio. ACS-K has a good solubility in water, and therefore, highly concentrated solutions suitable for household use can be manufactured. No problems of stability have to be anticipated for solutions in normal storage conditions. Similarly, no problems have been reported for the dissolution of tablets or powders. When blending ACS-K with other intense sweeteners for beverage applications, the blend ratio may depend on different factors including the flavor or flavor type. For this reason, in orange-flavored beverages, considering the time intensity curves of sweetness and fruitiness similar to sucrose-sweetened beverages, blends of ACS-K and ASP (40:60) have been used. In raspberry-flavored beverages containing natural flavors, 40:60 to 25:75 (ACS-K/ASP) blend ratios are considered optimum, whereas in beverages with artificial raspberry flavors, blend ratios of 50:50 to 20:80 are considered optimum. Nowadays, emerging trends in beverages include replacement of sugar in fully sugared beverages with intense sweeteners like ACS-K. These two sweeteners are authorized for use in the European Union (EU) and the United States.

ASP is referred to as a "first-generation sweetener." This generation was followed by a second generation of sweeteners; ACS-K belongs in this second generation. However, the new sweeteners have similar limitations to the older ones. The taste is often accompanied by a bitter and metallic aftertaste and does not provide the same taste of regular sugar. A key attribute that distinguishes sweeteners from other ingredients is their characteristic and pleasurable sweet taste and intensity. The standard for comparing sweetness is sucrose. The sweetness value assigned to sucrose is either 100 or 1.0, depending upon the scale or scoring system used (Helstad 2006). ASP–acesulfame dissolves completely in saliva and gastric juice. Although this salt mainly consists of the two approved sweeteners, it is considered as a separate compound, which requires specific approval in certain countries. In the EU, it is part of the Proposed Amendment to the Sweeteners Directive that was adopted in 2004. It is approved in the United States, Canada, United Kingdom, Mexico, Russia, and China.

4.6 CYCLAMATE

Among artificial sweeteners are the modern noncaloric alternatives to sugars that are widely used throughout the world. Cyclamate (cyclohexyl sulfamic acid monosodium salt) is an artificial sweetener that is 35 times sweeter than sugar. It has been widely used in low-calorie foods and beverages. The usage of sodium cyclamate has been the center of controversy during the last few decades

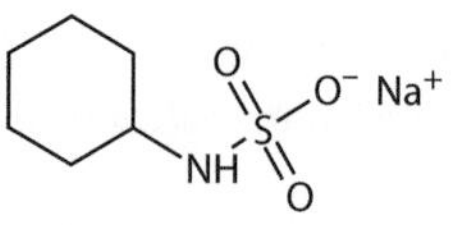

Figure 4.4 Chemical structure of sodium cyclamate.

due to its possible carcinogenic effects. Owing to a study by Wagner (1970) who found an increased incidence of bladder carcinomas in rats, the use of cyclamate was prohibited in the United States and the United Kingdom. However, further evaluations by the Cancer Assessment Committee of the Centre for Food Safety and Applied Nutrition of the FDA, by the SCF of the European Union, and by the World Health Organization (WHO) concluded that cyclamate is not a carcinogen (Takayama et al. 2000). Nowadays, cyclamate is approved for use in more than 50 countries worldwide. The ADI value for cyclamate has been set at 11 mg/kg body weight by the JECFA and at 7 mg/kg body weight by the SCF (Armenta et al. 2004a,b). Because the safety of cyclamate to humans is not clear completely, the restricted content level in foods is different in different countries (Wagner 1970). Cyclamate (Figure 4.4) is generally used in the form of a sodium salt because it is more soluble in water than the free acid. The calcium salt is also used as a sweetener, but, for some applications, it is not suitable as it can cause gelation and precipitation. Sodium cyclamate exhibits good stability in the solid form and is also stable in soft drink formulations within the pH range 2–10. Cyclamate is permitted in several countries (EU, Australia, Canada, New Zealand, etc.). However, it has been banned in the United States after controversial toxicity studies.

4.7 GLYCYRRHIZIN

Glycyrrhizin is found in the licorice root of a small leguminous shrub, *Glycyrrhiza glabra* L., from Europe and Central Asia. Glycyrrhizin or 20-β-carboxy-11-oxo-30-norolean-12-en-3β-yl-2-*O*-β-D-glucopyranurosyl-α-D-glucopyranosiduronic acid ($C_{42}H_{62}O_{16}$, MW 822.92) is a triterpenoide glycoside (saponin) with glycyrrhetinic acid, which is condensed with *O*-β-D-glucuronosyl-(1′→2)-β-D-glucuronic acid (Figure 4.5). After harvest, the roots are dried to 10% moisture, shredded, extracted with aqueous ammonia, concentrated in vacuum evaporators, precipitated with sulfuric acid, and crystallized with 95% alcohol providing a crude ammonium glycyrrhizin (AG). Further treatment yields a white, crystalline mono-ammonium glycyrrhizin (MAG). Both derivatives have

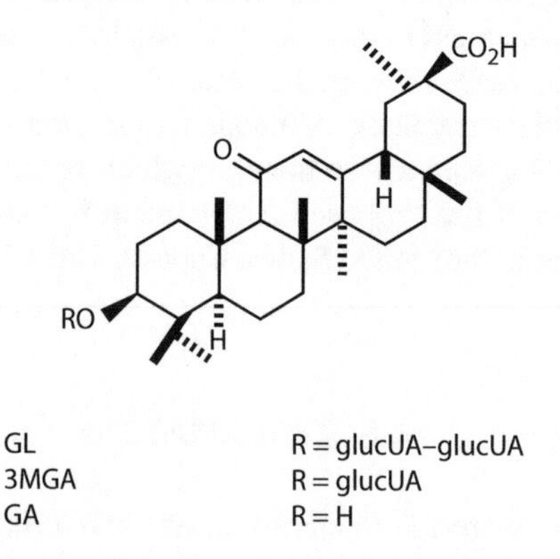

Figure 4.5 Chemical structures of glycyrrhizin (GL) and its metabolites: 3-monoglucuronyl-glycyrrhetinic acid (3MGA) and glycyrrhetinic acid (GA).

the same sweetness but differ in solubility and sensitivity to pH. AG is relatively stable and highly soluble in hot or cold water and in alcohol. It withstands temperatures above 105°C for a short period of time and precipitates at pH values below 4.5. MAG is used in applications where low pH and color rule out AG.

Glycyrrhizin is 50–100 times sweeter than sucrose and has a slow onset of sweetness followed by a lingering licorice-like aftertaste (Table 4.1). It exhibits a sweet woody flavor, which limits its use as a pure sweetener. Glycyrrhizin enhances food flavors, masks bitter flavors, and increases the perceived sweetness level of sucrose. It has the potential for providing functional characteristics including foaming, viscosity control, gel formation, and possibly antioxidant characteristics.

Studies have focused on the pharmacological effects of glycyrrhizin as antiulcer, antiinflammatory, antiviral, anticariogenic, and antispasmodic. It also has corticoid activity, influencing steroid metabolism to maintain blood pressure and volume and to regulate glucose/glycogen balance. Glycyrrhizin can be hydrolyzed by human intestinal microflora to 18-β-glycyrrhetinic acid and two molecules of glucuronic acid. After release of the acids, the compound binds to plasma protein, enters the enterohepatic circulation, and is almost completely metabolized. However, side effects, typically involving cardiac dysfunction, edema, and hypertension, have been reported among subjects receiving high doses of glycyrrhizin-based pharmaceuticals or consuming large amounts of licorice-containing confectionery or health products over a prolonged period.

The ammonium salt of glycyrrhizin is approved as a flavoring and flavor enhancer in the United States. It is on the FDA "generally recognized as safe" (GRAS) list. The use of glycyrrhizin is permitted in Japan and Taiwan. In Japan, it is used as a flavoring for hydrolyzed vegetable protein, soy sauce, and bean paste to control saltiness. At levels of 30–300 mg/kg, it enhances the flavor of cocoa and chocolate-flavored products, flavors and sweetens candy, confectionery, and beverages, and masks the bitter taste of pharmaceuticals. Because of its pharmacological action, it should be used in moderate amounts as a sweetener. Glycyrrhizin is used in Japan and in other countries as a sweetening agent. In the United States, it is approved for use as a flavor and flavor enhancer below 4.5.

4.8 NEOHESPERIDIN DIHYDROCHALCONE

Citrus fruits contain bitter flavanone glycosides, all derivatives of the disaccharide 2-*O*-α-L-rhamnopyranosyl-β-D-glucopyranose, neohesperidose. In 1963, Horowitz and Gentili found that catalytic hydrogenation of the chalcone form gave dihydrochalcone neohesperidosides, several of which were intensely sweet. Numerous dihydrochalcone derivatives were synthesized for taste and toxicity trials, from which NHDC emerged as a promising sweetener. It is prepared by alkaline hydrogenation of the biflavanoid neohesperidin present in Seville (bitter) oranges (*Citrus aurantium*).

NHDC is a semisynthetic nonnutritive intense sweetener. Chemically, it is 1-[4-[[2-*O*-(6-deoxy-α-L-mannopyranosyl)-β-D-glucopyranosyl]oxy]-2,6-dihydroxyphenyl]-3-(3-hydroxy-4-methoxyphenyl)-1-propanone with the molecular formula $C_{28}H_{36}O_{15}$ and a molecular weight of 612.60 (Figure 4.6). NHDC shows a slow buildup of sweetness, rising from 250 to 2000 times that of a 50 g/L sucrose solution, but more persistent (Table 4.1). It has a pleasant taste, flavor-modifying properties, ability to improve the sweetness quality and profile, and remarkable synergistic effects. Its flavor enhancement has been perceived in several products, especially fruit flavors. NHDC has the ability to decrease the perception of bitterness, saltiness, sharp, and spicy attributes. The sweetness intensity of NHDC depends on many factors such as concentration, pH, and the product to which it is added. As the concentration increases, the sweetness of NHDC decreases relative to the level of sucrose. Caffeine significantly enhances the sweetness of NHDC in certain soft drinks. At higher concentrations, NHDC has a lingering menthol or licorice-like aftertaste and a cooling sensation, which distinguishes it from other sweeteners. However, modifications of the sensorial

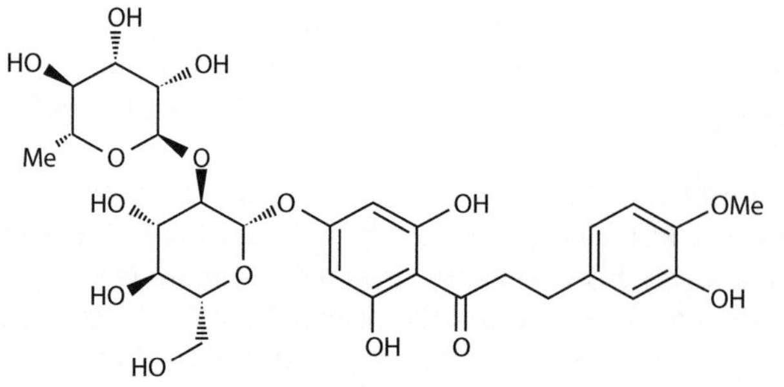

Figure 4.6 Chemical structure of NHDC.

properties of NHDC are possible by the admixture of bulk sweeteners, certain flavors, and other taste-modifying food additives such as gluconates, amino acids, or nucleotides. It also shows synergism with saccharin, ASP, cyclamate, sucralose, ACS-K, and sugar alcohol.

NHDC is a nonhygroscopic colorless crystalline solid. It is sparingly soluble in water (0.50 g/L) at 20°C but is highly soluble at 80°C (650 g/L). It is also soluble in alcohol and aqueous alkali, but a higher solubility is achieved in ethanol–water mixtures than in water or ethanol alone. Where a higher solubility is required, monobasic salts may be used that are freely soluble in water and exhibit a shorter duration of sweetness than the parent compound. The solubility of NHDC may also be enhanced by dissolving it in glycerol and propylene glycol, as well as by using it in mixtures with readily water-soluble polyols such as sorbitol. Interestingly, these bulk sweeteners also act as taste modifiers of NHDC by reducing its menthol-like aftertaste.

NHDC presents high stability at pH 2–6. It is stable under most food-processing and storage conditions and withstands pasteurization, ultra high temperature processes, and the normal shelf life of soft drinks. It is stable during fermentation of yogurt but undergoes hydrolysis at high acidity and elevated temperatures, yielding hesperetin dihydrochalcone, hesperetin dihydrochalcone-4′-β-D-glucoside, rhamnose, and glucose.

NHDC is noncariogenic and has a caloric value of 2 kcal g^{-1}. Little of the compound is absorbed unchanged from the small intestine. After cleavage of the glycosidic side chain by intestinal mucosal or bacterial glycosidases, the residual primary metabolites are partly excreted unchanged in the bile and partly metabolized further. Standard toxicity tests have suggested its safety. In 1987, the Scientific Committee for Food of the Commission of the European Communities allocated an ADI of 0–5 mg per kilogram of body weight. It is currently approved for use in the European Communities, Sweden, Switzerland, Morocco, and Tunisia.

Owing to its highly intense and long-lasting sweetness, NHDC is normally used at concentrations of less than 100 mg/kg. Only in chewing gum are higher levels required because of its slow release from the gum base. For use in soft drinks, NHDC has been recommended in combination with other sweeteners at a concentration of 20 mg/kg. Owing to its ability to reduce bitterness and to its flavor-enhancing properties, NHDC is an ideal sweetener for grapefruit or orange juice. Promising results have been reported from its use in fruit-flavored yogurts. In tabletop products, the addition of small amounts of NHDC may result in significant savings because of its synergistic, sweetness-enhancing effect. It has been used in juice, soft drinks, dairy products, desserts, confectionery, spreads, jams, chewing gum, chocolate-based products, and ice cream.

NHDC is a semisynthetic sweetener prepared from neohesperidin or naringin, two flavanones extracted from citrus peel. In aqueous solutions, NHDC is stable in the pH range 2.5–3.5. NHDC is currently allowed for many applications within the EU. In the United States, it is approved for flavoring food products.

4.9 NEOTAME

Neotame is a new high-potency nonnutritive sweetener that is considered as the potential successor of ASP. As a close derivative of ASP, it has the intrinsic qualities of this compound, which were at the root of the ASP commercial success, notably a very clean sweet taste, close to sucrose, with no undesirable bitter or metallic taste that occur in other well-known artificial sweeteners. Moreover, NTM offers additional salient advantages such as the following: a status, at use levels, of a no-calorie sweetener; an increase in stability in the neutral pH range, which strongly improves or widens the ASP applications (e.g., in baked goods); chemical inertness toward reducing sugars and aldehydic derivatives allowing its association with reducing sugars (glucose, fructose, high-fructose corn syrup, maltose, lactose, etc.) and flavoring agents based on aldehydic constituents (vanillin, ethyl vanillin, cinnamaldehyde, benzaldehyde, citral, etc.); an insignificant release of methanol and phenylalanine into the organism after intake (with, in particular, no possible hazard for phenylketonuric subjects); and a foreseeable highly competitive relative cost (cost per sucrose equivalent) as a result of its high sweetness potency.

NTM is a derivative of the dipeptide composed of the amino acids aspartic acid and phenylalanine (Figure 4.7). The optimum pH for maximum stability is ~4.5. NTM has been approved in the United States, Australia, and New Zealand.

NTM is the generic name for N-[N-(3,3-dimethylbutyl)-l±a-aspartyl]-l-phenylalanine 1-methyl ester. Its structural formula is given in Figure 4.7; its molecular formula is $C_{20}H_{30}N_2O_5$, and its molecular weight is 378.47. NTM's solubility is 740 times greater than necessary to obtain a sweetness level matching a 10% sucrose solution (17 g/L of NTM).

NTM is conveniently prepared from ASP and 3,3-dimethylbutyraldehyde; in a one-step high-yield process, NTM is an odorless white crystalline compound, which may be obtained anhydrous or, more usually, as a hydrate (4.5% hydration water; empirical formula $C_{20}H_{30}N_2O_5.H_2O$; formula weight 396.48). The melting point of the NTM hydrate is 80.9°C ± 83.4°C (without decomposition of the molecule below 200°C).

The crystal structure of NTM was obtained from single crystal X-ray diffraction analysis; the two hydrophobic groups of the NTM molecule are superposed in the crystal in a U-shaped conformation. NTM is stable under dry storage conditions; in dry conditions, its estimated shelf life is several years at ambient temperature. The monohydrated form is not hygroscopic.

The solubility in water of NTM is 12.6 g/L (10 g/L for airborn particulate matter [APM]) at 25°C. Its MRS is 10% (i.e., the multiple of the required solubility to match the sweetness intensity of NTM is an amphoteric compound). At 25°C, its pK_1 is 3.01, and its pK_2 is 8.02 (3.1 and 7.9, respectively, for APM). The pH of its isoelectric point (minimal charge and, in general, minimal solubility) is 5.5 (close to that of APM). NTM, at use levels, has negligible effects on viscosity. Its insignificant viscosity should not give rise to any mixing problems, and the negligible lowering of

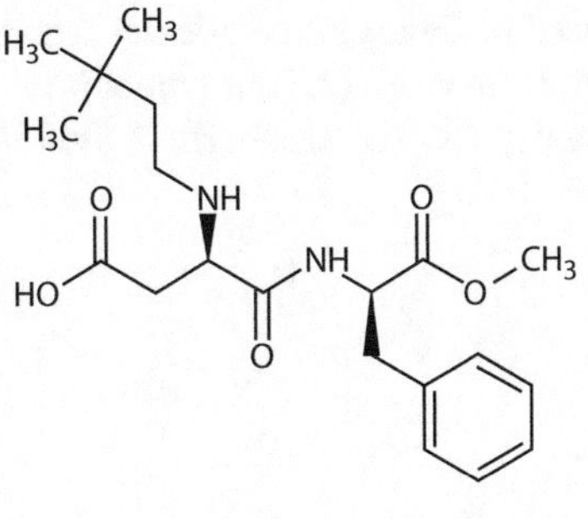

Figure 4.7 Chemical structure of NTM.

the surface tension and pH of its solutions should not lead to excessive foaming, for example, in carbonated soft drinks. In aqueous solution, the stability of NTM varies strongly with the pH and the temperature. Like APM, NTM is relatively stable at pH from 3 to 5.5. The optimal pH for NTM stability is 4.5. The degradation of NTM follows a pseudo–first-order kinetics. At pH 4.5, the half-life of NTM is 30 weeks at 25°C, 45 days at 40°C, and 40 h at 80°C in 0.1 M phosphate buffers. At pH 3, the half-life of NTM is 11 weeks at 25°C, 22 days at 40°C, and 24 h at 80°C. The stability of NTM at 80°C in pH range of 3–5.5 implies that high-temperature short-time processing is possible with food products sweetened with NTM; for example, after 30 min at 80°C, NTM percentage remaining in a pH 3 solution is 98.6%, which indicates that there is practically no loss of NTM within 30 min at 80°C. At pH 7, the half-life of NTM is 2 weeks at 25°C, 3 days at 40°C, and 4 h at 80°C.

4.10 SACCHARIN

Saccharin (SAC) is characterized as an artificial high-intensity sweetener. It is also called non-nutritive sweetener, and today is commonly used in the food industry. However, in the past, saccharin was used in a variety of applications apart from its major known use today as a sweetener. It was first used as an antiseptic and preservative to retard fermentation in food. Later on, saccharin was used in the plastic industry as an antistatic agent and as a modifier. Saccharin has been even used as a brightener in nickel-plated automobile bumpers (Arnold et al. 1983). Saccharin is the oldest high-intensity sweetener. It is commercially available in three forms: acid saccharin, sodium saccharin (Figure 4.8), and calcium saccharin. Sodium saccharin is the most commonly used form because of its high solubility and stability. Saccharin and its salts in their solid form show good stability under conditions present in soft drinks. However, at low pH, they can slowly hydrolyze to 2-sulfobenzoic acid and 2-sulfamoylbenzoic acid. Nowadays, it is exclusively used for low-calorie food intake that helps consumers to maintain their weight. It is also medically suggested to diabetics to consume foods containing saccharin instead of sugar.

Saccharin (1,2-benzisothiazol-3(2H)-on-1,1-dioxide) is the oldest sweetener on the market. It was discovered accidentally in 1879 by Constantin Fahlberg, a graduate student at Johns Hopkins University. He was working on the synthesis of toluene derivatives. One day, while having his lunch, he tasted something sweet in his bread (Sardesai and Waldshan 1991). This sweet taste was attributed to a compound known today as saccharin.

It is a white crystalline powder with an intensely sweet taste and is available in three forms (acid saccharin, sodium saccharin, and calcium saccharin). Aqueous solutions of saccharin are slightly acidic. Calcium and sodium salts of saccharin are neutral and slightly basic. The melting point of acid saccharin is 228.8°C–229.7°C, whereas the salts decompose above 300°C. Saccharin is characterized by its high solubility, stability, and bitter metallic aftertaste. Saccharin is about 300–500 times sweeter than sugar. The excellent stability of saccharin under food processing makes it ideally suited in many different products. It is used in a wide range of cases where heat processing is required (e.g., jams, canned products). However, at low pH (2.5), it can slowly hydrolyze to 2-sulfobenzoic acid and 2-sulfoamylobenzoic acid (Pearson 1991). Saccharin is used in a wide variety of food products

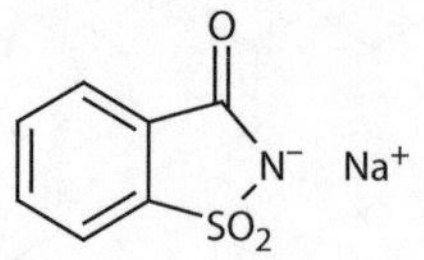

O
N^- Na^+
SO_2

Figure 4.8 Chemical structure of saccharin (sodium salt).

including baked goods (dry bases for mixes), beverages (dairy beverages, instant tea, instant coffee, fruit-based beverages), soft drinks (colas, citrus-flavored drinks, fruit-based soft drinks), sugar preserves and confectionery (calorie-free dustings, frostings, icings, toppings, fillings, syrups), alcoholic drinks (beer), vinegar, pickles and sauces (sandwich spreads, salad dressings), dairy products (yogurt and yogurt-type products, puddings, desserts and dairy analogues, sugar-free ice cream), fruits, vegetables, nut products, sugar-free jams and marmalades, low-calorie preserves, and other food products (i.e., chewing gums, liquid concentrates, frozen and refrigerated desserts).

Even today, even if saccharin is widely used, there is a controversy over its safety. Numerous toxicological studies have taken place with a variety of animal species. Saccharin is excreted unchanged except for a small percentage that is being accumulated in the bladder (Renwick 1985). It is shown that the compound is not genotoxic, and it does not bind to DNA. The greatest concern relating to metabolic effects of saccharin has been the uncertainty about its carcinogenicity to humans (Whysner and Williams 1996). Several early carcinogenicity studies in rats that included *in utero* exposure had indicated that feeding of saccharin may lead to bladder tumors (Cohen-Addad et al. 1986). On the contrary, studies in monkeys have shown that saccharin does not pose carcinogenic effects on the primate urinary tract (Takayama et al. 1998). A number of extensive epidemiological studies in humans have shown no association between saccharin consumption and urinary bladder cancer (Weihrauch and Diehl 2004). Saccharin may reasonably be anticipated to be a carcinogen, whereas a lethal dose in mice and rats is set for this compound (The Merck Index 1996).

The U.S. Food and Drug Administration delisted saccharin in 1972 because of the uncertainty of the safety of saccharin, and its use in foods and beverages was proposed to be banned in 1977. However, public protest led to imposing a moratorium on the ban, which has been extended up to the present (Pearson 1991; Kroger et al. 2006). In Canada, saccharin was banned in 1977 for use in foods (Arnold 1984). However, it is permitted to be sold in pharmacies as a tabletop sweetener. In other countries, saccharin is permitted, although its use is restricted to varying degrees. EU Directives 94/35/EC (European Commission 1994), 96/83/EC (European Commission 1996), 2003/115/EC (European Commission 2003), and 2006/52/EC (European Commission 2006) define in which food products and in what quantity saccharin can be used.

Consumer safety demands the monitoring of saccharin intake. A number of analytical methods based on different principles are available for the determination of saccharin in a broad range of food matrices. The aim here is to present the available methodologies for sample pretreatment and the available protocols of analysis.

4.11 STEVIOSIDE

Stevioside or stevia (Figure 4.9) is the name given to a group of sweet diterpene glycosides extracted from the leaves of *Stevia rebaudiana* plant (native of South America). Steviosides show good stability in the solid form. They are also quite stable in acidic condition beverages at 22°C. Steviosides are approved for food use in several South American and Asian countries, but lack approval in Europe and North America.

Stevioside is a natural sweetener extracted from the leaves of *S. rebaudiana* (Bertoni) Bertoni. *S. rebaudiana* (Bertoni) Bertoni is a perennial shrub of the Asteraceae (Compositae) family native to certain regions of South America (Paraguay and Brazil). It is often referred to as "the sweet herb of Paraguay." *S. rebaudiana* Bertoni (Asteraceae), an herb native to South America, is well known for sweet-tasting glycosides of the diterpene derivative steviol (ent-I 3-hydroxykaur-I 6-en-19-oic acid; Kohda et al. 1976). The major constituent is a triglycosylated steviol (stevioside) constituting 5%–10% in dry leaves. Other main constituents are rebaudioside A (tetraglucosylated steviol), rebaudioside C, and dulcoside A. The plant has several medicinal claims including that for antidiabetic activity (Goyal and Goyal 2010).

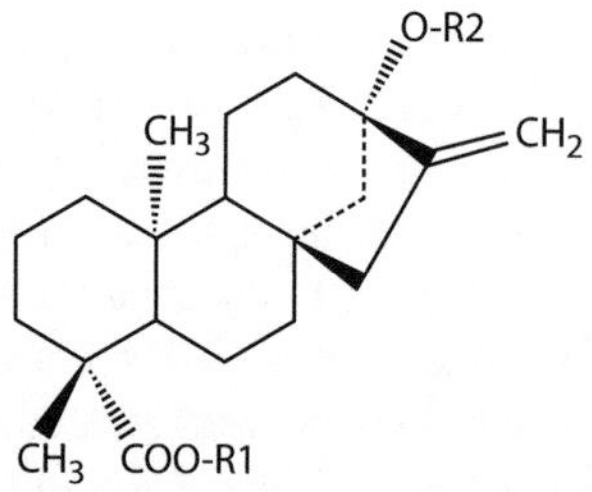

	Compound name	R1	R2
1	Steviol	H	H
2	Steviolbioside	H	β-Glc-β-Glc(2→1)
3	Stevioside	β-Glc	β-Glc-β-Glc(2→1)
4	Rebaudioside A	β-Glc	β-Glc-β-Glc(2→1) β-Glc(3→1)
5	Rebaudioside B	H	β-Glc-β-Glc(2→1) β-Glc(3→1)
6	Rebaudioside C (Dulcoside B)	β-Glc	β-Glc-α-Rha(2→1) β-Glc(3→1)
7	Rebaudioside D	β-Glc-β-Glc(2→1)	β-Glc-β-Glc(2→1) β-Glc(3→1)
8	Rebaudioside E	β-Glc-β-Glc(2→1)	β-Glc-β-Glc(2→1)
9	Rebaudioside F	β-Glc	β-Glc-β-Xyl(2→1) β-Glc(3→1)
10	Dulcoside A	β-Glc	β-Glc-α-Rha(2→1)

Figure 4.9 Chemical structure of steviosides. In rebaudioside D and E, R1 is composed of 2 b-Glc-b-Glc(2→1). In rebaudioside A, B, C, D, E, and F in group R2, an additional sugar moiety is added on carbon 3 of the first b-Glc. In rebaudioside F, one b-Glc is substituted for by-b-Xyl.

Acute and subacute toxicity studies revealed a very low toxicity of stevia and stevioside. Stevia and stevioside are safe when used as a sweetener. They are suited for both diabetics and PKU patients, as well as for obese persons intending to lose weight by avoiding sugar supplements in the diet. No allergic reactions to it seem to exist. Stevioside, the main sweet component in the leaves of *S. rebaudiana* (Bertoni) Bertoni, tastes about 300 times sweeter than sucrose (0.4% solution). Structures of the sweet components of Stevia occurring mainly in the leaves are given in Figure 4.9. Their content varies between 4% and 20% of the dry weight of the leaves depending on the cultivar and growing conditions. Stevioside 3 is the main sweet component. The advantages of stevioside as a dietary supplement for human subjects are manifold: it is stable, is noncalorific, maintains good dental health by reducing the intake of sugar, and opens the possibility for use by diabetic and PKU patients and obese persons.

There has been considerable commercial interest in using *Stevia* as a food/beverage sweetener. In December 2008, the FDA gave a "no objection" approval for GRAS status to Truvia (developed by Cargill and the Coca-Cola Company) and PureVia (developed by PepsiCo and the Whole Earth Sweetener Company, a subsidiary of Merisant), both of which are wholly derived from the *Stevia* plant. A bitter aftertaste associated with *Stevia* has been a problem for companies wishing to use the sweetener, and flavor companies have been trying to find ways to mask it without detracting from the perceived benefits of its natural status. As a result of research investigating the bitter component, we now report on a new water-soluble iminosugar alkaloid in *Stevia*. Alkaloids have not been previously reported from *Stevia* species. In an ongoing search for new iminosugars, *Veltheimia capensis* Hyacinthaceae (Sand Lily), a native to Western Cape, South Africa, was also found coincidentally to contain the same alkaloid.

Steviamine (1) has a molecular formula $C_{10}H_{20}NO_3$. The positive response to chlorine-*o*-tolidine reagent suggested that steviamine was an alkaloid. The ^{13}C nuclear magnetic resonance (NMR) spectroscopic data revealed the presence of a single methyl (δ 19.2), four methylenes (δ 23.7, 29.2, 33.3, and 56.5), and five methine (δ 52.8, 61.3, 66.7, 69.1, and 73.8) carbon atoms. The connectivity of the carbon and hydrogen atoms was defined from COSY and HMBC spectroscopic data. Three methine signals (δ 52.8, 61.3, and 66.7) with relatively down-field chemical shifts were suggestive of being bonded to the nitrogen of the indolizidine ring. The methylene signal at δ 56.5 (C-1) was attributed to the hydroxymethyl carbon, and this showed HMBC correlations to δ 61.3 (C-2) and 69.1 (C-3). The methine signal at δ 61.3 showed HMBC correlations to δ 69.1 (C-3), 73.8 (C-4), and 66.7 (C-5), suggesting a five-membered ring.

Iminosugar indolizidine alkaloids such as swainsonine and castanospermine have strong biological activities (Watson et al. 2001), including glycosidase inhibition, immune modulation, and antiviral and anticancer activity. The biological activity of steviamine and the synthesis of other 5-methyl-indolizidine alkaloids are under investigation. No study has been reported about the effect of iminosugars on taste in humans, but we consider it possible that polyhydroxylated iminosugars like steviamine may have an effect on the sweet taste of stevioside and rebaudosides. Since iminosugars are difficult to resolve and detect by most commonly used analytical systems and have similar solubility to the steviol glycosides, it is very likely that they occur in some *Stevia* products.

4.12 SUCRALOSE

Sucralose, 1,6-dichloro-1, 6-dideoxy-β-D-fructofuranosyl 4-chloro-4-deoxy-α-D-galacto-pyranoside, or 4,1′,6′-trichloro-4,1′,6′-trideoxy-*galacto*-sucrose (Figure 4.10), is a chlorinated derivative of sucrose, discovered in 1976 by carbohydrate research chemists at Queen Elizabeth College and Tate and Lyle, U.K. It is derived from a patented multistep process, involving selective chlorination of sugar at the 4, 1′, and 6′ positions substituting three hydroxyl groups on the sucrose molecule. It is the result of a study on a large number of related compounds, carefully synthesized and evaluated to determine the spatial structure and molecular configuration required for sweetness perception.

Ingredients and tabletop forms of sucralose are being marketed under the brand name Splenda. Its chemical formula is $C_{12}H_{19}O_8Cl_3$ (MW 397.35). Sucralose is a white, odorless crystalline powder and is readily dispersible and soluble in water, methanol, and ethanol. At 20°C, a 280 g/L solution of sucralose in water is possible. Sucralose presents Newtonian viscosity characteristics, a negligible

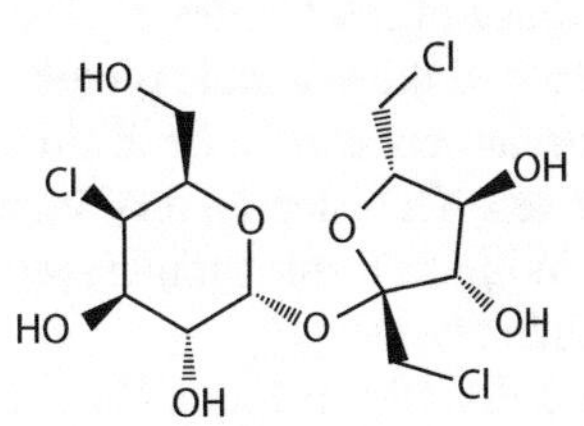

Figure 4.10 Chemical structure of sucralose.

lowering of surface tension, and no pH effects, and its solubility increases with increasing temperature. In ethanol, the solubility ranges from approximately 110 g/L at 20°C to 220 g/L at 60°C, and sucralose's solubility in ethanol facilitates in formulating alcoholic beverages and flavor systems.

Sucralose is 400–800 times sweeter than sucrose. Although its sweetness varies with pH, sucralose has a clean sugar-like taste and a time–intensity profile like that of sucrose, albeit more persistent. It has an excellent taste profile and no bitter or objectionable aftertaste. It is a flavor enhancer and shows sweetness synergism with cyclamate, ACS-K, and NHDC.

Sucralose offers a broad pH, aqueous solution, thermal processing, and shelf stability. It does not interact with food ingredients and is stable in the dry form (4 years at 20°C). It withstands high temperatures, thus making it well suited for use in pasteurized, aseptic processing, sterilized, cooked, and baked foods. However, under extreme conditions of pH, temperature, and time, sucralose may be hydrolyzed, producing 4-chloro-deoxy-D-galactose and 1,6-dideoxy-1,6-dichloro-D-fructose, or degraded with elimination of hydrogen chloride in basic medium.

Sucralose is noncariogenic. It resists hydrolysis in the human digestive tract, being excreted unchanged in the feces, and the very small portion absorbed is rapidly eliminated in the urine. Therefore, it produces no glycemic response and is virtually noncaloric. Following safety testing and toxicological studies in humans and animals, the FDA concluded that sucralose does not pose any carcinogenic, reproductive, or neurological risk. The JECFA reviewed it favorably and in 1990 recommended an ADI of 0–15 mg per kilogram of body weight. Sucralose is approved for use in a wide range of food products in Canada, United States, Australia, Mexico, Russia, Romania, China, the European Union, and Mercosur. It has been used as a tabletop sweetener and in carbonated, still, and alcoholic beverages, frozen desserts, confectionery, bakery products, canned fruits and vegetables, fruit spreads, chewing gum, dry-mix products, dairy products, condiments, dressings, and breakfast cereals.

4.13 THAUMATIN

Thaumatin is a group of intensely sweet basic proteins isolated from the fruit of *Thaumatococcus danielli* (West African Katemfe fruit). It consists essentially of the proteins Thaumatin I and Thaumatin II. Thaumatin is a taste-modifying protein that functions as natural sweetener or flavor enhancer. It is stable in aqueous solutions between pH 2.0 and 10 at room temperature. As occurs with ASP, it is nutritive, containing 4 kcal g^{-1}, but due to its intense sweetness, the amounts used are small enough for thaumatin to be considered and classified as a nonnutritive sweetener. Thaumatin is approved for a number of uses in the United Kingdom, Japan, Australia, the EU, and in many other countries. In the United States, it is approved as a flavor enhancer.

4.14 SAMPLE PRETREATMENT

A number of analytical methods based on different principles are available for the determination of ACS-K and ASP in a broad range of food matrices. The aim here is to present the available methodology for sample pretreatment and the available protocols of analysis.

Generally speaking, sample pretreatment cannot be avoided in most of the analytical methods for the determination of food additives. This step is very important because without it, the food sample cannot be directly analyzed. With the term "sample pretreatment," we refer to sample preparation or/and sample cleanup prior to analysis.

The determination of sweeteners directly in foods often cannot be achieved due to interferences. Often, sample pretreatment is the most time-consuming step of the method of analysis. For most assays, the weight of the sample taken for quantitative analysis has to be known. Then, some preparative operations are likely necessary before an extraction can be performed. For the extraction of sweeteners, these operations that make the complete extraction easier are usually the change of volume and the change of pH prior to extraction.

Food samples are characterized as difficult matrices. The food matrix presents a great variability in its composition. Carbohydrates, proteins, lipids, minerals, preservatives, colors, thickeners, and vitamins may stand alone or coexist in a food matrix. All of the components can interfere in the determination of sweeteners. Sample pretreatment procedures must be suitable with the method that the analyst will use, considering the instrumentation that he or she has available in the laboratory. The success of the method is often totally owed to the effectiveness of sample pretreatment, and it depends on the accuracy (quantitative or/and qualitative) that the analyst wants to obtain ("fit for purpose").

The chemical and the physical properties of food vary. The variability in composition of a given food sample can be minimized with proper sample preparation. Generally speaking, preparation of food samples is achieved in four steps: (1) homogenization, (2) extraction, (3) cleanup, and (4) preconcentration.

Analysis of liquid samples does not require the homogenization step because of their liquid state. After the sample preparation, some matrix components may still be present interfering in the analysis. They may coextract with analytes due to similar solubility in the solvents used for extraction. The presence of matrix interferences in a sample extract can contribute to problems on the accuracy of the method. The only way to resolve this problem is to further clean up the sample for analysis. Cleanup is achieved commonly by (1) solid-phase extraction (SPE), (2) dialysis, (3) liquid–liquid extraction (LLE), (4) precipitation, and (5) filtration.

Appropriate extraction and cleanup procedures maximize recovery of the analytes. Optimal sample preparation can reduce analysis time, enhance sensitivity, and enable confirmation and quantification of analytes. Extraction, cleanup, and/or purification might be necessary, depending on the complexity of the sample and the sensitivity and selectivity of the method used (Self 2005).

The approach of determination of ACS-K and ASP in a simple matrix is much easier and less time consuming than the determination of these two sweeteners in a more complicated food sample matrix. Bulk samples of sweeteners have a much more simple pretreatment stage than complex food matrices. The interference of other food additives in the determination of sweeteners is more common in a complicated food sample matrix. These additives may be in the same level of magnitude as the sweeteners or in a much higher level of magnitude causing interference in the determination. In bulk samples (e.g., tabletop solid tablets), additive compounds exist in a much lower quantity. Usually three types of additives are added in bulk powders in order to deal with possible problems during the preparation of tablets. Glidants are added for dealing with a poorly flowing material. Pure lubricants give more effective mixing, and antiadherents prevent the tablet adhering to the die. When we want to determine sweeteners in tabletop solid tablets, these are turned into powder. A portion of the powder is being weighed, directly dissolved in ultrapure water, and transferred into volumetric flasks. Then according to the methods protocol, the next steps of determining ACS-K or/and ASP follow. Samples characterized by a relatively simple matrix, like liquid sweeteners and beverages, can simply be diluted or dissolved in deionized water or in an appropriate buffer. In the case of carbonated drinks, the samples have to be degassed. Degassing is being done by sonication, by sparging with nitrogen, or under vacuum.

All samples are filtered prior to the analysis, and the extracts may need centrifugation. This simple sample preparation procedure is found in published procedures. It is very quick and cheap. To preconcentrate the analytes or/and remove the chemical interferences, we must take advantage of the SPE technique. It fractions the compounds of our choice based on the affinity of the compound or a group of compounds to the stationary phase. The most frequently used SPE cartridges are the nonpolar C_{18}. Their stationary packing material consists of nonpolar C_{18} chains. The protocol of an SPE procedure consists of four steps: (1) cartridge conditioning, (2) sample load, (3) cartridge wash, and (4) elution of analytes.

In the most common mode of SPE, an aliquot of the sample extract is loaded onto a previously conditioned SPE cartridge. The type of SPE packing material, solvents, pH, and the flow rates need to be properly selected in order to retain analytes effectively within the cartridge. The interfering substances should be retained very strongly or not retained at all. As a result, weakly retained substances are readily removed from the cartridge during sample load and/or cartridge wash. Analytes are eluted during the elution step, and interfering substances having a strong affinity to the sorbent stay adsorbed within the cartridge. The sensitivity of a final determination can easily be enhanced by evaporating the final SPE extract to dryness and reconstituting it with a smaller amount of a solvent of choice (preconcentration). SPE-based sample-preparation protocols seem to be the best available choice. They are simple, reproducible, reasonably quick, and inexpensive. They are universal and compatible with the most popular techniques used in food analysis (Tunick 2005).

Sweetened beverages include two types of beverages with carbon dioxide and two types of beverages without carbon dioxide. Beverages with carbon dioxide include soft drinks "light" and soft drinks sugar-sweetened. Beverages without carbon dioxide also include the same types of beverages: soft drinks "light" and soft drinks sugar-sweetened. In soft drinks "light" with carbon dioxide, a mixture of ASP and ACS-K is more often used to sweeten the products (Leth et al. 2007). Micellar electrokinetic capillary chromatography (MEKC) is a rapid method for the determination of artificial sweeteners in low-calorie soft drinks and is often used. The sample solutions of "light" soft drinks are prepared by diluting the products with an appropriate amount of deionized water. The solutions are then simply filtered through a 0.45-μm cellulose acetate filter before analysis (Thompson et al. 1995).

Today, the most common technique for the determination of artificial sweeteners in soft drinks is high-performance liquid chromatography (HPLC) analysis. The sample preparation for HPLC analysis includes firstly filtering through a 0.45-mm membrane filter, and then the sample is ultrasonicated before the analysis. Nectars are first centrifuged, filtered through membrane filters, and then ultrasonicated (Lino et al. 2008). For the determination in cola drink, first a volume is accurately weighed into a volumetric flask and then degassed in an ultrasonic bath. If the determination is obtained through HPLC analysis, the volume of cola drink is directly diluted with mobile phase (Demiralay et al. 2006).

Gum samples are prepared by placing the sample in a flask and extracting with a mixture of glacial acetic acid, water, and chloroform. Hard or soft candy samples are shaken with water until they dissolve (Biemer 1989). Milk and dairy products are homogenized prior to analysis, and an aliquot of a homogeneous sample is transferred to a flask followed by the addition of distilled water. The mixture is thoroughly stirred and allowed to stand and then filtered (Ni et al. 2009).

Sweeteners are determined in diet jams by mixing the jam with water and sonicating the mixture. The mixture is made up to volume and then filtered through a 0.45-mm filter (Boyce 1999). Preserved fruit is grounded and homogenized. Then it is weighed into a volumetric flask, and water is added. This mixture is extracted ultrasonically and diluted to volume with water after cooling to room temperature. A volume of supernatant is applied to a conditioned Sep-Pak C_{18} cartridge (Chen and Wang 2001), and the subsequent SPE follows according to the previous protocol.

The determination of sweeteners with flow injection analysis (FIA) methods is very common. It is reported that beverages, juices, strawberry sweets, and tomato sauce containing ACS-K and/or ASP can be analyzed with the use of a FIA setup coupled with an ultraviolet (UV) detector. An adequate amount of beverage is taken, degassed, and diluted, adjusting the same conditions as the carrier. Finally, it is filtered through a 0.2-mm Millipore filter. For the analysis in juices before the step of filtering, a centrifuge step may be necessary. In the case of strawberry sweets, an adequate amount is weighed and thoroughly crushed in a glass mortar, and then dissolved in water with the aid of an ultrasonic bath, adjusting to the same conditions as the carrier. After that, it is centrifuged and filtered as above. Finally, in the case of tomato sauce, the amount is suspended in water; then a portion is diluted and centrifuged, and then the same procedure is followed as for the beverages (Jiménez et al. 2009).

Fourier transform infrared (FTIR) spectroscopy is reported as a quick method for the determination of ACS-K in commercial diet food samples without the use of even an extraction procedure prior to analysis. Samples that are characterized as difficult food matrices like chocolate syrup, coffee drink, coffee creamer, cranberry juice, ice cream, and instant chocolate milk can be easily analyzed. The ice cream sample is converted to mixture at 60°C in a water bath with continuous agitation. All other samples are used without any pretreatments. A portion of each food sample is mixed thoroughly with ultrapure water. The biggest advantages of direct measurement through FTIR spectroscopy are speed and lack of time-consuming sample pretreatment. The samples are then maintained at 40°C in an incubator followed by ultrasonication. Carrez I (3.6 g potassium hexacyanoferrate trihydrate dissolved and made up to 100 mL with distilled water) and Carrez II reagent (7.2 g zinc sulfate heptahydrate dissolved to 100 mL with distilled water) solutions are added to the samples, followed by centrifugation at 4°C. The supernatant is carefully separated using a syringe to avoid the fat layer for each sample and filtered. The water-soluble extract obtained is directly poured onto the surface of attenuated total reflectance (ATR-FTIR; Shim et al. 2008).

The simultaneous determination of other additives including ACS-K and ASP in some food samples is very common nowadays. For example, a simultaneous determination in a food sample of soy sauce has been reported. For the pretreatment of this sample, it is placed in a volumetric flask and then diluted with water. After mixing thoroughly, a portion of the solution is added to a Sep-Pak C_{18} cartridge. Some additives that are not adsorbed by the packing material come through the cartridge and are collected as eluate A. Then water is passed through the cartridge to remove interfering substances. The other additives that are absorbed onto the packing material were then eluted with acetonitrile–water (2:3, v/v) and collected as eluate B. Eluates A and B are combined and poured into a volumetric flask and diluted to volume with acetonitrile–H_2O (2:3, v/v). The same procedure can be followed for food sample of dried roast beef and sugared fruit. The only difference is that they are previously ground into fines with a grinder (Chen and Fu 1995).

Appropriate extraction and cleanup procedures maximize recovery of saccharin. For its efficient extraction, the analyst should optimize the solvent volume and its pH prior to extraction. Optimal sample preparation can reduce analysis time, enhance sensitivity, and optimize confirmation and quantification of saccharin. The most common sample preparation procedures that are found nowadays in published procedures use the SPE technique. It is simple, reproducible, quick, and inexpensive. The SPE technique fractions the compounds of our choice based on the affinity of the compound or a group of compounds to the stationary phase. The protocol of an SPE procedure consists mainly of four steps. First, the conditioning of the cartridge precedes, followed by the loading of the food extract; washing of the cartridge follows, and finally, elution of the analytes is performed by adequate solvent(s). The type of SPE packing material, solvents, pH, and flow rates need to be properly selected in order to retain analytes effectively within the cartridge. The interfering substances should be retained very strongly or not retained at all. As a result, weakly retained substances are removed from the cartridge during sample load and/or cartridge wash. Analytes are eluted during the elution step, and interfering substances having a strong affinity to the sorbent stay adsorbed within the cartridge. SPE procedures are universal and compatible with the most popular techniques used in food analysis (Tunick 2005). In published procedures for the determination of saccharin, one can witness a variety of food matrices being suitably treated prior to analysis. Cola drinks are usually degassed in an ultrasonic bath (Demiralay et al. 2006). The same applies to every carbonated drink. Degassing is performed by sonication, by sparging with nitrogen, or under vacuum. Chewing gum samples are prepared by placing them in a flask and extracting with a mixture of glacial acetic acid, water, and chloroform. Hard or soft candy samples are shaken with water until dissolved (Biemer 1989). Milk and dairy products are homogenized prior to analysis, and an aliquot of a homogeneous sample is transferred to a flask followed by the addition of distilled water. The mixture is thoroughly stirred and then filtered (Ni et al. 2009). Saccharin is determined in diet jams by mixing the jam with water and sonicating the mixture. The mixture is then made up to a certain volume and filtered (Boyce 1999). Preserved fruits are grounded and homogenized. Then water is added, and the formed mixture is extracted ultrasonically and diluted to a certain volume with water (Chen and Wang 2001). For the analysis in juice before filtration, a centrifuge step may be necessary. In the case of strawberry sweets, an adequate amount is weighed and thoroughly crushed in a glass mortar and then dissolved in water with the aid of an ultrasonic bath. After that, it is centrifuged and filtered. In the case of tomato sauce, the amount is suspended in water, and then a portion is diluted and centrifuged (Jiménez et al. 2009). When the determination of saccharin in tabletop solid tablets is needed, tablets are turned into powder. A portion of the powder is being weighed and directly dissolved in ultrapure water. Samples characterized by a relatively simple matrix like liquid sweeteners can simply be diluted or dissolved in deionized water or in an appropriate buffer. The approaches for the determination of saccharin in simple matrices are much easier and less time consuming than the determination of saccharin in more complicated food sample matrices.

4.15 ANALYTICAL METHODOLOGY

The number of methods for the determination of ASP and ACS-K in food samples is large and is classified according to the detection method. The method is selected as a compromise between the following factors: (1) accuracy, (2) precision, (3) cost, (4) detection limits, (5) selectivity, (6) safety, (7) sample throughput, (8) consumption of sample and reagents, (9) simple operation (automated or not), and (10) contamination risks.

However, three main parameters are important in selecting the method of analysis: (1) The reagents and the apparatus of the method must have a low cost. (2) We do not need a very sensitive method with a low detection limit since the sweeteners fluctuate in the micromolar range and sensitive methods often are costly. (3) The sample throughput and the consumption of sample and reagents must be relatively low. The following analytical and detection techniques meet these criteria:

- Biosensors
- Spectrophotometry
- Electroanalysis
- Chromatography

These detection techniques have been applied to the simultaneous determination of several kinds of sweeteners in foods (Cantarelli et al. 2009). The magnitude of sensitivity and selectivity of the method (including sample preparation prior to analysis) used is different when only one of these two sweeteners in foodstuffs is determined and different when the two sweeteners simultaneously with other food additives are determined. When analysts want to determine simultaneously a number of compounds, they must compromise the best analytical parameters of each compound. In this case, it is common to lose magnitude of sensitivity and selectivity of the majority of the compounds analyzed simultaneously. On the other hand, this does not occur when a method for the determination of one and only compound is developed.

For routine analysis of complex food matrix samples, MEKC, capillary zone electrophoresis, HPLC, and ion chromatography (IC) are preferred. Even thin-layer chromatography (TLC; Baranowska et al. 2004) has been reported for the determination of the two sweeteners. These instrumental methods are important reference methods for food sample analysis and are based on expensive analytical instruments and reagents. Liquid chromatographic determination in foods is simple because beverages or aqueous extracts from foods can often be injected into the columns immediately after filtration. The most commonly used methods are basically reversed-phase HPLC separations coupled with UV light detector.

NMR spectroscopy and mass spectrometry (MS) are useful techniques for structural elucidation of unknown compounds, but the results obtained are difficult to quantify. Liquid chromatography (LC) methods have been used extensively for the determination of highly intense sweeteners because, in many instances, the sample matrices from which they are to be determined may be complex. In addition, a sweetener may be used in combination with other sweetener(s). Nevertheless, in recent years, capillary electrophoresis (CE) methods have been developed that compete successfully with LC methodologies. There are no methodologies proposed for the determination of NTM and only a few methods for alitame and thaumatin. Most methods for the determination of thaumatin involve immunochemical assays (IC) and measurement in an enzyme-linked immunosorbent assay reader. On the other hand, the largest number of methodologies has been proposed for the determination of saccharin because it is the oldest known and used high-intensity sweetener. Automated continuous determination by FIA has been developed for ACS-K, ASP, cyclamate, and saccharin; generally, these procedures involve spectrophotometric and electroanalytical detections.

4.15.1 Determination of ACS-K

In the last few years, there has been an interest in the development of analytical devices for the detection and monitoring of various biological and chemical analytes. For several decades, analytical chemists were inspired from biological sciences, and nowadays, the detection of analytes using biosensors is very common. Biosensors offer the capability to develop a method for the rapid screening of these sweeteners with a respective low cost (Nikolelis and Pantoulias 2000). Lipid films can be used for the rapid detection or continuous monitoring of a wide range of compounds in foods and in the environment. Such electrochemical detectors are simple to fabricate and can provide a fast response and high sensitivity. Investigations have taken place for the interactions of the artificial sweeteners ACS-K with freely suspended bilayer lipid membranes (BLMs) that can be used for the direct sensing of these sweeteners. The interactions of the sweeteners with these BLMs produce transient electrochemical current signals, the magnitude of which could be used to quantify the concentration of the sweetener. Determination of the mechanism of signal generation involves analysis of the structural effects caused by interactions of the sweeteners with model lipid membranes. Differential scanning calorimetry studies have shown that the interactions of sweeteners with lipid vesicles stabilize the gel phase of lipid films. Monolayer compression techniques at an air–water interface have revealed an increase in the molecular area of lipids when sweeteners are added to the aqueous subphase. Such structural changes are known to cause electrostatic and permeability changes with lipid membranes.

A conductometric method based on the use of surface stabilized BLMs (s-BLMs) composed from egg phosphatidylcholine is developed for monitoring ACS-K and other sweeteners. The interactions of sweeteners with s-BLMs produce a reproducible electrochemical ion current signal increase that appears within a few seconds after exposure of the membranes to the sweetener. The current signal increases relatively to the concentration of the sweetener in bulk solution in the micromolar range. The alterations in the current signal are observed even from low concentrations between 0.4 and 7 μM in electrolyte solutions. The s-BLM–based biosensor is stable for long periods of time (over 48 h) and can be easily constructed at low cost (and therefore can be used as a disposable sensor) with fast response times on the order of a few seconds. The lipid layer is deposited into a nascent metallic surface while immersing the metal wire into the lipid solution. The wire with the lipid layer is immersed into KCl aqueous solution, and ionic current is stabilized over a period of 10 to 15 min depending on the diameter of the silver wire. Calibrations are subsequently done by stepwise additions of 0.1 or 1.0 mM sweetener standard solution added to the KCl electrolyte while continuously stirring. Once the calibration plot or its equation is set up, the unknown sweetener concentration of a solution can be independently determined using a fresh BLM on a nascent metallic surface, and the procedure of immersing the wire with BLM into a KCl solution is repeated. The detection limit of ACS-K is 1 μm, and the reproducibility is on the order of ±4% to 8% in a 95% confidence level. The recovery ranged between 96% and 106% and shows no interferences from the matrix (Nikolelis et al. 2001a).

Enzyme electrodes coupled with FIA have also been reported. An ammonia-sensitive electrode L-aspartase or carboxypeptidase A with aspartate ammonia lyase immobilized and a three-enzyme system including aspartate, aminotransferase, glutamate oxidase, and ASP hydrolyzing enzyme coupled with an H_2O_2 probe have been reported. The major drawback of these biosensors is the interferences from L-aspartate, which is usually present in samples (Compagnone et al. 1997).

Potentiometry is the procedure in which a single measurement of electrode potential is employed to determine the concentration of an ionic species in a solution. A potentiometric method is developed for the titration of ASP. It makes use of a selective polyvinyl chloride membrane electrode. This electrode is based on the electroactive species of cetylpyridinium (CP^+) and 2,4,6-trinitrobenzene-sulfonate ($TNBS^-$). The electrode exhibits a rapid and Nernstian response to $TNBS^-$ from 5.0×10^{-5} M to 1.0×10^{-2} M at 25°C ± 0.1°C. The response is unaffected by the change of pH over the range 2 to 12. The electrode is successfully applied to the determination of ASP in pure solutions with a precision and accuracy of 0.9% to 1.3% without interference from other constituents. The selectivity of ion-pair–based membrane electrodes depends on the selectivity of the ion-exchange process at the membrane–test solution interface and the mobilities of the respective ions in the membrane. The CP-TNBS electrode is highly selective for the TNBS anion. The organic and inorganic anions do not interfere due to the differences in their mobilities and permeabilities as compared with the TNBS anions. It is found that the periodate ion and picric acid interfere with the CP-TNBS electrode. In the case of sugars, high selectivity is mainly attributed to the difference in polarity and the lipophilic nature of their molecules relative to the TNBS anion or CP cation. This method has a high degree of accuracy, a mean recovery of 99.8% to 101%, and excellent precision by the small values of the relative standard deviation (RSD). It is found that TNBS reacts with the investigated molecules in the ratio 1:1. The analysis indicates the high accuracy and precision depending on less complicated instrumentation and time-consuming pretreatment steps. The combination of sensitivity, selectivity, and simplicity of ion-selective electrode potentiometry makes it an excellent and versatile analysis technique (Buduwy et al. 1996). An established potentiometric assay for the indication of ACS-K is performed by nonaqueous titration with 0.1 N perchloric acid in glacial acetic acid (Joint FAO/WHO 2001).

Spectrophotometry is a detection technique widely used for the determination of sweeteners. This technique is not commonly used in food samples because it is sensitive to interference due to food additives. Spectrophotometric procedures for determining ASP involve different chemical reagents such as ninhydrin, chloroanilic acid, diethyldithiocarbamate, and p-dimethylaminobenzaldehyde (Fatibello-Filho et al. 1999).

The determination of ACS-K by FTIR spectra has been achieved using an advanced spectrometer equipped with a multibounce horizontal attenuated reflectance accessory. The instrument can produce 12 reflections with a penetration depth (infrared beam) of 2.0 μm. The accessory is composed of a ZnSe crystal mounted in a shallow trough for sample containment with an aperture angle of 45° and a refractive index (RI) of 2.4 at 1000 cm^{-1}. Single beam spectra (4000 to 400 cm^{-1}) of the ACS-K extract from real samples are obtained and corrected against the background spectrum of pure Milli-Q water. All spectra are collected in triplicate and averaged before subjecting to multivariate analysis (Shim et al. 2008).

4.15.2 Determination of Alitame

Analysis of ACS-K, alitame, ASP, caffeine, sorbic acid, theobromine, theophylline, and vanillin in tabletop sweeteners, candy, liquid beverages, and other foods using a μ-Bondapak C-18 column

Table 4.3 Simpler Methods for Analysis of Food Products for Nonnutritive Sweeteners

Cola, pudding, chocolate	Saccharin, cyclamate, alitame	μ Bondapak C-18 or Supelcosil LC-18	Phosphate buffer 20 mM (pH 3.5)–acetonitrile 97:3	RI, UV, 200 nm
Candy, soft drinks, yogurt, custard, fruit juice, nectar, biscuit, chocolate	Aspartame and its decomposition products, saccharin, alitame, acesulfame-K	μ Bondapak C-18	Phosphate buffer 125 mM (pH 3.5)–acetonitrile 90:10, 85:15, 98:2	UV, 220 nm

and a mobile phase of acetonitrile–0.0125 M potassium dihydrogen phosphate (10:90, v/v) at pH 3.5 and UV detection at 220 nm has been advocated. This method allows for the simultaneous determination of theobromine, theophylline, caffeine, vanillin, dulcin, sorbic acid, saccharin, alitame, ASP, and their degradation products in a single run of 60-min duration. Tabletop sweeteners, candies, soft drinks, fruit juices, fruit nectars, yogurts, creams, custards, chocolates, and biscuits have been analyzed by simple extraction or just by dilution using this method.

Some of the simpler LC methods for sweetener analysis are given in Table 4.3.

4.15.3 Determination of ASP

A screening flow-injection spectrophotometric method in tabletop sweeteners and in food samples (pudding, gelatin, and refreshment) using ninhydrin as a colorimetric reagent is reported. The reaction is conducted in a 1:1 v/v methanol–isopropanol medium that also contains potassium hydroxide. The absorbance measurements are made at 603 nm. The results obtained for the determination of ASP have a good correlation coefficient: $r = 0.998$. Thirty-six samples can be analyzed per hour, and the RSD is less than 3.5% ($n = 6$) for all samples. The detection limit is 0.0381 mM of ASP (Nobrega et al. 1994).

Another flow-injection spectrophotometric method is developed for determining ASP in tabletop sweeteners without interference of saccharin, ACS-K, and cyclamate. Samples are dissolved in water, and a portion of the solution is injected into a carrier stream of 5.0×10^{-3} M sodium borate solution (pH 9.0). The sample flow through a column packed with $Cu_3(PO_4)_2$ immobilized in a matrix of polyester resin. Then Cu(II) ions are released from the solid-phase reactor by the formation of $Cu(II)(ASP)_2$ complex. The mixture is merged with a stream of borate buffer solution (pH 9.0) containing 0.02% (w/w) alizarin red S, and the Cu(II)–alizarin red complex formed is measured spectrophotometrically at 550 nm. The calibration graph for ASP is linear in the 20 to 80 μg/mL concentration range, with a detection limit of 2 μg/mL of ASP. The RSD is 0.2% for a solution containing 40 μg/mL ASP ($n = 10$), and 70 measurements are obtained per hour. The column is stable for at least 8 h of continuous use (500 injections) at 25°C (Fatibello-Filho et al. 1999).

The establishment of analytical conditions for the determination of ASP by means of HPLC–electrospray ionization–MS (HPLC–ESI–MS) has been studied. MS is a powerful qualitative and quantitative analytical technique that has been introduced in many analytical and research laboratories in the last 10 years. The combination of HPLC with tandem MS yields a particularly powerful tool, and it is now the method of choice for analysis. However, HPLC–ESI–MS methods are not completely without problems that can compromise the quality of the results. ASP is analyzed in the positive-ion mode at 3.05-kV probe voltage with a methanol/water mobile phase (30:70, pH 3.00 by addition of acetic acid). Mass spectra at different spraying capillary voltages are reported and discussed. A detection limit of 5 pg and a linear response for ASP in aqueous solution in the range of 1 to 200 pg are obtained. The method is applied to the detection of ASP in three soft drinks commercially available in Italy. HPLC–ESI–MS provides an additional tool for the sensitive and selective detection of ASP and for the improvement of possibly critical separations such as that of ASP from caffeine (Galletti et al. 1996).

ASP can be determined by HPLC with electrochemical detection. ASP, which is electrochemically inactive, is made oxidizable in the range 0.1–1.1 V after postcolumn irradiation at 254 nm. A detection limit of 0.5 mg/L (signal-to-noise ratio 3:1) is attained using a coulometric detector with the working cell set at 0.8 V and a C_6 column (150 × 4.6 mm I.D.) operated under isocratic conditions with 0.1% perchloric acid–methanol (85:15, v/v) as the eluent at a flow rate of 1 mL/min. A linear response for aqueous solutions of ASP in the range 1 to 20 mg/L and a 5% standard deviation for five replicate injections were obtained. The method was applied to the determination of ASP in two diet colas (Galletti and Bocchini 1996).

The separation and determination of the sweetener ASP by IC coupled with electrochemical amperometric detection are reported. Typically in IC, electrochemical detection is employed such as an amperometric or conductivity detector. On the other hand, UV absorbing excipients such as flavors and dyes may not give any electrochemical response and can therefore be eliminated as an interferent in quantitation of the analyte. Thus, IC offers the opportunity to streamline method development and increase sample throughput. Sodium saccharin, ACS-K, and ASP were separated using 27.5 mM NaOH isocratic elution on a Dionex IonPac AS4A-SC separation column. ASP can be determined by integrated amperometric detection without interference from the other two sweeteners. The method can be applied to the determination of ASP in tabletop, fruit juice, and carbonated beverage samples, and the results obtained by integrated amperometry are in agreement with those obtained using a UV detection method. The recoveries for samples ranged from 77.4% to 94.5%. The peak area response for ASP is linear in the range 0.1–10 μg/mL. For seven consecutive injections of a standard solution with a concentration of 5 μg/mL, the RSD is 1.29%, and the detection limit (signal-to-noise ratio of 3:1) is 0.031 μg/mL for ASP (Qu et al. 1999).

4.15.4 Determination of Cyclamate

The common HPLC-UV detection mode is not suitable for determination of cyclamate because of the lack of UV chromophore in its molecule. Some complicated and time-consuming procedures are necessary for the absorbance detection of this sweetener in HPLC. This problem has been solved by using indirect UV photometry, postcolumn ion-pair extraction, and precolumn derivatization. One of the first approaches employed postcolumn ion-pair extraction with absorbance detection for the LC determination of cyclamate. After chromatographic analysis, the sweetener was mixed with an appropriate dye (methyl violet or crystal violet) and detected by absorption in the visible (Vis) range. In this method, the eluted sweetener is mixed with an appropriate dye (methyl violet or crystal violet) being detected by absorption in the Vis range (Lawrence 1987).

Hydrolysis of cyclamate to cyclohexylamine has been employed for the analysis of tabletop sweeteners. The hydrolysis step was performed batchwise by treating cyclamate with hydrogen peroxide and hydrochloric acid. The cyclohexylamine was derivatized with 1,2-naphthoquione-4-sulfonate (NQS) in an flow injection system. The NQS derivative was monitored at 480 nm (Cabero et al. 1999).

Cyclamate was determined in soft drinks using RP-HPLC combined with indirect Vis photometry at 433 nm. The analytical signal was derived from changes in absorbance of mobile phase with addition of chromogenic dye (Methyl Red; Choi et al. 2000).

Cyclamates in tabletop sweeteners and some low-calorie soft drinks were determined by an FIA spectrophotometric method. The procedure utilized the reaction between nitrite and cyclamate in medium containing phosphoric acid. The excess of nitrite was determined by the measurement of absorbance of Griess reaction product at 535 nm (Gouveia et al. 1995).

An HPLC–ESI–MS method using tris(hydroxymethyl) aminomethane as an ion-pair–forming agent was employed for analysis of cyclamate in foods. Cyclamate was separated on a C8 column in isocratic mode with 100% aqueous mobile phase. MS was operated in negative, selected ion (m/z = 178) recording mode. The method was found to be highly sensitive, specific, and simple (Huang et al. 2006).

An HPLC-tandem MS (HPLC-MS^2) method characterized by minimal sample preparation, high sensitivity, and selectivity was developed for the determination of cyclamate in foods. Under negative ESI conditions, parent ions of $m/z = 177.9$ and product ions of $m/z = 79.9$ were collected and used for quantitation (Sheridan and King 2008).

4.15.5 Determination of NHDC

The HPLC-UV method seems to be the most popular option for the determination of NHDC. NHDC can be analyzed using a standard C_{18} column in gradient mode. The optimal analytical wavelength is 282 nm, since common food components (e.g., sweeteners and preservatives) are transparent or absorb very slightly at this wavelength (Nakazato et al. 2001).

NHDC is a minor component of sweetener blends. It was separated from other commonly added sweeteners using a 100-mM borate buffer (pH 8.3). Direct UV detection was applied, and a wavelength of 282 nm was found optimal to avoid interferences from other sweeteners and sugars. The method was successfully applied to the determination of low levels of NHDC in soft drinks, fruit juices, and yogurts (Pérez-Ruiz et al. 2000).

4.15.6 Determination of Saccharin

In the last few years, there has been an interest in the development of analytical devices for the detection and monitoring of various biological and chemical analytes. For several decades, analytical chemists were inspired from the biological sciences, and nowadays, the detection of analytes using biosensors is very common. Biosensors offer the capability to develop methods for the rapid screening of saccharin with a respective low cost (Nikolelis and Pantoulias 2000). Lipid films can be used for the rapid detection or continuous monitoring of a wide range of compounds in foods. Such electrochemical detectors are simple to fabricate and can provide a fast response and high sensitivity. The major interference from proteins can be eliminated by modulation of the carrier solution that does not allow adsorption of these compounds in BLMs (Nikolelis et al. 2001b). A conductometric method based on the use of surface s-BLMs is developed for monitoring saccharin and other sweeteners. The interactions of sweeteners with s-BLMs produce a reproducible electrochemical ion current signal increase that appears within a few seconds after exposure of the membranes to the sweetener. The current signal increases relatively to the concentration of the sweetener in bulk solution in the micromolar range (Nikolelis et al. 2001a,b). Recently, a nanohybrid membrane sensor has been proposed (artificial BLMs) and has been compared with the surface s-BLMs offering higher stability (Chalkias and Giannelis 2007).

Potentiometry is a technique in which a single measurement of electrode potential is employed to determine the concentration of an ionic species in a solution. A potentiometric technique is characterized by the selectivity, the stability, and the response time of the electrode used. The selectivity of the electrode is generally evaluated considering some important ionic species normally found as components in food. Stabilization and response times are evaluated by recording the potential response of the membrane versus time, as the concentration of saccharin increases in solution. Drift is determined by measuring changes of electrode potential with time, in a fixed saccharin concentration solution. In literature, a number of electrodes are proposed for potentiometric determination of saccharin in foodstuff with specific selectivity, stability, and response time properties in each case. A polymer (silsesquioxane 3-n-propylpyridinium chloride) coated graphite rod ion-selective electrode with fast response times has been proposed. The electrode response was based on the ion pair formed between saccharinate acid and the 3-n-propylpyridinium cation from the silsesquioxane polymer (Alfaya et al. 2000). A silver wire electrode (Assumpcao et al. 2008) has also been proposed demonstrating great stability and lifetime. An electrode based on a polypyrrole-doped membrane is developed, exhibiting high selectivity toward saccharin in the presence of other compounds commonly found in food (Álvarez-Romero et al. 2010).

In literature, a number of sensitive methods have been proposed overcoming the drawbacks of interferences. A rapid, sensitive, and selective spectrophotometric method is developed using a formation of saccharin with Nile Blue (Cordoba et al. 1985). The formation of saccharin with dyes is a smart way to enhance sensitivity. Spectrophotometry coupled with an FIA system offers reproducible and accurate results (Capitán-Vallvey et al. 2004b). Recently, a sensitive spectrophotometric method has been proposed, taking advantage of the different kinetic rates of saccharin and other sweeteners in their oxidative reaction with $KMnO_4$. The data obtained from the kinetic rates of each analyte were processed with the help of chemometrics (Ni et al. 2009).

For routine analysis of complex food matrix samples, MEKC, HPLC, and IC are preferred (Table 4.2). These instrumental methods are important reference methods for food sample analysis. HPLC is commonly used for food sample analysis. Liquid chromatographic determination of saccharin in food is simple because beverages or aqueous extracts from foods can often be injected into a column immediately after filtration. The most applicable method for the determination of saccharin is reverse-phase HPLC coupled with UV light detector. The most popular reference method that everybody can come across in bibliography is the HPLC-diode array detector (HPLC-DAD) method proposed by Lawrence and Charbonneau (1988). A 5-μm C_8 bonded silica in a 150 mm × 4.6 mm column was used as a stationary phase in this method, with a mobile phase gradient ranging from 3% acetonitrile in 0.02 M KH_2PO_4 (pH 5) to 20% acetonitrile in 0.02 M KH_2PO_4 (pH 3.5) at a constant flow rate of 1.0 mL/min. The chromatograms were obtained at a wavelength of 210 nm (Lawrence et al. 1988).

The analytical conditions for the determination of saccharin by means of HPLC–ESI–MS have been studied recently. MS is a powerful qualitative and quantitative analytical technique that has been introduced in many analytical and research laboratories in the last 10 years. The combination of HPLC with tandem MS yields a particularly powerful tool, and it is the future method of choice for the determination of sweeteners. A simple and rapid method for the simultaneous determination of nine sweeteners, including saccharin, in various foods by HPLC–ESI–MS is developed. Mass spectral acquisition is done in the negative ionization mode by applying selected ion monitoring. The sweeteners are extracted from foods with 0.08 mol/l phosphate buffer (pH 7.0)–ethanol (1:1), and the extract is cleaned up on a Sep-pak Vac C_{18} cartridge after the addition of tetrabutylammonium bromide and phosphate buffer (pH 3.0). The quantification limit of saccharin is 1 mg/kg (Koyama et al. 2005). Another simultaneous determination method of saccharin and other additives in foods by LC–ESI(-)–MS/MS is proposed. A mixture of acetonitrile–water (1:1) was used to extract these additives from solid food matrices, and acetonitrile was used to extract them from liquid food matrices. Saccharin was identified and determined in the negative ionization mode using single reaction monitoring (SRM) of the product ion (m/z = 106) from its precursor ion (m/z = 182; Ujiie et al. 2007).

CE is an interesting alternative to HPLC. The resolving power of this technique is comparable with that of HPLC. Different types of CE have been used. MEKC (Boyce 1999) and capillary isotachophoresis (Herrmannová et al. 2006) are the types of CE most used for the determination of saccharin.

In the past, the preferred technique used for screening saccharin and other sweeteners was TLC, and for quantification, it was gas chromatography (GC). TLC methods were developed for separating sweeteners from other impurities using generally layers of polyamide. A TLC method that was developed in 1970 detected saccharin in a quantity of 2 μg in various foods (Takeshita 1972). In literature, there are a number of methods that determine saccharin and other sweeteners by TLC–UV and are characterized by a sensitivity that fluctuates in the microgram range (Nagasawa et al. 1970). Artificial sweeteners including saccharin have been determined by GC methods, which are usually sensitive and selective. The only main drawback of GC methods is the time-consuming derivatization step. This step is necessary for the conversion of sweeteners in volatile compounds, so that we can determine them with GC. The derivatization of saccharin is commonly achieved with trimethylsilylation (Dickes 1979).

The determination of saccharin individually or simultaneously with other sweeteners in mixtures is very important for consumer safety. Researchers focus their efforts on developing analytical methods for simple, rapid, and low-cost sensitive determination. Sensitive and robust analytical methods are essential to meet the needs of growing markets in quality control and consumer safety. Scientists have applied a wide variety of instrumental techniques. Today, the method of choice for the determination of artificial sweeteners in different complex food matrices is HPLC because of its multianalyte capability, compatibility with the physicochemical properties of sweeteners, high sensitivity, and robustness. However, biosensor technology has offered some methods of direct detection in simpler and cheaper matrices, without sample preparation. Due to a constant rising demand for alternative methods for determination of saccharin, and due to the increasing development of MS or tandem mass spectrometric methods, a number of procedures based on MS coupled with LC for the determination of saccharin are expected to appear.

4.15.7 Determination of Sucralose

Since sucralose does not absorb in the usable UV/Vis range, making sensitive and specific detection by direct UV absorption difficult, a derivatization procedure is necessary. This task can be accomplished using p-nitrobenzoyl chloride (PNBCl). Sucralose treated with PNBCl is converted into a strongly UV-absorbing derivative, having strong absorption at 260 nm, which allows for its sensitive, direct UV detection. Another solution is to use an RI or MS detector (Kobayashi et al. 2001).

A simple, fast TLC method was proposed for the quantification of sucralose in various food matrices. The method requires little or no sample preparation to isolate or to concentrate the analyte. The separation of sucralose was performed on amino-bonded silica-gel high performance thin layer chromatography (HPTLC) plate. The use of the DAD resulted in excellent limit of detection (Spangenberg et al. 2003).

A few procedures for determination of sucralose in various food and beverage products were reported. In all cases, high performance anion exchange chromatography (HPAEC) with pulsed amperometric detection (PAD) was applied. The high resolving power of HPAEC and the specificity of PAD allow the determination of sucralose with little interference from other ingredients. High precision, method ruggedness, and high spike recovery are possible for these complex sample matrices (Hanko and Rohrer 2004).

CE with indirect absorption measurement is also a suitable tool for monitoring the content of sucralose in various foodstuffs. Sucralose determination in low-calorie soft drinks can be achieved without any sample cleanup using 3,5-dinitrobenzoate buffer at pH 12.1 and indirect UV detection at 238 nm. The scope of the method has been extended to include the possibility of analysis of yogurts and candies by introducing a sample cleanup step (centrifugation, filtration, and SPE on Alumina A cartridges; McCourt et al. 2005).

4.16 MULTIANALYTE ANALYSIS

However, these methods are time consuming or do not have the selectivity required for ASP determination in some commercial samples. Spectrophotometry is usually coupled with an FIA system. The analysis of analyte mixtures by means of FIA systems has been accomplished in different ways:

1. Use of a microcolumn after flow cell to retain one analyte preferentially, the other being transiently retained in the solid support placed in the flow-through cell (Ruiz-Medina et al. 2001).
2. Use of differences in transient retention for both analytes at the flow cell. The transient retention of one analyte in the upper part of the flow cell, away from the measuring area, makes it possible to measure what is less retained (Capitán-Vallvey et al. 2004a).

3. Use of retention of only one analyte in the solid phase that fills the flow cell, while the second analyte is measured when it flows along the interstitial solution through the particles.
4. Use of a chemometric approach without separation of the analytes prior to the detection step (Jiménez et al. 2009).

A multianalyte flow-through sensor spectrophotometric method achieved the simultaneous determination of ASP and ACS-K in tabletop sweeteners. The procedure is based on the transient retention of ACS-K in the ion exchanger Sephadex DEAE A-25 placed in the flow-through cell of a monochannel FIA setup using a pH 2.7 composed from orthophosphoric acid/sodium dihydrogen phosphate buffer (0.06M) as a carrier. In these conditions, ASP is very weakly retained, which makes it possible to measure the intrinsic UV absorbance of first ASP at 226 nm and then ACS-K at 205 nm after desorption by the carrier itself. The linear concentration range for ASP is from 10 to 100 μg/mL, the detection limit is 5.65 μg/mL, and the RSD is 3.4% (at 50 μg/mL). The linear concentration range for ACS-K is from 40 to 100 μg/mL, the detection limit is 11.9 μg/mL, and the RSD is 1.61% (at 50 μg/mL). No interference is caused by glucose, sucrose, lactose, maltose, fructose, glycine, and leucine even when present in concentrations higher than those commonly found in the tabletop sweeteners analyzed. The level of the solid phase in the flow cell should be that needed to fill it up to a sufficient height, allowing the radiation beam to pass completely through the solid layer. The height of the solid support considerably influences the separation of both sweeteners (Jiménez et al. 2006).

A multianalyte flow-through method is proposed for the simultaneous determination of ASP, ACS-K, and saccharin in several food and soft drink samples. The procedure is based on the transient retention of the three sweeteners in a commercial quaternary amine ion exchanger monolithic column, placed in its specific holder, and allocated in a monochannel FIA setup using (pH 9.0) composed from Tris buffer 0.03 M, NaCl 0.4 M, and $NaClO_4$ 0.005M as a carrier. In these conditions, ASP is very weakly retained, while ACS-K is more strongly retained, making it possible to measure the intrinsic UV absorbance of ASP and then ACS-K after desorption by the carrier itself. The linear concentration range for ASP is from 9.5 to 130.0 μg/mL, the detection limit is 2.87 μg/mL, and the RSD is 1.46% (at 65 μg/mL). The linear concentration range for ACS-K is from 2.2 to 600.0 μg/mL, the detection limit is 1.0 μg/mL, and the RSD is 0.08% (at 300 μg/mL). The method is applied and validated satisfactorily for the determination of ASP and ACS-K in foods and soft drink samples, comparing the results with an HPLC reference method (Jiménez et al. 2009).

A new method to determine mixtures of the sweeteners ASP and ACS-K in commercial sweeteners is proposed making use of chemometrics. A classical 5^2 full factorial design for standards is used for calibration in the concentration matrix. Salicylic acid is used as the internal standard in order to evaluate the adjustment of the real samples in the PLS-2 model. This model is obtained from UV spectral data, validated by internal cross-validation, and used to find the concentration of analytes in the sweetener samples. The mean value of recovery degree is 99.2% with standard deviation of 3.2%. The proposed procedure is applied successfully to the determination of mixtures of ASP and ACS-K in bulk samples (Cantarelli et al. 2009).

HPLC and GC are commonly used for food sample analysis (Table 4.4). A common HPLC method that is used nowadays as a basic procedure for sweetener determination and used as a reference method for validation is the following: The method uses a UV light detector by using absorbance measurements at 205 nm. Seven standard solutions and five replicates are prepared for both ASP and ACS-K. The HPLC carrier flow is 0.75 mL/min. Every sample is accurately weighted in a volumetric flask and diluted to the volume with 0.02 M KH_2PO_4: acetonitrile (90:10 v/v). Then every sample is sonicated for 5 min in an ultrasonic water bath to extract sweeteners from the matrix. One milliliter of the extract is diluted and filtered through a 0.22-μm nylon membrane. The RSD (RSD%) values are better than 0.3% for ASP and 0.1% for ACS-K (Armenta et al. 2004a,b). For the estimation intake of intense sweeteners from nonalcoholic beverages in

Table 4.4 Chromatographic Techniques for Multianalyte Analysis in Functional Foods

ACS-K[a], ASP[b], SAC[c]	Soft drinks, juice, tomato sauce, strawberry sweets	FIA with online monolithic element	Water (0.4 M in NaCl, 5×10^{-3} M $NaClO_4$) (pH = 9)	Quaternary amine ion exchanger monolithic column	0.9 µg/mL	0.09%	97.6%–103.4% and 3–600 µg/mL	Jiménez et al. 2009
ACS-K, ASP, SAC, vanillin, sorbic acid, benzoic acid	Cola drinks, instant-powder drinks	HPLC-UV	ACN-Ammonium acetate buffer (pH = 4)	YMC-ODS Pack AM (5 µm × 250 mm × 4 mm I.D.)	0.2–3.1 µg/g	1.0%–2.2%	99%–101% and N/A	Demiralay et al. 2006
ACS-K, SAC, CYC[d]	Gum	IC-UV	300 mg/L Sodium carbonate	Dionex AS4A Separator column	N/A	1.1%	99.8%–102.5% 0.0262 mg/mL –0.0022 mg/mL	Biemer et al. 1989
DUL[e], ACS-K, SAC, preservatives, antioxidants	Soy sauce, sugared fruits, dried roast beef	Ion-paired LC-UV	ACN-aqueous a-hydroxy-iso butyric acid solution (pH = 4.5) containing hexadecyltrimethylammonium bromide	Stainless-steel Shoko $5C_{18}$ column (5 µm × 25 mm × 4.6 mm I.D.)	0.5 µg/g	N/A	81.9%–89.64% and N/A	Chen et al. 1995
ACS-K, SAC, ASP, antioxidants, preservatives	Beverages, low joule jam	MEKC-UV	Sodium tetraborate solution (pH = 9.5) with Na cholate, dodecyl sulfate-10% ACN or isopropanol or MeOH	Fused silica capillary (60 cm × 75 µm I.D.)	N/A	N/A	98.9%–100.86% and N/A	Boyce 1999
SAC, ASP, ACS-K, sorbic acid, benzoic acid, caffeine, theobromine, theophylline	Cola drinks, preserved fruits, tablets, fermented milk, fruit juice	IC-UV	Aqueous NaH_2PO_4 (pH = 8.20) –4% (v/v) ACN	IC-A3 Shim-Pac (5 µm × 150 mm × 4.6 mm I.D.)	20 ng/mL	1.5%	85%–104% and N/A	Chen et al. 2001
SAC, ASP, ACS-K, CYC, citric acid	Drinks, powdered tabletop sweeteners	IC–UV–ELCD	Na_2CO_3 solution	Dionex IonPac AS4A-SC (250 mm × 4 mm I.D.)	0.26 µg/mL	0.84%	93%–107% and 2–100 mg/mL	Chen et al. 1997

(*continued*)

Table 4.4 (Continued) Chromatographic Techniques for Multianalyte Analysis in Functional Foods

ACS-K, SAC, benzoic acid, sorbic acid	Beverages, jams	HPLC–UV	8%MeOH in phosphate buffer (pH = 6.7)	Spherisorb C_{18} ODS-1 (5 µm × 250 mm × 4.6 mm I.D.)	<0.1 mg/100 mL	N/A	100.4%–103.2% and 0–100 mg/L	Hannisdal 1992
SAC, ASP, CYC, ACS-K	Carbonated cola drinks, fruit juice, preserved fruit	IC-suppressed conductivity detector	KOH	Dionex Ion Pack AS 11 (250 mm × 2 mm I.D.)	0.045 mg/l	N/A	98.5%–102.4% and N/A	Zhu et al. 2005
SAC, ASP	Carbonated beverages, soft drinks, strawberry jam	FIA with online SPE	Dihydrogen phosphate buffer 3.75 × 10^{-3} mol/L	N/A	1.4 µg/mL	1.6%	99%–101% and 10–200 µg/mL	Capitán-Vallvey et al. 2006
ACS-K, SAC, ASP, CYC, sorbitol, mannitol, lactitol, xylitol	Chewing gum, candy	Capillary isotachophoresis	Two electrolytes used: HCl–Tris (pH = 7.7) (E1) L-histidine–Tris (pH = 8.3) (E2)	Capillary ethylene propylene copolymer (90-mm length)	0.052 mM	2.6%	98.2%–102.5% and N/A	Herrmannová et al. 2006
SAC, preservatives	Coffee drink	Ion pair Chromatography	ACN-Water-0.2M phosphate buffer (pH = 3.6) (7:12:1)	Nucleosil5C_{18}	10 µg/g	1.18%	102.4% and N/A	Terada et al. 1985
SAC, ASP, CYC, ALI[f], ACS-K, DUL, NEO[g], SCL[h], NHDC[i]	Soft drinks, canned and bottled fruits, yogurt	HPLC–ELSD	TEA format buffer–methanol–acetone	Nucleodur C_{18} Pyramid (5 µm × 250 mm × 3 mm I.D.)	15 µg/g	0.9%–4.5%	93%–109% and N/A	Wasik et al. 2007
ACS-K, SAC, ASP, benzoic acid, sorbic acid, ponceau 4R, sunset yellow, tartrazine	Soft drinks	HPLC–UV	MeOH– phosphate buffer (pH = 4)	Lichrosorb RP_{18} (10 µM, 250 × 4.6 mm I.D.)	0.1–3 mg/L	N/A	98.6%–102.3% and N/A	Dossi et al. 2006

SAC, ASP, benzoic acid, sorbic acid	Soft drinks	HPLC	Phosphate buffer (pH = 4.5)–ACN	Lichrosorb C_{18} (5 µm × 25 cm × 4 mm I.D.)	N/A	<3.2%	>95% and N/A	Moors et al. 1991
AK, SCL, SA, CYC, ASP, DUL, GA[j], STV[k], REB[l]	Solid and liquid food matrices	LC–MS	Acetonitrile–water (8:2)	Zorbax Eclipse XDB-C_{18} (150 mm × 2.1 mm I.D.)	N/A	N/A	75.7%–109.2% and N/A	Koyama et al. 2005
SA, sorbic acid, benzoic acid, p-HBA[m] ethyl, p-HBA isopropyl, p-HBA propyl, p-HBA isobutyl, p-HBA butyl	Solid and liquid food matrices	LC–MS–MS	0.01% formic acid solution–acetonitrile	TSK gel ODS8OTs (150 mm × 4.6 mm I.D.)	10 µg/g	N/A	78%–120% and N/A	Ujiie et al. 2007
ASP, SAC	Dietary foods	HPLC–UV	Solution TEA–phosphate buffer (pH = 3)–MeOH–THF	Hypersil C_{18} (10 µm, 250 mm × 4 mm I.D.)	N/A	N/A	95%–97% and N/A	Di Pietra et al. 1990
AK, SCL, SA, CYC, ASP, DUL, GA, STV, REB	Beverages, canned fruits, cakes	HPLC–ESI–MS	TEA formate buffer–MeOH–ACN	Spherogel C_{18} (5 µm × 250 mm × 4.5 mm I.D.)	<0.10 µg/mL	N/A	95.4%–104.3% and 0.05–5.00 µg/mL	Yang and Chen 2009

Note: a. ACS-K, acesulfame-k; b. ASP, aspartame; c. SAC, saccharin; d. CYC, cyclamate; e. DUL, dulcin; f. ALI, alitame; g. NEO, neotame; h. SCL, sucralose; i. NHDC, neohesperidin dihydrochalcone; j. GA, glycyrrhizic acid; k. STV, stevioside; l. REB, rebaudioside; m. p-HBA, p-hydroxybenzoic acid.

Denmark, sweeteners including ACS-K and ASP are separated by HPLC on a C-18 column (5 mm, 250 mm × 4.6 mm) equipped with a guard column, eluted isocratically with a mixture of methanol and water buffered with potassium hydrogen phosphate, and measured spectrophotometrically at 220 nm (Leth et al. 2007). For the estimation of the intake of ACS-K and ASP from soft drinks for a group of Portuguese teenage students, reversed-phase LC was used with a Hichrom C_{18} column (5 μm, 250 mm × 4.6 mm) and a buffered mobile phase [KH_2PO_4 0.02 M/acetonitrile (90:10, v/v)/phosphoric acid] at 1 mL/min. The pH was rigorously controlled at 4.2–4.3 for an adequate resolution between ASP and benzoic acid, which is present in most analyzed samples. Detection was performed with a UV detector at 220 nm. An external standard method is used for quantification (Lino et al. 2008). An isocratic separation of some food additives with HPLC analysis was carried out on a Shimadzu class LC-VP HPLC system with an autosampler (SIL-10ADVP) and a diode-array detector (SPD-M 10AVP). A 5-μm YMC-ODS Pack AM column (250 mm × 4.6 mm I.D.) was used for the analysis. In this study, three different values of the mobile phase acetonitrile content (15%, 20%, and 25%, v/v) were prepared. The pH values were adjusted to 4.0 with glacial acetic acid. The flow rate was 0.8 mL/min, and the volume injected was 20 μL. For ACS-K, detection followed at 230 nm, and for ASP, detection followed at 203 nm. This HPLC-UV procedure was able to separate ACS-K and ASP from other additives in cola drinks and in instant powder drinks. The recovery achieved is 99%–101%, the LOD is 0.2 to 3.1 μg/g, and the RSD is 1.0% to 2.2% for all analytes determined (Demiralay et al. 2006).

An HPLC-UV procedure was able to separate ASP and ACS-K from saccharin, benzoic acid, sorbic acid, Ponceau 4R, Sunset Yellow, and Tartrazine in soft drinks using a LiChrosorb C_{18} column (10 μm, 250 mm × 4.6 mm) and a mobile phase that consisted of MeOH–phosphate buffer (pH 4). The recovery achieved is 98.6% to 102.3% with a limit of detection of 0.1 to 3 mg/l for all analytes determined (Dossi et al. 2006).

An HPLC-evaporative light scanning detector (HPLC-ELSD) procedure was able to separate ASP and ACS-K from saccharin, cyclamate, sucralose, dulcin, alitame, NTM, and NHDC in non-carbonated soft drinks, canned or bottled fruits, and yogurts using C_{18} stationary phase and a mobile phase that consisted of TEA formate buffer–MeOH–acetonitrile (ACN). The limit of detection is 15 μg/g, and the RSD is 0.9% to 4.5% for all analytes determined (Wasik et al. 2007). An ion paired HPLC-UV procedure is able to separate ACS-K from saccharin, dulcin, preservatives, antioxidants in sugared fruits, soy sauces, and dried roast beef. A Shoko stainless-steel $5C_{18}$ (5 μm, 250 mm × 4.6 mm) was used as the stationary phase and the mobile phase consisted of ACN–aqueous a-hydroxyisobutyric acid solution containing hexadecyltrimethylammonium bromide. The recovery is 81.9% to 103.27%, the limit of detection achieved is 0.15 to 3 μg/g, and the RSD is 0.3% to 5.69% for all analytes determined (Chen and Fu 1995). A high-performance IC–UV–electrolytic conductivity detector (HPIC–UV–ELCD) procedure is able to separate ACS-K and ASP from saccharin, cyclamate, and citric acid in drinks and powdered tabletop sweeteners. A Dionex Ion Pac AS4A-SC (254 mm × 4 mm) was used as the stationary phase and the mobile phase consisted of Na_2CO_3. The recovery is 93% to 107%, the limit of detection is 0.019 to 0.044 mg/L, and the RSD is 0.84% to 1.38% for all analytes determined (Chen et al. 1997). An HPIC–UV procedure is able to separate ACS-K and ASP from saccharin, benzoic acid, sorbic acid, caffeine, theobromine and theophyline in drinks, juices, fermented milk drinks, preserved fruits, tablet drinks, and powdered tabletop sweeteners. A Shim-pack IC-A3 (5 μm, 150 mm × 4.6 mm) was used as the stationary phase and the mobile phase consisted of NaH_2PO_4 (pH 8.20)–ACN. The recovery is 85% to 104%, the limit of detection is 4 to 30 mg/L, and the RSD is 1% to 5% (Chen and Wang 2001). Finally, an HPIC suppressed conductivity detector procedure is able to determine ACS-K and ASP in the presence of saccharin and cyclamate in carbonated cola drinks, fruit juice drinks, and preserved fruits. A Dionex Ionpac AS11 (250 mm × 2 mm) was used as the stationary phase and the mobile phase consisted of KOH. The recovery is 97.96% to 105.42%, and the limit of detection is 0.019 to 0.89 mg/L for all the analytes determined (Zhu et al. 2005) (Table 4.4).

A MEKC–UV procedure is able to determine ASP and ACS-K with saccharin, dulcin, alitame, caffeine, benzoic acid, and sorbic acid in low-energy soft drinks, cordials, tomato sauce, marmalades, jams, and tabletop sweeteners. An uncoated fused-silica capillary (75 cm × 75 μm) was used as the stationary phase and the mobile phase consisted of a buffer comprising sodium deoxycholate, potassium–dihydrogen orthophosphate, and sodium borate (pH 8.6). The recovery is 104% to 112%, and the RSD is 0.63% to 2.6% for all analytes determined (Thompson et al. 1995). Another MEKC-UV procedure is able to determine ASP and ACS-K with saccharin, preservatives, and antioxidants in cola beverages and low-energy jams. A fused-silica capillary (52 cm × 75 μm) was used as the stationary phase and the mobile phase consisted of borate buffer with Na cholate, dodecyl sulfate, and MeOH (pH 9.3). The recovery is 98.9% to 100.86%, and the RSD is 0.9% to 1.5% for all analytes determined (Boyce 1999). Finally, a MEKC–UV procedure is able to determine ASP and ACS-K with saccharin, preservatives, and colors in soft drinks. An uncoated fused-silica capillary (48.5 cm × 50 μm) is used as the stationary phase and the mobile phase consisted of carbonate buffer (pH 9.5) with sodium dodecyl sulfate. The LOD is 0.005 mg/mL for all analytes determined (Frazier et al. 2000).

A simple method for the simultaneous determination of five artificial sweeteners, alitame, ACS-K, saccharin, ASP, and dulcin, in various foods by HPLC and detecting the species at 210 nm was reported. The recoveries of the five sweeteners from various kinds of foods spiked at 200 mg/g ranged from 77% to 102%. The detection limits of the five sweeteners were 10 mg/g (Kobayashi et al. 1999).

The most popular reference method used in bibliography is the HPLC-DAD method proposed by Lawrence and Charbonneau. A 5-mm C_8 silica in a 150 mm × 4.6 mm column was used as the stationary phase in this method, with a mobile phase gradient ranging from 3% acetonitrile in 0.02 M KH_2PO_4 (pH 5) to 20% acetonitrile in 0.02 M KH_2PO_4 (pH 3.5) at a constant flow rate of 1.0 mL/min. The chromatograms were obtained at a wavelength of 210 nm. In order to obtain the calibration function, six different concentration levels and three replicates of each one of the standard solutions are analyzed using peak area as the analytical parameter (Lawrence et al. 1988).

Quantitative determination by GC is impossible due to the low volatility of ACS-K and due to the fact that methylation produces differing ratios of methyl derivatives. A method of determining ASP and its degradation products was reported in 1975 by GC (Furda et al. 1975).

CE, as mentioned above, is an interesting alternative to HPLC, and the resolving power of this technique is, in many cases, comparable with that of HPLC; frequently, their running costs are lower. High-pressure CE appears to be a viable method for the determination of ASP in real commercial products. The analysis time is significantly faster than that reported for HPLC methods, and no interferences were detected in soft drink samples tested. Caffeine has a different migration time from ASP, and its low sensitivity at the detection wavelength precludes the appearance of a peak in the electropherogram. The linear calibration curve developed is compatible with the range of ASP concentrations or amounts in the products tested. It appears that a small amount of ASP adsorption occurs on the capillary walls, but this affects quantitative determinations well below the useful range for typical commercial samples. A linear calibration curve between 25 and 150 μg/mL for the analyte solution is established, which can be used for quantitative determinations of ASP in typical food and beverage products. Six commercial samples are analyzed, and one diet cola with a known ASP concentration gives an RSD of 2.6% from the manufacturer's value (Qu et al. 1999).

A method for isotachophoretic determination of sweeteners of different character in candies and chewing gums is developed. A capillary made of fluorinated ethylene–propylene copolymer with an internal diameter of 0.8 mm and an effective length of 90 mm is filled with an electrolyte system consisting of a leading electrolyte (10 mM HCl with 14 mM Tris, pH 7.7) and a terminating electrolyte (5 mM L-histidine with 5 mM Tris, pH 8.3). The analysis is performed at a driving current of 200 μA, and for detection current, the magnitude is decreased to 100 μA. Boric acid is added to the aqueous sample solution to form borate complexes with substances of polyhydroxyl nature and make them to migrate isotachophoretically. Using conductivity detection, the calibration curves in the tested concentration range up to 2.5 mM and are linear for all components of interest: ACS-K, saccharin, ASP, cyclamate, sorbitol, mannitol, lactitol, and xylitol. The concentration detection limits for the determined compounds range between 0.024 and 0.081 mM. Good precision of the method is evidenced by favorable RSD values for all compounds ranging from 0.8% to 2.8% obtained at the analyte concentration of 1.0 mM ($n = 6$). The analysis time is about 20 min. Simplicity, accuracy, and low cost of analyses make this an alternative procedure to methods used so far for the determination of ionizable sweeteners (Herrmannová et al. 2006).

TLC methods have been developed for analysis of sweeteners. ASP, ACS-K, sodium cyclamine, and benzoic acid are separated on thin layers of silica gel G with 10% (v/v) ethanol–40% (v/v) isopropanol–1% (v/v) (12.5%) aqueous ammonia. This chromatographic system was applied to the analysis of sweeteners in 23 sparkling and nonsparkling drinks (Baranowska et al. 2004).

A new chromatographic modality that does not require high pressures and also allows renewal of the stationary phase as desired is reported. The technique is based on a thin-layer paramagnetic stationary phase (Fe_3O_4–SiO_2) retained on the inner wall of a minicolumn through the action of an external magnetic field, which also plays an important role in separating the analytes. Accordingly, the name "renewable stationary phase liquid magnetochromatography" has been proposed for it. The technique is used to separate and quantify the sugar substitute ASP and its constituent amino acids (hydrolysis products), L-aspartic acid and L-phenylalanine, in diet fizzy soft drinks. When the results obtained for ASP were compared with those obtained using the HPLC reference method of Lawrence and Charbonneau, no significant differences were observed. The system proposed is fully automated, making it an economic and competitive alternative to conventional methods of determining ASP and its amino acid components (Barrado et al. 2006).

The determination of ASP and ACS-K in tabletop samples as was previously described has been achieved by Fourier transform middle-infrared (FTIR) spectrometry. With the use of a fully mechanized online extraction, the contact of the operator with toxic solvents was avoided and differentiates between samples that contain ASP and ACS-K and those that include only ASP, reducing the time needed for the analysis of the last kind of samples to 5 min. The method involves the extraction of both active principles by sonication of samples with 25:75 v/v $CHCl_3/CH_3OH$ and direct measurement of the peak height values at 1751 cm^{-1}, corrected using a baseline defined at 1850 cm^{-1} for ASP, and measurement of the peak height at 1170 cm^{-1} in the first-order derivative spectra, corrected by using a horizontal baseline established at 1850 cm^{-1} for ACS-K. Limit of detection values of 0.10% and 0.9% w/w and RSDs of 0.17% and 0.5% are found for ASP and ACS-K, respectively On the other hand, an HPLC method needs approximately 35 min for completion (Armenta et al. 2004a,b).

Electrochemistry at the liquid–liquid interface enables the detection of non-redox-active species with electroanalytical techniques, and the electrochemical behavior of two food additives, ASP and ACS-K, can be investigated. Both ions were found to undergo ion-transfer voltammetry at the liquid–liquid interface. Liquid–liquid electrochemistry as an analytical approach in food analysis is very suitable. Differential pulse voltammetry was used for the preparation of calibration curves over the concentration range of 30–350 μM with a detection limit of 30 μM. The standard addition method was applied to the determination of their concentrations in food and beverage samples such as sweeteners and sugar-free beverages. Selective electrochemically modulated LLE of these species in both laboratory solutions and in beverage samples was achieved (Herzog et al. 2008).

4.17 CONCLUSIONS

The determination of artificial sweeteners individually or simultaneously in mixtures is very important for legal aspects. Researchers focus their efforts on developing analytical methods for simple, rapid, and low-cost sensitive determination. Sensitive and robust analytical methods are essential to meet the needs of growing markets in quality control and consumer safety. Scientists have applied a wide variety of instrumental techniques. Today, the method of choice for the determination of artificial sweeteners in different food matrices is HPLC because of its multianalyte capability, compatibility with the physicochemical properties of sweeteners, high sensitivity, and robustness. Due to a constant rising demand for alternative methods for determination of sweeteners, and due to the rising development of mass spectrometry or tandem mass spectrometry methods, we will witness a number of procedures based on MS coupled with LC for the determination of ACS-K and ASP.

CE and IC are both interesting alternatives to HPLC. The resolving power of these techniques is, in many cases, comparable with that of HPLC, and frequently, their running costs are lower. However, it seems that due to limited robustness, in the case of CE methods, and the modest choice of separation mechanisms, in the case of IC, these methods are less popular. TLC and GC have been applied occasionally to analysis of artificial sweeteners. TLC methods are characterized by poor separation efficiency, and GC methods require derivatization that is time consuming and labor intensive. Due to a demand for simple, rapid, and low-cost alternative methods for determination of sweeteners, in many instances, chromatographic methods can be replaced by electroanalytical, spectroscopic, or FI procedures. Some of them are even more sensitive and selective and require very little sample preparation. Unfortunately, their applications are limited to one or two sweeteners only.

However, there still remains the challenge of developing stable, reliable, and robust methods for the determination of artificial sweeteners in difficult food matrices. Robust and reliable analytical methods are essential to meet the needs of growing markets in quality control and consumer safety.

REFERENCES

Alfaya, R. V. S., Alfaya, A. A. S., Gushikem, Y. et al. 2000. Ion selective electrode for potentiometric determination of saccharin using a thin film of silsesquioxane 3-n-propylpyridinium chloride polymer coated graphite rod. *Anal. Lett.* 33: 2859–71.

Álvarez-Romero, G. A., Lozada-Ascencio, S. M., Rodriguez-Ávila, J. A., Galán-Vidal, C. A., Páez-Hernandez, M. E. 2010. Potentiometric quantification of saccharin by using selective formed by pyrrole electropolymerization. *Food Chem.* 120(4): 1250–4.

Armenta, S., Garrigues, S., de la Guardia, M. 2004. Sweeteners determination in table top formulations using FT-Raman spectrometry and chemometric analysis. *Anal. Chim. Acta* 521: 149–55.

Armenta, S., Garrigues, S., de la Guardia M. 2004. FTIR determination of aspartame and acesulfame-K in tabletop sweeteners. *J. Agric. Food Chem.* 52(26): 7798–803.

Arnold, D. L. 1984. Toxicology of saccharin. *Fundam. Appl. Toxicol.* 4: 674–85.

Arnold, D. L., Krewski, D., Munro, I. C. 1983. Saccharin: A toxicological and historical perspective. *Toxicology* 27: 179–256.

Assumpcao, M. H. M. T., Medeiros, R. A., Madi, A. et al. 2008. Development of a biamperometric procedure for the determination of saccharin in dietary products. *Quimica Nova* 31(7): 1743–6.

Baranowska, I., Zydron, M., Szczepanik, K. 2004. TLC in the analysis of food additives. *J. Planar Chromatogr.-Mod. TLC* 17(1): 54–7.

Baron, R. F., Hanger, L. Y. 1998. Using acid level, acesulfame potassium aspartame blend ratio and flavor type to determine optimum flavor profiles of fruit flavored beverages. *J. Sensory Studies* 13: 269–83.

Barrado, E., Rodriguez, J. A., Castrillejo, Y. 2006. Renewable stationary phase liquid magnetochromatography: Determining aspartame and its hydrolysis products in diet soft drinks. *Anal. Bioanal. Chem.* 385: 1233–40.

Biemer, T. A. 1989. Analysis of saccharin, acesulfame-K and sodium cyclamate by high performance ion chromatography. *J. Chromatogr.* 463: 463–8.

Boyce, M. C. 1999. Simultaneous determination of antioxidants, preservatives and sweeteners permitted as additives in food by mixed micellar electrokinetic chromatography. *J. Chromatogr. A* 847: 369–75.

Buduwy, S. S., Issa, Y. M., Tag-Eldin, A. S. 1996. Potentiometric determination of L-dopa, carbidopa, methyldopa and aspartame using a new trinitrobenzenesulfonate selective electrode. *Electroanalysis* 8(11): 1060–4.

Cabero, C., Saurina, J. Hernandez-Cassou, S. 1999. Flow-injection spectrophotometric determination of cyclamate in sweetener products with sodium 1,2-naphthoquinone-4-sulfonate. *Anal. Chim. Acta* 381: 307–13.

Cantarelli, M. A., Pellerano, R. G., Marchevsky, E. J. et al. 2009. Simultaneous determination of aspartame and acesulfame-K by molecular absorption spectrophotometry using multivariate calibration and validation by high performance liquid chromatography. *Food Chem.* 115: 1128–32.

Capitán-Vallvey, F., Valencia, M. C., Nicolas, E. A. 2004a. Flow-through spectrophotometric sensor for the determination of aspartame in low-calorie and dietary products. *Anal. Sci.* 20(10): 1437–42.

Capitán-Vallvey, L. F., Valencia, M. C., Arana Nicolás, E. et al. 2006. Resolution of an intense sweetener mixture by use of a flow injection sensor with on-line solid-phase extraction—Application to saccharin and aspartame in sweets and drinks. *Anal. Bioanal. Chem.* 385: 385–91.

Capitán-Vallvey, L. F., Valencia, M. C., Nicolás, E. A. et al. 2004b. Flow-through spectrophotometric sensor for the determination of saccharin in the text low-calorie products. *Food Add. Contam.* 21: 32–41.

Chalkias, N. G. and Giannelis, E. P. 2007. A nanohybrid membrane with lipid bilayer-like properties utilized as a conductimetric saccharin sensor. *Biosens. Bioelectron.* 23: 370–6.

Chen, B. H., Fu, S. C. 1995. Simultaneous determination of preservatives, sweeteners and antioxidants in foods by paired-ion liquid chromatography. *Chromatographia* 41(1/2): 43–50.

Chen, Q.-C., Wang, J. 2001. Simultaneous determination of artificial sweeteners, preservatives, caffeine, theobromine and theophylline in food and pharmaceutical preparations by ion chromatography. *J. Chromatogr. A* 937: 57–64.

Chen, Q. C., Mou S. F., Liu, K. N. et al. 1997. Separation and determination of four artificial sweeteners and citric acid by high-performance anion-exchange chromatography. *J. Chromatogr. A* 771: 135–43.

Choi, M. M. F., Hsu, M. Y., Wong, S. L. 2000. Determination of cyclamate in low-calorie foods by high-performance liquid chromatography with indirect visible photometry. *Analyst* 125: 217–20.

Cohen-Addad, N., Chatterjee, M., Bekersky, I. et al. 1986. In utero-exposure to saccharin: A threat? *Cancer Lett.* 32: 151–4.

Compagnone, D., O'Sullivan, D., Guilbault, G. G. 1997. Amperometric bienzymic sensor for aspartame. *Analyst* 122: 487–90.

Cordoba, M. H., Garcia, I. L., Sanchez-Pedreño, C. et al. 1985. Spectrophotometric determination of saccharin in different materials by a solvent extraction method using Nile blue as reagent. *Talanta* 32: 325–7.

Demiralay, E., Çubuk, Ö. G., Guzel-Seydim, Z. 2006. Isocratic separation of some food additives by reversed phase liquid chromatography. *Chromatographia* 63: 91–6.

Di Pietra, A. M., Cavrini, V., Bonazzi, D. et al. 1990. HPLC analysis of aspartame and saccharin in pharmaceutical and dietary formulations. *Chromatographia* 30: 226–33.

Dickes, G. J. 1979. The application of gas chromatography to food analysis. *Talanta* 26: 1065–99.

Dossi, N., Toniolo, R., Susmel, S. et al. 2006. Simultaneous RP-LC determination of additives in soft drinks. *Chromatographia* 63: 557–62.

EFSA. 2009. Updated opinion on a request from the European Commission related to the 2nd ERF carcinogenicity study on aspartame, taking into consideration study data submitted by the Ramazzini Foundation in February 2009. *EFSA J.* 1015: 1–3.

European Commission. 1994. Directive 94/35/EC of European Parliament and of the Council of 30 June 1994 on sweeteners for use in foodstuff. *Off. J. Eur. Un.* L237(13).

European Commission. 1996. Directive 96/83/EC of European Parliament and of the Council of 19 December 1996 on sweeteners for use in foodstuffs. *Off. J. Eur. Un.* L048(16).

European Commission. 2003. Directive 2003/115/EC of European Parliament and of the Council of 22 December 2003 amending Directive 94/35/EC on sweeteners for use in Foodstuffs. *Off. J. Eur. Un.* L024(65).

European Commission. 2006. Directive 2006/52/EC of European Parliament and of the Council of 5 July 2006 amending directive 95/2/EC on food additives other than colors and sweeteners and Directive 94/35/EC on sweeteners for use in foodstuffs. *Off. J. Eur. Un.* L024(10).

Fatibello-Filho, O., Marcolino Jr., L. H., Pereira, A. V. 1999. Solid-phase reactor with copper(II) phosphate for flow-injection spectrophotometric determination of aspartame in tabletop sweeteners. *Anal. Chim. Acta* 384: 167–74.

Frazier, R. A., Inns, E. L., Dossi, N. et al. 2000. Development of a capillary electrophoresis method for the simultaneous analysis of artificial sweeteners, preservatives and colours in soft drinks. *J. Chromatogr. A* 876: 213–20.

Furda, I., Malizia, P. D., Kolor, M. G. et al. 1975. Decomposition products of L-aspartyl-L-phenylalanine methylester and their identification by gas-liquid–chromatography. *J. Agric. Food Chem.* 23(2): 340–3.

Galletti, G. C., Bocchini, P. 1996. High-performance liquid chromatography with electrochemical detection of aspartame with a post-column photochemical reactor. *J. Chromatogr. A* 729: 393–8.

Galletti, G. C., Bocchini, P., Gioacchini, A. M. et al. 1996. Analysis of the artificial sweetener aspartame by means of liquid chromatography/electrospray mass spectrometry. *Rapid Commun. Mass Spectrom.* 10: 1153–5.

Gouveia, S. T., Fatibello-Fihlo, O., Norbega, J. A. 1995. Flow injection spectrophotometric determination of cyclamate in low calorie soft drinks and sweeteners. *Analyst* 120: 2009–12.

Goyal, R. K., Goyal, S. K. 2010. Stevia (*Stevia rebaudiana*) a bio-sweetener: A review. *Int. J. Food Sci. Nutr.* 61: 1–10.

Hanko, V. P., Rohrer, J. S. 2004. Determination of sucralose in splenda and a sugar-free beverage using high-performance anion-exchange chromatography with pulsed amperometric detection. *J. Agric. Food Chem.* 52: 4375–9.

Hannisdal, A. 1992. Analysis of acesulfame-K, saccharin and preservatives in beverages and jams by HPLC. *Z. Lebensm. Unters. Forsch.* 194: 517–9.

Helstad, S. 2006. *Ingredient Interactions Effects on Food Quality*, 2nd ed. Anilkumar G. Gaonkar and Andrew McPherson (eds.), pp. 167–94. Boca Raton, FL: CRC Press, Taylor & Francis.

Herrmannová, M., Křivánková, L., Bartoš, M. et al. 2006. Direct simultaneous determination of eight sweeteners in foods by capillary isotachophoresis. *J. Sep. Sci.* 29: 1132–7.

Herzog, G., Kam, V., Berduque, A. et al. 2008. Detection of food additives by voltammetry at the liquid–liquid interface. *J. Agric. Food Chem.* 56(12): 4304–10.

Huang, Z., Ma, J., Chen, B. et al. 2006. Determination of cyclamate in foods by high performance liquid chromatography-electrospray ionization mass spectrometry. *Anal. Chim. Acta* 555: 233–7.

Ilbäck, N.-G., Alzin, M., Jahrl, S. et al. 2003. Estimated intake of the artificial sweeteners acesulfame-K, aspartame, cyclamate and saccharin in a group of Swedish diabetics. *Food Add. Contam.* 20(2): 99–114.

Jiménez, G. J. F., Valencia, M. C., Capitán-Vallvey, L. F. 2006. Improved multianalyte determination of the intense sweeteners aspartame and acesulfame-K with a solid sensing zone implemented in an FIA scheme. *Anal. Lett.* 39: 1333–47.

Jiménez, G. J. F., Valencia, M. C., Capitán-Vallvey, L. F. 2009. Intense sweetener mixture resolution by flow injection method with on-line monolithic element. *J. Liquid Chromatogr. Relat. Technol.* 32: 1152–68.

Joint FAO/WHO Expert Committee on Food Additives (JECFA). 2001. In Toxicological evaluation of certain food additives specifications, 57th Session, FAO Food and Nutrition Paper 52 (9).

Kobayashi, C., Nakazato, M., Ushiyama, H. et al. 1999. Simultaneous determination of five sweeteners in foods by HPLC. *J. Food Hyg. Soc. Japan* 40(2): 166–71.

Kobayashi, C., Nakazato, M., Yamajima, Y. et al. 2001. Determination of sucralose in foods by HPLC. *Shokuhin Eiseigaku Zasshi (J. Food Hyg. Soc. Japan)* 42: 139–43.

Kohda, H., Kasai, R., Yamasaki, K. et al. 1976. New sweet diterpene glucosides from *Stevia rebaudiana*. *Phytochemistry* 15: 981–3.

Koyama, M., Yoshida, K., Uchibori, N. et al. 2005. Analysis of nine kinds of sweeteners in foods by LC/MS. *J. Food Hyg. Soc. Jpn.* 46: 72–8.

Kroger, M., Meister, K., Kava, R. 2006. Low-calorie sweeteners and other sugar substitutes: A review of the safety issues. *Compr. Rev. Food Sci. Food Saf.* 5: 35–47.

Lawrence, J. F. 1987. Use of post-column ion-pair extraction with absorbance detection for the liquid chromatographic determination of cyclamate and other artificial sweeteners in diet beverages. *Analyst* 112: 879–81.

Lawrence, J. F., Charbonneau, C. F. 1988. Determination of seven artificial sweeteners in diet food preparations by reverse phase liquid chromatography with absorbance detection. *J. AOAC* 71: 934–7.

Leth, T., Fabricius, N., Fagt, S. 2007. Estimated intake of intense sweeteners from non-alcoholic beverages in Denmark. *Food Add. Contam. A* 24(3): 227–35.

Lino, C. M., Costa, I. M., Pena, A. 2008. Estimated intake of the sweeteners, acesulfame-K and aspartame, from soft drinks, soft drinks based on mineral waters and nectars for a group of Portuguese teenage students. *Food Add. Contam. A* 25(11): 1291–6.

MacKinnon, D. K. 2003. *Food Additives Data Book*. Jim Smith and Lily Hong-Shum (eds.), pp. 901–1003. UK: Blackwell Science.

McCourt, J., Storka, J., Anklam, E. 2005. Experimental design-based development and single laboratory validation of a capillary zone electrophoresis method for the determination of the artificial sweetener sucralose in food matrices. *Anal. Bioanal. Chem.* 382: 1269–76.

Moors, M., Teixeira, C. R. R. R., Jimidar, M. et al. 1991. Solid-phase extraction of the preservatives sorbic acid and benzoic acid and the artificial sweeteners aspartame and saccharin. *Anal. Chim. Acta* 255: 177–86.

Nagasawa, K., Yoswidome, H., and Anryu, K. 1970. Separation and detection of synthetic sweetners by thin layer chromatography. *J. Chromatogr.* 52: 173–6.

Nakazato, M., Kobayashi, C., Yamajima, Y. et al. 2001. Determination of neohesperidin dihydrochalcone in foods. *Shokuhin Eiseigaku Zasshi (J. Food Hyg. Soc. Japan)* 42: 400–4.

Ni, Y., Xiao, W., Kokot, S. 2009. Differential kinetic spectrophotometric method for determination of three sulphanilamide artificial sweeteners with the aid of chemometrics. *Food Chem.* 113: 1339–45.

Nikolelis, D. P., Pantoulias, S. 2000. A minisensor for the rapid screening of acesulfame-K, cyclamate, and saccharin based on surface stabilized bilayer lipid membranes. *Electroanalysis* 12(10): 786–90.

Nikolelis, D. P. and Pantoulias, S. 2001b. Selective continuous monitoring and analysis of mixtures of acesulfame-K, cyclamate, and saccharin in artificial sweetener tablets, diet softdrinks, yogurts, and wines using filter-supported bilayer lipid membranes. *Anal. Chem.* 73: 5945–52.

Nikolelis, D. P., Pantoulias, S., Krull, U. J. et al. 2001a. Electrochemical transduction of the interactions of the sweeteners acesulfame-K, saccharin and cyclamate with bilayer lipid membranes (BLMs). *Electrochim. Acta* 46: 1025–31.

Nobrega, J. D., Fatibello, O., Vieira, I. D. 1994. Flow-injection spectrophotometric determination of aspartame in dietary products. *Analyst* 119(9): 2101–4.

Pearson, R. L. 1991. *Alternative Sweeteners.* Lyn O'Brien Nabors (ed.), pp. 147–65. USA: Marcel Dekker.

Pérez-Ruiz, C., Martinem-Lozano, V., Tomás, A. et al. 2000. Quantitative assay for neohesperidin dihydrochalcone in foodstuffs by capillary electrophoresis. *Chromatographia* 51: 385–9.

Qu, F., Qi, Z. H., Liu, K.-N. et al. 1999. Determination of aspartame by ion chromatography with electrochemical integrated amperometric detection. *J. Chromatogr. A* 850: 277–81.

Renwick, A. G. 1985. The disposition of saccharin in animals and man—A review. *Food Chem. Toxicol.* 23: 429–35.

Ruiz-Medina, A., Fernández-de Córdova, M. L., Ortega-Barrales, P. et al. 2001. Flow-through UV spectrophotometric sensor for determination of (acetyl) salicylic acid in pharmaceutical preparations. *Int. J. Pharm.* 216: 95–104.

Sardesai, V. M., Waldshan, T. H. 1991. Natural and synthetic intense sweeteners. *J. Nutr. Biochem.* 2: 236–44.

Self, R. 2005. *Extraction of Organic Analytes from Foods, a Manual of Methods.* UK: The Royal Society of Chemistry.

Sheridan, R., King, T. 2008. Determination of cyclamate in foods by ultraperformance liquid chromatography/tandem mass spectrometry. *J. AOAC Int.* 91: 1095–102.

Shim, J. Y., Cho, I. K., Khurana, H. K. et al. 2008. Attenuated total reflectance–Fourier transform infrared spectroscopy coupled with multivariate analysis for measurement of acesulfame-K in diet foods. *J. Food Sci.* 73(5): C426–31.

Soffritti, M., Belpoggi, F., Tibaldi, E. et al. 2007. Life-span exposure to low doses of aspartame beginning during prenatal life increases cancer effects in rats. *Environ. Health Perspect.* 115: 1293–7.

Spangenberg, B., Storka, J., Arranz I. et al. 2003. A simple and reliable HPTLC method for the quantification of the intense sweetener Sucralose®. *J. Liquid Chromatogr. Relat. Technol.* 26: 2729–39.

Takayama, S., Renwick, A. G., Johansson, S. L. et al. 2000. Long-term toxicity and carcinogenicity study of cyclamate in nonhuman primates. *Toxicol. Sci.* 53: 33–39.

Takayama, S., Sieber, S. M., Adamson, R. H. et al. 1998. Long-term feeding of sodium saccharin to nonhuman primates: Implications for urinary tract cancer. *JNCI* 90: 19–25.

Takeshita, R. 1972. Application of column and thin-layer chromatography to the detection of artificial sweeteners in foods. *J. Chromatogr.* 66: 283–93.

Terada, H., Sakabe, Y. 1985. Simultaneous determination of preservatives and saccharin in foods by ion-pair chromatography. *J. Chromatogr.* 346: 333–40.

The Merck Index. 1996. 12th ed., New Jersey, USA: Merck & Co.

Thompson, C. O., Trenerry, V. C., Kemmery, B. 1995. Micellar electrokinetic capillary chromatographic determination of artificial sweeteners in low-Joule soft drinks and other foods. *J. Chromatogr. A* 694: 507–14.

Tunick, M. H. 2005. *Methods of Analysis of Food Components and Additives*. Semih Ötleş (ed.), pp. 1–14. USA: CRC Press Taylor & Francis Group.

Ujiie, A., Hasebe, H., Chiba, Y. et al. 2007. Simultaneous determination of seven kinds of preservatives and saccharin in foods with HPLC, and identification with LC/MS/MS. *J. Food Hyg. Soc. Japan* 48: 163–9.

Wagner, M. W. 1970. Cyclamate acceptance. *Science* 168: 1605.

Wasik, A., McCourt, J., Buchgraber, M. 2007. Simultaneous determination of nine intense sweeteners in foodstuffs by high performance liquid chromatography and evaporative light scattering detection—Development and single laboratory validation. *J. Chromatogr. A* 1157: 187–96.

Watson, A. A., Fleet, G. W. J., Asano, N. et al. 2001. Polyhydroxylated alkaloids—Natural occurrence and therapeutic applications. *Phytochemistry* 56: 265–95.

Weihrauch, M. R., Diehl, V. 2004. Artificial sweeteners—Do they bear a carcinogenic risk? *Ann. Oncol.* 15: 1460–5.

Whysner, J., Williams, G. M. 1996. Saccharin mechanistic data and risk assessment: Urine composition, enhanced cell proliferation, and tumor promotion. *Pharmacol. Ther.* 71: 225–52.

Wilson, L. A., Wilkinson, K., Crews, H. M. et al. 1999. Urinary monitoring of saccharin and acesulfame-K as biomarkers of exposure to these additives. *Food Add. Contam.* 16(6): 227–38.

Yang, D.-J., Chen, B. 2009. Simultaneous determination of nonnutritive sweeteners in foods by HPLC/ESI-MS. *J. Agric. Food Chem.* 57: 3022–7.

Yebra-Biurrun, M.C. 2000. Flow injection determinations of artificial sweeteners: A review. *Food Add. Contam.* 17(9): 733–8.

Zhu, Y., Guo, Y., Ye, M. et al. 2005. Separation and simultaneous determination of four artificial sweeteners in food and beverages by ion chromatography. *J. Chromatogr. A* 1085: 143–6.

Zygler, A., Wasik, A., Namieśnik, J. 2009. Analytical methodologies for determination of artificial sweeteners in foodstuffs. *Trends Anal. Chem.* 28(9): 1082–102.

CHAPTER 5

Honey

Athanasios Labropoulos and Stylianos Anestis

CONTENTS

5.1 INTRODUCTION

Honey is a popular sweet product that goes back to linguistic borders. Nowadays, more than a million tons of honey is produced per year all over the world.

Honey is most simply defined as a sugar-like sweetener collected by bees from flowers and live plants and modified in their wax combs. It also invokes our memories from a very young age and can be thought of as the most outstanding of the hive products.

Honey is a 100% natural product often labeled according to American honey legislation as "pure" honey (National Honey Board 2011a; Tonelli et al. 1990). It takes its pure originality from the flowers' nectar which is pure and natural.

The nectar is an outstanding energy source for the bees, because it contains simple sugars, such as glucose, fructose, and sucrose. It also bursts with vitamins, minerals, enzymes, amino acids, and compounds such as organic acids and aromatic matters.

Honey is the main everyday energy source in the hive during winter and also has a role as an insulator in order to help the bees overcome any adverse climatic variations.

The first historical references to honey date back to the Neolithic or Paleolithic periods approximately 12,000 years ago, as certified by a painting of a super found in 1921 near Valencia, Spain. In Egypt, dating from approximately 2400 B.C., one encounters the oldest evidence of beekeeping practice. In ancient Greece and Rome, it reached its highest levels (vanEngelsdorp and Meixner 2010).

The honey-producing countries did not slow any major revolution in beekeeping. However, these days, extracting cells from hives when they have 80%–90% of their honey cells sealed after a major nectar flow from May to October yields approximately 10–15 kg of honey per hive per annum. There are many techniques of removing honey from the combs and bees from the supers. Care must be taken to avoid contamination to supers and to the honey. The combs removed from the supers have to be uncapped. Then, the combs are put into an extractor to remove the honey by centrifugation.

After extraction, honey has to be put in a honey-ripening (settling) tank for purification at 30°C–35°C for 3–5 days (which delays further crystallization). The impurities and the air are eliminated by filtration through a strainer of 0.1-mm mesh. Then, appropriate packaging keeps the honey safe from air and humidity (to avoid fermentation) and away from strong light to preserve its nutritional and antimicrobial properties.

To avoid crystallization, honey producers have to reliquefy the honey through pasteurization, which is also used to avoid honey fermentation (usually for honey > 19% moisture). For some marketing purposes, honey can be encouraged to crystallize uniformly by mixing it with approximately 10% of very finely crystallized honey.

To maintain the best quality, honey requires complex care regarding its preservation, and the best conditions to store honey are at 14°C in a dry, dark, and airy place.

5.2 USES

Honey has been used throughout history (beyond its use as a human food) as a talisman and as a sweetness symbol. Archaeologists have found that humans started hunting for honey more than 10,000 years ago (as seen in Mesolithic rock paintings in Valencia, Spain). Ancient Egyptians were using honey in cakes, and in the Middle East, it was used for embalming the dead. We find the art of beekeeping mentioned in ancient China, as well as the importance of beekeeping box quality.

Honeybees were also found to be cultivated in ancient Maya (Central America) for culinary purposes and health uses, for example, as ointment for rashes, burns, and sore throats. Honey also makes appearances in the history of many of the world's religions:

- Honey is a symbol for the New Year among Jewish people. The Hebrew Bible refers to honey, and the Book of Judges (Samson) describes a swarm of bees and honey in the carcass of lion.
- In the festival of Madhu Purnima celebrated by the Buddhists, honey plays an important role, because a monkey brought Buddha honey to eat in the wilderness.
- According to the Christian New Testament, John the Baptist has long lived in the wilderness eating wild honey and locusts.
- In Islam, the Qur'an, through the Prophet Muhammad, recommends honey as a nutritional and healthy food and contains an entire chapter called An Nahl (the honey bee).

In Western civilization, "honey" is a term of endearment that can be used for casual acquaintances or for loved ones. In Russian and other European languages and books, bears are depicted as eating honey, the word is coined from the noun meaning honey, and that is the reason honey is often sold in bear-shaped jars. Honey is often used in bakeries, confectioneries, commercial beverages, wines, and beers.

5.3 HONEY FORMATION

Honey is stored by bees as a food in order to be used when fresh food sources are scarce during bad weather. There are three kinds of bees in the hives (Whitmyre 2006):

- A single female queen
- A seasonally changing number of male drones that fertilize the queen
- Approximately 30,000 female worker bees

The worker bees collect the nectar and bring it to the hive, where it is partially digested in the bees' stomachs until the product reaches a desired quality. Then, it is stored in honeycomb cells, which are either created by the bees or artificially made. Inside the hive, a process is taking place: bees fan their wings, enhancing the evaporation of the water in the nectar, which concentrates the sugar and greatly increases the shelf life by preventing fermentation.

5.4 HONEY FORMS

In general, honey is found in various forms (National Honey Board 2011a), such as liquid honey, cut-comb honey, crystallized natural honey, and creamed honey. These forms are described as follows:

- Liquid honey is extracted from combs by centrifuge, gravity, and/or straining and is easily used in bakery and confectionary products.

- Comb honey is an edible honey found as it is produced in the beeswax comb and packaged in various-size containers.
- Cut-comb or a liquid and cut-comb combination is a liquid-style honey in which chunks of honeycomb have been added to the jar.
- Crystallized or granulated honey is a honey in a semisolid state in which part of its sugar glucose has been crystallized by losing water and becoming glucose monohydrate. It immobilizes other honey components in a suspension that forms a semisolid body with an orderly structure.
- Creamed or whipped honey is a supersaturated solution made by controlling the crystallization process to produce fine crystals, resulting in a smooth and spreadable product. It is also known as churned honey, honey fondant, candied honey, granulated honey, and honey spread.

Honey is brought into the market and usually sold as either pure honey or a honey product. Some honey products do not meet the compositional criteria of pure honey:

- Dried honey is dehydrated by heating honey mixed with starches or sugars.
- Flavored or fruited honey is a honey in which fruit, color, or flavor has been added.
- Infused honey is honey in which herbs or peels and other flavoring is added by steeping.

Kosher and organic honeys are produced and packaged according to specific rules and regulations concerning ethnic Jewish dietary regulations and organic productions regulations, respectively. In general, honey is bought and sold by supermarkets, farmer's markets, or directly from beekeepers by variety or by color and taste/flavor. However, most consumers will typically buy a pure honey or a particular honey variety (e.g., clover honey), and industrial users (bakers and food and beverage processors) usually buy honey by color and taste/flavor, because they are linked. Honey that is darker usually has a stronger, more robust taste/flavor. On the other hand, lighter color honey is often more delicate and sweeter.

5.5 NUTRITION

Honey, of course, consists mainly of sugars, but it also contains other components such as vitamins, minerals, antioxidants (chrysin and pinobanksin), and enzymes (such as catalase). Its glycemic index ranking ranges from 31 to 78, depending on the variety of honey (Krell 1996). A nutritional profile of the varieties of honey shown in Table 5.1 consists of many elements, some of which can be summarized as follows:

- Carbohydrates: Glucose, fructose, sucrose, maltose, isomaltose, erlose, panose, cellobiose, dextrin, raffinose, isomaltotriose, isomaltopentaose, maltulose, trehalose, isomaltotriose, isopanose, and formic
- Amino acids: Lysine, histidine, asparagine, proline, cystine, methionine, leucine, phenylalanine, arginine, threonine, glutamic acid, and valine
- Aromatic substances: Hexyl acetate, octyl acetate, terpined, β-phenylethyl, p-cymol, aldehyde, ketone, geraldol, and ester
- Vitamins: Thiamine (B1), ascorbic acid (C), pyridoxine (B6), phylloquinone (K), riboflavin (B2), pantothenic acid (B5), nicotinic acid, retinol (A), and tocopherol (E)
- Enzymes: Diastase, invertase, catalase, phosphate, peroxide, and inulase
- Metallic elements: Lead, sulfur, chlorine, boron, zinc, iridium, chromium, potassium, copper, iron, calcium, phosphorus, silver, magnesium, manganese, nickel, and sodium
- Color-affecting compounds: Carotenoids, polyphenols, tyrosine, and tryptophan
- Fatty acids: Linoleic, stearic, oleic, palmitoleic, and linolenic
- Antibiotic compounds

Table 5.1 Average Nutritional Values per 100 g of Various Honeys

Proximate Analysis	Quantity
Energy	1.272 kJ
Water	17.2 g
Carbohydrates	82.4 g
Sugars	81.1 g
Protein	0.3 g
Fiber	0.2 g
Fat	0.0 g
Ash	0.2 g
Other components	3.2 g
Vitamins	
Riboflavin (vitamin B2)	0.04 mg
Niacin (vitamin B3)	0.12 mg
Pantothenic acid (vitamin B5)	0.07 mg
Vitamin B6	0.03 mg
Vitamin C	0.52 mg
Folate (vitamin B9)	2.00 mg
Minerals	
Potassium	52 mg
Calcium	6 mg
Sodium	4 mg
Phosphorus	4 mg
Magnesium	2 mg
Iron	0.4 mg
Zinc	0.2 mg

5.6 HONEY CLASSIFICATION

Generally, honey is classified into main categories and subcategories. Usually, it is classified by its source (floral and secretions), region, and processing and packaging, and it is also graded by color and optical density. For example, the United States Department of Agriculture (USDA) grades honey on a scale ranging from 0 for "white" honey to more than 114 for dark honey (Pridal and Vorlova 2002).

5.6.1 Classification by Floral Sources

Monofloral honey is made from the nectar of one type of flower, for example, thyme honey. In order to get this type of honey, beekeepers bring the beehives to an area where the bees have access to a particular flower. In practice, however, it is difficult to keep bees from visiting other types of flowers in the area (Devillers et al. 2004). Typical examples of monofloral honeys include clover, orange blossom, sage, buckwheat (North America); thyme, acacia, dandelion, sunflower, and chestnut (Europe); and cotton and citrus (Africa).

Monofloral honeys from different floral and geographical origins can be classified by pattern recognition rheometric techniques (Wei, Wang, and Wang 2010). It has been stated that heather honeys exhibit non-Newtonian, shear-thinning behaviors with a tendency to yield stress and were also thixotropic (Witczak, Juszczak, and Galkowska 2011). An electronic tongue was used by Wei, Wang, and Liao (2009) to classify honey samples of different floral and geographical origins.

Polyfloral honey is well known as wildflower honey derived from the nectars of various flowers. The taste (aroma and flavor) may be varied, depending on the prevalent flower source. However, many commercially available honeys are different in color, flavor, and density, and their properties depend on their geographic origin, for example, a variety of honey flavors from Texas versus honey from the mountains of Greece.

5.6.2 Classification by Nonfloral Secretions

When, instead of nectar as food, bees are taking the sweet secretions of aphids and other plant sap-sucking insects, the honey is called "honeydew." This type of honey is dark or brown in color with the rich flavor of stewed fruit, and it is not as sweet as the flower honeys. Such honeys come from pine, fir, and other similar resin trees found in Germany, Bulgaria, Northern California, Greece, and other countries (Gounari 2006). In Greece, honeydew constitutes the largest percentage of the country's annual honey production. Honeydew, although popular, has some complications and dangers to the bees as a result of its larger proportion of indigestibles, which can cause dysentery in the bees and the death of their colonies.

5.6.3 Classification by Processing and Packaging

Honey can be subjected to various processing methods and packaging techniques before it is sold to marketplaces under the following names:

- *Crystallized honey (granulated honey).* This is honey in which glucose has been crystallized over time and can be brought back to solution in liquid form by heating at a low temperature, approximately 50°C (120°F; Tosi et al. 2004).
- *Pasteurized honey.* This honey is pasteurized to destroy yeast cells and avoid microcrystallization in the honey; however, excessive heat may result in quality deterioration because of increases in hydroxymethylfurfural levels and decreases in enzyme (e.g., diastase) activity, which darkens the natural color and deteriorates odor, flavor, and overall acceptability (Subramanian, Hebbar, and Rastogi 2007).
- *Raw honey.* This honey is called raw, because it is obtained by extraction from the beehives without any additional processing treatment or is minimally processed. It may contain some small particles of wax and pollen (Prescott, Harley, and Klein 1999).
- *Strained honey.* This type of honey has passed through mesh to remove pieces of wax, propolis, and defect materials without removing minerals or enzymes.
- *Ultrafiltered honey.* This honey has been processed by filtration under pressure and low heating (65°C–77°C or 150°F–170°F) to recover all the extraneous solids. It is clear and has a longer shelf life (LaGrange 1991).
- *Ultrasonicated honey.* This type has been processed by ultrasonication (a nonthermal processing). This type of honey reduces the danger of fermentation as a result of yeast cell destruction.
- *Whipped honey.* This is also called creamed, churned, spun, candied, and fondant honey. This honey has been processed to create a large number of small crystals to prevent crystallization.
- *Dried honey.* Moisture is extracted from the liquid to create a solid honey, which is usually used in garnishing desserts.
- *Comb or cut-comb honey.* This is the honey in the honeybees' wax comb collected in wooden frames. The comb is then cut out in chunks and packaged.
- *Chunk honey.* This is honey that is packaged as pieces of comb immersed in liquid honey.

5.7 NUTRACEUTICAL ASPECTS

Honey grading is based on various factors, including water content, aroma, flavor clarity, and absence of defects, as indicated by USDA standards. The quality of honey can be graded by the

Table 5.2 Honey Grading Scale

Grade	Moisture	Taste	Clarity	Quality Defects
A	<18.6%	Very good	Very clear	Free of quality defects
B	<18.6%	Good	Clear	Fairly free of quality defects
C	<20.0%	Fairly good	Fairly clear	Some quality defects
Below standard	>20.0%	Not good	Not clear	Quality defects

Source: Modified from Grout, R. A. *The Hive and the Honey Bee, Revised Edition*, Dadant and Sons, Hamilton, IL, 1992.

above criteria, as shown in Table 5.2, according to how the honey flows without breaking into separate drops. Honey with low water content is suitable for long preservation because of its unique composition and physical, biological, and chemical properties.

Excessive heat (37°C–50°C or 98°F–122°F) causes the loss of important antibacterial, nutritional, and enzymatic components. On the other hand, crystallization does not affect the nutritional aspects of the honey and is a function of factors such as temperature and other specific compounds (sugars and traces) in the honey.

In medicine, honey has been used to treat various ailments to humans through its antiseptic and antimicrobial properties, which are the result of low water activity (causing osmosis), a hydrogen peroxide effect, high acid, and methylglyoxal (MGO; an active antimicrobial).

Honey is a mixture of two primarily saturated monosaccharides (fructose and glucose). Its low water activity is associated with contained sugars, and therefore, few water molecules remain available for microbial growth.

When honey is applied to a wound, hydrogen peroxide is released by the enzyme glucose oxidase (present in honey) by the dilution of the honey with body fluids and acts as an antiseptic agent. This is described in the following equation, which shows the glucose oxidase reaction:

$$C_6H_{12}O_6 + H_2O + O_2 \rightarrow C_6H_{12}O_7 + H_2O_2 \tag{5.1}$$

In addition, the relatively low acidic pH of honey (3.2–4.5) prevents the growth of many bacteria, yeast, and molds (Waikato Honey Research Unit 2011). The antibiotic activity of honey is a result of MGO and a synergistic unknown component. In addition, honey has functional effects through its antioxidants implicated in reducing damages from colon problems (disease colitis).

5.8 MICROBIOLOGICAL ASPECTS

The quality and safety of honey is influenced by microbes, but of most concern are the following: (1) microbes commonly formed postharvest (yeasts and spore-forming bacteria); (2) microbes that indicate the commercial quality of honey (coliforms and yeasts); and (3) microbes that cause human illness (Snowdon and Cliver 1996).

Sources of honey's microbial contamination are both primary (e.g., pollen, digestive bee tracts, and nectar air pollution) and secondary (e.g., air, handling, and equipment). Microbes often found in various honeys are primarily of concern; they are yeasts (from very low to very high numbers) and spore-forming bacteria (usually of the Bacillus genus and less often of the Clostridium genus). However, no such vegetative disease-caused bacterial species have been found because of the antimicrobial properties of honeys. Several tests, such as standard plate count (SPC) and other specialized tests (yeast and coliform counts), may be useful as an indicator of honey's sanitary quality (Waikato Honey Research Unit 2011).

Botulism spores are naturally present in honey, and therefore, it is not safe to give unprocessed honey to children under 1 yr of age because of their low digestive and immune system developments.

Honey produced from flower sources of oleanders, azaleas, mountain laurels, and others may cause health symptoms to humans, such as dizziness, weakness, and nausea (FDA 2007). Honey may be also considered toxic when bees are proximate to tutu bushes (Coriaria arborea) and when bees gather honeydew produced by the insect known as the passionvine hopper (*Scolypopa australis*), which is often found in New Zealand.

Microbes of concern in honey handling are those that are commonly found in honey (yeasts and spore-forming bacteria), those that indicate the commercial quality of honey (coliforms and yeasts), and those that cause illness under certain conditions (Snowdon and Cliver 1996). Primary sources of microbial contamination include pollen, bee's digestive tracts, dust, air, earth, and nectar, and secondary sources include food handling, cross contamination, equipment, and buildings.

On the other hand, during storage, honey is also susceptible to chemical, biological, and physical changes influencing its shelf life maintained for at least a couple of years. A recommended storage temperature should be between 18°C–24°C (64°F–75°F) for processed honey and below 10°C (50°F) for unprocessed honey. Pasteurization (170°F/s or 145°F/30 min) is well known to affect yeast and the microbial cells that are responsible for honey fermentation.

5.9 GENERAL PHYSICOCHEMICAL VIEW

Honey, a sweet food, is produced by bees (genus Apis) using flower nectar and other sweet stuff through a process of regurgitation, and is stored as a primary food source in wax honeycombs inside beehives. Honey mainly owes its sweetness to fructose and glucose and has a relative sweetness that approaches that of granulated sugar (National Honey Board 2011b).

Honey has long been used by humans in various foods and beverages as a flavoring and sweetening agent. It has distinct properties in generating heat, creating and replacing energy, from certain tissues and enzymes to promote oxidation. For a healthy body, honey as an already-digested sugar deserves consideration as the most assimilative carbohydrate compound.

Natural honey contains natural sugars and other components produced by honeybees that take nectar from flowers and/or secretions of plants. Scientists looking at honey's components found a complex of natural sugars, enzymes, minerals, vitamins, and amino acids. The color and flavor of honey depend on the nectar and secretion sources. However, there are more than 300 kinds of honey produced in the United States alone, including mainly clover, eucalyptus, orange blossom, etc., but the top producers of natural honey include China, Argentina, Turkey, Ukraine, and Mexico.

The effect of temperature and time of constant heating on the rheological properties of light and dark types of honey was examined by Abu-Jdayil et al. (2002). They revealed that a light-colored, low water–content, heat-treated honey showed a change in viscosity only at higher temperatures and a dark-colored, heat-treated honey showed a change in viscosity at all levels of heating temperatures.

The viscosity of two honeydew (pine and fir) and four monofloral nectar (thymus, orange, helianthus, and cotton) honeys at their initial and other (17%–21%) water content at 25°C–45°C was studied by Yanniotis, Skaitsi, and Karaburnioti (2006). They claimed a viscosity variation between 0.421 and 23.405 Pas, which was time independent and indicating Newtonian behavior (a shear stress varied linearly with the shear rate). Another experiment on the viscosity of monofloral honey and supersaturated sugar solutions measured at –5°C and 70°C indicated Newtonian behavior in all systems with reducing viscosity as temperature increased (Recondo, Elizalde, and Buera 2006). The high-frequency dynamic shear rheology of honey using an ultrasonic spectrometer showed that it can be used for the quality control of honey and other viscous food products (Kulmyrzaev and McClements 2000).

The processing temperature and water activity or the initial reactant concentration has a significant impact on browning and the color change of honey (Vaikousi, Koutsoumanis, and Biliaderis 2009).

Ultrasonic longitudinal attenuation and velocity technique could be useful for honey quality control by measuring various frequencies (e.g., 0.5–13.5 MHz) versus temperatures in honey with various moisture content (e.g., 15%–19%; Laux, Camara, and Rosenkrantz 2011).

Studies have shown that an ultrasound treatment speeds up the liquefaction of honey (especially at lower, <50°C, temperatures), creates clearer and more transparent honey than heat-treated-only honey, and results in a smaller amount and size of crystals (Kabbani, Sepulcre, and Wedekind 2011).

Camara and Laux (2010) applied an ultrasonic technique based on the measurement of the complex reflexion coefficient to distinguish two honeys with moisture content of a difference less than 0.2%.

Guo et al. (2010) suggested that microwave dielectric properties could be used in developing sensors to determine sugar and water content for distinguishing pure and water-added honey because there were strong linear correlations between the dielectric constant and the total soluble solids and water content. In addition, Chirife, Zamora, and Motto (2006) examined some fundamental aspects of the relationship between water activity and moisture in honey and found a very good straight-line relationship between both parameters in the range of 15%–21% moisture.

Results of the study of Yao et al. (2004) revealed that the high-performance liquid chromatography analysis of the phytotechnical constituents (e.g., phenolic acids) of honey could be used for the authentication of the botanical origin of honey. However, a common profile of phenolic acids such as gallic, chlorogenic, coumaric, and other acids can be found in honey.

A managed population of honeybees is influenced by various factors, including diseases, parasites, pesticides, the environment, and other social and economic factors (vanEngelsdorp and Meixner 2010). Although a large number of literature cover the evolution, behavior, and physiology of the bee genome, honeybees are uniquely suited to integrative studies of the genetic mechanisms of the behavioral transition, which includes large-scale changes in hormonal activity, metabolism, circadian rhythms, sensory perception, and gene expression.

5.9.1 Carbohydrates of Honey

Honey's carbohydrate profile is mainly composed of fructose (38.5%) and glucose (31.0%) sugars, sucrose or saccharose (1.5%), and higher sugars (7.2%), as shown in Table 5.3. Higher sugars are referred to as oligosaccharides and are medium-size carbohydrates containing more than three simple sugar subunits of monosaccharides and disaccharides such as erlose, maltotriose, panose, lecrose, and kestose (Cotte et al. 2003). In general, honey is approximately 1.5 times sweeter than sugar on a dry basis and contains an average of 82.5 g of carbohydrates per 100 g, providing approximately 300 kcal per 100 g of product.

Table 5.3 Main Sugars in Honey

Sugar	Average Values (%)
Fructose	38.5
Glucose	31.0
Sucrose	1.5
Maltose	7.2
Other carbohydrates[a]	4.2

Source: White, J. W. Jr., *J. Assoc. Off. Anal. Chem.*, 63(1), 11–18, 1980. With permission. In White, J. W., Jr. et al., *Composition of American Honeys*. Tech. Bull. 1261, Agricultural Research Service, U.S. Department of Agriculture, Washington, DC, 1962. With permission.

[a] Trisaccharides and other sugars, e.g., isomaltose, turanose, and kojibiose.

Table 5.4 Sensory Evaluation of Certain Solutions for Honey Sweetness Intensity

Concentrated Honey Solution (%)	Clover (%)	Orange Blossom (%)	Wildflower (%)
25	98.3	99.6	95.5
50	129.5	124.5	120.5
75	137.1	137.3	132.0
100	130.0	134	130.0

Source: Neumann, P.E., and Chambers, E., *Cereal Foods World* 38 (6) 418, 1993.

Table 5.5 Honey Substitution Based on Carbohydrate Equivalences

Sweetener	Solids (%)	Water (%)	Sweetener Replacement (kg honey/kg liquid)
Honey	82	18	—
Sucrose	100	0.0	Add 1.2 kg/0.2 kg
HFCS[a]	70	30	Add 0.85 kg/0.14 kg
CS[b]	80	20	Add 0.98 kg/0.03 kg
Molasses	72	28	Add 0.88 kg/0.03 kg

Source: NHB. Carbohydrates and the Sweetness of Honey. http://www.honey.com/images/downloads/carb.pdf.
[a] High-fructose corn syrup.
[b] Corn syrup.

Carbohydrate composition plays a key role in the crystallization functionality of honey, and therefore, it can be used to predict honey's tendency to crystallize and/or to avoid it. Honey may be varied in sweetness, flavor, and aroma, resulting from its floral type (monofloral and polyfloral) and the manufacturer's blend. Researchers have shown the sweetness intensity of honey increases with concentration as shown in Table 5.4. For example, the addition of 25% honey to a 10% sucrose solution increased the sweetness intensity by 16%, and it doubles the intensity by the addition of 25% honey to a 5% sucrose solution. Manufacturers of bakery products often substitute more than 10%–15% of total sugar with an equal amount of honey. A common practice on a sweetness/moisture basis is to replace 1.8 lbs of water with 10 lbs of honey in the original formula (Shin and Ustunol 2005). A substitution based on carbohydrate equivalences using honey as a substitute is shown in Table 5.5. On the other hand, a 68% honey solution freezes at 21.5°F (–12°C), and a 15% honey solution freezes at 29.5°F (–1.4°C), which has an impact on product application, for example, in ice cream, where the freezing point is very important (Bachmann 1995).

In addition to sweetness, honey carbohydrates offer many other properties, such as its ability to (1) hold moisture and/or extend shelf life because of its high reducing sugar content, (2) enhance flavors because of its sugar combination, and (3) promote healthy aspects to consumers because of its nutrient variety.

5.9.2 Honey Acidity

A number of organic acids (0.6%) and amino acids (0.05%) were found in honey with an average pH of 4.0 (3.5–6.0 range) and a typical acidity of 29 mEq/kg. The major organic acids of honey are gluconic (most pronounced), acetic, butyric, citric, formic, lactic, malic, succinic, and pyroglutamic acids. A number of 18 free amino acids were also found present in honey but in small amounts, with proline being the most abundant.

Among other acids found in honey, there are a range of aliphatic and aromatic acids with an important contribution to honey's flavor and aroma profiles. The low pH of honey has characteristics compatible with many low-acidity food products, inhibiting the presence and growth of microorganisms. On the other hand, honey has a favorable application in sauces, dressings, beverages, condiments, and other acidic or sour manufactured food products and the ability to smooth the flavor of acidic products such as lemon juice and vinegar.

5.9.3 Honey Enzymes

Honey, unlike other sweeteners, contains a variety of enzymes in low concentrations. Diastase (amylase), invertase (α-glucosidase), and glucose oxidase are predominant (Gilliam and Jackson 1972). Most of these enzymes play a virtual role in nectar transferred into honey, resulting in an array of various components that contribute to honey's functionality (Table 5.6), e.g., a transglycosylation activity.

Levels of diastase (Gothe scale average value of 21) or α-amylase vary in honey, depending on factors such as floral origin, pH, foraging patterns, and storage temperature exposure. Invertase enzyme inactivated by heating hydrolyzes sucrose to fructose and glucose. Glucose oxidase originates in bees, plays a role in honey formation, and oxidizes glucose, yielding gluconolactones (gluconic acid formation) and hydrogen peroxide.

5.9.4 Honey Color

Color is an important characteristic upon which honey is classified by governmental agencies, organizations, producers, and customers (Gonzales-Miret et al. 2007). For example, the USDA classifies honey into seven color categories, which can be determined by the Pfund color (PC) scale (in millimeter) and optical density (OD) as follows (USDA Agricultural Marketing Service 1985):

- Water white (PC > 8 and OD = 0.0945)
- Extra white (PC = 9–17 and OD = 0.189)
- White (PC = 18–34 and OD = 0.378)
- Extra light amber (PC = 35–50 and OD = 0.595)
- Light amber (PC = 51–85 and OD = 1.389)
- Amber (PC = 86–114 and OD = 3.008)
- Dark amber (PC > 114 and OD > 3.008)

The color of honey can be assessed by a number of methods, including the PC grader, CIE 1976, CIELAB, and AOAC methods. For example, the AOAC uses a Lovebird 2000 visual comparator, and the PC grader compares a standard amber-colored glass wedge with liquid honey. The color

Table 5.6 Enzymes and Their Functionality in Honey

Enzyme	Category	Functionality
Diastase	α-Amylase group	Converts starch to other carbohydrates
Invertase	α-Glucosidase group	Converts sucrose to glucose and fructose
Glucose oxidase	Peroxidases group	Converts glucose to gluconic acid and peroxide
Catalase	Oxidoreductases group	Converts peroxide to water and oxygen
Acid phosphatase	—	Removes phosphate from organic phosphate
β-Glucosidase	—	Converts β-glucans to glucose and oligosaccharides
Esterase	Hydrolases group	Breaks down esters to other components
Protease	Protein hydrolases group	Hydrolyzes proteins to other peptides

Source: White, J.W. Jr., *Adv. Food Res.* 24, 288, 1978; Crane, E., *A Book of Honey*, Charles Scribners' Sons, New York, 1980.

intensity of honey is usually expressed as the distance along the amber wedge ranging from 1 to 140 mm (Gonzales et al. 1999). However, honey changes its natural/initial color, darkening as a result of storage temperature and composition. Lighter honeys (e.g., clover and alfalfa) usually have a milder flavor and are much preferable than the darker ones, with a few exceptions (e.g., basswood honey is light in color with a strong flavor).

5.9.5 Quality Assurance

The total phenolic flavonoid and carotenoid content varies among honeys, with the highest values to be obtained in amber honeys (Alvarez-Suarez et al. 2010).

The use of multivariate analysis on physicochemical parameters—moisture, water activity, electric conductivity, color, hydroxymethylfurfural, acidity, pH, proline, diastase and invertase, and sugar composition (fructose, glucose, sucrose, maltose, isomaltose, trehalose, turanose, and melezitose)—can be determined in blossom and suspected floral honeys or honeydews in order to differentiate them (Bentabol-Manzanares et al. 2011).

With regard to the botanical and geographical origin of honeys, there are suitable methods based on the analysis of specific components, investigating flavors, patterns, distribution of pollen, aroma compounds, and special marker compounds for the detection of botanical origin and some other profiles of oligosaccharides, amino acids, and trace elements for the detection of geographical origin (Anklam 1998).

Studies also revealed it is possible to produce clarified honeys and/or enzyme-enriched honeys using a combination of microfiltration and ultrafiltration membrane processing, which removes yeast cells, resulting in improved stability of the processed honey (Barhate et al. 2003).

Sucrose syrup is a common additive in honey adulteration for developing a cheap, simple, convenient, and rapid sucrose-adulterated honey. Dielectric properties can be used to detect sucrose in adulterated honey or to sense sucrose content in honey, because the dielectric constant of pure honey is higher than honey sucrose–syrup mixtures (Guo et al. 2011a).

5.10 MICROWAVE APPLICATIONS

In the food industry, honey is accepted as a sweetening and a browning agent with many applications in various products, such as sauces, dressings, beverages, glazes, spreads, jellies, baked goods, and confections. Honey as a lightly reactive ingredient for browning is also used widely in microwave-processed products because of the electrolyte effects of various sugars; the ionic conductance effects of phosphates, gluconic acid, and others; and the presence of salts of organic acids. Therefore, honey has superior microwave reactivity compared to other sweeteners (e.g., sugar and corn syrup) because of its special attributes of fructose, glucose, and maltose, which are favorable to browning reaction, and the presence of organic acids. Researchers have shown that honey heats twice as fast as water and significantly faster than corn syrup because of the concentration of honey solids and other physical properties, as shown in Table 5.7 (National Honey Board 2011c). A load factor curve measures a load of honey or a blend that absorbs maximum power in a microwave oven and can determine the efficiency of the oven. A dielectric constant reflects the reduction in microwave wavelength as it passes through foods. The loss factor determines the rate of heating and the dissipation of microwave energy, which represents the complex permittivity and is viewed as a shunt resistance. The loss tangent or dissipation factor determines absorption and penetration depth in the microwave processing of honey products and can be expressed as a loss factor and dielectric constant. For example, a 20% sugar solution has a dielectric constant close to water and a loss factor value higher than for honey and syrups. The microwave frequencies used for industrial and household microwaves are at 915 and 2450 MHz, respectively.

Table 5.7 Average Values of Physical Properties of Honey at 20°C

Physicochemical Properties	Value
Specific weight (moisture 17%)	1.425(kg/L)
Specific gravity	1.423 kg
Specific heat (moisture 17%)	(0.57 cal/g°C (1.825 kJ/Kg°C)
Thermal conductivity	130 × 10^{-5} cal/cm s°C
Viscosity at 25°C, 16% H_2O, clover	90.0 poise
Freezing point	−1.42°C to −1.53°C (15% honey solution) −5.8°C (68% honey solution)
Refractive index	1.490
Sugars (Brix)	81.5
Sourness (acidity)	Fairly low (pH 3.9)
Color—Pfund color scale (mm)	Water white <8 to Dark amber >114

Source: Adapted from National Honey Board, Storage and usage of tips for honey. Nature's Simple Sweetener. http://www.honey.com/images/downloads/broch-honey-simplified.pdf, 2011; White, J.W., 1992; Krell, R., *Value-Added Products from Beekeeping: FAO Agricultural Services Bulletin No. 124*, Food and Agriculture Organization of the United Nations, Rome. http://www.fao.org/docrep/w0076E/w0076e04.htm, 1996.

Researchers on honey composition and its dielectric properties claim salts and amino acids could be responsible for honey's added microwave reactivity in food applications. Studies on the penetration depth of honey indicate that the penetration depth falls under 1 cm in baked goods with less than 55% honey solids and decreases as the concentration of honey solids increases. The loss factor decreases to almost half the value as the temperature increases from 20°C to 65°C in a 50% aqueous solution of honey at 2400 MHz, while the dielectric constant remains the same as temperature changes (Figure 5.1). Tables 5.8 and 5.9 show data on dielectric properties for various honey-based compositions and different dilutions (National Honey Board 2011c; Guo et al. 2011a). The units for penetration depth, absorption, and conductance are centimeters, Napier per centimeter, and siemens per meter, respectively.

There are not much published data on the above dielectric properties of honey, but honey is more microwave reactive than other similar syrup types, for example, high-fructose corn syrup (HFCS) and other sugar solutions. In microwave baking and/or cooking, it is preferable for the foodstuffs to be circular, with a thickness less than 1 cm and a diameter equal to multiples of 112 wavelength (Cui et al. 2008).

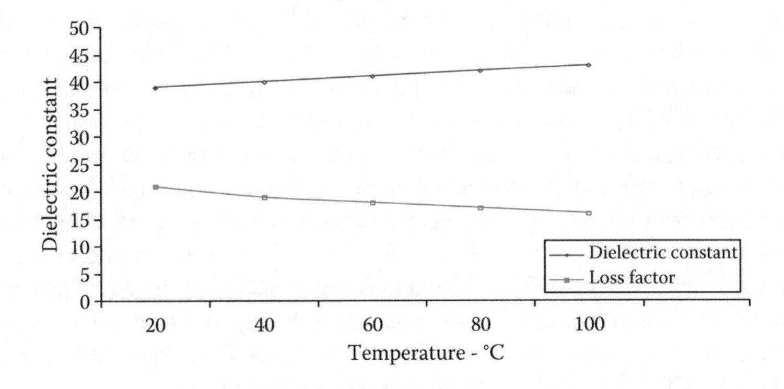

Figure 5.1 Temperature dependence of dielectric properties of honey. (Data from National Honey Board, Different white technical papers, http://www.honey.com/nhb/technical/technical-reference/, 2011.)

Table 5.8 Dielectric Properties at Various Honey Compositions at 21°C and 2400-MHz Frequency

Honey Composition	Temperature (°C)	Penetration Depth (cm)	Absorptivity (N/cm)	Conductance (S/m)	Loss Factor	Loss Tangent	Reflectivity (%)
Honey 21%	59	0.25	3.98	1.85	13.5	0.23	60
Honey 50%	22	0.42	2.43	1.24	9.2	0.42	44
Blend honey + sugar (20% + 20%)	51	0.27	3.71	2.00	14.7	0.29	58
Blend honey + sugar + HFCS (20:16:4%)	44	0.29	3.45	1.81	13.2	0.30	56

Source: Adapted from Guo, W. et al., *J. Food Eng.*, 97, 2, 2010. Guo, W. et al., *J. Food Eng.*, 102, 3, 2011. With permission. Guo, W. et al., *J. Food Eng.*, 107, 1, 2011. With permission.

Table 5.9 Dielectric Properties of Honey (Clove) at Different Dilutions, 21°C Temperatures, and 240-MHz Frequency

Honey Dilution (%)	Temperature (°C)	Penetration (cm)	Absorption (N/cm)	Conductance (S/m)	Loss Factor	Loss Tangent	Reflectivity (%)
5	75	0.23	4.40	1.16	8.68	0.12	63
25	64	0.25	4.10	1.98	14.88	0.24	62
55	35	0.32	3.07	2.45	18.29	0.52	54
75	12	0.54	1.84	1.04	7.80	0.64	36

Source: Guo, W. et al., *J. Food Eng.*, 107, 1, 1–7, 2011. With permission. Guo, W. et al., *J. Food Eng.*, 102, 3, 209–216, 2011b. With permission.

5.11 CRYSTALLIZATION

Honey sometimes takes a semisolid granulated form known as "crystallized" when one of the main sugars, glucose, precipitates out of the supersaturated solution, becoming glucose monohydrate by water losses and having the solid body of a crystal (Gleiter, Horn, and Isengard 2006).

The tendency of honey to crystallize depends on factors such as glucose content, moisture level, and other compositional substances of the honey, including sugars, minerals, acids, and proteins. However, crystallization can also be stimulated by any small particles existing in the honey, such as dust, pollen, bits of wax, propolis, and air bubbles. All these possibilities are related to handling, processing, storing, and the type of honey (Assil et al. 1991). The temperature of bottling and the kind of container may also have an impact on crystallization. Temperatures of bottling at 104°F–130°F (40°C–60°C) reduces crystallization significantly because of the crystals dissolving and expelling incorporated air into the honey. Filtering under pressure influences some nuclei (undissolved glucose crystals and pollen particles) and eliminates potential crystallization.

Some researchers try to predict the tendency to crystallize by applying ratios of glucose/water (e.g., ratios of <1.7 indicate honey that will stay liquid for a long time, and ratios >2.1 show quick-crystallization tendencies; Doner 1977). A high percentage of fructose causes honey to remain in a liquid condition for a significant period of time. Thus, honey at various fructose/glucose ratios plays a significant role in the crystallization process, which could be prevented by storing honey at temperatures below 52°F (10°C) and by avoiding moisture absorption.

Fermentation problems in honey depend on the initial count of microbes, the time and temperature of storage, and the moisture content. For example, honey with <17% moisture avoids fermentation as well as pasteurization (reduction of microbial content). Processes such as pasteurization at 66°C for 30 min usually delays crystallization and reduces fermentation by affecting yeast cells, and filtering removes particles that initiate crystallization. Most honeys containing <30% glucose resist granulation tendencies controlled mainly through proper storage, heating, and filtering. For example, mild heating to 60°C–70°C dissolves nuclei crystals, thus delaying crystallization. Cool temperatures (<10°C) avoid crystallization, while moderate ones (10°C–20°C) encourage it. High temperatures (20°C–27°C) degrade honey and discourage crystallization, while higher temperatures (>27°C) prevent crystallization, degrade the honey's quality, and encourage spoilage. However, unprocessed honey avoids crystallization at <10°C, and processed honey avoids crystallization at 18°C–24°C storage conditions (Assil, Sterling, and Sporns 1991).

5.12 DRIED AND OTHER HONEY PRODUCTS

Dried honeys are available commercially for various uses. These products are produced by special drying processes, and aids are added at the time to facilitate product processing and quality stability. Dried honey uses are limited in the industry, but its various forms, such as honey powders, flakes, and granulated honey, can be useful in special food applications (Sun et al. 2008).

Dried honey products are produced through technologies such as drum spray drying, microwave vacuum (MWV), and freeze drying and have a tendency to cake as a result of their high fructose content. The honey content in these products ranges from 50% to 70% according to the desired product.

Ingredients commonly used in the drying process are sweeteners (e.g., corn syrup, maltol, and sugar) and aids such as bulking and anticaking agents (e.g., calcium stearate, dextrins, starch, bran, and lecithin). Furthermore, some dried honeys are customized with fibers and vitamins, and others contain only honey and flour bran with no processed additives (Sun et al. 2008). However, dried honey products do not have a comparable aroma to liquid monofloral honeys, and their color may vary from light or golden yellow to tan or brown.

Advantages of dried honey products include low moisture content (2%–3.5%), which allows its easy blending into dry mixes (seasonings and coatings); consistency in texture, flavor, and color; and convenience (free flow). The honey's density is characteristically varied with the moisture content; for example, a standard U.S. gallon (3.8 L) of liquid honey weighs 4.4 kg (National Honey Board 2010).

The shelf life and functionality of dried honey products depends on the type and quantity of added ingredients, storage, and packaging conditions. Dried honeys can be used in injection and tumble–processed poultry products, development of peanut butter and honey products, quality improvement of oil-free potato chips, and in frozen dough.

MWV drying was investigated by Cui et al. (2008) as a potential method for obtaining high-quality dried honey. They heated liquid honey in a MWV dryer (30°C–50°C for 10 min) to a moisture content of approximately 2.5% and found that (1) there are no significant changes in the content of fructose, glucose, maltose, and sucrose in the honey, (2) the volatile acids, alcohols, aldehydes, and esters changed slightly, and (3) the acids decreased markedly, whereas the aldehydes and the ketones increased remarkably.

5.13 HONEY FUNCTIONALITY

5.13.1 Effect of Honey on Sweetness Intensity

The use of honey as a sweetness enhancer is helpful in low-sugar product formulations in which the addition of 25% honey to a 5% sucrose solution doubled the sweetness intensity of the solution, especially with the orange blossom and clover honeys (National Honey Board 2011d).

5.13.2 Effect of Honey on the Sourness of Solutions

The sourness perception is linearly related to the acid concentration of the product in which the addition of 25% honey at a 0.08% citric acid solution resulted in an approximately 75% decrease in sourness (from a rating of 100–25). However, citrus-based and other sour fruit-based beverages will benefit from the addition of honey by masking the flavor of the fruit. For example, the success in yogurt (Europe) and citrus beverages (United States, Japan, and Australia) can partially be a result of honey's ability to decrease sourness and improve the flavor profile of the product (National Honey Board 2011a).

5.13.3 Effect of Honey on Bitterness Intensity

Bitterness in food products is often the result of molecular breakdown and bitter peptide formation during processing, and it is usually associated with off-flavor developments. A study at Georgia University (according to the National Honey Board) measuring the ability of honey to decrease the perception of bitterness using 0.16% caffeine solutions resulted in a decrease of 100% bitterness intensity to approximately 30% when 25% orange blossom and/or wildflower honey was added. However, food manufacturers may be using honey to improve and/or mask the bitterness of some otherwise desirable ingredients of savory products.

5.13.4 Effect of Honey on Saltiness Perception

Because of honey's sugar composition, it has a potential ability to decrease saltiness in foods. Experiments at Georgia University (according to the National Honey Board) showed that 25% clover, wildflower, or orange blossom honey that was added to 0.35% sodium chloride solution reduced the saltiness intensity from 50 to 18, 15, and 28, respectively. Such honey attributes have been used by ham and bacon manufacturers for their salt-cured products, and other product developers have used honey to adjust the saltiness of their savory products.

5.13.5 Effects of Honey on Frozen Dough

Several studies, according to the National Honey Board, that have included honey as a functional ingredient in frozen and nonfrozen doughs revealed the following:

- A study at the University of Kentucky (1995) indicates honey at 4%–6% (flour basis) has the following effects:
    - Improves the rheological properties
    - Protects the dough from freezing damage
    - Improves dough strength
    - Decreases staling
    - Has a desirable effect on the crust and crumb color of frozen dough
    - Increases consumer acceptability

Thus, manufacturers of frozen dough will benefit in terms of the shelf life of the products by adding liquid or dried honey to their product formulations (Tong et al. 2010).

- A baking study, as shown in Table 5.10, indicated that freezing dough decreased its rheological properties, but the addition of 4% liquid or dry honey increased the resistance to the extension of the frozen dough, improving the strength of the dough. Thus, dough containing liquid honey showed an increased extendibility and loaf volume (Figure 5.2 and Table 5.11) compared to dough with dry honey and more so to dough with no added ingredients.

Table 5.10 Base Dough Formulation

Ingredients	Quantity (%)
Flour	100
Water	100
Honey	4, 6, 8, 10, and 12
Yeast (compressed)	5.3
Salt	1.5
Shortening	3.0
Ascorbic acid	100 ppm
Potassium sorbate	40 ppm

Source: Adapted from NHB, *Honey in Frozen Dough*, National Honey Board, Longmont, CO, 1995.

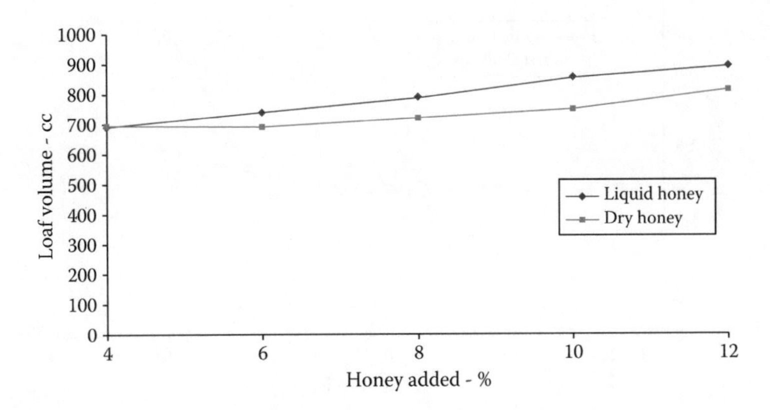

Figure 5.2 Loaf volume of frozen dough as a function of added honey. (Adapted from NHB, *Honey in Frozen Dough*, National Honey Board, Longmont, CO, 1995.)

Table 5.11 Loaf Volume of Frozen Doughs with Various Added Levels of Honey (Liquid and Dry)

Honey Added (%)	Liquid Honey (cc)	Dry Honey (cc)
4	690	695
6	740	690
8	790	720
10	855	750
12	895	815

Source: Adapted from NHB, *Honey in Frozen Dough*, National Honey Board, Longmont, CO, 1995.

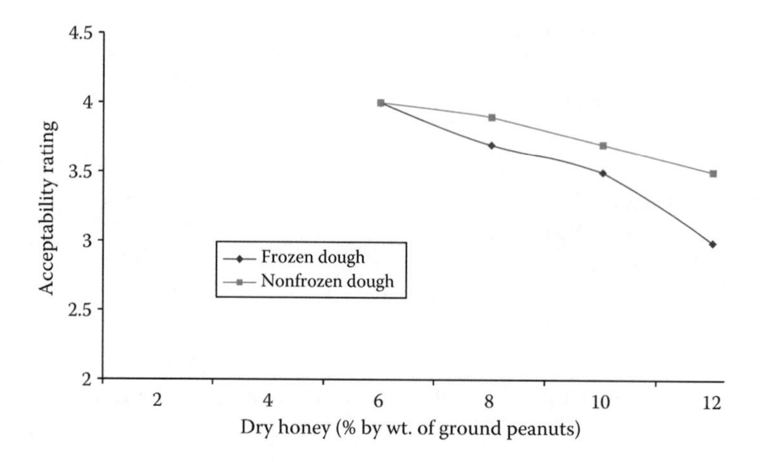

Figure 5.3 Consumer rating (1 being the lowest and 6 being the highest) of freshness of added dry honey to frozen and nonfrozen dough. (Adapted from NHB, *Honey in Frozen Dough*, National Honey Board, Longmont, CO, 1995.)

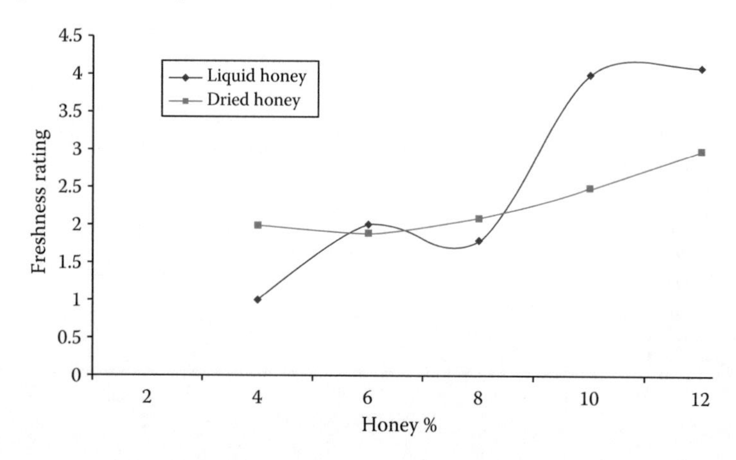

Figure 5.4 Consumer rating (1 being the lowest and 6 being the highest) for freshness of adding dry honey to frozen and nonfrozen dough. (From National Honey Board, Storage and usage tips for honey. Nature's Simple Sweetener. http://www.honey.com/images/downloads/broch-honey-simplified.pdf, 2011.

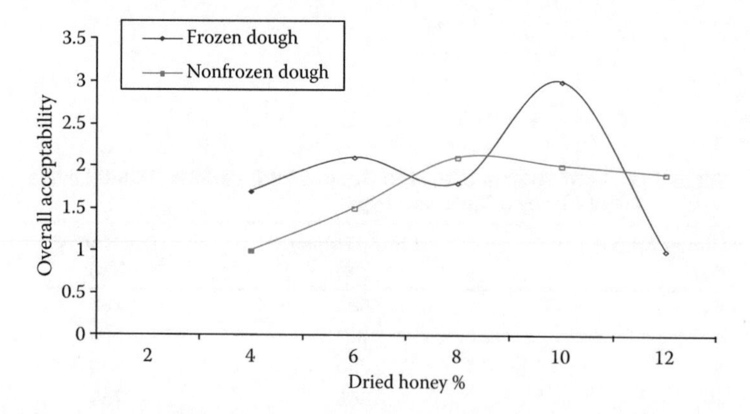

Figure 5.5 Consumer rating (1 being the lowest and 6 being the highest) for the overall acceptability of adding dry honey to frozen and nonfrozen dough. (Adapted from NHB, *Honey in Frozen Dough*, National Honey Board, Longmont, CO, 1995.)

- A sensory evaluation study showed an almost-equal preference for frozen and nonfrozen dough when they contained up to 10% liquid or 4%–8% dry honey. Figures 5.3 through 5.5 show no significant differences in freshness, flavor, and overall acceptability between frozen and nonfrozen dough.

5.14 HONEY APPLICATIONS

5.14.1 Honey in Drinks (Beverages and Beers)

Each variety of honey contributes differently to the color, flavor, and aroma of drinks (beverages and beers). The addition of selected honeys, especially wildflower, sage, or citrus, to beer adds quality and appeal to consumers, but it should not be exposed to high temperatures during beer processing in order to retain the flavor and aroma of the honey in the beer. Ultrafiltered and pasteurized honeys yield the best results for a wide variety of drinking products because of the sweetening properties and distinctive flavor components of honey. Brewers usually applied heat at 78°C for 2.5 h. Because of its sugars, honey's contribution to fermented beverages (e.g., beers, ales, and lagers) increases the popularity of these fermented products (e.g., honey beer), creating a dry, smooth, and crisp taste, color, and consumer appeal (LaGrange 1994).

However, adding 3%–10% honey is recommended for a mild aromatic flavor. Lately, new honey forms have been developed for high-quality and clear-colored needs, such as natural honey extracts for beverages such as black and herbal teas and flavored instant coffees. Ice teas as healthy alternatives to carbonated soft drinks are sweetened with honey to promote their natural flavor. In other vegetable-based beverages, honey has a role in decreasing the bitterness perception. In addition, forms of coprocessed dry honeys and milks are used for creamers and/or sweeteners. In sport drinks, honey provides the monosaccharides (fructose and/or glucose), which are very important nutrients for these types of beverages (Crane and Visscher 2009).

5.14.2 Honey in Frozen Desserts

The uses of liquid and/or dry honey in frozen desserts have been directed toward the following objectives:

- To develop more stable freeze-thaw products such as honey-based ice creams and frozen yogurts
- To develop good taste and extended melting time of the products
- To enhance the fruit flavors of the desserts

5.14.3 Honey in Fruit Gels

Many commercial fruit gels using honey and/or other sugars (as sugar sources) lack stability and break down while in storage or distribution channels. Thus, several research studies have undertaken to optimize production conditions using honey as an additive with the following objectives:

- To determine the proper quantity of honey as the major sweetener
- To optimize the reduction of discoloration and loss of flavor in the market
- To choose proper stabilizers (e.g., pectins and gums) for stable gels
- To adjust moisture levels in the formulation during processing
- To develop formulations with desirable pH, Brix %, gel set, flavor, color, and consistency

For example, the processing line and formulation ingredients, as shown in Table 5.12, for producing honey-apple gels includes the following steps:

Table 5.12 Honey-Apple Fruit Gel Formulations

Ingredients	Formulations (Honey Solids)	
Pectin/Sucrose Solution	**50%**	**100%**
Pectin	0.35	0.65
Sucrose	0.35	0.65
Water	12	12.5
Other ingredients		
Apple juice concentrate	14	14.5
Citric acid	0.2	0.2
Sucrose	33.3	0
Honey	35	70
Water	0	5.5

Source: Adapted from NHB, *Honey Fruit Spreads: Formulation, Production and Stability of Commercially Viable Honey Fruit Spreads*, National Honey Board, Longmont, CO, 1995.

- Water is boiled in a kettle.
- A pectin/sucrose mixture is dissolved in the boiling water.
- Concentrated apple juice is added.
- Honey and sucrose are mixed, and citric acid is added.
- The solution is boiled for 1 min and then poured into containers for cooling and storage.

Sensory evaluation indicates the addition of honey solids to the formula caused a darker coloration, increased sweetness and intensity expressed in flavor tartness, and a slight gel breakdown. The higher honey content differed minimally in quality, stability, and acceptability compared to products containing lower honey levels.

5.14.4 Honey in Spreads

Several spreads lack stability during storage and distribution. Thus, research tends to develop spread products (e.g., fruit and peanut) with improved stability and acceptable sensory characteristics by utilizing various types of honeys based on the following objectives:

- To optimize honey content as a major sweetener
- To reduce browning and flavor loss during production and storage
- To determine natural stabilizers (e.g., gums and pectins) for the production of desirable consistency
- To optimize the moisture level of the finished products with minimal processing

Honey (6%–20%) is often used in peanut butter spread formulations to improve consistency, accentuate the flavor, and increase consumer appeal. Experimental findings, according to the National Honey Board, have shown honey–peanut butter spread containing various levels of honey and three different levels of salt (1%, 1.5%, and 2%) showed no significant difference at the honey level (Figure 5.6).

Another type of spread is "gourmet," which is used for tasty toppings (bagels and ice creams) and for flavorful fillings (baked products). Most of these honey spreads are usually preferable when they contain nuts and not only flavor.

5.14.5 Reduced-Fat Honey Spreads

Reduced-fat honey spreads are favorable in many food categories, including bagels, special breads, and other baked products. Although these products are popular accompaniments to many

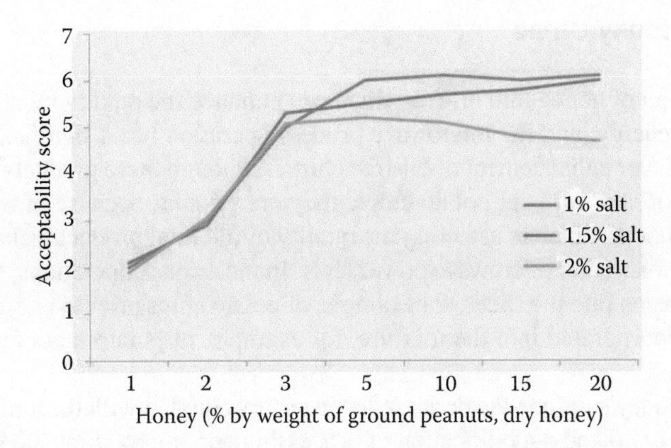

Figure 5.6 Changes in the acceptability of honey peanut butter spread as a function of honey and salt. (From NHB, *Honey Peanut Butter: Development, Optimization and Shelf-Stability of a Product Containing Peanut Butter and Honey*, National Honey Board, Longmont, CO, 1995.)

foodstuffs, they contain a notable level of fat, which can be a deterrent to consumers who are looking for low-fat or fat-free honey spread products. Studies, according to the National Honey Board, based on the formulation and shown in Table 5.13, have claimed the following processing steps:

- Liquefy the honey (heating to 66°C).
- Add the thickening agent and homogenize the mixture.
- Add a blend of premelted fat (oil/vegetable shortening, 1:1) and an emulsifier.
- Homogenize the fat blend and honey spread (66°C for 4 min).
- Cool the spread to room temperature.
- Package and properly store the product.

A commercial spread contains 65% fat and 20% honey, and the reduced-fat spread contains 70% honey and 13.5% fat. Other components used in these spreads include triglycerides, water, salt, lecithin, citric acid, flavors, beta carotene, and preservatives. Conclusions based on studies, according to the National Honey Board, indicate the following:

- Reduced-fat honey spread (13.5% shortening) contains 27% fewer calories than the commercial ones.
- Reduced-fat honey spreads are preferred over those of regular honey spreads.
- Reduced-fat and reduced-calorie honey spreads showed similar texture quality to commercial honey spreads.
- Adding pieces of fruit to reduced-fat honey spreads had a significant effect on mouthfeel but not on the overall acceptability.
- In general, 84% of consumers preferred to buy reduced-fat honey spreads.

Table 5.13 Composition Percentage of Various Honey Fat Spreads

Ingredients	Commercial	Reduced Fat	Reduced-Fat Calories
Honey	20	70	70
Oil/fat	65	27	13.5
Reduced fat	—	—	13.5
Monoglycerides	3	3	3
Calories (kcal/100 g)	66	480	425

Source: NHB, *Honey Fruit Spreads: Formulation, Production and Stability of Commercially Viable Honey Fruit Spreads*, National Honey Board, Longmont, Colorado, 1995. With permission.

5.14.6 Oil-Free Honey Chips

The incorporation of honey into oil-free chips can enhance the quality by utilizing a novel process of microwave energy, and the microwave process operation has a significant potential for the color, texture, and flavor enhancement of oil-free chips. Although these products may lack the color, flavor, and texture of conventional potato chips, they are popular because of their nutritional and healthy aspects. Other parameters affecting the quality of the final product include product variety, types and levels of honey, and microwave power level. In most snack operations, liquid honey is usually utilized for infusion into the slices, for example, of potato chips prior to microwave dehydration, and dry honey is incorporated into the mixture, for example, of potato mass for the production of formed potato chips.

Except for the honey's unique flavor, its reduced sugars coupled with the amino-bearing protein compounds of the potato and a modification of processing conditions stimulate browning reactions for a desired golden-brown color in the product. The liquid honey preferred for this type of product is clover honey (82% Brix) in an amount of 4%–6% of the brine weight, and the dry (powdered) honey is the drum-dried with 70% solids in levels 2%–3% of the brine weight. Quality objective criteria for this product category include moisture content, hardness, thickness, salt, color, and diameter of the product, and subjective criteria include mouthfeel and flavor.

5.14.7 Honey Snacks and Cereals

Researchers at the University of Illinois, according to the National Honey Board, have developed and tested extruded honey snacks that possess advanced-quality characteristics compared to other established snacks (e.g., tortilla snacks). Consumer sensor tests shown in Table 5.14 indicate differences in certain sensory characteristics. For example, snacks with increased honey levels tend to receive higher acceptability scores for most sensory parameters (flavor, hardness, intensity, and sweetness).

Manufacturers use honey in other extruded snacks (puffed, baked, curls, multigrain, and potato) in order to promote healthier and better quality snacks (improved surface browning and added flavor).

Multigrain snacks. Three formulas (Tables 5.15 and 5.16) with 8% honey, 4% brown sugar, and a control (without sugar) show that honey significantly increased browning in the multigrain snacks after being fried. The snacks with 8% honey produced medium-brown pellets after extrusion compared to the formulas with brown sugar and without sugar.

Breakfast cereals. In general, honey is well recognized as an important substitute for sugar in ready-to-eat cereals to satisfy the consumer's breakfast appetites. Extrusion is a common continuously used processing method for breakfast cereal manufacturing, in which a preformed dough material is cooked, shaped, and puffed by mechanical (pressure) and thermal (heating) energy

Table 5.14 Acceptability and Intensity Scores of Honey Low-Fat Chips on 1–7 Scale Points

Honey Level (%)	Overall Acceptability	Appearance	Flavor	Texture	Hardness	Sweetness
0	4.9	5.6	4.5	5.7	5.3	2.5
6	5.1	6.0	4.8	5.6	5.5	3.0
12	5.4	5.5	5.3	5.6	5.5	3.9
18	5.3	5.5	5.6	5.6	5.6	4.6

Source: LaGrange, V. and Sanders, S.W., *Cereal Foods World*, 33(10), 833–838, 1988.

Table 5.15 Extruded Multigrain Chips with and without Honey

	Amount of Treatments (%)		
Ingredient	**Honey**	**No Sugar**	**Brown Sugar**
Whole wheat flour	27	35	31
Rice flour	25	25	25
Corn flour	25	25	25
Wheat bran	7	7	7
Oat flour	5	5	5
Salt	2	2	2
Honey	8	—	—
Brown sugar	—	—	4

Source: LaGrange, V. and Sanders, S.W., *Cereal Foods World*, 33(10), 833–838, 1988.

inputs. Research findings (LaGrange and Sanders 1988) indicate that a combination of high levels of honey (15%) and low extrusion speeds (280 rpm) produced the most desirable results for flavor, sweetness, crispy texture, and color (golden appearance). It was also found that honey levels (>10%) improved the structural integrity of the extruded honey cereals. These data indicate that higher honey levels in cereals increased the bulk density and radial expansion of cereals but lowered the thickness of pieces and the number of broken pieces. In addition, the decrease in mechanical energy input (lower screw speeds) during extrusion was attributed to the effect of broken pieces imparted by the honey in high honey cereals. Thus extrusion conditions were evidently effective in the final product quality as follows:

- The higher honey (10%) extruded cereals were preferred by consumers and ranked as the best in overall acceptability.
- No off-flavor was detected in the higher honey level (10%–15%).

Based on the results above and other studies, the following conclusions can be made for extruded honey cereals:

- High honey levels impact perceivable honey flavors in extruded honey cereals.
- Higher extrusion screw speeds (>200 rpm) tend to lower honey flavor intensity.
- The extrusion process allows a uniform dispersion of the hygroscopic honey components in the cereal.

Table 5.16 Base Formula of Extruded Honey Cereal Products (on a Dry Basis)

Ingredient	**Quantity (%)**
Whole-grain hard wheat flour	50.0
Straight grade hard wheat flour	29.5
Wheat starch	12.5
Honey/water (2:1) dilution	6.0
Salt	1.0
Sodium bicarbonate	0.8
Cinnamon	0.2
Total	100.0

Source: National Honey Board, Carbohydrates and the sweetness of honey, http://www.honey.com/images/downloads/carb.pdf, 2011. USDA Agricultural Marketing Service, United States Standards for Grades of Extracted Honey, May 23, USDA, Washington, DC, 1985.

- The equilibrium of moisture content increased more rapidly at a higher level of honey content in the cereal.
- Consumers showed a preference for extruded honey cereals containing 10%–15% honey, because they exhibit higher levels of crispness than the control ones.
- Extruded cereals with higher honey levels could help delay rancidity development and thus minimize the need for antioxidants.

5.14.8 Honey Continents (Salsas and Marinades)

Honey is used in many condiments (dressings, sauces, mustards, salsas, and marinades). Honey flavor and sweetness increased, in general, with increased honey levels, and sourness response decreased despite increasing acidity. A consumer evaluation (Waikato Honey Research Unit 2011) of heat-processed and fresh salsas for overall acceptability, sweetness, and sourness is shown in Table 5.17. Fresh salsas received significantly lower scores in overall acceptability than the heat-processed ones, possibly because consumers have experience (familiarity) with heat-processed salsa (commercial) products. Heat-processed salsas received considerably higher scores than fresh salsas in sweetness, because heat-processed tomatoes in the salsas release sweet aromatics and flavors from the caramelization of the natural sugar in the tomatoes. Sourness was significantly affected (decreased in both salsas with increasing honey levels) by honey level and heat processing. Increasing honey levels reduced water activity and moisture in both fresh and processed salsas and resulted in extended shelf life.

Some other studies claim that honey in meat salsas and/or marinades imparts oxidative stability and antimicrobial effects (Johnston et al. 2005). However, manufacturers are using marinades to improve the flavor and moisture of their products, because they are adding seasoning (mixtures of spices, oil, and acids), flavor, and tenderization to meats (e.g., turkey, chicken, and ham). Thus, the addition of honey has a significant effect on marinade retention, depending on the marination methodology (immersion versus injection).

5.14.9 Dairy Honey Products

Honey is a favorite ingredient in yogurt products such as drinkable shakes, desserts, and dips because of its healthy aspects. Research at Michigan State University on drinkable yogurts formulated with 1%–2% milk fat, 0%–6% nonfat dry milk powder, 0%–2% stabilizer, and 5% honey mixed and pasteurized at 75°C for 30 min, cooled to 40°C, incubated with 2% yogurt culture at 40°C for 4–6 h, and cooled to 4°C for further experimentation (National Honey Board 2011a,b; Waikato

Table 5.17 Consumer Sensory Evaluation Test on a Nine-Point Hedonic Scale for Fresh and Heat-Processed Salsas

		Grading		
Treatment	**Honey Level (%)**	**Overall Acceptability**	**Sweetness**	**Sourness**
Fresh	0	3.8	2.5	3.5
Processed	0	5.2	3.0	3.0
Fresh	5	4.0	4.5	3.0
Processed	5	5.0	5.8	2.5
Fresh	10	3.8	6.2	2.7
Processed	10	4.5	7.2	2.3

Source: National Honey Board, Storage and usage tips for honey, Nature's Simple Sweetener, http://www.honey.com/images/downloads/broch-honey-simplified.pdf, 2011. National Honey Board, Honey and bees. http://www.honey.com/nhb/about-honey/honey-and-bees/, 2011. Waikato Honey Research Unit, Honey as an antimicrobial agent, http://bio.waikato.ac.nz/honey/honey_intro.shtml, 2011.

Honey Research Unit 2011) revealed that sweetness, flavor intensity, viscosity, and smoothness were influenced primarily by the honey content and the type (honey and/or sucrose) of the sweetener.

5.14.10 Bakery Honey Products

Honey offers unique functional properties to baked goods (baking mixes and breads) such as extending their shelf life and reducing crumbliness. A baking cycle (a total mixing and baking time of approximately 4 h) consists of the following stages:

- First kneading
- Resting
- Second kneading
- First racing
- Stirring down
- Second racing
- Baking
- Cooling

Some studies indicate that liquid honey in bread making produces high-quality products, because the addition of honey increases color changes and improves bread moisture and flavor. The dry honey mixes compared to mixes that include liquid honey and sugar (sucrose) were slightly tougher and more brittle. For example, bakery products such as bagels with 25% fat, 75% hard wheat flour, and 6% honey had the highest preference rating for aroma, flavor, texture, and overall quality characteristics.

5.15 CONCLUSION

In conclusion, the references in this chapter relate to honey production, honey composition, its physical properties, good formulations, and applications.

Honey is a liquid sweetener containing primarily glucose and fructose sugars and numerous other secondary types of sugars (e.g., oligosaccharides), as well as acids, proteins, vitamins, and minerals. Although honeys may have various different flavors and aromas, their sweetening power is similar and can be successfully incorporated at various levels into a wide number of food and drink products.

Honey contains a number of acids, including amino acids (0.05%–0.10%) and organic acids (0.17%–1.17%), with an average pH of 3.9 or typical range of 3.5–6.5, which inhibits the growth of microorganisms, makes honey compatible with many food and drink products in terms of overall acceptability, and contributes to a manufactured product's nutrition, texture, flavor, aroma, and consumer preference profile.

The enzymes in honey—predominantly the diastase (amylase), invertase, and glucose oxidase—do not generally influence the final food products, because they are present in relatively low concentrations.

The color of honey is related to its mineral content, is characteristic of its source, and is an important characteristic upon which honey is classified by honey producers and end users.

Because of its antimicrobial properties (e.g., low water activity, low moisture, low pH, and high sugar continent), honey discourages the persistent growth of microbes, which are primary yeasts and spore-forming bacteria.

Honey that is properly processed, packed, and stored retains its quality for a long time, but for practical purposes, a shelf life of two years is often stated.

REFERENCES

Abu-Jdayil, B., A. A. M. Ghzawi, K. I. M. Al-Malah, and S. Zaitoun. 2002. Heat effect on rheology of light and dark colored honey. *J. Food Eng., 51: 1*, 33–38.

Alvarez-Suarez, J. M., S. Tulipani, D. Díaz, Y. Estevez, S. Romandini, F. Giampieri, E. Damiani, P. Astolfi, S. Bompadre, and M. Battino. 2010. Antioxidant and antimicrobial capacity of several monofloral Cuban honeys and their correlation with color, polyphenol content and other chemical compounds. *Food and Chemical Toxicology, 48: 9*, 2490–2499.

Anklam, E. 1998. A review of the analytical methods to determine the geographical and botanical origin of honey. *Food Chemistry, 63: 4*, 549–562.

Assil, H. I., R. Sterling, and P. Sporns. 1991. Crystal control in processed liquid honey. *J. of Food Science, 56: 4*, 1034–1038.

Bachmann, L. 1995. Research shows consumer acceptability of frozen desserts with honey. *The National Dipper, 11: 4*, 30–34.

Barhate, R. S., R. Subramanian, K. E. Nandini, and H. U. Hebbar. 2003. Processing of honey using polymeric microfiltration and ultrafiltration membranes. *J. Food Eng., 60: 1*, 49–54.

Bentabol-Manzanares, A., Z. Hernández García, B. Rodríguez Galdón, E. Rodríguez Rodríguez, and C. Díaz Romero. 2011. Differentiation of blossom and honeydew honeys using multivariate analysis on the physicochemical parameters and sugar composition. *Food Chemistry, 126: 2*, 664–672.

Camara, V. C., and D. Laux. 2010. Moisture content in honey, determination with a shear ultrasonic reflect meter. *J. Food Eng., 96: 1*, 93–96.

Chirife, J., M. C. Zamora, and A. Motto. 2006. The correlation between water activity and % moisture in honey: Fundamental aspects and application to Argentine honeys. *J. Food Eng., 72: 3*, 287–292.

Cotte, J. F., H. Casabianca, S. Chardon, J. Lheritier, and M. F. Grenier-Loustalot. 2003. Application of carbohydrate analysis to verify honey authenticity. *J. of Chromatography A, 1021: 1*, 145–155.

Crane, E. 1980. *A book of honey*. New York: Charles Scribners' Sons.

Crane, E., and P. K. Visscher. 2009. Honey. In: V. H. Resh and R. T. Cardé (eds.), *Encyclopedia of Insects, Second Edition* (459–460). Burlington, MA: Academic Press.

Cui, Z. W., L. J. Sun, W. Chen, and D. W. Sun. 2008. Preparation of dry honey by microwave-vacuum drying. *J. of Food Eng., 84: 4*, 582–590.

Devillers, J., M. Morlot, M. H. Pham-Delègue, and J. C. Doré. 2004. Classification of monofloral honeys based on their quality control data. *Food Chemistry, 86: 2*, 305–312.

Doner, L. W. 1977. The sugars of honey–A review. *J. of the Science of Food and Agriculture, 28: 5*, 443–456.

FDA. 2007. Grayanotoxin. *U.S. Food and Drug Administration Bad Bug Book: Foodborne Pathogenic Microorganisms and Natural Toxins Handbook.* http://www.fda.gov/Food/FoodSafety/FoodborneIllness/FoodborneIllnessFoodbornePathogensNaturalToxins/BadBugBook/ucm071128.htm.

Gilliam, M., and K. K. Jackson. 1972. Enzymes in honey bee (*Apis mellifera* L.) hemolymph. *Comp. Biochem. and Physiol. B, 42: 3*, 423–427.

Gleiter, R. A., H. Hom, and H.-D. Isengard. 2006. Influence of type and state of crystallization on the water activity of honey. *Food Chemistry, 96: 3*, 441–445.

Gonzales, A. P., L. Burin, and M. del Pilar Buera. 1999. Color changes during storage of honeys in relation to their composition and initial color. *Food Res. Int., 32: 3*, 185–191.

González-Miret, M. L., F. Ayala, A. Terrab, J. F. Echávarri, A. I. Negueruela, and F. J. Heredia. 2007. Simplified method for calculating color of honey by application of the characteristic vector method. *Food Res. Int., 40: 8*, 1080–1086.

Gounari, S. 2006. Studies on the phenology of *Marchalina hellenica* in relation to honeydew flow. *Journal of Apicultural Research, 45: 1*, 8–12.

Grout, R. A. 1992. *The hive and the honey bee. Revised edition*. Hamilton, IL: Dadant and Sons.

Guo, W., X. Zhu, Y. Liu, and H. Zhuang. 2010. Sugar and water contents of honey with dielectric property sensing. *J. Food Eng., 97: 2*, 275–281.

Guo, W., Y. Liu, X. Zhu, and S. Wang. 2011b. Temperature-dependent dielectric properties of honey associated with dielectric heating. *J. Food Eng., 102: 3*, 209–216.

Guo, W., Y. Liu, X. Zhu, and S. Wang. 2011a. Dielectric properties of honey adulterated with sucrose syrup. *J. Food Eng., 107: 1*, 1–7.

Kabbani, D., F. Sepulcre, and J. Wedekind. 2011. Ultrasound-assisted liquefaction of rosemary honey: Influence on rheology and crystal content. *J. Food Eng., 107: 2*, 173–178.

Krell, R. 1996. *Value-added products from beekeeping: FAO agricultural services bulletin No. 124*. Rome: Food and Agriculture Organization of the United Nations. http://www.fao.org/docrep/w0076E/w0076e04.htm (accessed July 2011).

Kulmyrzaev, A., and D. J. McClements. 2000. High-frequency dynamic shear rheology of honey. *J. Food Eng., 45: 4*, 219–224.

LaGrange, V. 1991. Ultrafiltration of honey. *American Bee Journal, 131: 7*, 453–458.

LaGrange, V., and S. W. Sanders. 1988. Honey in cereal-based new food products. *Cereal Foods World, 33: 10*, 833–838.

Laux, D., V. C. Camara, and E. Rosenkrantz. 2011. α-Relaxation in honey study versus moisture content: High-frequency ultrasonic investigation around room temperature. *J. Food Eng., 103: 2*, 165–169.

National Honey Board (NHB).1995a. *Honey fruit spreads: Formulation, production and stability of commercially viable honey fruit spreads*. Longmont, CO: National Honey Board.

National Honey Board (NHB). 1995b. *Honey peanut butter: Development, optimization and shelf-stability of a product containing peanut butter and honey*. Longmont, CO: National Honey Board.

National Honey Board (NHB). 1995c. *Honey in frozen dough*. Longmont, CO: National Honey Board.

National Honey Board (NHB). 2011a. Storage and usage tips for honey. Nature's Simple Sweetener. http://www.honey.com/images/downloads/broch-honey-simplified.pdf (accessed July 2011).

National Honey Board (NHB). 2011b. Honey and bees. http://www.honey.com/nhb/about-honey/honey-and-bees (accessed July 2011).

National Honey Board (NHB). 2011c. Honey and microwaveable foods. (http://www.honey.com/nhb/technical/ (accessed September 2011).

National Honey Board (NHB). 2011d. Carbohydrates and the sweetness of honey. Retrieved from http://www.honey.com/images/downloads/carb.pdf (accessed September 2, 2011).

Neumann, P. E., and Chambers, E. 1993. Effects of honey type and level on the sensory and physical properties of an extruded honey-graham formula breakfast cereal. *Cereal Foods World, 38: 6*, 418.

Prescott, L. M., J. P. Harley, and D. A. Klein. 1999. *Microbiology*. Boston: McGraw-Hill.

Pridal, A., and L. Vorlona. 2002. Honey and its physical parameters. *Czech J. Anim. Sci., 47: 10*, 439–444.

Recondo, M. P., B. E. Elizalde, and M. P. Buera. 2006. Modeling temperature dependence of honey viscosity and of related supersaturated model carbohydrate systems. *J. Food Eng., 77: 1*, 126–134.

Shin, H. S., and Z. Ustunol. 2005. Carbohydrate composition of honey from different floral sources and their influence on growth of selected intestinal bacteria: An in vitro comparison. *Food Research Int., 38: 6*, 721–728.

Snowdon, J. A., and D. O. Cliver. 1996. Microorganisms in honey. *International J. Food Microbiology, 31: 1–3*, 1–26.

Subramanian, R., H. U. Hebbar, and N. K. Rastogi. 2007. Processing of honey: A review. *Int. J. of Food Properties, 10: 1*, 127–143.

Tonelli D., E., Gattavecchia, S., Ghini, C., Porrini, G., Celli, and A. M. Mercuri. 1990. Honey bees and their products as indicators of environmental radioactive pollution. *J. Radioanal. Nucl. Chem., 141: 2*, 427–436.

Tong, Q., X. Zhang, F. Wu, J. Tong, P. Zhang, and J. Zhang. 2010. Effects of honey powder on dough rheology and bread quality. *Food Res. Int., 43: 9*, 2284–2288.

Tosi, E. A., E. Re, H. Lucero, and L. Bulacio. 2004. Effect of honey's high-temperature short-time heating on parameters related to quality, crystallization phenomena and fungal inhibition. *LWT Food Sci. and Technol., 37: 6*, 669–678.

USDA Agricultural Marketing Service. 1985. United States Standards for Grades of Extracted Honey. May 23. Washington, DC: USDA.

Vaikousi, H., K. Koutsoumanis, and C. G. Biliaderis. 2009. Kinetic modeling of nonenzymatic browning in honey and diluted honey systems subjected to isothermal and dynamic heating protocols. *J. Food Eng., 95: 4*, 541–550.

vanEngelsdorp, D., and M. D. Meixner. 2010. A historical review of managed honey bee populations in Europe and the United States and the factors that may affect them. *J. Invertebr. Pathol., 103*(Suppl. 1), S80–S95.

Waikato Honey Research Unit. 2011. Honey as an antimicrobial agent. Retrieved from http://bio.waikato.ac.nz/honey/honey_intro.shtml.

Wei, Z., J. Wang, and W. Liao. 2009. Technique potential for classification of honey by electronic tongue. *J. Food Eng., 94*(3–4), 260–266.

Wei, Z., J. Wang, and Y. Wang. 2010. Classification of monofloral honeys from different floral origins and geographical origins based on rheometer. *J. Food Eng., 96*, 469–479.

White, J. W. 1992. Honey. In: J. M. Graham eds, "The Hive and the Honey Bee." Dadant & Sons, Hamilton, Illinois. 1324 pp.

White J. W. Jr. 1962. Reducing Sugars and pH: Calculated from data in White. Composition of American Honeys. Tech. Bull. 1261, Agricultural Research Service, U.S. Department of Agriculture, Washington, DC.

White J. W. Jr. 1978. Honey. *Adv. Food Res., 24:* 288.

White J. W. Jr. 1980. F:G ratio, fructose, glucose, sucrose: White. Detection of noney adulteration by carbohydrate analysis. *J. Assoc. Off. Anal. Chem, 63: 1*, 11–18.

Whitmyre, V. 2006. The plight of the honeybees. Retrieved from http://groups.ucanr.org/mgnapa/Articles/Honeybees.htm.

Witczak, M., L. Juszczak, and D. Galkowska. 2011. Non-Newtonian behavior of heather honey. *J. Food Eng., 104: 4*, 532–537.

Yanniotis, S., S. Skaitsi, and S. Karaburnioti. 2006. Effect of moisture content on the viscosity of honey at different temperatures. *J. Food Eng., 72: 4*, 372–377.

Yao, L. H., Y. M. Jiang, R. Singanusong, N. Datta, and K. Raymont. 2004. Phenolic acids and abscisic acid in Australian eucalyptus honeys and their potential for floral authentication. *Food Chemistry, 86: 2*, 169–177.

CHAPTER 6

Syrups

Athanasios Labropoulos and Stylianos Anestis

CONTENTS

6.1 INTRODUCTION

Sugar is an important ingredient in many food and drink applications, and liquid sweeteners or syrups possess additional applications, benefits, and properties.

Among common viscous carbohydrate-based liquids, syrups include honey, corn syrup, inverted sugar, and molasses. In order to understand how liquid sweeteners differ from solid sweeteners, such as granulated sugars, it is imperative to understand their chemistry. Of course, sucrose as a disaccharide of glucose and fructose possesses a relative sweetness of 100 as a base for all other sweeteners. However, liquid sugars or syrups are different in chemical structure and functional properties from the competitive granulated sugar (Anderson 1997).

Honey, being the oldest all-natural liquid sweetener, is used extensively in bakery goods, extending their self life as a result of its humidity absorbance and slow release of moisture.

Corn syrup refers to a sweetener group that differs in glucose concentration (20%–98%), but when it is blended with fructose, it results in syrups of 42%–95% fructose.

Inverted sugar is a liquid product that is created through an inversion process of sucrose into equal amounts of glucose (relative sweetness of 70) and fructose (relative sweetness of 120).

Molasses is a refined by-product of cane and beet sugar manufacturing. Its color and flavor depends on the number of extraction processes, and its acidity may be neutralized by the addition (1%) of baking soda (El-Refai et al. 1992).

Some of the most common liquid sweeteners include the following:

- Maple syrup is obtained by collecting tree sap, boiling it, and concentrating it into a syrup.
- Golden syrup is made from evaporated cane juice and has a smooth consistency and a unique rich, toasty flavor.
- Sorghum syrup is processed from sorghum stalks and gives color to cereals, crackers, snacks, and baked goods.

- Malt syrup is made from mashing barley, converting it to sugars, and concentrating the extract into a thick syrup with 78%–82% solids with a malty flavor and a malty to caramel color.
- Brown rice syrup (BRS) is made from brown rice converted by fermentation to a maltose disaccharide liquid with the consistency of honey and half the sweetness of table sugar.
- Concentrated juice syrup is made by heating fruit juices, treating with enzymes, filtering, and concentrating it into liquid sweeteners.

Liquid sweeteners or syrups are important ingredients in many products of the food and beverage industry, where they can be used as food for yeast (rice of selected baked goods), as a sweetening agent, or for browning. Food manufacturers usually do not overlook the many kinds of syrups (commercially existing in the marketplace) that possess additional benefits beyond these three reasons for adding a sweetener to a product formulation (Mitchell 2006). However, note that food processors cannot simply make a 1:1 substitution of a liquid sweeter (syrup) for sugar in the formulation of a product. Each product and syrup has different properties, which influence the needs for sweetener levels and other ingredients as well.

The most common liquid sweeteners are inverted syrup, corn syrup, honey, molasses, maple syrup, malt and sorghum syrup, and fruit cereal syrup. All these types of syrup (viscous liquid carbohydrates) provide unique flavor profiles and a number of functional properties. However, to understand syrup's functionality, it is imperative to examine the chemical structure of each variety. For example, cane sugar syrup is a disaccharide composed of two monosaccharides, glucose and fructose, possessing the base for sweetness comparison with other sweeteners, such as inverted syrup. Through a process called inversion, sucrose is split into its two sugars, glucose and fructose. Thus, a traditional inverted sugar (syrup) is a liquid sweetener that contains an equal amount of glucose with sweetness, relative to table sugar, approximately equal to 70 and fructose with a relative sweetness approximately equal to 120. This difference of the two sugar components in inverted sugar syrup results in higher sweetness levels than table sugar. However, commercial inverted sugar syrup is available in a variety of sweetness levels and sugar profiles useful for any food and drink formulation (Mitchell 2006).

6.1.1 Corn Syrup

Corn syrup is a group of liquid sweeteners that differ in glucose concentration (solids) levels, which may range commercially from 20% to 98% glucose. Thus, corn syrup enriched with fructose is produced from corn syrup containing as much glucose as possible and is called high-fructose corn syrup (HFCS). However, it is a blend of fructose and glucose syrup with an increased fructose level ranging from 42% to 95% by weight. HFCS is usually added to product formulations for its sweetening power, because the more the fructose in the HFCS blend is, the sweeter the syrup becomes. Some corn syrup with a relative sweetness of approximately 30 and high-maltose syrup with a relative sweetness of approximately 34 are usually preferred in certain bakery confections and beverage applications as a body and bulk agent with a little sweetness (Watson and Ramstrad 1987).

In general, corn syrup contributes smoothness, moisture, and chewiness to baked goods (e.g., cakes, cookies, and pies), makes ideal glazes, and assists in any topping's quality.

6.1.2 Honey

Honey is the oldest liquid sweetener known to humans and animals and is characterized as the most all-natural sweetener (commercially). It is approximately 1.5 times sweeter than table sugar. The very important characteristics of color and flavor of honey varies with the feed source gathered by the bees producing the honey. The darker the honey, the stronger the flavor.

Honey helps extend the shelf life of bakery goods because of its slow moisture release and humidity absorbance while adding fructose and color (browning).

6.1.3 Molasses Syrup

Molasses is a by-product of the cane and/or beet sugar industry. The dark color and caramelized flavor of the resulting molasses syrup after the sugar refining process depends on the extraction procedure when cane or beet juice is boiled into syrup before the sugar crystals are removed. The first extraction always yields the lighter and sweeter molasses syrup, the second extraction makes a darker molasses syrup with moderate sweetness, and the third extraction produces a very dark, strong-tasting and hardly sweet blackstrap molasses syrup, which is not suitable to the food industry.

Molasses syrup can be substituted with sugar to a suggested (e.g., bakery goods) ratio of 1.3 to 1.0 on a weight basis. Finally, in its applications, molasses may need to be neutralized by the addition of baking soda, because it is more acidic than sugar syrup.

6.1.4 Maple Syrup

Maple syrup, a common liquid sweetener produced by tapping trees, collecting sap, and boiling and heat concentrating the watery sap to caramelize the natural sugars.

Various grades of maple syrup are assigned by European and American agencies, such as the one by the United States Department of Agriculture (USDA), according to which maple syrups are classified as grade A light amber (GALA), grade A medium amber (GAMA), grade A dark amber (GADA), and Grade B. GALA has a delicate flavor and a light color made early in the season (in colder weather). GAMA has a darker color with easily discernible maple flavor. GADA is a very dark, strong maple flavor syrup used most often in baked goods. Finally, Grade B, which is also called cooking syrup, is extremely dark in color and has a strong maple flavor and hints of caramel. In general, granular sugar can be substituted with maple syrup in a ratio of 1 to 1.5 for bakery product applications but with the addition of baking soda to neutralize the acidity.

6.1.5 Golden Syrup

Golden syrup, a popular English sweetener, is made from sugar cane juice through evaporation and has high clarity with a golden hue, the smooth consistency of corn syrup, and a unique rich, toasty, butterscotch flavor. In most food applications, such as bakery goods, pies, and nut bars, it substitutes for corn syrup with a great success.

6.1.6 Sorghum Syrup

Sweet sorghum syrup or sorghum molasses produced from sorghum has been used widely in the United States since 1950, especially in southern Appalachia, as a traditional breakfast with hot biscuits.

6.1.7 Malt Extract Syrup (Malt Syrup)

Malt syrup is very interesting to the brewing and other drink industries. It is made by mashing barley, converting starches to sugars, rendering soluble proteins by enzymes, and concentrating the extracted juice into a thick liquid (78%–82% solids). The color of the final product is controlled by the kilning or roasting temperature applied to the malt before its mashing stage and ranges from light amber to deep brown.

Malt syrup is also nutritious, because its processing procedure preserves most of the barley grain's natural health characteristics.

Although its main application is in brewing products, malt syrup in baked foods binds ingredients, adds bulk, acts as a humectant, and adds malty flavor and color profiles ranging from mild to caramel. It works very well in particular breads as a yeast food source and as a liquid or powdered light sweetener (approximately 65% as sweet as sugar) in many food and drink products.

6.1.8 Brown Rice Syrup

BRS is a natural sweetener similar to honey in consistency and can readily substitute for honey in many food and drink formulations. It can also substitute—1.3 parts BRS for one part table sugar—in baked food formulations, as long as baking soda is also added for the acidic adjustment of the product.

The production involves converting brown rice starch through fermentation into maltose, a disaccharide consisting of two units of glucose, and is approximately half as sweet as table sugar.

The most common application of BRS is in baked foods intended to have a hard or crispy texture and a unique caramel-like flavor, such as granola bars and cookies.

6.1.9 Fruit Juice Syrup

Another syrup in the liquid sweetener category comes from fruit juices. Different genuses and plant species enable syrup manufacturers to produce an array of syrup with different flavor profiles.

One of the latest arrivals of fruit syrup is produced from the sap developed in the heart of the blue agave cactus by crushing the plant, extracting the syrup in raw form, and filtering and heating it until the raw sugars are broken down to fructose. Then, the liquid is either further refined to produce a pale, amber-colored syrup or bottled as is in its dark (chestnut) color (Ashurst 2005).

Agave syrup appears as a golden syrup composed of 90% fructose and 10% glucose, containing 25% water and making it 1.5 times sweeter than table sugar. Recently, it has been used as sweetener, and its increased popularity is a result of its low glycemic quality. Agave syrup that is 25% less than sugar can be substituted in most baked goods, resulting in moisture retention properties similar to those of honey.

In general, these types of syrup are made by heating fruit juices (water removing), treating them with enzymes, filtering (to strip all color and flavor), and concentrating them to the desired final products. Among various fruits, grapes, and pears are the primary sources of fruit juice concentrates (syrups) used as sugar replacement in food and beverage products.

6.2 GLUCOSE AND FRUCTOSE SYRUP

6.2.1 Syrup Nomenclature

The term "syrup" comes from the ancient languages of Latin, *siropus*, and Persian, *sharab*, and it can be described as a viscous liquid consisting of a dissolved sugar–water solution with little tendency toward crystallization. Multiple hydrogen bonds between sugar and water are responsible for the viscous consistency of the solution, which should not be close to a supersaturation point (65%–67% by weight). Most types of syrup are usually sugar–water solutions and are prepared by solubilizing of sugar ingredients to purified water, mixing with or without heating, and straining.

6.2.2 Definition, Production, and Classification

Glucose syrup is mainly a solution of monosaccharides, which are produced enzymatically from cornstarch with a solid content depended to the dextrose equivalent (DE) and solid content according to the proposed syrup type.

Fructose syrup is produced from glucose syrup through partial isomerization. For glucose and fructose syrup, corn (maize) starch is the most common raw material used in the United States, whereas in the European Commission, potato starch and soft wheat starch (grade A) are used in a small amount (approximately 10%) of the total production (Schenck 2006).

The classification of glucose and fructose syrup is based on DE, which is the amount of reducing sugars (dextrose) in the syrup and expressed as a percentage of the dry total solids (Schenck 2006). For example, syrups are classified as types I–IV, depending on their DE:

- Type I: DE 20–38
- Type II: DE 39–58
- Type III: DE 59–73
- Type IV: DE > 74

A separate classified syrup group is the high-fructose syrup, which can be extended to include high-maltose syrups with a wide variation of fructose and maltose and DE values similar to conventional syrup.

6.2.3 Glucose Syrup Production

Production methodology for glucose syrup derived from corn starch has been available for more than 100 years since the mid-nineteenth century. Glucose syrup processing has since moved toward highly automated continuous processes through new enzyme systems producing a variety of syrups for the food and beverage industries (Melanson et al. 2007).

The basic steps of glucose syrup production are (1) conversion, (2) purification, and (3) concentration. The starch hydrolysis defines the type of syrup produced.

Starch is defined as a chemical compound, i.e., a glycan, that contains two glucose polymers: amylase and amylopectin molecules. The different relative proportions of these components in various starches are not relevant for glucose syrup production, but they do have some minor differences in the conversion and the purification steps (Schenck 2006).

Conversion is the processing step during which starch is hydrolyzed with acid–enzyme aid to result in lower molecular weight products (glucose syrups). The systems used in the conversion process are acid, acid–enzyme, and enzyme–enzyme. In the processing steps for the first two conversion systems, starch slurry is acidified and hydrolyzed with acid or acid–enzyme aid and heated in converters. Then, the liquors are cooled, neutralized, and purified to remove impurities (suspended proteins and lipids) before their concentration to a determined solid level. The purification (discoloration) step is performed using carbon treatment, and the degree of conversion is controlled by acid, temperature, and reaction time.

Syrup types with DE values of 20–55 are usually acid-converted starches made into regular glucose syrups. Other types with lower DE values are considered maltodextrins, and those with DE values higher than 55 are bitter by-products. However, a secondary enzyme conversion stage is required for syrup types with higher DE values of up to 94. The solid content of glucose syrup varies from 70% to 83% according to product type and market request.

In the process flow of enzyme–enzyme converted syrup, the alpha amylase is used as a hydrolyzing agent, and the kind of syrup (e.g., high-maltose, dextrose, or fructose) to be produced depends on the substrate. For example, for high-dextrose syrup, an amyloglucosidase enzyme, combined often with pullulanase, is used to produce a level of more than 95% dextrose product.

After purification and demineralization, an immobilized enzyme system is commonly used to produce a 42% fructose syrup, which, by enrichment through a chromatographic process, can lead to very high fructose (>90%) syrup. However, a variety of syrup blends can be obtained by mixing different fructose and glucose syrup.

Glucose syrup possesses functional properties according to conversion degree, sugar composition, and processing system, as shown in Table 6.1. The viscosity of glucose syrup is influenced by

Table 6.1 Some Functional Properties of Glucose Syrup at Various Conversion DE Levels

	Type of Syrup	
Functional Properties of Glucose Syrup	**Low DE**	**High DE**
Flavor	–	+
Color stabilization	±	±
Moisture stabilization	±	±
Nutritive value	±	±
Sweetness	–	+
Viscosity	–	+
Crystallization	+	–
Osmotic pressure	–	+
Hygroscopicity	–	+
Cohesiveness	+	–
Thickening agent	+	–

Source: Adapted from Hull, P.: *Glucose Syrups: Technology and Applications*, 2010. Copyright Wiley-VCH Verlag GmbH & Co. KGaA. Reproduced with permission. Advanced Instruments, Inc., *The Physical Chemistry, Theory, and Technology of Freezing Point Determinations*, Advanced Instruments Inc., Norwood, MA, 1971. Aurand, L. W. et al., *Food Composition and Analysis*, Van Nostrand Reinhold, New York, 1987. Balston, J. N., and B. E. Talbot, *A Guide to Filter Paper and Cellulose Powder Chromatography*, H. Reeve Angel & Co., London, 1952. Browne, C. A., and F. W. Zerban: *Physical and Chemical Methods of Sugar Analysis*. 1955. Copyright Wiley-VCH Verlag GmbH & Co. KGaA. Reproduced with permission. Emerton, V., and E. Choi, *Essential Guide to Food Additives*, RSC Publishing, Cambridge, U.K., 2008. McCance, R. A., and E. M. Widdowson, *The Composition of Foods, Royal Society of Chemistry*, Cambridge, U.K., 1991. Kirk, R. S. et al., *Pearson's Composition and Analysis of Foods*, Longman, London, 1991.

temperature and solid content, whereas nutritional value and solubility remain constant for all syrup types. In general, the viscosity of syrup decreases with increased temperatures and increases with increased solids and acid conversion.

6.2.4 Syrup Appearance (Color)

The color of corn syrup (glucose and fructose) is water-clear and may be changed (discoloration) during storage by nonenzymatic browning through the Maillard reaction (proteinous material with reducing sugars) and acidic caramelization (chemical dextrose degradation).

The stability of color is decreased by increased sugar levels (glucose and fructose), temperature, and solids (proteinous material and impurities), while the demineralization of high-fructose syrups seems to improve color stability.

6.2.5 Sweetness Aspects

Table 6.2 shows the relative sweetness of various sugars compared to the standard value of 100 for sucrose sweetness. The relative sweetness of glucose and fructose syrup is related to the conversion rate and depends on the amount of dextrose and maltose content (O'Brien-Nabors 2001).

An interesting consideration is the fact that syrups with higher than 42% fructose have the same sweetness as sucrose, and sweetness becomes higher when glucose syrup is mixed with sucrose. It implies the possibility of using glucose syrup in specific foods and drinks (e.g., nonalcoholic beverages and juices).

Table 6.2 Relative Sweetness of Sweeteners Compared to 10% Sucrose Solution (Value 100)

Sweeteners	Relative Sweetness
Sucrose	100
Glucose syrup, 42 DE	41
Glucose syrup, 64 DE	59
Dextrose	72
Fructose	114
Inverted sugar	95
Maltose	46
Galactose	63
Lactose	39
Xylose	67
Mannose	59
Sorbitol	51
Mannitol	69

6.2.6 Fermentation and Moisture Aspects

Increased environmental humidity will create a product that absorbs moisture to the equilibrium relative humidity (ERH). Glucose syrup at a low conversion level requires low moisture content to reach the equilibrium (humectant syrup), while in high-conversion syrup, the moisture absorbed is higher (hygroscopic syrup; O'Brien-Nabors 2001).

6.2.7 Physical Properties

Physical properties such as freezing point (FP), boiling point (BP), and osmotic pressure (OP) have an increased effect on glucose and fructose syrup.

FP depression is relevant to the dextrose and fructose levels in the syrup. However, in order to obtain a soft-structure ice cream (fewer ice crystals), high fructose and dextrose syrups are required. On the other hand, for firmer ice cream structure (more ice crystals), low-DE-value glucose and maltose syrups are used because of their lower freezing point.

As far as temperature is concerned, BP is minimally increased with a reduction in browning, especially with low conversion (syrups of low DE value). However, an important application of high-conversion syrup is in preserves because of its high OP, which affects microorganism contamination.

6.2.8 Applications to Foods and Drinks

Glucose and fructose syrups are applied in many foods and drinks such as confections, preserves, baked goods, frozen products, and beverages because of their functional properties (texture, consistency, and extended shelf life). In both the United States and European Commission countries, these syrups have gained a wide application in the food industry (Nelson 2009).

Recently, new biotechnology processes have developed various glucose syrups with DE values ranging from 20 to 97 and, from them, other products such as gluconic and glutamic acids, xanthan gum, mycoproteins, ethanol, enzymes, nutritive sweeteners (sorbitol and hydrogenated syrups), ascorbic acid (vitamin C), and caramel.

6.2.9 Safety and Consumption

Many claims about HFCS are made, for example, that it is responsible for an increased incidence of obesity, suggesting that it is metabolized differently, but the American Medical Association (AMA) has stated that HFCS does not appear to contribute more to obesity than other caloric sweeteners. A study by Melanson et al. (2007) has found no important differences in insulin, leptin, fasting blood glucose, and ghrelin among the two sweeteners, i.e., HFCS and sucrose. Other researchers conducted studies on abnormally high levels of pure fructose, which may have led to confusion about the relationship of HFCS and obesity. However, no persuasive evidence supports the above claims, because HFCS, like other sweeteners such as glucose and sucrose, contributes calories and, in moderate consumption, does not contribute to obesity. The source of the added sugar in food and beverages should not be a concern, because it is the large amount of added sugars (total calories) that is important. Thus, sugars of various sources in foods and beverages can be enjoyed by consumers as part of their diet and lifestyle if they are consumed in moderation.

An expert panel led by Richard Forshee of the Center for Food, Nutrition, and Agriculture Policy published a report in the August 2007 issue of *Critical Reviews in Food Science and Nutrition*, which announced that "the currently available evidence is insufficient to implicate HFCS per se as a causal factor in the overweight and obesity problem in the United States." Once the combination of fructose and glucose found in HFCS and table sugar (sucrose) in nearly identical composition (50% fructose and 50% sucrose) is absorbed into the body, both (fructose and glucose) appear to be metabolized similarly. Although HFCS and sugar are identical in composition and caloric value, they differ in the chemical bonding of their sugars. The saccharide caloric composition of fructose and glucose and other sugars and components are shown in Table 6.3.

In general, all caloric sweeteners cause an insulin response to a greater (pure glucose) or lesser (fructose) extent. However, both HFCS and sugar (sucrose) cause about the same insulin release in the body, because they contain about the same amounts of glucose and fructose.

A study by Linda M. Zukley at the Rippe Lifestyle Institute on the effect of HFCS on uric acid and body triglycerides of lean women resulted in "no differences in the metabolic effects." The glycemic index (GI), which is based on how much blood sugar increases over 2–3 h after a meal, has a moderate value of 55–60 for HFCS and sugar compared to pure glucose (100) and fructose (20). Moreover, in response to claims of a food allergy effect from HFCS consumption, little immunological connection due to trace proteins that remained after HFCS production has been reported.

Table 6.4 shows some of the most common caloric and/or nutritive sweeteners and their general nutritional characteristics, which are summarized as follows:

- Each has approximately the same composition (fructose and glucose)
- Each offers roughly the same sweetness per gram (4 kcal/g)

Table 6.3 Composition of HFCS[a] versus Other Sugars

Sweeteners	HFCS 42 (%)	HFCS 55 (%)	Sucrose (%)	Inverted Sugar[b] (%)	Honey[c] (%)
Fructose	42	55	50	45	49
Glucose	53	42	50	45	43
Other sugars	5	3	0	10	5
Other substances	0	0	0	0	3

Source: Adapted from Hull, P.: *Glucose Syrups: Technology and Applications*, 2010. Copyright Wiley-VCH Verlag GmbH & Co. KGaA. Reproduced with permission. Reprinted from *Developments in Sweeteners: Applied Science*, Grenby, T. H., Copyright 1987, with permission from Elsevier. O'Brien-Nabors, L., *Alternative Sweeteners*, Marcel Dekker, Inc., New York, 2001.

[a] HFCS of two formulations, 42% and 55% fructose with the balance of glucose and higher sugar.
[b] Hydrolyzed sugar.
[c] Honey: 3% of proteins, amino acids, vitamins, and minerals.

Table 6.4 General Nutritional Characteristics of Common Caloric Sweeteners

Common Sweeteners	Nutritional Characteristics
HFCS	Monosaccharides of fructose and glucose made from corn
Glucose	Simple sugar
Fructose	Simple fruit sugar
Sucrose	Crystalline white sugar made from cane and sugar beets composed of fructose and glucose bonded together
Inverted sugar	Fructose and glucose in liquid form made from sugar hydrolyses
Honey	Liquid sweetener of fructose, glucose, and other higher sugars, vitamins, minerals, and enzymes
Fruit juice concentrate syrups	Liquid sweeteners of concentrated fruit juices

- Each is absorbed by the body at about the same rate
- HFCS is nutritionally the same as sugar (sucrose)

6.3 MAPLE SYRUP

Maple syrup is a sweetener made from the xylem-exuded sap of maple trees, which has been concentrated by water evaporation. It is often used as a sweetener or topping for waffles, pancakes, oatmeal, and French toast and for other uses such as flavoring agents in baking, pastry, dressings, and sauces. Maple syrup consists of 66% sucrose, and it is produced entirely from maple sap called sweet water or "sinzibukwud" (drawn from tree) by the aboriginal people of North America (archeological evidence), who started the maple sugaring operation by boring holes in the trunks of maple trees, inverting wooden spouts in the holes, and collecting sap in wooden buckets hung from the protruding end of each spout (Hopkins 2007). Then, the harvested sap was transported to the base camp and poured into large metal vessels, where it was boiled until it reached the desired consistency. The production has remained basically the same since colonial days. Maple syrup is made by boiling sap (20–50 L) until 1 L of maple syrup is obtained. It was around the American Civil War when cane sugar replaced maple sugar in popularity in the United States. Later, modern maple processing methods, such as reverse osmosis, improvements in tubing, the use of vacuum, new filtering systems, new preheaters, and better storage conditions, were perfected to better product quality (Davenport and Staats 1998).

Canada produces more than 80% of the world's maple syrup, with the vast majority coming from Quebec, Ontario, and Nova Scotia, and exports more than 30,000 tons per year. Maple syrup has also been produced in the United States (e.g., Vermont, Maine, Ohio, and Michigan) and in other countries such as Japan and South Korea on a small scale (USDA 2010).

Maple syrup in Canada is divided into several major grades (with a density of 66 Brix), including extra light (AA), light (A), medium (B), amber (C), and dark (D). Maple syrup in the United States is classified into two major grades (depending on its translucence), including grade A (light amber, medium amber, and dark amber) and grade B (darker). The nutrition profile of maple syrup is shown in Table 6.5 (Elliot 2006).

The basic ingredients of maple syrup consist primarily of sucrose and water with small amounts of other sugars (glucose and fructose), organic acids (malic acid), amino acids, and relevant amounts of minerals (potassium, calcium, zinc, and manganese). Furthermore, it contains a variety of volatile organic compounds (e.g., vanillin, hydroxybutanol, and propionaldehyde), which are primarily responsible for its distinctive maple flavor.

In some instances, off-flavor is found in maple syrup as a result of various causes, including contaminants from the boiling apparatus, fermentation (the sap sitting too long), or "buddy" sap (late in the season's harvest). Maple syrup is similar to sugar calorie-wise, but compared to honey, it has 15 times more calcium.

Table 6.5 Nutrition Profile of Maple Syrup

Nutritional Ingredients, Vitamins, and Minerals	Nutritional Value per 100 g	RDA (%)
Energy	1.1 kJ	
Carbohydrates	67.1 g	
Sugars	595 g	
Dietary fiber	0.0 g	
Fat	0.2 g	
Protein	0.0 g	
Thiamine (vitamin B1)	0.005 mg	0
Riboflavin (vitamin B2)	0.010 mg	1
Niacin (vitamin B3)	0.030 mg	0
Pantothenic acid (vitamin B5)	0.035 mg	1
Vitamin B6	0.002 mg	0
Potassium	204 mg	4
Calcium	67 mg	7
Magnesium	14 mg	4
Zinc	4.2 mg	42
Manganese	3.30 mg	165
Phosphorous	2 mg	0
Iron	1.20 mg	10

Source: United States Department of Agriculture, Maple Syrup: Production, Price and Value, 2000–2004, U.S. and Canadian Provinces, 2010.
Note: RDA: Recommended Daily Allowance.

British culinary expert Delia Smith has described maple syrup as a unique sweetener, different from any other, smooth and silky in texture, and with a taste hinting caramel to coffee (Werner 2011). Maple syrup applications vary and include products, such as apple sauce, baked goods, candies, cakes, breads, and teas, and maple syrup is used as a replacement for other flavoring (e.g., for honey in wine). Finally, labeling laws in the United States prohibit the name "maple" on imitation maple syrups, and Canadian laws allow only syrup with a density of 66 Brix to be marketed as maple syrup.

6.4 RICE SYRUP

Rice syrup is a natural product made enzymatically from white, brown, and/or certified organic rice. According to its application, rice syrup is classified into the following categories, with a DE ranging from 26 to 70:

- White rice syrup (WRS) is a traditional product filtered lightly to produce a translucent syrup with a light butter flavor.
- Clarified white rice syrup (CWRS) is a clear amber syrup produced by further filtration to a low-flavor-profile product for delicately flavored and/or frozen applications.
- BRS is a traditional syrup made from brown rice filtered lightly to an opaque syrup with a buttery flavor.
- Clarified brown rice syrup (CBRS) is a golden-colored product produced by further filtration to a lower flavor profile and preferred in nonfat product formulations as a low-conversion syrup (26 DE).
- Organic rice syrup (ORS) is a syrup made from organically grown white or brown rice with the same clarity and DE ranges as the nonorganic certified ones.

A typical composition of rice syrup and its carbohydrate profile are shown in Tables 6.6 and 6.7. The viscosity of various rice syrups shows different patterns at 60% solids (Brix) and above, and it

Table 6.6 Typical Composition of Various Rice Syrups

Nutritional Ingredients	WRS (%)	CWRS (%)	BRS (%)	CBRS (%)
Water	21.0	20.5	21.0	21.0
Fat	0.6	0.2	1.8	0.4
Protein	0.7	0.8	0.9	1.1
Ash	0.6	0.4	0.7	0.6
Carbohydrate	77.0	78.0	75.5	77.0

Table 6.7 Typical Carbohydrate Profiles (%) for a Variety of Rice Syrups According to Their DE

	Dextrose Equivalent (DE)					
Carbohydrate (%)	26	42	50	55	60	70
Glucose	5	5	18	22	32	56
Maltose	14	44	43	38	29	23
Maltotriose	16	15	8	3	3	3
Maltotetraose	8	4	2	2	1	3
Maltopentaose	17	5	2	2	3	3
Maltohexaose	10	2	2	2	2	2
Others alone	30	25	25	31	30	10

is also diminished at 60°C and above. The viscosity of different DE rice syrups varies with solids or sugar content (Brix) at 25°C, with CWRS of 26 DE obtaining the highest viscosity (10,000 cps) at 80 Brix and CBRS the lowest viscosity (2,000 cps) (Spiler and Mitchell 1992).

The syrup's solid content (%) on the viscosity starts at 60 Brix. On the other hand, the temperature effect on rice syrups of viscosity at 80 Brix starts to be equal at temperatures of 60°C and above. The effect of temperatures from 25°C to 60°C on the viscosity of rice syrups is shown to be much higher for the CWRS than for the CBRS.

6.4.1 Brown Rice Syrup

The rice syrup sweetener known as BRS is derived by cooking brown rice (dried rice sprouts) through enzymatic processing to break down the rice starches to sugars (45% maltose, 3% glucose, and 52% maltotriose; www.brownricesyrup.info). The final product is adjusted according to the desired sweetness and application (e.g., drinks and rice milks).

6.5 FRUIT SYRUP

Fruit syrup is made of concentrated juices (e.g., apple, pear, and pineapple) that are used as sweeteners and extendable quantity agents.

In recent years, the commercial availability of several alternative sweeteners (nutritive and intense) has focused attention on their possible use in food products and beverages as a partial or complete replacement for sugar. One such nutritive sweetener with a caloric value similar to table sugar is marketed under the trade name "fruit source." This product is prepared from fruit juice (grape) and rice syrup, with promising results for the food and drink industry because of its blend of simple and complex carbohydrates. It has a distinctly bland taste and a sweetness of 80 (dry basis) compared to sugar (100). Table 6.8 lists the composition of fruit source syrup in grams per 100 g of

Table 6.8 Composition of Fruit Source Sweetener Component

Nutritional Ingredients	Nutritional Values (g)
Moisture	22
Carbohydrates	75
Protein	1
Ash	1.4
Vitamins and minerals	—
Vitamin C	5.3
Riboflavin	0.02
Niacin	1.1
Thiamin	0.02
Calcium	26.8
Iron	0.7
Phosphorus	40
Potassium	238.0
Sodium	39

liquid (syrup) for macrocomponents (proximate analysis elements) and in milligrams per 100 g for micronutrients (vitamins and minerals; O'Brien-Nabors 2001).

The carbohydrate composition of fruit source sweetener is listed in Table 6.9. It is a potentially very effective all-natural sweetener with all these carbohydrates. The natural humectant qualities and low water activity of fruit source sweetener make it useful for a broad range of applications in confections, baked goods, and sports endurance items. More importantly, it functions as a very effective fat replacer in many applications, so fat-free, low-fat, or reduced-fat versions of the items in a product line can be created using fruit source syrup as a replacement for sweeteners and fats. In addition, it provides the added benefits of improved mixing and baking times as well as often eliminating the need for emulsifiers, antioxidants, and preservatives in food and drink formulations.

Because of the unique properties of this sweetener, it is necessary to mention some of the important benefits, including reduced fat and salt levels, which are appealing to health-conscious consumers, as well as lower cost and extended shelf life. Fruit source syrup made from all-natural ingredients (rice syrup and grape juice) with no artificial flavors and colors has a broad application in bakery products. It is a unique sweetener acting as a binder, an emulsifier, and an antioxidant. On the other hand, it adds a pleasant, sweet taste without the aftertaste found in many other sugar substitutes and artificial sweeteners.

A fat reduction of at least 25% is needed for a reduced- or less-fat claim on packaging, and converting a formulation to fruit source syrup also reduces the salt requirements in food and beverage formulations by 25% or more. At the same time, excellent taste, texture, and mouthfeel can

Table 6.9 Carbohydrate Profile of Fruit Source Sweetener

Carbohydrates	Value (%)
Glucose	25
Fructose	22
Maltose	9
Sucrose	2.5
Complex sugars	41.5

Source: Adapted from Meyers, T., *AIB Techn. Bul.*, 25, 11, 1993.

be achieved, which is unlikely with gum-based, starch-based, and other protein fat replacers. Reductions can range up to as much as 100% in some cases.

The natural humectant qualities of fruit source sweetener together with its low water activity have been found to significantly increase the shelf life of bakery products compared to similar products containing preservatives. Fruit source syrup is an advanced sweetener, because it provides good sweetening power and a better taste, texture, and overall pleasure than most other common sweeteners. Also, it controls water activity very effectively, eliminates the need for stabilizers and preservatives, and needs no refrigeration. The reduced-fat and reduced-salt claims, together with the all-natural and fruit-and-grain-sweetener aspects, are powerful consumer acceptance motivators. As an unrefined sweetener, it contains important nutrients (vitamins and minerals) for a good nutrition and metabolization in contrast to refined sweeteners (sucrose).

6.6 INVERTED SYRUP

Inverted sugar syrup is a mixture of fructose and glucose made by splitting table sugar (sucrose) into its simple monosaccharides through hydrolytic processing. This methodology involves a hydrolysis reaction induced either by heating sucrose solution or through the use of catalysts called surceases (animal origin) and invertases (plant origin). Both are a kind of glycoside hydrolase enzyme. In addition, some acids (e.g., lemon juice) are added to accelerate the conversion of sucrose to inverted syrup. "Inverted" is a word that comes from the polarimetric methodology used to measure the concentration of sugar syrup by optical rotation to the right of polarized light. This rotation is reduced until the direction has changed (inverted) from right to left, meaning the sucrose solution is fully converted, according to the following reactions by Equation 6.1:

$$\begin{array}{lclclcl} C_{12}H_{22}O_{11} & + & H_2O & \rightarrow & C_6H_{12}O_6 & + & C_6H_{12}O_6 \\ \textit{Sucrose} & + & \textit{water} & \rightarrow & \textit{Glucose} & + & \textit{fructose} \\ \text{(spec. rotation} + 66.5^\circ) & + & \text{water} & \rightarrow & \text{(spec. rotation} + 52.7^\circ) & + & \text{(spec. rotation} - 92^\circ) \end{array} \quad (6.1)$$

This type of sucrose hydrolysis (net rotation of 66.5°, which is converted to −39°) yields 85% glucose and fructose when the temperature is maintained at 50°C–60°C. The procedure to produce inverted syrup is simply adding 1 g of citric acid (ascorbic acid) per kilogram of sugar or lemon juice at approximately 10 mL per kilogram of sugar. The mixture is boiled for 20 min until enough sucrose is converted to inverted syrup (Aider et al. 2007).

Another way, aside from the use of acids and/or enzymes, of producing inverted syrup is by thermal means (simmer 2:1 parts of sucrose to water for 5 min). Possible crystallization can be restored back to liquid form by light heating applications. Examples of products using inverted syrups include jams, granola bars, and candies.

6.7 MALT (BARLEY) SYRUP

Barley or malt syrup is a sweetener made from sprouted barley through enzymes that convert the staves to liquid syrup, which contains 65% maltose, 30% other complex syrups, and 3% protein. It is viscous and dark brown in color and has a distinctive flavor with approximately half the sweetness of table sugar.

Standard malt syrup is also available in diastolic and nondiastolic forms. The diastolic form contains natural active barley enzymes and is used primarily by bakers for color and other functionalities. The nondiastolic malt syrup does not contain active enzymes and is used as a flavoring agent or a humectant and to improve color and sweetness (Edney and Izydorczyk 2003; Bamforth 1998).

6.8 LIQUID CORN SWEETENERS

6.8.1 Corn Syrup

Corn syrup, also known as glucose syrup, is a complex syrup made from maize starch through enzymatic processing, which produces a sweet sugar compound containing varying amounts of glucose, maltose, and other oligosaccharides. It can be used to soften texture, add volume, prevent crystallization, and enhance the flavor of foods. Finally, it is distinct from HFCS. Glucose syrup comes from liquid starch hydrolysis of monosaccharides, disaccharides, and higher saccharides made from common starches such as wheat, rice, and potatoes (Levenson and Hartel 2005).

In the production of corn syrup, 2.3 L of corn yields approximately 950 g of starch to produce 1 kg of glucose syrup. Commercially, corn syrup is mainly produced by enzyme α-amylase addition to a mixture of cornstarch and water. This enzyme breaks down the cornstarch to oligosaccharides, which are broken down to glucose by the glucoamylase enzyme (γ-amylase). Then, the glucose is transformed to fructose using the enzyme D-xylose isomerase. Corn syrup's viscosity and sweetness depends on the hydrolysis level, and its grade is rated according to DE. In commercial foods, corn syrup is used as a thickener, sweetener, and humectant (maintains freshness) (Kearsley and Dziedzic 1995).

Due to the increase in the price of cane sugar, corn syrup and HFCS are less expensive alternatives that are used to produce foods, candied fruit, and beverages. Glucose syrup usage was later expanded to a variant of HFCS in which other enzymes are used to convert some glucose to fructose.

6.8.2 High-Fructose Corn Syrup

HFCS is made from cornstarch that does not contain artificial or synthetic ingredients. HFCS is nearly identical in composition to table sugar, containing approximately 50% glucose and 50% fructose, and has the same number of calories. USDA data show the U.S. consumption of HFCS is lately on the decline even though obesity and diabetes continue to rise as in other continents (Europe, South America, and Australia).

Glucose syrup mainly consists of glucose and is used as a nonsweet thickener. HFCS is predominately made of fructose and is used as a sweetener (Tsai 2005).

HFCS plays a key role in the integrity and functionality of the following products:

- *Baked goods.* HFCS gives a pleasing brown crust to breads and cakes, contributes to the fermentation processing of raised products, reduces sugar crystallization during baking for soft, moist textures, and enhances the flavor of fruit fillings.
- *Yogurt.* HFCS provides fermentable sugars that enhance fruit and spice flavors, control moisture to prevent separation, and regulate tartness.
- *Canned and frozen fruits.* HFCS protects the firm texture of canned fruits and reduces freezer burn in frozen fruits.
- *Sauces, salad dressings, ketchups, and other condiments.* HFCS enhances flavor and balances variables such as tartness and acidity of tomatoes.
- *Beverages.* HFCS provides greater stability in acidic carbonated sodas than sucrose, so the flavor remains consistent and stable over the entire shelf life of the product.

The absorption and metabolism of HFCS is similar to that of sugar (sucrose) as indicated by Schorin (2005). HFCS was introduced into the food supply of North America in 1968, but its popularity as a sweetener grew after the U.S. Food and Drug Administration (FDA)'s decision in 1983 that HFCS is generally recognized as safe. The first part of the decision explains the composition, consumption, and metabolism, and the second part explores the healthy impact of HFCS consumption. Schorin (2006) concluded that it is difficult to identify a plausible physiological explanation for how approximately equal amounts of fructose and glucose should have differential effects when chemically bonded (sucrose) or not bonded (HFCS). Therefore, the current evidence does not support claims of a specific unique effect of HFCS on health effects.

The prevalence of the overweight/obesity problem in the U.S. population has increased with the consumption of HFCS per capita since the early 1980s (Hein et al. 2005). Current research published in scientific journals does not support a cause–effect relationship between HFCS consumption and overweight/obesity although some public health reports hypothesized that these two trends are directly related (Hein et al. 2005).

6.8.3 Glucose Syrup

Glucose syrup is produced worldwide from maize or corn under various names, including corn syrup, glucose syrup, high-fructose glucose syrup (HFGS), HFCS, and high-fructose (HF) syrups. When one sugar in the glucose syrup is predominant, e.g., dextrose or maltose, the syrup is referred to as dextrose or maltose syrup.

The glucose industry has been trying to find a product similar to and as cheap as sucrose until it found fructose, which is 1.2–1.7 times sweeter than sucrose. Glucose may be produced by one of the following methods (Hull 2010):

- Acid or enzyme hydrolysis of starch slurry
- Fructose separation from inverted sucrose
- Heating dextrose with alkali

Table 6.10 shows the differences in sugars of glucose syrups produced either by acid or acid–enzyme hydrolysis but with the same DE value equal to 42. These differences are a result of a higher dextrose proportion in the process using only acid. It has an impact on the dark color of applied products, e.g., darker toffees, and in the viscosity of syrup because of its higher sugar content. Some functional properties of the glucose syrups are given as follows.

- *Body agent.* Higher sugars increase the viscosity of glucose syrup, which improved the mouthfeel of drinks by giving them body, especially to those using high-intensity sweetness.
- *Browning.* The browning reaction, well known as the Maillard reaction, occurs when reduced sugars of glucose syrups are heated in the presence of proteins.
- *Cohesiveness.* Glucose syrup is viscous and sticky because of its low DE and higher sugar content; therefore, it is ideal as a binder for various products such as cereal bars.

Table 6.10 Sugar Spectra of Glucose Syrups Produced by Different Methods of Hydrolysis

Process	Dextrose (%)	Maltose (%)	Maltotriose (%)	Higher Sugar (%)
Acid	19.0	14.0	11.0	56.0
Acid + enzyme	6.0	44.0	13.0	36.0

Source: Hull, P., *Glucose Syrups: Technology and Applications*, Wiley-Blackwell Publ., West Sussex, U.K., 2010; Emerton, V., and E. Choi, *Essential Guide to Food Additives*, RSC Publishing, Cambridge, U.K., 2008; McCance, R. A., and E. M. Widdowson, *The Composition of Foods*, Royal Society of Chemistry, Cambridge, U.K., 1991; Kirk, R. S., R. Sawyer, and H. Egan, *Pearson's Composition and Analysis of Foods*, Longman, London, 1991. With permission.

- *Fermentability.* Glucose syrups such as 95 DE HFGS and maltose syrups, because of their high content of easily fermented sugars (dextrose, fructose, and maltose), are excellent energy sources for microorganisms (yeast and bacteria) to utilize in fermentation processes.
- *Flavor enhancement.* Glucose syrup can enhance or change fruity, cereal, and other flavors due to its molecular structure, having a joint occupancy in the taste sensations of sweet, sour, salt, and bitter.
- *Foam stabilizer.* Glucose syrups of low DE are viscous and, when added to foamed products, will make the walls of the bubbles thicker and stronger, resulting in more stable foam in aerated products such as marshmallows and ice cream.
- *Freezing point.* FP depression depends on the concentration of molecules in the liquid (glucose syrup). This ability is used in the formulation of frozen desserts (ice cream) and is advantageous to product quality. A similar term is the freezing point factor (FPF), which is defined as the molecular weight of sucrose divided by the molecular weight of a sweetener.
- *Humectancy.* Medium- and high-DE glucose syrups are good moisture retainers, and therefore, they can be used in baking to prevent products from drying out, crystallizing, and hard icing.
- *Nutritive value.* Glucose syrups have a similar caloric value to sugar, because they are composed of sugars (regardless of DE) that, on a dry basis, give energy of 4 kg/g (17 kJ/g). Because glucose syrups are soluble and easily assimilated, they are a good source of nutritive solids and, therefore, ideal energy providers to human bodies.
- *Osmotic pressure.* In general, the osmotic pressure of glucose syrup is related to its molecular weight. Thus, the lower the molecular weight, the greater the osmotic pressure.
- *Preservative value.* Glucose syrup exerts an osmotic pull on dilute solutions because of sugar concentration. As a result, the syrup pulls out the liquid and dehydrates the microorganism cells. This ability makes glucose syrup an effective preserving agent in products where it is applied.
- *Crystallization.* Glucose syrup of low DE (35 and 42) contains high amounts of higher sugars, which, because of its increased viscosity, will slow down the crystallization process and prevent the formation of large crystals (Ostwald ripening) by creating an insulation around individual sucrose crystals. Thus, glucose syrup is useful in products where smooth texture is required, e.g., fondant formulation with a smooth, white, opaque luster, popsicles, and other similar frozen desserts.
- *Viscosity.* Glucose syrup varies in viscosity with DE value, solid content, and temperature. The flexibility of viscosity in glucose syrup makes it more manageable in terms of processing, handling, storage, and application. The viscosity of glucose syrup plays an important role in the texture and mouthfeel of the end products. For example, when a low DE is used, the high viscosity will affect the end product, giving a bland taste because this type of syrup is unable to carry the flavor to tastebud receptors. Table 6.11 shows the tremendous differences in viscosity of glucose syrup as affected by DE values, solids, and temperatures.

Table 6.11 Effects of Temperature and Solids on the Viscosity of Glucose Syrups

	Viscosity (cps)		
Temperature (°C)	**42 DE (81% Solids)**	**63 DE (82% Solids)**	**95 DE (75% Solids)**
20	250,000	90,000	1,000
40	18,000	7,000	200
80	550	250	20

Source: Adapted from Hull, P.: *Glucose Syrups: Technology and Applications*, 2010. Copyright Wiley-VCH Verlag GmbH & Co. KGaA. Reproduced with permission. Advanced Instruments, Inc., *The Physical Chemistry, Theory, and Technology of Freezing Point Determinations*, Advanced Instruments Inc., Norwood, MA, 1971. Aurand, L. W. et al., *Food Composition and Analysis*, Van Nostrand Reinhold, New York, 1987. Balston, J. N., and B. E. Talbot, *A Guide to Filter Paper and Cellulose Powder Chromatography*, H. Reeve Angel & Co., London, 1952. Browne, C. A., and F. W. Zerban: *Physical and Chemical Methods of Sugar Analysis*. 1955. Copyright Wiley-VCH Verlag GmbH & Co. KGaA. Reproduced with permission. Emerton, V., and E. Choi, *Essential Guide to Food Additives*, RSC Publishing, Cambridge, U.K., 2008. McCance, R. A., and E. M. Widdowson, *The Composition of Foods*, Royal Society of Chemistry, Cambridge, U.K., 1991. Kirk, R. S. et al., *Pearson's Composition and Analysis of Foods*, Longman, London, 1991.

Table 6.12 Typical Sweetness of Various Glucose Syrups Compared to Sucrose, Dextrose, and Fructose

Sweetener	Sweetness Value
Sucrose	100
Fructose	150
Dextrose	80
42 DE glucose syrup	50
63 DE glucose syrup	70
HFGS	95

Source: Adapted from Hull, P.: *Glucose Syrups: Technology and Applications*, 2010. Copyright Wiley-VCH Verlag GmbH & Co. KGaA. Reproduced with permission. Matz, S. A., *The Chemistry and Technology of Cereals as Food and Feed*, Van Nostrand Reinhold/AVI, New York, 1991. Harris N., et al., *A Formulary of Candy Products*, Chemical Publishing Co., Revere, MA, 1991. Grenby, T. H.: *Developments in SWEETENERs: Applied Science*. 1987. Copyright Wiley-VCH Verlag GmbH & Co. KGaA. Reproduced with permission. O'Brien-Nabors, L., *Alternative Sweeteners*, Marcel Dekker, Inc., New York, 2001.

- *Sweetness*. The sweetness of glucose syrup is very subjective, because there is no chemical test and/or instrument to measure it. However, it is dependent on various factors such as concentration (solids), pH and acidity, temperature, and combined sweetness and ingredients. Usually, the sweetness of glucose syrup increases as the DE value increases but always less than sucrose sweetness. However, by blending glucose syrup of different DEs, the sweetness of the glucose syrup blend product can be substantially increased to make it a versatile ingredient. Typical sweetness levels of various glucose syrup compared to sucrose, dextrose, and fructose are shown in Table 6.12. Thus, by blending, for example, fructose syrup with 63 DE glucose syrup to a certain degree, the sweetness profile of the blend can be close to sucrose. However, sweetness is perceived in the mouth differently for every sweetener or a blend of sweeteners. For example, sucrose has a lower perceived sweetness than fructose and a little higher than dextrose, but it lasts for a longer time.

6.8.4 Glucose Syrup Applications

Because glucose syrup is very versatile, it is difficult to mention every application, but in this section an attempt will be made to indicate typical applications.

6.8.4.1 Glucose Syrup (42 DE)

This clear and viscous syrup, containing dextrose (19%), maltose (14%), maltotriose (11%), and higher sugars (56%), has many applications in the confectionary industry for products such as caramels and toffees, fudge and fondants, and glazes.

6.8.4.2 Lower than 42 DE Glucose Syrup

These types of syrup are characterized by their high viscosity resulting from the high content of higher sugars, as shown in Table 6.13 (Hull 2010). The 35 DE and 28 DE syrups are produced by acid and enzyme–enzyme hydrolysis, respectively, and because they contain less dextrose than the 42 DE syrups and are therefore less humectant, they are useful in confections where moisture absorbance is a problem.

These lower DE syrups are also useful in the ham processing industry to help in water retention and hence decrease cost (increased weight) and enhance red color-reducing nitrates and nitrites in the meat curing.

Table 6.13 Typical Sugar Analysis of 42 DE and Lower Syrups

Sugars	28 DE Syrup (%)	35 DE Syrup (%)	42 DE Syrup (%)
Dextrose	2	11	14
Maltose	10	12	14
Maltotriose	16	14	11
Higher sugars	72	68	56

Source: Adapted from Hull, P.: *Glucose Syrups: Technology and Applications*, 2010. Copyright Wiley-VCH Verlag GmbH & Co. KGaA. Reproduced with permission. Grenby, T. H.: *Developments in SWEETENERs: Applied Science*. 1987. Copyright Wiley-VCH Verlag GmbH & Co. KGaA. Reproduced with permission. O'Brien-Nabors, L., *Alternative Sweeteners*, Marcel Dekker, Inc., New York, 2001.

These low DE and viscous syrups also have certain other properties in the food industry because of their higher sugar content (responsible for high viscosity), resulting in excellent spray- and cospray-dried products (stable droplet formation) with a good application in coffee whiteners, which usually contain 57% 28 DE glucose syrup, 34% vegetable oil, and other stabilizers and emulsifiers (Hull 2010).

6.8.4.3 Glucose Syrup (63 DE)

Glucose syrup is produced by the acid–enzyme treatment of 42 DE syrup and contains 34% dextrose, 33% maltose, 10% maltotriose, and 23% higher sugars.

The 63 DE syrups are characterized by their balanced sugar spectrum and medium viscosity, which make them an ideal ingredient in a wide range of products, such as soft sweetness dextrose (sugar) confectionery and bakery products, sauces and dressings, jams, brewing, and soft drinks, as described below (Hull 2010):

- *Confections.* A 63 DE syrup offers an acceptable viscosity with a reasonable sweetness and minimal risk of graining to confectionary products; for example, a softer check will result when it is used in pectin jellies.
- *Bakery products.* A typical inclusion rate (5% dry solids) will contribute to sweetness and improvement of the brown crust color of bakery products as well as humectants in cakes.
- *Fermentation.* A 63 DE glucose syrup, because of its balanced combination of fermentable and nonfermentable sugars, is ideal for use in browning and other fermented drinks (Russian Kvass). In addition, it can be used in liqueur (distilled spirits) manufacturing, providing sufficient sweetness to mask the hardness of the spirits, enhance the flavor, and increase the viscosity. Generally, a 63 DE syrup is similar to brewer's yeast in terms of fermentability, offering 70% of its sugars to be fermented by the yeast while 30% are unfermented sugars, which provide body and mouthfeel to the end product (Hull 2010).
- *Jams.* The use of 63 DE syrup in jams has the following advantages: (1) contributes to microbiological stability because of its molecular weight (high osmotic pressure); (2) helps reduce crystallization; (3) reduces apparent sweetness, allowing fruit flavor to come through; and (4) enhances the sheen of jams, making them more appealing products.
- *Soft drinks.* Glucose 63 DE syrup, which is less sweet than sucrose, plays an important role in the soft drink industry, especially when used in combination with high-intensity sweeteners. This type of 63 DE syrup masks the usual harsh flavor associated with high-intensity sweeteners, enhances body and mouthfeel, and gives the overall sweetness a more round character similar to sucrose.

6.8.4.4 Glucose Syrup (95 DE)

A 95 DE glucose syrup mainly contains dextrose (94%) with maltose (4%) and a minimum amount of maltotriose (1%) and higher sugars (1%). It can be spray dried to produce a free-flowing powder that is interchangeable with dextrose. There is an advantage of using 95 DE syrup when large volumes are involved, especially in industrial fermentations where processes need to be automated. Typical applications include the following:

- *Baked goods.* A 95 DE syrup can be used in bread making and other fermented dough as an energy source for yeast and, with the high reducing sugar present, for crust browning and a distinctive fresh aroma.
- *Confections.* A 95 DE syrup is avoided in mixes with chocolate because of the water present, but in spray-dried applications, it has no problem with incompatibility.
- *Fermented products.* A 95 DE syrup is suitable for the production of fermented products such as citric acid, lactic acid, and xanthan gum.
- *Other uses.* A 95 DE syrup can also be used for the production of antibiotics as an ideal carbohydrate source and for growing meat-free mycoprotein products used in vegetarian meat substitutes.

6.8.4.5 High-Fructose Glucose Syrup (HFGS)

HFGS includes syrup of 42%, 55%, and 80% fructose with 71%, 77%, and 79% solids, respectively.

Only HFGS containing 42% and 80% fructose is available within Europe because of European Union (EU) regulations. The 55% fructose is available for the rest of the world where, by blending, it is possible to have any fructose level.

The sugar profile of HFGS is very important, and sweetness levels compared to sucrose are shown in Table 6.14.

Although sweetness is a common aspect of HFGS and sucrose, they are completely different. HFGS is composed of dextrose and fructose (both reduced sugars), which can react with protein to give a brown color to products. In addition, HFGS more effectively lowers the freezing point of a formulation because of its lower molecular weight compared to sucrose and acts as an excellent preservative because of its higher osmotic pressure. In general, some of the most interesting applications of HFGS in food are the following:

- *Bakery products.* HFGS is a good source of easily fermentable sugars for the yeast in breads and other yeast-raised bakery goods.
- *Confections.* HFGS has limited use in confections because of its high hygroscopic effect.
- *Fruit preparation (fillings).* The use of HFGS in fruit preparations (fillings) offers the following advantages: (1) increases the sweetness of fructose in lower temperatures; (2) increases osmotic

Table 6.14 Sweetness Values of HFGS Compared to Sugar

Sweeteners	Sweetness
Sucrose	100
Fructose	150
HFGS 42 DE	95
HFGS 55 DE	105
80% fructose syrup	130

Source: Adapted from Hull, P.: *Glucose Syrups: Technology and Applications*, 2010. Copyright Wiley-VCH Verlag GmbH & Co. KGaA. Reproduced with permission. Matz, S. A., *The Chemistry and Technology of Cereals as Food and Feed*, Van Nostrand Reinhold/AVI, New York, 1991. Harris N., et al., *A Formulary of Candy Products*, Chemical Publishing Co., Revere, MA, 1991. Grenby, T. H.: *Developments in SWEETENERs: Applied Science*. 1987. Copyright Wiley-VCH Verlag GmbH & Co. KGaA. Reproduced with permission. O'Brien-Nabors, L., *Alternative Sweeteners*, Marcel Dekker, Inc., New York, 2001.

pressure in products, improving the prevention of microbial growth; (3) reduces the freezing point of a mix, allowing end products to remain soft; and (4) enhances the fruity flavor of products.

- *Soft drinks.* HFGS 42 and 55 have a major use in soft drinks, with a 100% sucrose replacement achievable. The reason is the fruit flavor enhancement and the reduction of soft drink calories.

6.8.4.6 Low-DE (<20) Glucose Syrups or Maltodextrins

Low-glucose syrups (DE < 20) are usually available as spray-dried, free-flowing, white powders with a little sweetness and a bland flavor. They are normally sold to DE specifications of 1, 5, 10, 15, and 18.

Maltodextrins are made from very low DE syrups in a process similar to glucose, with subtle variations in functionality depending on the type of starch (corn, potato, rice, or tapioca) used in the production and method of hydrolysis (acid, acid–enzyme, or enzyme–enzyme). Their functional properties make them suitable for many different applications and include low dextrose content, low browning, lack of sweetness, low hydroscopic quality, good moisture control, and a makeup of complex, soluble, and nutritional carbohydrates. The above properties of maltodextrins make them applicable to various products:

- *Confectionery agent.* The high viscosity of maltodextrins can be used to change the texture of confections, to act as a foam stabilizer, to control crystallization, to increase the chewiness of soft confections, or to act as a binder in granola bars and in tableting.
- *Balking agent.* Maltodextrins are good balking agents in dry formulations because of their bland taste and solubility. They are blended with high-intensity artificial sweeteners, acting as a balking agent that effectively dilutes the high-intensity sweeteners to an acceptable level.
- *Calorie reduction agent.* Maltodextrins can be used as a partial (30%–40%) fat replacer, because they are bland and act as emulsifiers and stabilizers. By replacing part of the fat in a product, they reduce the consumption of calories, resulting in more healthy products.
- *Flavor carrier.* Maltodextrins are ideal carriers for flavors because of their bland taste and large surface area. Flavors can be lost during the drying process if they are added to the maltodextrin syrup prior to spray drying, but maltodextrins will aid the encapsulation because of their high viscosity, which makes them good film formers.

6.9 SORGHUM SYRUP

Sorghum syrup is a natural sweetener made from juice squeezed from stalks of sweet sorghum through milling, extraction of juice, and an evaporation process.

Lately, grain sorghum has also been enzymatically converted to an all-natural and nutritious grain-based sweetener because of its high protein, amino acid, vitamin, and mineral content. It can be substituted for malted barley extract syrup, which has wide applications in food products, especially in baked goods as a sweetener and/or browning agent.

Sorghum syrup is used in the food and beverage industry as sorghum molasses usually blended with sugar cane molasses and other types of syrup.

Sorghum syrup, because it is gluten-free, offers a solution to the challenge of many products formulated to be gluten-free, such as cereals, crackers, snack foods, and other food products.

It promotes browning and flavor enhancement like malt extract syrup, because it contains the same amounts of reducing sugars and amino acids. In addition, it can be substituted 1:1 with other malt extract syrup, because it has other similar characteristics, including moderate- to long-lasting sweetness, moisture retention, and medium viscosity.

6.10 CANE OR BEET SYRUP

Cane or beet syrup is an unrefined, thick, dark syrup that is directly produced from sugar cane or sugar beets by cooking, squeezing, pressing, and concentrating the resulting juice until it has the desired consistency. Beet syrup, called "zuckerrüben sirup" in Germany, has been used particularly in the Rhine area as a spread for sandwiches and as a sweetener in sauces, cakes, and desserts (Mudoga et al. 2008). This syrup can be hydrolyzed and converted to HF syrup similar to HFCS when it has DE > 30.

Current studies reveal that sugar syrup from glyphosate-resistant sugar beets has the same nutritional value [genetically modified organism (GMO)] as sugar from conventional sugar beets (non-GMO). By 2010, in the United States, 95% of sugar beet syrups came from glyphosate-resistant sugar beets. The weeds can be controlled using glyphosate herbicide without hurting the sugar beet crops, which allows the achievement of higher yields and more recoverable sugar than conventional crops (Gyura et al. 2002; Gyura et al. 2005).

6.11 GOLDEN SYRUP

Golden syrup is a thick, amber-colored form of inverted sugar cane syrup produced by the acid treatment of a sugar solution and used as substitute for corn syrup. It is a combination of by-products at the crystallization stage of cane sugar refining like similar products from the beet sugar refining process, during which the disaccharide sucrose is broken down to glucose and fructose by acid hydrolysis or by enzymes (invertase). The high fructose content gives golden syrup a sweeter taste than sugar and less crystallization than sucrose syrup (Clarke 2003).

Golden syrup or syrup, defined as uncrystallized syrup, is the syrup produced from sugar cane refining. It is a dark syrup with a distinctive strong flavor and slightly bitter taste. The term "theriac" has also been referred to as treacle, which is composed of many ingredients and was used as an antidote for poisons (snake bites) and other ailments. "Treacle" comes from the Latinization of the Greek *theriake*, which is derived from *therion* (wild animal).

Treacle is made when raw sugars are processed by affination, a treatment that causes the dark-colored washings to be boiled until sugar precipitates, forming a mash, which is centrifuged, yielding a brown sugar and a fluid by-product known as treacle.

6.12 MOLASSES SYRUP

Molasses is a dark and viscous by-product of the processing of sugar cane and/or sugar beets into the table sugar. Its name comes from the Portuguese word *melaço* and/or the Latin world *mel*, which means honey (Hogan 2008). Factors such as the sugar cane maturity and the method of extraction affect the molasses's quality, which is different between beet and cane molasses.

There are sulfured and unsulfured (SO_2 used as a preservative) molasses and three grades: mild or first molasses, second molasses, and blackstrap molasses are commercially known in the food industry. The first molasses comes from the first boiling toward the production of sugar crystals and contains the highest sugar content. The second molasses comes from the second boiling of sugar extraction and has a slightly bitter taste. The blackstrap molasses comes from the third boiling, which contains the highest levels of sugars. The last one contains significant amounts of vitamins and minerals (calcium, magnesium, potassium, and iron), which makes it a significant source of nutrients.

Molasses coming from sugar beets contains predominantly sucrose (50%) and lesser amounts of glucose and fructose. It also contains salts (calcium, potassium, and chloride), limited amounts of biotin (vitamin H or B7), and other compounds (betaine, trisaccharide raffinose) as a result of the concentration process and/or other reasons.

Some of the many substitutions made for molasses with a great degree of success are the following: black treacle, dark corn syrup, maple syrup, and honey.

In addition, because of its special properties, molasses has wide applications in the following food and nonfood products (Olbrich 1963):

- Food and consumption derivatives
  - As the main base for fermentation purposes (e.g., rum)
  - As an additive to dark brewed drinks (e.g., dark ales)
  - As a flavoring agent to special tobacco products (e.g., those smoked in a narghile or hookah, such as shisha)
  - As an iron supplement
  - As an additive in livestock grains
  - As ground bait in fishing
- Chemical and industrial conditions
  - As a carbon source for in situ remediation of chlorinated hydrocarbons
  - As an alternative source base (fermentation) for ethanol production (used as fuel in motor vehicles)
  - As a chelating agent for removing rust in a ratio of 1:10 molasses to water
  - As a component of mortar for brickwork
  - As a microbial active agent in the soil of most plants
  - As a flowering enhancer—does not apply to hydroponic gardening because molasses contains sucrose, which cannot be used by the plant for cellular production like other sugar boosters (xylose, ribose, and lyxose), which deliver usable energy to plants

6.13 BEVERAGE SYRUP

Beverages and other drinks usually use various sweeteners to offset the tartness of drink recipes. Syrups as liquids are more easily used in mixed drinks and beverages than granulated sugar. However, syrups for beverages are varied according to their usage and can be classified as follows:

- Simple syrup is a sugar-water sweetener made by dissolving granulated sugar in hot water (1:1 or 1:2) by stirring; it is also known as sugar syrup, simple sugar syrup, and bar syrup.
- Flavored syrup is a type of syrup that is made by adding flavors to simple syrup, such as cinnamon, vanilla, and chocolate.
- Gomme syrup is like simple syrup but it has added gum arabic, which prevents crystallization while adding a smooth consistency.
- Agave syrup is a sweetener found in Mexico and South Africa—made from various species of the agave plant such as the blue, green, grey, salmiana, or thorny agave—that is sweeter than sugar (1.5 times) and honey but less viscous (Kamozawa and Talbot 2009).

Agave syrup consists mainly of fructose (56%) and glucose (20%; Johannes 2009). Its GI rating is comparable to fructose and much lower than sucrose. Thus, large consumption could trigger fructose malabsorption, decreased glucose tolerance, and hyperinsulinemia. Its physical properties allow it to be used in cold beverages (ice teas) and as a vegan alternative to honey in cooking. Agave syrup is found in many color forms such as light, amber, dark, and raw (Eskander et al. 2010). The light one has a mild to neutral flavor and is used in delicate foods and drinks, whereas both the amber and dark forms have medium to strong caramel flavor and are used as a topping to pancakes or as an enhancer of flavor to meat and seafood dishes. Agave is produced from the *Agave tequilana* plant by juice extraction (Hocman 2009), filtration, heating for sugar hydrolyzation into simple fructose and glucose units, and concentration to a syrup liquid thinner than honey (Catalano 2008).

6.14 OTHER TYPES OF SYRUP

6.14.1 Raisin Juice Syrup

Raisin juice syrup (RJS) is a natural sweetener produced by the extraction of sugars from raisin by-products, employing processes that do not alter the nutritional properties of raisin components. The U.S. Food and Drink Directory defines it as a product derived through the concentration of raisin extract after distancing of quantities of existing acids (Pilando and Wrostland 1992). Other studies define it as the physical product derived though the steeping of raisins in hot water (various stages) and concentrating (vacuum) to a liquid product of approximately 70 Brix or soluble solids (70%).

Physicochemical analyses revealed that there is a slight variation in the composition of raisin syrup, depending on the variety and origin of the raisins. A proximate analysis of raisin syrup composition is given in Table 6.15. The main physical characteristics are pH = 4.5, specific weight = 1.4, and density = 41–42 Be. Raisin syrup is a 100% physical product with a number of advantages:

- Offers physical sweetness to products
- Gives a fruity flavor and enhances existing flavor in products
- Increases freshness because of its high quantity of reduced sugars
- Contributes to the physical coloring as it adds a caramel to dark brown color to special products (e.g., whole meal bread), enhancing their wholesome and attractive appearance
- Acts as a natural preservative, inhibiting microbial growth because of its propionic and other organic acids
- Substitutes for fat in products, giving a similar texture and taste feeling to products as real fat

Successful applications of raisin syrup found in many foods and beverages include various types of muffins, brownies, cookies, and chips, which, besides the addition of natural sweetness, involves other attractive quality attributes.

6.14.1.1 Fat Substitute

The demand for food products with lower fat or light products has continuously increased since the last century. Raisin syrup as a fat substitute gives body to products and contributes to their freshness, preservation, and overall sensory attributes. For example, bakers and other food producers can replace a great quantity of butter and margarines with raisin syrup and succeed in the reduction of fat.

Table 6.15 Raisin Syrup Composition per 100 g of Product

Nutritional Ingredients	Value (g)
Moisture	21.8
Carbohydrates	72
Fat	0.3
Protein	2.4
Ash	1.3
Crude fiber	0.05

6.14.1.2 Sugar and Salt Substitute

Raisin syrup can successfully substitute for sugar as a basic ingredient in the development of many ethnic cuisine recipes. One of the latest advances is the use of raisin syrup in ethnic (Indian, Chinese, and Malaysian) products, especially in sauces and dressings, due to its sweetness, flavor, and intensifying characteristics on spicy ingredients and citrus fruits.

A raisin syrup level of 5%–10% in ethnic cuisine products is not distinguishable in flavor and sweetness, but it is enough to intensify the flavor and improve the overall quality of the product (e.g., Chinese spring rolls). The use of large (>20%) quantities of raisin syrup counterbalances the sour taste of vinegar, contributing to the cohesiveness and consistency of sauces, dressings, and sweet or sour fillings while prolonging their preservation.

6.14.1.3 Food Additives

The expansion of the food market and changes in the lifestyle and nutritional habits of consumers have led to the use of additional substances for the improvement of the production and quality of food products. In this case, raisin syrup may be used as an additive to improve the quality of a product. Raisin syrup can be used successfully as an ingredient to enhance the color, flavor, texture, and freshness of products, such as multigrain breads, cakes, snacks, brownies, sauces, dressings, and ice cream.

6.14.2 Must Syrup

Must syrup is a product concentrated by heat from grape juice. Must syrup production follows a traditional methodology in which grape juice is boiled with wooden ash, sugar, or maribor powder to neutralize the acids. After boiling, the partially concentrated syrup liquid is filtered and then boiled again to desired viscous, dark syrup. It usually is used for flavoring, sweetening, and coloring food products (for particular reasons) such as sauces, dressings, and bakery or dairy products with distinguished flavors.

Must syrup is a natural liquid food without preservatives or other additives. Its nutritional value is shown in Table 6.16. Calcium, phosphorous, and iron together with other elements, such as anthocyanins, tannins, salts, and acids, make it a valuable ingredient for many food applications. In addition, must syrup is considered a medicine for diseases like anemia and depression.

6.14.3 Date Syrup

Dates contain up to 70% total sugar, from which 50% is inverted sugar, and have a water activity of 0.6 with a pH of 5.5. Dates are processed into a concentrated liquid form called date syrup, which is convenient for industrial uses. This syrupy date liquid brings its sweetness and flavor to baked goods and to all types of liquid food and beverage applications (Al-Farsi et al. 2007).

Table 6.16 Nutritional Ingredients of Must Syrup per 100 g

Nutritional Ingredients	Value
Carbohydrates	81
Fat	0.4
Proteins	0.9
Calcium	74
Phosphorous	40
Iron	1.2

Source: Hockenhull, D. J. D., *Progress in Industrial Microbiology*, J. & A. Churchill, Ltd., London, 1968. With permission.

To produce date syrup, all soluble solids are extracted from the fruit. Then, it is filtered and concentrated to 70 Brix with a pH of 4.8–5.0. The natural concentrated date syrup with no preservatives added has up to 1 year of shelf life and blends easily into batters for sweet breads, cakes, muffins, cookies, and other bakery products, providing sweetness and moisture retention, which helps retard staling.

Date syrup liquid offers opportunities for the food industry to create unique liquid products, e.g., juice blends, yogurt drinks, and other beverages. In addition, it can be used in spirits, wines, and vinegars through the fermentation of the sweet syrup liquid.

6.14.4 Birch Syrup

Birch syrup is a sweetener made from the sap (1%–2% sugar) of birch trees, found mostly in Alaska, Russia, Ukraine, and Scandinavia, resembling maple syrup (Bauman 2005). The product contains approximately 42%–52% fructose, 44% glucose, and other sugars (e.g., sucrose and galactose), and it is used like maple syrup as flavoring for pancakes, candies, sauces, dressings, and for beers, wines, and beverages.

6.14.5 Palm Syrup

Palm syrup or palm honey is a sweet syrup that is produced in the Canary Islands and on the coast of South Africa from the sap of *Phoenix canariensis* trees (or *Jubaea chilensis* in Chile). The process involves collecting this sap, known as guarapo, and concentrating it into syrup by boiling for several hours until a reduction in volume by 90% to a dark brown, rich, sweet syrup, and then, it is packaged. Four to six months after harvesting, palm syrup tends to thicken and crystallize.

Palm syrup is widely used in southern Asian cooking and baking (pastries and desserts) as a flavor enhancer in spicy curries, as a refreshing agent in drinks, and in the production of various alcohols such as arrack by fermentation. Finally, it is used for general medicinal purposes.

6.14.6 Medicated Syrup

Medicated syrups are aqueous solutions that contain some other recipients, except for sugars, such as the following:

- Sugar polyols (glycerol, sorbitol, and maltitol)
- Acids (citric acid)
- Preservatives (parabens, benzoates, and antioxidants)
- Chelating agents [sodium ethylenediaminetetraacetic acid (EDTA)]
- Flavorings
- Bufferings
- Colorings

Sugar-free syrups are those in which sugar has been replaced by polyols (e.g., glycerol and sorbitol) or by artificial sweeteners (e.g., aspartame, neotame, and sucralose), polysaccharides (e.g., carrageenan and xanthan gum), and others. Syrups are usually sugar–water solutions and are prepared by the following:

- Solubilization of a sugar ingredient into purified water
- Mixing with/without heating
- Strain where it is needed
- Addition of water to the right volume

REFERENCES

Advanced Instruments, Inc. 1971. *The Physical Chemistry, Theory, and Technology of Freezing Point Determinations*. Norwood, MA: Advanced Instruments Inc.

Aider, M., D. de Halleux, and K. Belkacemi. 2007. Production of granulated sugar from maple syrup with high content of inverted sugar. *J. Food Eng., 80*: 3, 791–797.

Al-Farsi, M., C. Alasalvar, M. Al-Abid, K. Al-Shoaily, M. Al-Amry, and F. Al-Rawahy. 2007. Compositional and functional characteristics of dates, syrups, and their by-products. *Food Chem., 104*: 3, 943–947.

Anderson, H. G. 1997. Sugars and health: A review. *Nutrition Research, 17*: 9, 1485–1498.

Ashurst, P. R. (Ed.). 2005. *Chemistry and Technology of Soft Drinks and Fruit Juices*. Oxford, U.K.: Blackwell Publishing.

Aurand, L. W., A. E. Woods, and M. R. Wells. 1987. *Food Composition and Analysis*. New York: Van Nostrand Reinhold.

Balston, J. N., and B. E. Talbot. 1952. *A Guide to Filter Paper and Cellulose Powder Chromatography*. London: H. Reeve Angel & Co.

Bamforth, C. 1998. *Beer: Tap into the Art and Science of Brewing*. New York: Oxford University Press.

Bauman, M. 2005. Haines birch syrups attract gourmet following. *Alaska Journal of Commerce* (May).

Browne, C. A., and F. W. Zerban. 1955. *Physical and Chemical Methods of Sugar Analysis*. Hoboken, NJ: John Wiley & Sons, Inc.

Catalano, A. 2008. *Baking with Agave Nectar: Over 100 Recipes Using Nature's Ultimate Sweetener*. Berkeley, CA: Celestial Arts.

Clarke, M. A. 2003. Syrups. In: B. Caballero, L. Trugo, and P. M. Finglas (eds.), *Encyclopedia of Food Sciences and Nutrition* (5711). Burlington, MA: Elsevier, Inc.

Davenport, A. L., and L. J. Staats. 1998. Maple syrup production for the beginner. Retrieved from http://www.dnr.cornell.edu. May 2011.

Edney, M. J., and M. S. Izydorczyk. 2003. Malt types and products. In: B. Caballero, L. Trugo, and P. M. Finglas (eds.), *Encyclopedia of Food Sciences and Nutrition* (3671). Burlington, MA: Elsevier, Inc.

Elliot, E. 2006. *Maple Syrup: Recipes from Canada's Best Chefs*. Halifax, Nova Scotia: Formac Publ. Co.

El-Refai, A. H., M. S. El-Abyad, A. I. El-Diwany, L. A. Sallam, and R. F. Allam. 1992. Some physiological parameters for ethanol production from beet molasses by *Saccharomyces cerevisiae* Y-7. *Bioresource Tech., 42*: 3, 183–189.

Emerton, V., and E. Choi. 2008. *Essential Guide to Food Additives*. Cambridge, U.K.: RSC Publishing.

Eskander, J., C. Lavaud, and D. Harakat. 2010. Steroidal saponinds from the leaves of Agave macroacantha. *Fitoterapia, 81:* 5, 371–374.

Grenby, T. H. 1987. *Developments in Sweeteners: Applied Science*. London: Elsevier Applied Science.

Gyura J., Z. Šereš, G. Vatai, and E. B. Molnár. 2002. Separation of nonsucrose compounds from the syrup of sugar-beet processing by ultra- and nano-filtration using polymer membranes. *Desalination, 148*: 1–3, 49–52.

Gyura, J., Z. Šereš, and M. Eszterle. 2005. Influence of operating parameters on separation of green syrup colored matter from sugar beet by ultra and nanofiltration. *J. Food Eng., 66*: 1, 89–96.

Harris N., M. Peterson, and S. Crespo. 1991. *A Formulary of Candy Products*. Revere, MA: Chemical Publishing Co.

Hein, G. L., M. L. Storey, J. White, and D. R. Lineback. 2005. Highs and lows of high fructose corn syrup: A report from the Center for Food and Nutrition Policy and its Ceres Workshop. *Nutrition Today, 40*(6), 253–256.

Hockenhull, D. J. D. 1968. *Progress in Industrial Microbiology*. London: J. & A. Churchill, Ltd.

Hocman, K. 2009. Agave nectar aka agave syrup. Retrieved from www.thenibble.com/reviews. May 2011.

Hogan, C. M. 2008. Chilean wine palm: *Jubaea Chilensis*. Retrieved from http://globaltwitcher.auderis.se/artspec_information.asp?thingid=82831. May 2011.

Hopkins, K. 2007. Maple syrup quality control manual. Retrieved from http:/www.extension.umaine.edu. May 2011.

Hull, P. 2010. *Glucose Syrups: Technology and Applications*. West Sussex, U.K.: Wiley-Blackwell Publ.

Johannes, L. 2009. Looking at health claims of agave nectar. *The Wall Street Journal*. October 27.

Kamozawa, A., and H. A. Talbot. 2009. Agave nectar, a sweetener for any occasion. Retrieved from www.popsci.com. June 2011.

Kearsley, M. W. and S. Z. Dziedzic. 1995. *Handbook of Starch Hydrolysis Products and Their Derivatives*. London: Blackie Academic Publ.

Kirk, R. S., R. Sawyer, and H. Egan. 1991. *Pearson's Composition and Analysis of Foods*. London: Longman.

Levenson, D. A., and R. W. Hartel. 2005. Nucleation of amorphous sucrose-corn syrup mixtures. *J. Food Eng., 69: 1*, 9–15.

Matz, S. A. 1991. *The Chemistry and Technology of Cereals as Food and Feed*. New York: Van Nostrand Reinhold/AVI.

McCance, R. A., and E. M. Widdowson. 1991. *The Composition of Foods*. Cambridge, U.K.: Royal Society of Chemistry.

Melanson, K. J., L. Zukley, J. Lowndes, V. Nguyen, T. J. Angelopoulos, and J. M. Rippe. 2007. Effects of high-fructose corn syrup and sucrose consumption on circulating glucose, insulin, leptin and ghrelin and on appetite in normal-weight women. *Nutrition, 23*: 2, 103–112.

Meyers, T. 1993. Bakery application of a patented sweetener prepared from fruit juice and rice syrup. *AIB Techn. Bul., 25:* 11–15.

Mitchell, H. 2006. *Sweetness and Sugar Alternatives in Food Technology*. Oxford, U.K.: Blackwell Publ.

Mudoga, H. L., H. Yucel, and N. S. Kincal. 2008. Decolorization of sugar syrup using commercial and sugar beet pulp-based activated carbons. *Bioresource Technol., 99:* 9, 3528–3533.

Nelson, A. L. 2009. *Sweeteners Alternative*. St. Paul, MN: Eagan Press.

O'Brien-Nabors, L. 2001. *Alternative Sweeteners*. New York: Marcel Dekker, Inc.

Olbrich, H. 1963. *The Molasses*. Berlin: Fermentation Technologist Institut fur Zuckerindustrie.

Pilando, L. S., and R. E. Wrostland. 1992. Compositional profiles of fruit juice concentrates and sweetness. *Food Chem., 44:* 1, 19–27.

Schenck, F. W. 2006. *Glucose and Glucose-Containing Syrups*. Published online by Wiley Online Library, doi: 10.1002/14356007.a12_457.pub2.

Schorin, M. D. 2006. High fructose corn syrups—Part 2: Health effects. *Nutrition Today, 41:* 2, 70–77.

Schorin, M. D. 2005. High fructose corn syrups—Part 1: Composition, consumption and metabolism. *Nutrition Today, 40:* 6, 248–252.

Spiler, G. A., and C. R Mitchell. 1992. *Glycerin Response of Normal Children to Fructose, Sucrose and Fruit Rice Concentrate Sweetener*. Los Altos, CA: Health Res. and Stud. Center.

Tsai, W.-T. 2005. An overview of environmental hazards and exposure risk of HFCS. *Chemosphere, 61:* 11, 1539–1547.

United States Department of Agriculture (USDA). 2010. Maple Syrup: Production, Price and Value, 2000–2004, U.S. and Canadian provinces. p. 12 (Sept.).

Wainwright, T. 2000. *Basic Brewing Science*. Braunschweig, Germany: PTB Publ.

Watson, S. A., and P. E. Ramstrad. 1987. *Corn: Chemistry and Technology*. St. Paul, MN: American Association of Cereal Chemists, Inc.

Werner L. H. 2011. Maple sugar industry. *Canadian Encyclopedia.* Retrieved from http://www.thecanadianencyclopedia.com/index.cfm?PgNm=TCE&Params=A1ARTA0005095. May 2011.

CHAPTER 7

Other Sweeteners

Theodoros Varzakas and Athanasios Labropoulos

CONTENTS

7.1 RAISIN JUICE CONCENTRATE: STAFIDIN

Stafidin or raisin juice concentrate is a product produced by raisin manufacturers. It is made in a natural way that alters neither the natural character of the initial product nor the taste and nutritional properties of the raisins.

The Greek Food Code defines stafidin as "the concentrated aqueous extract of dry raisins, which has been treated for the partial removal of the highest percentage of dissolved acids."

It contains about 70% w/w inverted sugar and around 2% w/w proteins. It is also rich in trace elements (especially potassium, sodium, phosphorus, magnesium, and calcium) and vitamins (especially C, B3, and A).

Because the carbohydrates of raisin juice are in the form of glucose and fructose, they easily pass into the blood without digestion. This is of nutritional importance, especially for babies, children, people with celiac disease, and sportsmen and in situations demanding immediate energy (Batu 2005). Concentrated raisin juice has lower caloric content than sucrose (2250 and 3900 kcal/kg, respectively), whereas their sweetness is similar.

Raisin juice is a pure extract of dry raisins in the form of a dark brown syrup (Karathanos, Kostaropoulos, and Saravacos 1994; Camire and Dougherty 2003; Simsek, Artik, and Raspinar 2004) produced by boiling without the addition of sugar or other food additives (see Table 7.1).

Raisin juice concentrate contains acids that are beneficial in bringing out and rounding flavors and complementing other ingredients. It also contains high levels of propionic acid and a high content of reducing sugars, which, in combination with the lowered pH caused by the tartaric acid present (a natural flavor enhancer), is a mold inhibitor.

Sanders (1991) reported that by adding raisin juice concentrate to whole wheat bread at a proportion of 9%–12% w/w to the weight of flour, an extension to the mold-free life of the bread by 3–4 days was achieved. Raisin juice concentrate is used in bakery (Matz 1996; Matz 1989) and confectionery products (Riedel 1976), brown sauces, and salad dressings as a sweetener supplement and a color, volume, and shelf life enhancer.

The incorporation of sucrose and raisin juice (in concentrated or dried form) at 3% and 5% of flour weight to commercial wheat starch (Codex Alimentarius Commission 2000) in gluten-free flour was carried out to examine the effects on baking and the textural and sensory properties of bread. Breads made with gluten-free flour are usually characterized by poor color and baking characteristics, as well as short shelf life. Sabanis, Tzia, and Papadakis (2007) conducted a study to help solve these problems by using raisin juice, a natural sweetener that contains no preservatives, has lower caloric content than sucrose, and includes a number of important vitamins and minerals that are very important for people with celiac disease. The study showed that 3% raisin juice in concentrated form contributes to a great improvement in loaf volume, color, and hardness in gluten-free bread during the first day after baking but had a higher staling rate because of its high moisture content. Dried raisin juice gave bread higher loaf volume and better color compared to the control gluten-free bread and also increased its shelf life because of its moisture absorption properties.

Table 7.1 Composition of Raisin Juice Concentrate (Stafidin)

Nutritional Energy	Quantity per 100 g
Calories	225 kcal
Total soluble solids	70 g
Protein (N*6.25)	2.10 g
Ash	0.60 g
Fat	0.20 g
Dietary fiber	0.02 g
Total sugars	68 g
Reducing sugars	65 g
Acidity (expressed as tartaric acid)	2 g
Sodium	0.020 g
Potassium	0.095 g
Calcium	0.035 g
Magnesium	0.035 g
Phosphorous	0.095 g
Iron	0.002 g
Copper	0.0002 g
Zinc	0.0002 g
Vitamin C	0.0030 g
Niacin	0.0006 g
Thiamine	0.0001 g
Riboflavin	0.00002 g
Vitamin A	5 IU

Ziemke (1977) defines raisin juice concentrate as a natural product manufactured by raisin extraction with water in different stages. Following extraction, the liquid part is evaporated under vacuum to a final soluble solids content of 70°Bx.

The following characteristics of stafidin are given:

- Composition should vary between a pulpy and a solid state.
- Sugar concentration expressed as inverted sugar should not fall below 71% w/w.
- Discoloration of stafidin is allowed.
- The use of stafidin is allowed, as long as adequate labeling is followed.
- Manufactured stafidin (bakery stafidin) should have a concentration of inverted sugar not lower than 70% w/w.

7.1.1 Uses of Stafidin

7.1.1.1 Fat Substitute

Low fat or light products are in great demand lately. More particularly, raisin syrup as a fat substitute contributes to the freshness, preservation and overall organoleptic characteristics of food products giving them body. It can easily replace butter and margarines in bakery products and lead to fat reduction.

7.1.1.2 Sugar and Salt Substitute

Producers can develop products with satisfactory sweetness and promote them as products with no added sugar by substituting sugar to a high degree and salt to lower degree with raisin syrup.

More specifically, producers reduce and/or eliminate salt in some bakery products with the addition of tartaric acid (a flavor intensifier found in raisin syrups). The latter is added in chocolate cakes to strengthen the cocoa flavor and thus smaller amount of salt are added.

7.1.1.3 Ethnic Cuisine Recipes

Raisin syrup acts as a basic ingredient in the development of many ethnic cuisine recipes. The use of raisin syrup as a sweetening, flavoring and intensifying agent on spicy ingredients and citrus fruits is advanced in ethnic dishes (Indian, Chinese, Malaysian) and more particularly in sauces and dressings.

A raisin syrup level of 5–10% in ethnic dishes intensifies the flavor and improves the overall quality of the food (as in Chinese spring rolls) with the major advantage not to be distinguished in flavor and sweetness. Uses of larger (>20%) quantities of raisin syrup counter balance the sour taste of vinegar, contributing to the cohesiveness and consistency of sauces, dressings, sweat or sour fillings, and prolonging their preservation simultaneously.

7.1.1.4 Food Additive

The expansion of the food market and changes in the lifestyle and nutritional habits of consumers have led to the use of additional substances for the improvement of production and quality of food products. In this case, raisin syrup may be used as an additive to improve the quality characteristics of these products. Raisin syrup can be used successfully as an ingredient to enhance the color, flavor, texture, and freshness of products such as multigrain breads, cakes, snacks, brownies, sauces, dressings, and ice cream.

7.1.2 Raisin Receive

The Autonomous Organization for Raisins in Greece is responsible for the protection, cultivation, and growth of Corinthian raisins. From these black raisins, they withhold 15% of the total raisin production and use it for winemaking, stafidin production, and threptin.

Raisins are collected from the broken plants to ensure their use as stafidin and not for other use.

In Greece, stafidin is manufactured mainly from Corinthian raisins, which have the lowest cost and lowest transport and processing costs. However, in places such as Crete, where sultanas are produced in large quantities, it is possible to manufacture stafidin from sultanas.

Raisin juice concentrate is a natural sweetener in syrup or paste form, and it is produced from second-grade dry raisins by leaching them with water. Dried raisin juice, although easier to handle and has more potential applications than the syrup, is not available in the market. Papadakis, Gardeli, and Tzia (2006) produced raisin juice powder with a lab-scale spray dryer. The problem of stickiness in the drying chamber was overcome through the use of maltodextrins of 21, 12, and 6 dextrose equivalent (DE) as drying aids. For each type of maltodextrin, the dryer operating conditions and the minimum concentration of maltodextrin in the feed, necessary for successful powder production, were determined. The maximum ratio of raisin juice solids to maltodextrin solids achieved was 67:33 and was made possible with the use of 6-DE maltodextrin. The inlet and outlet drying air temperatures were 110°C, and 77°C, respectively, and the feed contained 40% w/w total solids. The physical and sensory properties of all powders produced were determined and found to be satisfactory; the only exception was their high hygroscopicity.

7.1.3 Raisin Extraction

Raisins are placed in stainless-steel extraction tanks that are very well insulated and have a pseudobed functioning as a filter. The distance of the longitudinal plates that comprise the pseudobed is modified according to the raisin size. The extraction process has the following four steps (Payne 1994):

- First stage
  - Water and raisins are poured into each tank.
  - Cold water, 20°C–25°C, is used.
  - Raisins stay in the water for 12 h.
  - The sugar solution is collected from each tank.

The first sugar extract should have a concentration of 20–25°Be.

- Second stage
  - Cold water is poured only into the first tank.
  - The sugar extract collected is poured into the second tank.
  - Extraction continues, the sugar extract collected goes to the third tank, and so on.
  - The sugar extract from the second extraction should appear with a concentration of 10–15°Be.
- Third stage
  - Cold water is poured only into the first tank.
  - Reextraction of the sugar extract in the next tanks takes place.
  - Extractions continue until the sugar extract in the last tank has a concentration of 6–7°Be.
- Fourth stage
  - When the concentration reaches 6–7°Be, cold water is poured into the first extraction tank.
  - Extraction starts, and steam insertion is carried out in the extract appearing in the region between the pseudobottom (filter) and the bottom of the tank to achieve the best icing of raisins.
  - Reextraction with steam insertion follows at a lower temperature in the succeeding tanks. The final extract has a temperature of 50°C–60°C and 10°Be–20°Be.
  - This extraction under heat continues until the concentration of the final extract reaches 3°Be–5°Be.

The raisin extraction process is shown in Figures 7.1 and 7.2.

1st Stage

H_2O, RAISINS 20°C–25°C

Sugar extract concentration of 20–25°Be

2nd Stage

↓

Sugar extract from second extraction has a concentration of 10–15°Be

3rd Stage

↓

Sugar extract in the last tank has a concentration of 6–7°Be

4th Stage

↓

Reextraction with steam insertion follows (at a lower temperature in the following tanks). Final extract has a temperature of 50°C–60°C and 10–20°Be.

Figure 7.1 Flow diagram of raisin extraction.

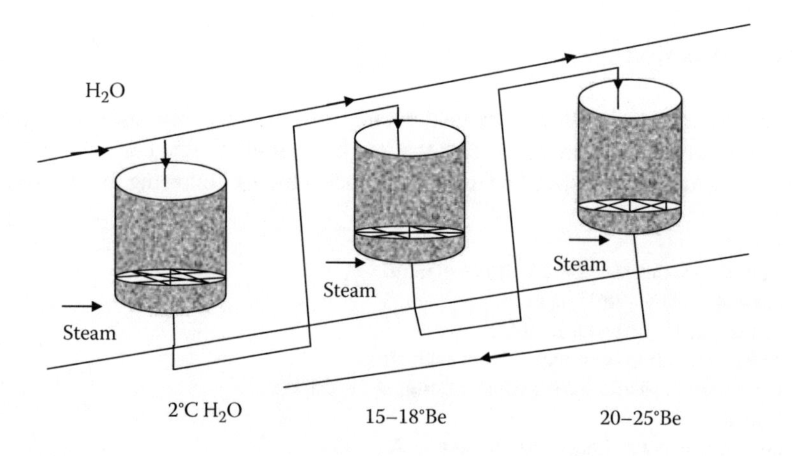

Figure 7.2 Basic steps in the flow diagram of raisin extraction.

7.1.3.1 Mixing of Extracts and Heating

Mixing of extracts is carried out to get an extract at 12°Be–15°Be. Heating follows at 45°C–50°C to retard mold growth and preserve the extract with no exposure to microbial spoilage.

7.1.3.2 Acid Removal

Partial removal of acids, mainly tartaric acid, is carried out to reduce the acidity of the final product. This is done with the addition of calcium emulsion; hence, tartaric acid is deposited on the bottom as insoluble calcium tartrate.

7.1.3.3 Denaturation

Sugar extract is stored in denaturation tanks, which are appropriately insulated to keep the temperature of the product constant.

Denaturation is carried out with the addition of 0.02% phenolphthalein. Phenolphthalein makes stafidin rapidly detectable when added to products where its use is not allowed, e.g., to wines, to increase the alcoholic grade.

7.1.3.4 Evaporation

Following denaturation, the extract is ready for evaporation. This is done in an evaporator in a thin layer under a vacuum. This type of evaporator is suitable for the evaporation of liquid products with high viscosity, such as raisin juice concentrate.

A schematic of the evaporator shows the introduction of the extract to the tank through a pump in the separator, which is filled with extract until it reaches two-thirds of its height. The extract passes through a tube to the heat exchanger, because the separator and heat exchanger are contact vessels. Steam insertion is carried out at a point, and condensates exit at the downward point of the exchanger. Inside the evaporator, vacuuming is done with a vacuum pump. The extract is heated in the heat exchanger and is boiled, discarding water in the form of vapors, which are collected on the upper side of the system. Products in the form of droplets during heating are carried through the tube to the separator, where they are mixed with the initial extract. The circulation of the evaporated product from the heat exchanger to the separator takes place as a result of the difference in vapor pressure existing between the separator and the evaporator.

When the level in the separator reaches a certain height so that the natural flow in the system is not possible, a new quantity of extract is added through a second valve. At another point, there is a glass valve through which level control is achieved. From the second valve, a sampling of the product is received to control the Baume (°Be), so the required evaporation is achieved. When the product reaches 42°Be, evaporation stops, and the final product is taken before packaging.

The characteristics of the product during evaporation are listed as follows:

- Initial temperature—40°C
- Final temperature—67°C
- Initial concentration—15°Be
- Final concentration—42°Be
- Degrees Brix of the final product—70°Bx

The product's physical characteristics are listed as follows:

- Color—Brown
- Taste—Fruity, raisin juice
- Acidity—pH 2.0–3.5, best for confections
- Sweetness—High concentration of monosaccharides, such as glucose and fructose (85%–95% dry solids)
- Viscosity—Approximately 260 cps or 42°Be at 42°C

7.1.4 Manufacture of Stafidin

In Figure 7.3, the manufacture of stafidin is shown. Initially, the product is a pure extract of stafidin. It is manufactured with the extraction of raisins using hot water (Figure 7.3b). Then, the product is evaporated under vacuum and is stored with no preservatives (Figure 7.3c). The final product with approximately 74% dissolved solids is packaged and distributed.

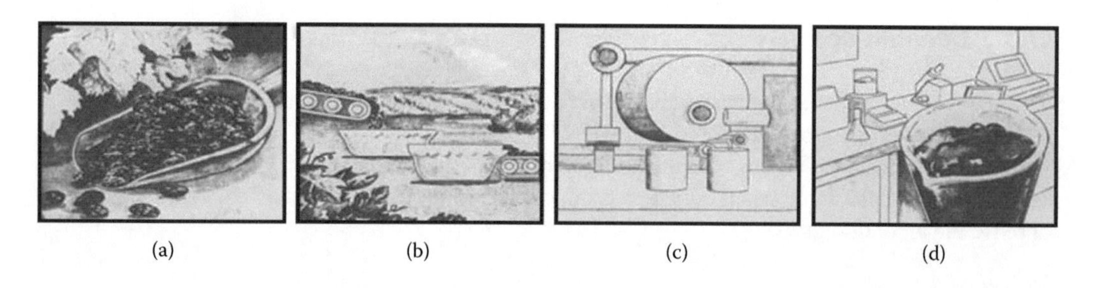

(a) (b) (c) (d)

Figure 7.3 Manufacture of stafidin. (a) Syrup is a raisin extraction product. (b) It is manufactured with raisin extraction using hot water. (c) This product is then concentrated under vacuum. (d) Final product with approximately 74% dissolved solids is packaged and distributed.

7.1.4.1 Advantages

Stafidin 42°Be or 70°Bx with a brown color can substitute for caramel color in foods as a natural product. This syrup is a taste enhancer that can smooth the most intense testability or bitterness of a food to make a product acceptable in terms of taste, even for the most demanding consumers. It is used in the manufacture of confectionery and bread products with the following quality issues:

- Taste—Improves the taste in whole meal breads and cellulosic substances
- Freshness—Increases the shelf life of foods and is used to reduce staling of bread products
- Coloring agent—Improves the color of the crust of bread products and confections
- Sweetness—Most acceptable sweetening agent compared to sugar, syrup, honey, or molasses
- Maintainability—Acceptable as a natural antifungal agent in foods
- Toasting—Improves toasting (richer color, improved aroma, and uniform toasting)
- Naturalness—Natural ingredient approved as healthy and hygienic by consumers
- Purchasing—Low cost, originates in natural products, and, in conjunction with multiple healthy attributes, lends a competitive advantage to its marketing

7.1.4.2 Storage and Packaging

Following evaporation, the product is transferred through stainless-steel tubes in a tank, where it remains for cooling at room temperature. Plastic containers of 5, 25, and 50 kg are used, as well as glass containers of 650 g.

7.1.4.3 Legislation for the Production of Stafidin in Manufacturing Plants

Processing of dry raisins to manufacture stafidin and similar products is only allowed with the concentration of the sugar liquid taken after the extraction of dry raisins or with the evaporation of the must of fresh raisins until it reaches a pulpy or solid state with or without the addition of aromatic substances.

Stafidin manufacture is only allowed in processing plants described as follows:

- Second-category distilleries that follow the legislation, according to which they are allowed to manufacture wines from fresh raisins
- Specific plants for vinegar manufacturing that implement the legislation
- Certain stafidin manufacturing plants that have permission to manufacture from the General Chemical State Laboratory (GCSL)

Requirements for certain plants to be granted permission to manufacture stafidin are described as follows:

1. Owners of specific stafidin manufacturing plants, as well as second-category distillers and vinegar producers, who intend to process dry or fresh raisins for stafidin manufacture should inform the GCSL before the first of August every year if they intend to start production and should list the following:
   - The name of the plant and the owner
   - The location of the plant
   - The machinery used for the production and all industrial installations, including the number and capacity of the tanks and containers used
   - The daily productivity and yield in ready-to-use stafidin
   - The method of production and the name of the product

2. The GCSL approves the license for the manufacture of stafidin from dry or fresh raisins if the following conditions apply:
   - The plant has the specific machinery for raisin extraction
   - The productivity is at least 300 kg daily
   - The plant adopts good manufacturing and hygienic practices

3. If the application is sent to the prefectures, then the GCSL sends chemists to conduct an audit and file a report for license permission.
4. The license can be withdrawn in case the rules mentioned above are not satisfied with a joint ministerial decision from the Ministries of Agriculture and Finance and an opinion by the Supreme Chemical Council.

7.1.4.4 Specific Mixing of Sugars

In the application filed with the GCSL, the amount (in liters) of liquid extract or must from fresh raisins and the quantity, which should not be below 1000 L, should be stated. The sugar percentage and the corresponding weight should also be stated in °Baume and in kilograms, as well as the total acidity, which should not be allowed to be higher than 1%, expressed as tartaric acid.

The specific mixing of dry raisin liquid extract or must from fresh raisins to produce stafidin is carried out with the addition of pure dextrin previously diluted in warm water and at a ratio of 1 kg of dextrin to 1000 L of mixed liquid.

7.2 SYKIN FROM FIGS

The common fig (*Ficus carica*) is a large, deciduous shrub or small tree native to southwest Asia and the Mediterranean region (from Afghanistan to Portugal). It grows to a height of 6.9–10 m (23–33 ft) tall with smooth, gray bark. The leaves are 12–25 cm (4.7–9.8 in) long and 10–18 cm (3.9–7.1 in) across and are deeply lobed with three or five lobes. The fruit is 3–5 cm (1.2–2 in) long with a green skin, sometimes ripening toward purple or brown (Figure 7.4). The sap of the fig's green parts is an irritant to human skin (Purdue University 2011).

The common fig is widely known for its edible fruit throughout its natural growing range in the Mediterranean and Middle Eastern regions, Iran, Pakistan, Greece, and northern India and also in other areas of the world with a similar climate, including Arkansas, Louisiana, California, Georgia, Oregon, Texas, South Carolina, and Washington in the United States; southwestern British Columbia in Canada; Durango, Nuevo León, and Coahuila in northeastern Mexico; as well as areas in Argentina, Australia, Chile, and South Africa.

Figs can also be found in continental climates that have hot summers, as far north as Hungary and Moravia, and can be harvested up to four times per year. Thousands of cultivars, most named, have been developed or come into existence as human migration brought the fig to many places outside its natural range. Fig has been an important food crop for thousands of years and was also thought to be highly beneficial in the diet.

Figure 7.4 Common fig fruit.

The edible fig is one of the first plants cultivated by humans. Nine subfossil figs of a parthenocarpic type dating to about 9400–9200 B.C. were found in the early Neolithic village Gilgal I (in the Jordan Valley, 13 km north of Jericho). The find predates the domestication of wheat, barley, and legumes and may thus be the first known instance of agriculture. It is proposed that they may have been planted and cultivated intentionally, 1000 years before the next crops were domesticated (wheat and rye; Kislev, Hartmann, and Bar-Yosef 2006a, 2006b; Lev-Yadun et al. 2006).

Figs can be eaten fresh or dried and made into jam. Most commercial production is in dried or otherwise processed forms, because the ripe fruit does not transport well and, once picked, does not keep well.

The Food and Agriculture Organization of the United Nations (FAO) reports that, in 2005, fig production was 1,057,000 metric tons. Turkey was the top fig producer (280,000 metric tons), followed by Egypt (170,000 metric tons) and other Mediterranean countries. The Aydın, İzmir and Muğla regions, which used to be called the ancient Caria region, are the top fig producers in Turkey.

7.2.1 Nutrition

Figs are one of the best plant sources of calcium and fiber. According to the United States Department of Agriculture (USDA) data for the mission variety, dried figs are the richest in fiber, copper, manganese, magnesium, potassium, calcium, and vitamin K, relative to human needs (Table 7.2). They have smaller amounts of many other nutrients. Figs have a laxative effect and contain many antioxidants. They are a good source of flavonoids and polyphenols (Vinson 1999), including gallic acid, chlorogenic acid, syringic acid, (+)–catechin, (–)–epicatechin, and rutin (Veberic, Colaric, and Stampar 2008). In one study, a 40-g portion of dried figs (two medium-size figs) produced a significant increase in plasma antioxidant capacity (Vinson et al. 2005).

7.2.2 Characteristics of Unprocessed Dried Figs

Unprocessed dried figs have been obtained from the ripe fruit of varieties of *Ficus carica domestica* L. dried naturally. According to Regulation 1573/1999, the following minimum requirements and tolerances for unprocessed dried figs should apply:

- Maximum moisture content of 24%
- Minimum size of 136 fruits per kilogram for small fruit varieties and 116 fruits per kilogram for other varieties
- A thin skin and a pulp of honey consistency
- A uniform color
- Clean and practically free from foreign matter

Table 7.2 Nutritional Value per 100 g (3.5 oz) of Dried, Uncooked Figs

Energy	1041 kJ (249 kcal)
Carbohydrates	63.87 g
Sugars	47.92 g
Dietary fiber	9.8 g
Fat	0.93 g
Protein	3.30 g
Thiamine (vitamin B_1)	0.085 mg (7%)
Riboflavin (vitamin B_2)	0.082 mg (5%)
Niacin (vitamin B_3)	0.619 mg (4%)
Pantothenic acid (vitamin B_5)	0.434 mg (9%)
Vitamin B_6	0.106 mg (8%)
Folate (vitamin B_9)	9 µg (2%)
Vitamin C	1.2 mg (2%)
Calcium	162 mg (16%)
Iron	2.03 mg (16%)
Magnesium	68 mg (18%)
Phosphorus	67 mg (10%)
Potassium	680 mg (14%)
Zinc	0.55 mg (6%)

Source: USDA Nutrient Database, 2009, Dried figs nutrition facts-dried figs calories. http://www.lose-weight-with-us.com/dried-figs-nutrition.html.
Note: Percentages are relative to U.S. recommendations for adults.

In each lot, the following tolerances shall be allowed:

- 30% by the number or weight of dried figs with internal or external damage from any cause, of which not more than 18% of figs are damaged by insects
- 3% by the number or weight of dried figs unsuitable for processing

7.2.3 Characteristics of Dried Figs

Dried figs shall have been obtained from the ripe fruit of varieties of *Ficus carica domestica* L. dried naturally. The following minimum requirements and tolerances for dried figs should apply:

- Maximum moisture content of 24%
- Minimum size of 136 fruits per kilogram for small fruit varieties and 116 fruits per kilogram for other varieties
- A thin skin and a pulp of honey consistency
- A uniform color
- Clean and free from foreign matter

In each lot, the following tolerance shall be allowed: 25% by the number or weight of dried figs with internal or external damage from any cause, of which not more than 15% of figs are damaged by insects.

7.2.4 Fig Syrup Concentrate

Fig syrup is delicious and nutritious. It is 100% natural and made from fresh, fully ripened black figs with no sugar added. Extremely nutritious, it can be added to milk as a substitute for chocolate, used as a coffee sweetener or a coffee substitute, diluted with water to make a fig juice energizer, or used in baking.

Figs provide more fiber than any other common fruit or vegetable and have a high quantity of polyphenol antioxidants. They contain no fat, no sodium, and no cholesterol. Figs have the highest overall mineral content of all common fruits. A 40-g serving provides 244 mg of potassium, along with calcium, iron, and zinc. Figs are also a natural source of serotonin, which stimulates the pineal gland. Figs contain vitamin B6, which is responsible for producing mood-boosting serotonin, which has been shown to lower cholesterol and prevent water retention. Recent research has determined that dried figs contain omega-3 and omega-6 essential fatty acids, as well as phytosterols. Phytosterols are credited with decreasing natural cholesterol synthesis in the body, thus lowering overall cholesterol counts.

Figs are one of the most abundant fruits in the Mediterranean diet (Solomon et al. 2006), and an artisan derivative of this fruit is fig syrup, a typical Calabrian food product made by boiling and concentrating fresh figs (generally, the cultivar is *Ficus carica*, also known as *fico dottato Calabrese*) in water without adding any other ingredient. The obtained product is a dense syrup characterized by its brown color and sweet taste and smell. It is widely found in the Mediterranean region, and in Italy, the production area matches the old Kingdom of the Two Sicilies. For a long time in Calabria, fig syrup has been a common ingredient in the preparation of typical foods and cakes in particular.

The efficacy of fig syrup, a Mediterranean fig derivative, as a nutraceutical supplement was demonstrated by Puoci et al. (2011). Fig syrup is a fruit concentrate used as a common ingredient in the preparation of typical foods, particularly cakes. *In vitro* assays were performed to determine the amount of nutraceutical ingredients, such as phenolic compounds (3.92-mg equivalent of gallic acid per gram) and flavonoids (0.35-mg equivalent of catechin per gram), and high-performance liquid chromatography (HPLC) analyses provided specific information about the composition of antioxidants in the syrup. Furthermore, total antioxidant activity, scavenging properties against 2,2-diphenyl-1-picrylhydrazyl and peroxyl radicals, and anticholinesterase activity clearly showed the efficacy of the syrup in preventing damage induced by free radicals and, thus, the applicability of this food derivative as a nutraceutical supplement.

Fig syrup concentrate is produced in the same way as stafidin.

7.3 PRUNE JUICE CONCENTRATE

Prune juice concentrate is similar to raisin and fig juice concentrates. It is extracted from dry prunes using osmotic pressure and evaporated at 70°Bx (70% sugars). It is dark brown in color and has soluble solids of 30%, of which reducing sugars are 85%–95%.

Prune juice concentrate can be diluted to prune juice. This juice in low doses (1%–5%) increases the product shelf life and is used against staling of bakery products, as a sweetener, for coloring of food products, and as a natural preservative in different products. It can be used as follows:

- Sugar substitute
- Binder, to hold up humidity in cakes and cookies
- Preservative agent in cereal bars
- Natural syrup in yogurt and ice cream
- Flavor and taste enhancer

Some of the advantages attributed to the product are the following:

- Natural product
- Nutritional benefits
- Natural sweetener
- Natural preservative

- Label benefit
- Increase in shelf life
- Natural color
- Variety in viscosity
- Easy use and easy preservation

Maintaining digestive health is vital to achieving overall wellness. Prunes, which are the dried fruit of various plum species, primarily the European plum or Prunus domestica, have been known since biblical times for their benefits in supporting digestive health.

Prune juice is richer in fiber—both soluble and insoluble—than plum juice and is often marketed as a remedy for constipation and for help with kidney stones. Insoluble fiber helps speed food through the digestive tract. Soluble fiber, on the other hand, is a slower moving fiber, good for lowering cholesterol and regulating blood-sugar levels.

Prune juice concentrate is made from the high-volume, low-temperature water extract of prunes. Prune concentrate is obtained by the vacuum evaporation of the water-soluble portion of prunes to a concentration of about 70% soluble to dissoluble matter.

Prunes are also high in polyphenols, which include antioxidants that protect cellular deoxyribonucleic acid against damage, decrease inflammation, and may help prevent cancer. Prune concentrate acts as a natural laxative by stimulatating peristalsis (movement of the muscles in the digestive tract) in the colon, thereby aiding the process of elimination. Prune concentrate promotes bowel movement and is believed to help regulate bowel function.

Interestingly, in a 1990 announcement that came as a shock to millions of Americans, the U.S. Food and Drug Administration (FDA) declared that the common prune was not an effective laxative. In reality, this was not to be interpreted to mean that prunes or prune concentrate do not have laxative properties when taken as foods; rather, the case here is that the FDA no longer regards them as effective as a drug (Ofiyeva 2011).

Canned or bottled juice is prepared from a water extract of dried plums and contains not less than 18.5% by the weight of water-soluble solids extracted from dried plums. Juice concentrate has been shown to be an effective natural mold inhibitor in bread while providing the added benefits of natural color, improved flavor, and better texture. The juice may contain one or more optional acidifying ingredients—lemon juice, lime juice, or citric acid—in a quantity sufficient to render the food slightly tart. It may also contain honey in a quantity not less than 2% or more than 3% by weight and may contain vitamin C, not less than 30 mg or more than 50 mg per 6-oz serving of the finished food.

Juice concentrate is a viscous form of juice, packed at a 70°Bx minimum, except for higher Brix packs for export shipments or on special orders. No preservatives are added to the juice concentrate, as the 70°Bx level product is self preserving. The weight of 70°Bx concentrate is 11.25 lb/gal (1.348 kg/L).

Some dried plum and fresh plum juice and concentrates can be used in meat and poultry applications to retain moisture, suppress the growth of some pathogens, and extend shelf life. The specifications of plum juice concentrate are shown in Table 7.3.

Prunes (a natural laxative) are plums that have been picked prior to full maturation and dehydrated. Although prunes are approximately 6% fiber, prune juice concentrate lacks the same fiber content according to Health Directories (http://www.health-directories.com/constipation-prunejuice.html. Interestingly, this lack of fiber in prune juice does not diminish its ability to promote regularity.

According to Health Directories, prunes promote the regularity of the bowels because of the active lithocholic acid. Prunes are high in sorbitol, which is not easily digested, thereby remaining undigested in the gut. The undigested sorbitol causes water retention, which softens the stool, making for an easier elimination. Although prunes aid in regularity, they are not the cure-all for severe constipation (http://www.ehow.com/about_6712505_prune-senna-chronic-constipation.html).

Table 7.3 Specifications of Plum Juice Concentrate

Dried plum (prune) juice	*Description*: Canned or bottled dried plum (prune) juice. Contains not less than 18.5% by the weight of water-soluble dried plum solids.
	Moisture Range: 18.5% min.
	Packaging: Glass:12/16 oz, 12/32 oz, 12/40 oz Tins: 4/6/5–1/2 oz, 6/48 oz, 8/6/5–1/2 oz, 24/12 oz, 12/46 oz, 6/No. 10.
	Product Use: Beverages
Dried plum (prune) juice concentrate	*Description*: Viscous form of dried plum juice. No preservatives added.
	Moisture Range: 70°Bx
	Acidity as citric acid: 6.7 ± 1.7
	pH: 2.8–4.0
	Specific gravity: 1.327–1.356
	Packaging: Steel drums, polyethylene pails.
	Product Use: Mold inhibitor in baked goods. Retains moisture in precooked meats. Can be injected into whole muscle meats. Inhibitor of food-borne pathogens/oxidation.
	Flavor: Full flavored and typical of fine-quality plum juice concentrate. Free from scorched, fermented, caramelized, or other undesirable flavor
	Shelf life: Refrigerated or frozen for approximately 2 years
Fresh plum juice concentrate	*Description*: Concentrate made from juice from mature fresh plums.
	Moisture Range: 70°Bx
	Packaging: Steel drums, polyethylene pails
	Product Use: Bakery products, syrups/toppings juice blends, meat/poultry sauces/marinades, snack foods/energy bars. Can be injected into whole muscle meats. Inhibitor of food-borne pathogens/oxidation.

Source: http://www.californiadriedplums.org/industrial/products/juice-concentrate, October 2011.

7.4 APPLE JUICE CONCENTRATE

Apple juice concentrate is a natural sweetener. Currently, it is mainly used as a juice, and we will focus on its production, as described in Figure 7.5. In supermarkets, frozen fruit juices are diluted at a ratio of 1:3 with water and consumed as fresh juices.

7.4.1 Fruit Juices

Fruit juices are classified into three main categories:

- Natural juices, which are juices extracted from fruits where a certain percentage of approximately 1.5% sweeteners is added (such as sugar, sorbitol, and fructose).
- Concentrated juices from natural juices, in which a sweetener is added according to a concentration degree.
- Nectar fruit juices, which are juices from fruits thickened with the addition of sweeteners.

Fruit juices constitute a very good source of nutrients, because they contain vitamins, inorganic compounds, and micronutrients, whereas they have a low percentage of fat. However, juices have an increased nutritional content in cases where sweeteners have been added.

Fruit juices are the most ancient drinks that contain sugars and were used by ancient Greeks because their gods required them to consume ambrosia and drink nectar, which was formed by fruit juices mixed with honey and pollen.

Fruit processing to make juices started with apple production at the end of the nineteenth century in Switzerland. Technological progress in recent decades has made a great number of good quality fruit juices.

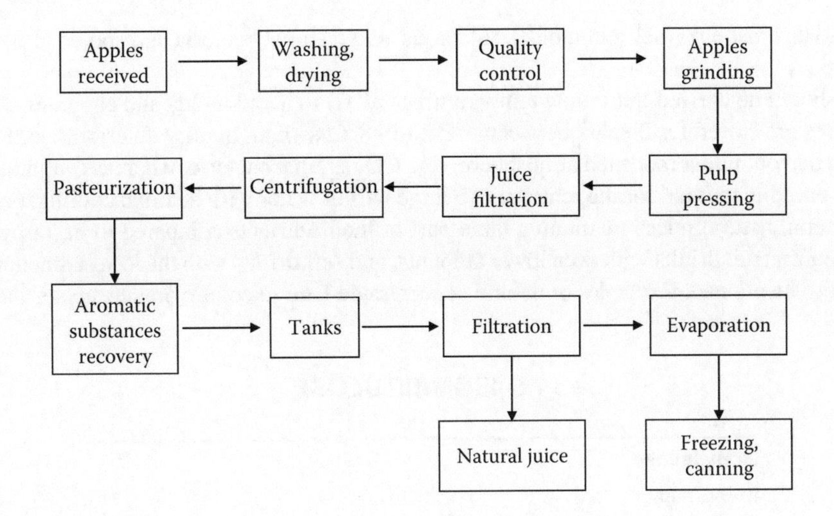

Figure 7.5 Flow diagram of apple juice manufacture.

The main constituents of fruit juices are sugars; inorganic salts and other elements, such as potassium, calcium, sodium, phosphorous, chlorine, and iron; and vitamins such as vitamins A, B, C, and D. The quality of juice is directly related to the quality of raw materials, i.e., the natural juices that will be added to the final product. Factors affecting the quality of the natural juices are the following:

- Ratio of sugars to acids, where the starch of unripe fruits during maturation is broken down to glucose, which, in conjunction with other soluble sugars such as fructose and saccharose, are responsible for the sweet taste.
- Organic acids of fruits, such as citric acid (citrus fruits), tartaric acid (grapes), and malic acid (apples), which have a high concentration when they are unripe and fall down during maturity in balance with sugars.
- Pectin substances, where protopectins during fruit maturation convert into soluble pectins, which affect the texture of flesh and fruit processing.
- Phenolic substances, such as tannins and coloring agents, that affect color and taste.
- Aromatic compounds, such as alcohols, esters, aldehydes, and ketones, which consist of a large number of volatile compounds (in trace quantities, they contribute in the formation of the final aroma of juices).
- Proteins in trace amounts, causing problems in juice clarity.

7.4.2 Manufacturing

Harvesting of mature fruits with a good balance of sugars and acids is followed by sampling for quality control, washing to remove impurities and pesticide residues, and mashing to remove the pulp. This breaking colonizes cellular structure and brings enzymes (phenol oxidases) in contact with their substrate, causing, with the effect of oxygen, the enzymic browning. Juice comes out of pressing, misty and muddy, and then is filtered with the addition of gelatin, albumin, or enzymes. Packaging is carried out in Tetra Pak machines aseptically.

7.4.3 Preservation and Storage

Pasteurization is carried out at 93°C for 30 s, which destructs pathogenic microorganisms and inactivates enzymes. Another preservation method is the addition of benzoic acid (antibacterial), sorbic acid (against enzymes and molds), sulfurous acid (against bacteria and spores of molds), and their salts.

This should be carried out in low temperatures (<4°C) to avoid molds and enzymes. Apple and grape juices are stored for 2 years between 5°C and −8°C without quality deterioration. Packaging can be carried out under modified atmosphere, e.g., CO_2 against bacteria, whereas evaporation is the main preservation method for the removal of a large part of water with heating, cooling, or freezing.

In general, natural juices maintain a large part of their nutrients compared to nectar, which has lower amounts, fruit drinks with even lower amounts, and soft drinks with the lowest amount of nutrients of those mentioned. Examples of natural concentrated fruit juices are orange, apple, and cherry.

7.5 ISOMALTULOSE

Isomaltulose

IUPAC name

6-O-α-D-glucopyranosyl-D-fructofuranose

Other names

Palatinose, lylose

Properties

Molecular formula	$C_{12}H_{22}O_{11}{\cdot}H_2O$
Molar weight	360.3 g/mol

Isomaltulose (chemical name: 6-O-α-D-glucopyranosyl-D-fructofuranose, monohydrate) also known by the trade name Palatinose, is a disaccharide that is commercially manufactured enzymatically from sucrose via bacterial fermentation (Figure 7.6). It is a natural constituent of honey and sugar cane and has a very natural sweet taste. It has been used as a sugar substitute in Japan since 1985. It is particularly suitable as a noncariogenic sucrose replacement (Lina, Jonker, and Kozianowski 2002).

Isomaltulose is fully absorbed in the small intestine as glucose and fructose. Like sucrose, it is fully digested and provides the same caloric value of approximately 4 kcal/g (http://en.wikipedia.org/wiki/Isomaltulose).

It is low glycemic and low insulinemic. The effect of isomaltulose is that the glucose enters the blood at a slow rate, avoiding high peaks and sudden drops in glucose levels and, therefore, insulin levels. This leads to a more balanced and prolonged energy supply in the form of glucose (Konig et al. 2007).

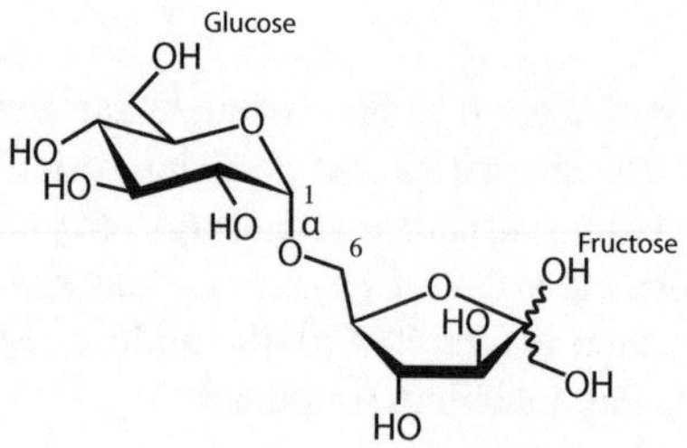

Figure 7.6 Chemical structure of isomaltulose. It is a white crystalline powder with not less than 98% isomaltulose as determined by HPLC, not more than 6.5% loss on drying as determined by Karl Fischer method and not more than 0.1 ppm lead content as determined by atomic absorption spectroscopy.

Being low insulinemic, isomaltulose also supports improved fat oxidation during physical activity, as high insulin levels hinder the use of lipids as an energy source. As such, isomaltulose can increase the amount of fat used as energy, thus enhancing performance endurance (Van Can et al. 2009).

Isomaltulose is tolerated like sucrose and not suitable for people with a preexisting intolerance to fructose and those who are unable to digest sucrose.

7.5.1 Isomaltulose and Disorders in Fructose and Sucrose Metabolism

Food Standards Australia New Zealand (FSANZ 2007) has recently approved the use of a sugar substitute called isomaltulose in food. Isomaltulose contains glucose and fructose and therefore has similar properties to traditional sugars. It provides the same amount of energy as sucrose but is digested more slowly, leading to lower and slower increases in blood glucose compared to sucrose. Isomaltulose is suitable for use as a total or partial replacement for sucrose in certain foods.

7.5.2 Where Does Isomaltulose Come From?

Isomaltulose is a white crystalline substance characterized by a sweetness quality similar to sucrose and a melting point of 123°C–124°C (Irwin and Sträter 1991).

Isomaltulose has about 42% of the sweetness of sucrose and is found naturally in very small quantities in honey and sugar cane juice (Siddiqua and Furhala 1967; Takazoe 1985; Eggleston and Grisham 2003). Commercial isomaltulose is manufactured from sucrose using enzymes.

Isomaltulose is a reducing disaccharide ($C_{12}H_{22}O_{11}$) produced by an enzymatic conversion of sucrose ($C_{12}H_{22}O_{11}$), whereby the 1,2-glycosidic linkage between glucose and fructose is rearranged to a 1,6-glycosidic linkage. It occurs as an intermediate in the production of isomalt (E953), permitted for use as a nutritive sweetener and marketed under the trade names Palatinit® and C*IsoMaltidex® by Südzucker AG (hereafter Südzucker) and Cerestar, respectively.

Specifically, the use of isomalt in food was considered acceptable by the Scientific Committee on Food (SCF) in 1984 (SCF 1984) (http://www.europa.eu.int/comm/food/fs/sc/scf/reports/scf_reports_16.pdf).

The production of isomaltulose by Cerestar is initiated by dissolving food-grade sucrose in water and subsequently treating the resulting solution with a biocatalyst obtained from a nonviable, nonpathogenic strain of *Protaminobacter rubrum* (Porter et al. 1991). Prior to the addition of the biocatalyst to the sucrose solution, *P. rubrum* cells are killed by treatment with formaldehyde. Following completion of the conversion of sucrose, residual *P. rubrum* material is removed by filtration. These steps prevent the presence of the production organism in the isomaltulose product. The crude isomaltulose product is then sequentially subjected to several stages of purification, including demineralization, crystallization, and washing. Drying and cooling of isomaltulose complete the production process, resulting typically in a product of 99% or greater purity.

Isomaltulose was first prepared by Südzucker as an intermediate compound in the production of isomalt. In particular, the production of isomalt is completed by catalytic hydrogenation following the enzymatic rearrangement of the sucrose to isomaltulose (Cargill-Cerestar 2003). Before progressing to the catalytic hydrogenation, isomaltulose is isolated by the concentration of the isomaltulose solution and subsequently purified by crystallization, which, in particular, removes residual sucrose. On an anhydrous basis, this purification method yields isomaltulose, which is typically 98% pure (Irwin and Sträter 1991; Südzucker 1996). Südzucker utilizes *P. rubrum* as the source of the biocatalyst required for the rearrangement of sucrose. The microorganism is grown on synthetic media consisting of sucrose, a nitrogen source, and inorganic salts. Subsequently, the cells are isolated, killed by treatment with formaldehyde, and, unlike Cerestar's method, which utilizes filtration to remove any residual cells, immobilized.

More recently, an alternative method for the purification of isomaltulose has been developed by Südzucker, allowing for the marketing of an additional grade of isomalt. Specifically, residual sucrose is removed by enzymatic conversion using *Saccharomyces cerevisiae*. The purity of isomaltulose typically obtained with this method is only approximately 82%.

This alternative method has been submitted for approval in the U.S. (Südzucker 1996), as well as in Europe, where it was accepted by the SCF (SCF 1997).

Additionally, isomaltulose is produced and made commercially available by the Shin Mitsui Sugar Company for the Japanese, Korean, and Taiwanese food markets. Furthermore, in Japan, the Shin Mitsui Sugar Company distributes Südzucker isomalt. The isomaltulose produced by Shin Mitsui is also prepared from food-grade sucrose via enzymatic rearrangement. Similar to Südzucker, Shin Mitsui immobilizes the biocatalyst (Nakajima 1984), and there are strong indications that the biocatalyst is obtained from *P. rubrum* (Okuda et al. 1986). This is corroborated by Shin Mitsui's role as the distributor of the isomalt produced by Südzucker in Japan. The resulting isomaltulose is purified through concentration and crystallization, providing a material of 99% or greater purity on a dry basis (Shin Mitsui Sugar Co. 2003).

7.5.3 Is Isomaltulose Safe?

Isomaltulose is considered safe. Safety of isomaltulose for human consumption has been evaluated by SCF 1984 and concluded that there are no public health and safety concerns for the general population associated with its use in foods.

According to a toxicological evaluation of the safety of isomaltulose by Cargill-Cerestar 2003 which was primarily based on (i) metabolic data in animals and humans (ii) clinical data pertaining to the glycemic response obtained following isomaltulose administration as compared to that obtained with either sucrose or glucose (iii) the results of human studies demonstrating that isomaltulose is well-tolerated; (iv) the results of a developmental toxicity study in rats (v) supportive animal sub-chronic and chronic toxicity data it was concluded that no treatment-related toxicological relevant effects were observed in any of the animal studies. Considering the safety of isomaltulose, clinical study data have been provided for fructose, one of the hydrolysis products of isomaltulose.

Metabolic data indicate that, prior to absorption, isomaltulose is almost completely hydrolyzed to fructose and glucose and both metabolites are subsequently utilized in well-characterized carbohydrate pathways. Nutritionally, the compound is therefore equivalent to sucrose, which has an extensive history of use in the European community. The safety of isomaltulose is confirmed by a series of published animal toxicity and human clinical studies, including human trials performed specifically using Cerestar isomaltulose, reporting no adverse toxicological effects relevant to the conditions of the intended uses in foods.

7.5.4 What Foods Could Potentially Contain Isomaltulose?

The following types of foods may contain isomaltulose:

- Beverages (soft drinks, instant drink preparations, teas, and fruit or vegetable juices or drinks)
- Breakfast cereals and cereal bars
- Confections and chewing gums
- Fondants and fillings
- Jams, marmalades, and sugar preserves
- Energy-reduced foods and meal replacements

7.5.5 Estimated Daily Isomaltulose Intake

In summary, on an all-user basis, the highest mean and 97.5th percentile intakes of isomaltulose by the U.K. population from all proposed food uses in the European Union as observed in male teenagers were estimated to be 37.8 g/person/day (0.7 g/kg body weight/day) and 97.8 g/person/day (1.9 g/kg body weight/day), respectively. On a body weight basis, children consumed the greatest amount of isomaltulose, with mean and 97.5th percentile all-user intakes of 1.6 and 4.0 g/kg body weight/day, respectively (Cargill-Cerestar 2003).

7.5.6 Labeling

If the food contains added isomaltulose, it will be declared as "isomaltulose" in the ingredient list.

7.5.7 Other Substances to Be Avoided by Individuals with Fructose Metabolism Disorders

Some other substances are permitted in foods but are unsuitable for individuals with disorders in fructose metabolism. Tagatose and sorbitol are sugar substitutes that are approved for use in certain foods in the *Australia New Zealand Food Standards Code*. These substances should also be avoided by people with disorders in fructose metabolism. Both tagatose and sorbitol can be readily identified in the ingredients list by the use of their common names, "tagatose" or "sorbitol."

It is recommended that people with disorders in fructose metabolism avoid foods containing fructose or sucrose (whether naturally present or added; Sanders 2009.

The FDA has ruled that the synthetic sweetener isomaltulose does not promote tooth decay, thereby giving the green light to manufacturers making certain claims on products that contain it. The FDA concluded that isomaltulose, marketed by the German company Palatint as Palatinose, is noncariogenic, meaning that it does not lead to tooth decay. Because of the strong molecular bonds holding the sweetener's molecules together, it cannot break down into its component sugars (including glucose) in the mouth but only in the more vigorous digestion processes that occur later (such as in the stomach). This means it does not provide any food source for plaque bacteria in the mouth. After eating sugar, plaque bacteria excrete certain acids as waste products, and these acids, in turn, degrade tooth enamel and lead to cavities. According to Palatint, isomaltulose is also a low-glycemic sweetener, meaning it causes a slow, sustained increase in blood-sugar levels rather than the spike caused by many sugars (Van Can et al. 2009, 2011). The company developed the sweetener primarily for people with diabetes or consumers concerned about preventing diabetes. The company touts many other properties of its artificial sweetener. Isomaltulose is nonhygroscopic, meaning it does not gather into lumps and dissolves easily into beverages. According to Palatint, this makes it ideal for use in powdered drinks. In addition to resisting breakdown by human saliva, isomaltulose also resists fermentation and digestion by the microbes in dairy products, including lactobacilli. This means it can be used in dairy beverages without the need for preservatives. Unlike some other artificial sweeteners, isomaltulose is actually a carbohydrate, meaning it does eventually break down into sugars in the body and provides an actual nutritional benefit. The FDA ruling is significant, because it allows the use of claims such as "does not promote tooth decay" or "may reduce the risk of dental caries" on products containing isomaltulose. This will make the sweetener more attractive to food manufacturers and consumers Gutierrez 2008).

7.5.8 Main Uses of Isomaltulose and Attributes

The main uses of isomaltulose and its attributes are listed as follows:

- Nutritive sweeteners
- Sport and fitness drinks
- Energy drinks and tablets

- Cereals
- Tooth-friendly confections
- Prolonged energy supply
- Low glycemic and low insulinemic response—Isomaltulose is fully digested and therefore provides the same caloric value of 4 kcal/g as sucrose. However, isomaltulose is digested more slowly than sucrose, leading to a low glycemic response. Because of this feature, isomaltulose is considered beneficial in products for diabetics and prediabetic dispositions
- Kind to teeth—FDA clinical data show isomaltulose is not fermented by oral bacterial flora to an extent that might lower dental plaque pH and generate the erosion of dental enamel. Therefore, the FDA has concluded isomaltulose does not promote tooth decay
- Exclusively derived from sugar
- Natural sweetness
- Improves taste and texture in foods and beverages

Individuals with type 1 diabetes mellitus (T1DM) are encouraged to consume CHO (consistent carbohydrates) to prevent hypoglycemia during or after exercise. However, research comparing specific types of CHO is limited. West et al. (2011) compared the alterations in metabolism and fuel oxidation in response to running after preexercise ingestion of isomaltulose or dextrose in people with T1DM.

After preliminary testing, on two occasions, eight individuals with T1DM consumed 75 g of either dextrose (DEX; GI = 96) or isomaltulose (ISO; GI = 32), 2 h before performing 45 min of treadmill running at 80% ± 1% $V{\cdot}O_{2peak}$. Blood glucose (BG) was measured for 2 and 3 h before and after exercise, respectively. Cardiorespiratory parameters were collected at rest and during exercise. Data (mean ± SEM) were analyzed using repeated measures of analysis of variance.

There was a smaller increase in BG in the preexercise period under ISO, with the peak BG occurring at 120 min after ingestion compared with 90 min under DEX (Δ + 4.5 ± 0.4 versus Δ + 9.1 ± 0.6 $mmol{\cdot}L^{-1}$, $P < 0.01$). During the final 10 min of exercise, there were lower CHO (ISO 2.85 ± 0.07 versus DEX 3.18 ± 0.08 $g{\cdot}min^{-1}$, $P < 0.05$) and greater lipid oxidation rates (ISO 0.33 ± 0.03 versus DEX 0.20 ± 0.03 $g{\cdot}min^{-1}$, $P < 0.05$) under ISO. After exercise, ISO BG was lower than DEX for the entire 180-min period, with the BG area under the curve and mean BG concentrations being 21% ± 3% and 3.0 ± 0.4 $mmol{\cdot}L^{-1}$ lower ($P < 0.05$), respectively.

The consumption of ISO improves BG responses during and after exercise through reduced CHO and improved lipid oxidation during the later stages of exercise.

The food industry is constantly seeking novel ingredients to improve existing products or to allow the introduction of new products. Such new materials must be safe, pure, and inexpensive; otherwise, they will be unacceptable in concept. They must also have organoleptic and textural properties that make them acceptable to the consumer. Sucrose is an extremely valuable food ingredient: Its best known property is its sweetness, but compared with high-intensity sweeteners such as aspartame and saccharin, sucrose is not a very sweet material. In many cases, the amount of sucrose required to provide the degree of sweetness required also provides significant bulk and texture to the product. Oftentimes, as in the cases of, for example, chocolate, cakes, and biscuits, consumers expect the bulk and texture provided by the sucrose. Public taste is changing, however, and the level of sweetness provided by the quantities of sucrose needed to give the desired bulk and texture are now excessive, at least to the tastes of many adults. Consequently, there is a demand for ingredients that provide bulk and the correct texture to foodstuffs but less sweetness than sucrose. Several disaccharides provide lower levels of sweetness than sucrose and provide similar bulking properties, but few of them have the appropriate textural properties. ISO is about one-third as sweet as sucrose but has a similar sweetness profile to that of sucrose (Cheetham and Bucke 1999).

7.6 GRAPE AND CLEAR GRAPE JUICE CONCENTRATES

Grape juice concentrate is an alternative natural sweetener used in the manufacture of homemade sweets. The raw material used in its manufacture is grapes, from which must is removed and evaporated using natural heating and then petimezi is produced. The latter should contain at least 70% w/w invert sugars.

Since ancient times, must from grapes has been very useful in the gastronomic area. Must was used in ancient Egypt as a sauce for fish instead of using lemon. Currently, it is used in fish soup in places such as Lefkada (an island in the Ionian sea), Greece, and in salad recipes with fried eggplant seasoned with sour grape made in Naoussa, north of Greece. However, what interests us is the role of grape juice as a sweetener. In ancient times, grape juice was the only available sweetener besides honey. Ancient Greeks used different types of this sweetener, depending on the concentration degree or boiling. There was the common must, a liquid must from ripe or overripe grapes, whether it was a product of must fermentation. In addition, there were other liquid musts in concentrated form, such as epsima, which was grape juice boiled until it became concentrated and used as a honey substitute, and siraion, another grape juice concentrate of different texture (not enough information, however, is provided).

Romans had a wide range of sweeteners based on grapes but with different boiling points. Some of these were caroenum, which was quite similar to boiled wine; sapa, a grape juice concentrate; and defrutum, which was quite similar to sapa. Greeks used must and its syrup in different recipes for foods and sweets. For example, master chef Apikios used boiled must in a cheese sauce for lettuce salad, in boiled turnips, as a seasoning in poultry and game, and as a sweetener in bakery products and desserts.

Grape juice currently has many applications in Greek cuisine. For example, moustalevria (from grape must) and must bread rolls are some of the most characteristic products. This sweet grape liquid has many applications throughout Greece either as a grape juice (must) or as a boiled syrup of grape juice commonly known as petimezi.

7.6.1 Physicochemical and Nutritional Properties

Grape juice, most specifically, grape juice from the roditis (concord) variety, has the highest antioxidant capacity compared to other fruit juices (e.g., apple juice and blackberry cocktail), red wine, and tea. Grape juice contains 124 mg of proanthocyanidins per 8 lb of juice (250 g), whereas cranberry juice has 91 mg per 5 lbs, tea has 32 mg per 8 lbs, and apple juice has 7 mg per 8 lbs. An older study showed double antioxidant quantities in grape juice from roditis compared to apple, orange, grapefruit, or tomato juice. Anthocyanins are flavonoids that protect human health from atherosclerosis or cancer. The physicochemical properties of red grape juice concentrate are shown in Table 7.4.

Table 7.4 Physicochemical Properties of a Grape Juice Concentrate

Color (Ratio of 520 nm/430 nm absorption)	2.1
Brix	8.0
Acidity (g/100 g as tartaric acid)	3.0
pH (at 68°Bx)	2.9
SO_2 (ppm/68°Bx)	5.0
Potassium (ppm)	2000
Sodium (ppm)	175
Calcium (ppm)	300

7.6.2 Characteristics

Pure sugars (mainly fructose–glucose) remain in grape juice, following the removal of all nonsugar constituents. It is a pure, colorless, odorless, and neutral product. Acidity is almost zero, as well as its concentration in sulphur and other substances (acids, tannins, coloring agents, and metals). It is stored safely without any fear of microbial spoilage because of its high concentration in sugars. Pure grape juice concentrate is a product with similar content to grape juice but without saccharose.

7.6.3 Manufacture of Pure Grape Juice Concentrate

During the manufacture of pure grape juice concentrate, clean and washed grapes are collected and taken to pneumatic presses, where they are crushed.

During crushing, SO_2 is added to prevent alcoholic fermentation. Next, grape juice is filtered to remove the grape solids. Filtered grape juice is then passed through an ion exchange resin column to become pure, i.e., to remove all nonsugar constituents (acids, tannins, colors, and metals).

Following purification, evaporation starts at 56°Bx under vacuum in three stages at 90°C, 65°C, and 45°C. Finally, the juice received consists of mainly sugars, such as fructose and glucose. The manufacture of pure grape juice concentrate is shown in Figure 7.7.

7.6.4 Quality Control

Each production lot is exposed to quality control analysis from the relevant authorities or the company producing it according to the specifications set up.

7.6.5 Use

Grape juice concentrate is used in juice manufacture, marmalades, confections, yogurt, ice cream, and canned fruit.

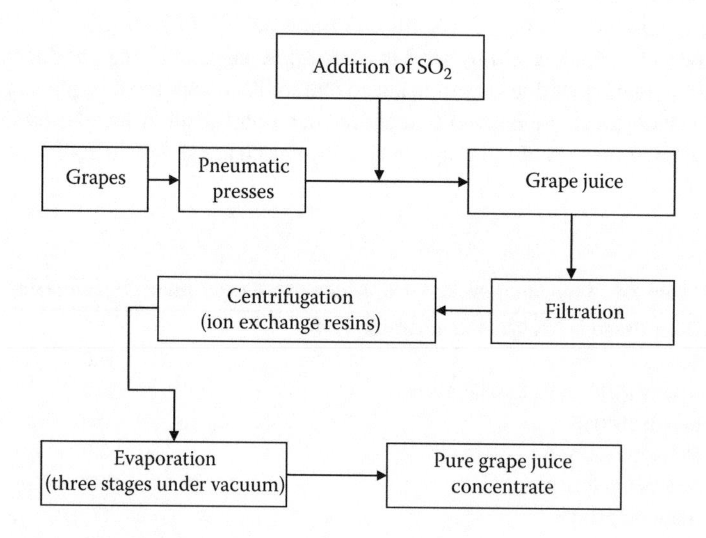

Figure 7.7 Manufacture of pure grape juice concentrate.

7.6.6 Sale

Grape juice concentrate is sold in liquid form at 56.0 ± 1°Bx in bulk or in aseptic vats of 250 kg.

7.6.7 Manufacture of Grape Juice Concentrate

In natural grape juice concentrate from selected grapes manufactured under vacuum, juice is transparent with a yellow-orange color. The final product is stabilized, filtered, pasteurized, and aseptically packaged (Figures 7.8 and 7.9). Processing guarantees top product quality and authenticity. Water addition brings it to its initial state.

During the manufacture of grape juice concentrate, grapes are collected clean after washing and taken to pneumatic presses, where they are crushed. During crushing, SO_2 is added to prevent alcoholic fermentation. Then, grape juice is filtered to remove the grape solids, and defumigation is carried out. Filtered grape juice is then evaporated at 65°Bx under vacuum in three stages at 90°C, 65°C, and 45°C and cooled at 10°C (Figure 7.10). Juice is then filtered again, pasteurized, and aseptically packaged. The manufacture of grape juice concentrate is described in Figure 7.11.

Figure 7.8 Pasteurizer.

Figure 7.9 Aseptic filler.

Figure 7.10 Grape juice concentrate.

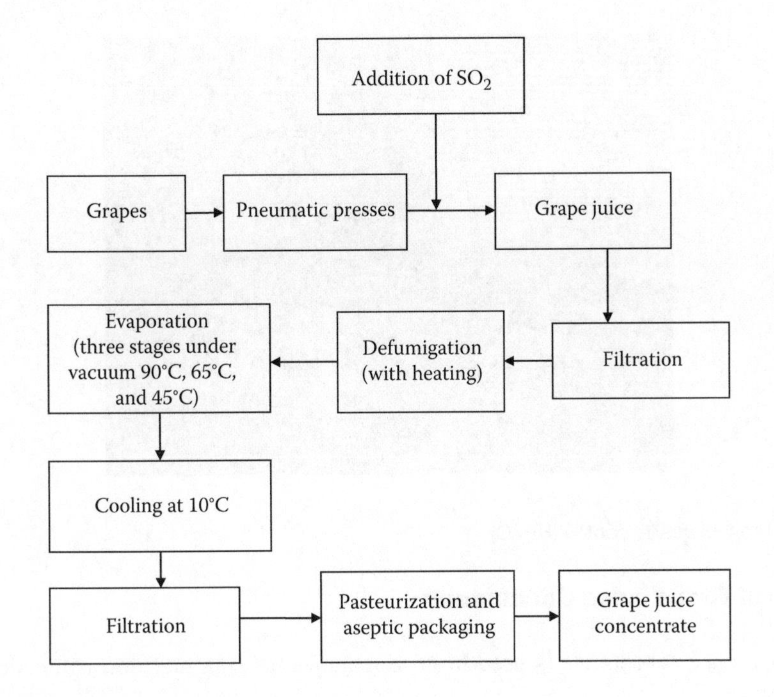

Figure 7.11 Manufacture of grape juice concentrate.

Figure 7.12 Typical evaporator unit.

7.6.8 Quality Control of Grape Juice Concentrate

A typical evaporation unit is shown in Figure 7.12. Each production lot is analyzed in the quality control department (Figure 7.13) and is checked according to the specifications set up by each company or producer.

Figure 7.13 Product quality control (HPLC).

7.6.9 Uses of Grape Juice Concentrate

Pure grape juice concentrate is used in juice manufacture, marmalades, confections, and the bakery industry.

7.6.10 Storage, Distribution and Sale of Grape Juice Concentrate

Pure grape juice concentrate is sold in liquid form at 65.0 ± 1°Bx in bulk or in aseptic sacks of 270 kg (Figure 7.14).

Figure 7.14 Storage tanks.

7.7 PETIMEZI

Petimezi (the origin of the word is Turkish) is a Turkish syrup-like liquid obtained after the evaporation of juices of (specifically) grape, fig, or mulberry by boiling with coagulants. It is used as a syrup or mixed with tahini for breakfast.

The same product with another name appears in ancient Greece. It was called *epsetos* (thin syrup) or *siraion* (concentrated syrup) with unknown uses. Romans called it *defrutum* (when must was halved during boiling), *caroenum* (when two-thirds was evaporated during boiling), or sapa (thick syrup). Finally, Italians named it *saba* or *vino cotto.*

7.7.1 Pekmez

Most of the grape products in Turkey are in the form of pekmez and raisins, particularly sultanas. There is also another kind of pekmez made from carob, called *keçiboynuzu pekmezi* or *harnup pekmezi* in the Turkish language (Ermurat 2006).

7.7.2 Production

Grapes and figs reserved for production are squeezed by a mortar or presser to produce grape juice. The juice is cooked at 50°C–60°C for 10–15 min. A special kind of sterile white soil, used to purify the juice from particles, should be added to the juice while it is being cooked. Soil is also useful for balancing the taste. The Ministry of Agriculture in Turkey advises producers to use 1–5 kg of soil for 100 kg of grape juice.

Grape juice is stored in containers for 4–5 h to ensure that all particles in the juice sink to the bottom. The liquid part of the juice is collected from the container and transferred to boilers for further heating. This boiling process gives pekmez its dark color and high viscosity. Pekmez has a dark color because of the caramelized sugar in the boiled grape juice. The juice loses its fluidity because of water loss by evaporation. Additionally, pekmez produces a special smell when prepared in cooking.

Petimezi is the thick, viscous syrup made by boiling must, usually found in areas making wine (Figure 7.15).

Athinaios of the ancient Greek gastronomy of the first century after Christ refers to the repulsion of humans to cannibalism because of the growth of cooking focusing on a dish with fish and

Figure 7.15 Petimezi, the modern revival of an ancient sweetener.

petimezi. This dark viscous elixir has sweetened the palate for many years now. For the current petimezi, this viscous syrup is made by boiling and removing the must (grape juice before the start of fermentation). It is found throughout Greece in towns and villages where people make their own wines and in selected shelves of supermarkets. Autumn is the time when cooks in villages use fresh must and petimezi to make biscuits, custards, cakes, breads, and halva even sweeter. In modern cooking, this sweetener has had a sort of revival, added beyond confectionery products to dishes with meats, fish, and vegetables and to sauces and dressings to satisfy the Greek appetite searching for something new and innovative. However, modern chefs know very little about this sweetener, which comes from Greek ancestors some thousands of years ago.

The ancient Greeks used several variations of sweet grape juice, distinguished by the degree to which they were boiled down, or concetrated (http://dianekochilas.com/889/grape-must-in-the-greek-cooking). There was the highly prized *gleukos*, which denoted either the fresh, free-flow must that was obtained from fully-ripe or overripe grapes before treading or the fermented product of this must; then there were several concentrated forms: *hepsema*, which was grape juice concentrated by boiling until rather thick, and which was a cheap substitute for honey; *hepsetos*, which was less concetrated; and, finally, *siraion*, yet another concentrated grape must, but to what degree or thickness it is hard to determine.

In Limnos, a fresh homemade pasta is boiled in grape must and served up as a hearty winter dessert. In Tinos, trays of moustalevria, once set, are left to dry in the sun and then dredged in cinnamon and sesame. A similar sweet is made in Roumeli called *moustopita*, and is sun-dried moustalevria sprinkled with sesame and seasoned with either bay leaf or basil.

Sutzuk lukum is another sweet, which has walnuts in a row dipped in moustalevria and then left to dry, a process repeated many times until a sweet with walnuts is made gelatinized and sausage-like.

Petimezi acts as a preservative in country cuisine, more specifically in sweets in a jar made with eggplant and pumpkin. Petimezi is poured on baked bread in tsaletia (pancakes from bobota, a sweet porridge made of flour and an old dish from Samos) and used in small spinach pies made on the island of Tinos like fesklopita, in most traditional dishes, and in at least one sweet and sour dish with fish, savoro, in different versions from the Cyclades.

To make 2 L of petimezi, we require 5 L of fresh must and two 5-cm cubes of stale bread. Must is poured in a big pan with bread.

We heat at a low temperature. When the must starts to boil, the foam is removed with a spoon. This is continuously done while the must is boiling slowly. Cooking is carried out until 5 L boil down to 2 L. A thick and viscous mixture will be made, with a density approximately between that of oil and honey. Then, it is left to cool and is stored in a clear jar in a cool and dry place.

The most common sweets made with must are moustalevria, a kind of custard, and moustokouloura, thick brown rolls that become soft and crispy with the taste of must (Kochilas 2009).

Cretan petimezi comes from the evaporation of must from selected mature grapes in Crete. It has a wonderful aroma and taste. It is made traditionally and used as a sweetener in homemade sweets, pancakes, doughnuts, and others. It is an excellent syrup for ice cream. This product was used from Greek ancestors as a heating agent during war and as a sweetener in confections during the period when sugar was a luxury because of its high price.

Traditional grape syrup can be used today at a rate of 1:1 instead of sugar in all sweets, even in bread and butter, whereas it is the main ingredient and preservative for sweets in a jar called *retselia*. In addition, throughout Greece and on many islands, we find pumpkin or quince preserved in petimezi. In Crete, petimezi is poured on figs before preserves are made. The most extraordinary recipe comes from Naoussa, where grape syrup is called honey and is made from a grape variety that is black and sour. One of the most traditional local sweets is *kores* in honey, a fruit preparation with roughly cut eggplant preserved in grape syrup.

Traditional grape syrup gives aroma, color, and many nutrients wherever it is used. It can also be used in the manufacture of moustalevria instead of must (Figure 7.16).

Figure 7.16 Moustalevria.

Petimezi is suitable for the following (http://www.petimezi.gr/petimezi.html):

- Moustalevria, must rolls, and moustokalamara (must in two to three parts water, depending on the sweetness required, and one part petimezi)
- Liquor petimezi (three parts petimezi and one part tsikoudia or Cretan raki)
- Petimezi shot (three parts tsikoudia and one part petimezi)
- Ice cream syrup (one spoonful in a bowl)
- Sweetener in yogurt (two spoonfuls in a bowl)
- Refreshing drink (one part petimezi and four parts cubed water)
- Iced drink (one part petimezi and four parts trimmed ice cubed water)

7.7.3 Nutritional Value

Petimezi is a natural syrup that has no additives or other preservatives and is rich in energy, calcium, iron, and vitamins with all the nutritional benefits of grape (tannins and anthocyanins). Thus, in the past, it was used as a medicine for anemia, a psychic energizer, and a soothing agent. However, its main use is as a sweetener, because it contains natural sugars metabolized easily to enhance taste and product freshness.

Nutritional ingredients and mineral values of petimezi compared to stafidin and sykin are shown in Table 7.5.

Table 7.5 Comparative Data on Nutritional Ingredients from Petimezi, Stafidin, and Sykin

Nutritional Ingredients and Minerals	Petimezi	Stafidin	Sykin
Carbohydrates (g)	80.9	70.5	91.5
Fat (g)	0.4	0.15	0.22
Proteins (g)	0.9	1.6	1.99
Calcium (mg)	74	30	-
Phosphorous (mg)	40	79	-
Iron (mg)	1.2	3	-
Energy (kcal)	330	245	385

7.7.4 Greek Petimezi Manufacture

Five kilograms of must from grape juice is transferred to a stainless-steel container and boiled for 5 min. Then, the heat is lowered, and sugar or ash from firewood is added, attached on a thin cloth. Bread or sugar might be added to neutralize the acids of the juice. Stirring of the must is carried out during boiling. Foam is continuously removed until the juice is clear and transparent (all dissolved solids are deposited). The pure juice is then boiled again until it is darkened and becomes thick syrup. Must is then allowed to rest for 3–4 h and passed through a thin cloth in a clean pot. Boiling continues until evaporation (like honey), and the color becomes dark. Storage in glass or clay containers is recommended for 3 years.

7.8 CAROB SYRUP

Made from the world's best varieties of carobs, carob syrup is a healthy nutritional product rich in calcium, phosphorous, iron, proteins, and carbohydrates. Carob syrup can be used as a sweetener for cakes, breakfast cereals, and hot drinks or beverages mixed with cold water or milk. Carob has natural therapeutic qualities and helps with stomach upsets and with the smooth functioning of the bowels.

Carob juice concentrate is derived from carob trees, a tree cultivated in Mediterranean countries, especially in Cyprus. Cypriot carob syrups are organic by nature, because they are cultivated and germinate solely in the wild. Marley soils favor carob growth, giving a pleasant and sweet taste in their fruits and in carob syrup. Sugar concentration is rich, exceeding 50%.

Carob syrup is manufactured after the extraction of carobs, which were the main natural sweeteners for years. There are reports in religious (the bread of John the Baptist) and historic (Spanish war and the First and Second World Wars) books. It is richer in calcium than milk (it contains 350 mg/100 g product, whereas milk contains 125 mg/100 g). It is rich in phosphorous (80 mg), protein (4%), carbohydrates (63%), vitamins (A and B), and low quantities of micronutrients (sodium and iron), and its density is 68°Bx. The nutritional analysis is shown in Table 7.6.

7.9 DATE SYRUP

Dates contain up to 70% total sugar, from which 50% is inverted sugar, and have a water activity of 0.6 with a pH of 5.5. Dates are processed into a concentrated liquid form of date syrup, which is convenient for industrial users. This date syrup liquid brings its sweetness and flavor to baked goods and to all types of liquid food and beverage applications (Al-Farsi et al. 2010).

To produce date syrup, all soluble solids are extracted from the fruit, filtered, and concentrated to 70°Bx with pH 4.8–5.0. The natural, no-preservatives-added, concentrated date syrup has up to a

Table 7.6 Nutritional Analysis of Carob Syrup

Parameters	Method	Unit	Results
Proteins (N × 6.25)	ISO1871: 75	%	1.9
Fat	Soxhlet	%	<0.5
Carbohydrates	HPLC	%	62.7
Energy (kcal)		kcal/100 g	259
Solid residue/marc	AOAC 32.1.03:99	%	67.3
Ash	AOAC 923.03:99	%	2.5

1-year shelf life and blends easily into batters for sweet breads, cakes, muffins, cookies, and bakery products, providing sweetness and humectancy, which helps retard staling.

Date syrup liquid offers opportunities to the food industry to create unique liquid products, e.g., juice blends, yogurt drinks, and other beverages. In addition, it can be used in spirits, wines, and vinegars by the fermentation of the sweet syrup liquid.

7.10 BIRCH SYRUP

Birch syrup is a sweetener made from the sap (1%–2% sugar) of birch trees found mostly in Alaska, Russia, Ukraine, and Scandinavia resembling maple syrup (Bauman 2005). The product contains approximately 42%–52% fructose, 44% glucose and other sugars (e.g., sucrose and galactose), and it is used like maple syrup as a flavoring for pancakes, candies, sauces, and dressings, as well as for beers, wines, and beverages.

7.11 PALM SYRUP

Palm syrup or palm honey is a sweet syrup that is produced in the Canary Islands and on the coast of South Africa from the sap of *Phoenix canariensis* trees (or *Jubaea chilensis* in Chile). The process involves collecting this sap, known as guarapo, and concentrating it into syrup by boiling for several hours until a reduction in volume by 90% to a dark brown, rich, sweet syrup, and then, the syrup is packaged (Millares 2011). Four to six months after harvesting, palm syrup tends to thicken and crystallize.

Palm syrup is widely used in southern Asian cooking and baking (pastries and desserts) as a flavor enhancer in spicy curries, as a refreshing agent in drinks, and in the production of various alcohols such as arrack by fermentation. It is also used for general medicinal purposes.

7.12 MEDICATED SYRUP

Medicated syrups are aqueous solutions that contain some other recipients, such as, except for sugars, the following:

- Sugar polyols (glycerol, sorbitol, and maltitol)
- Acids (citric acid)
- Preservatives (parabens, benzoates, and antioxidants)
- Chelating agents [sodium ethylenediaminetetraacetic acid (EDTA)]
- Flavorings
- Bufferings
- Colorings

Sugar-free syrups are those in which sugar has been replaced by polyols (e.g., glycerol or sorbitol) or by artificial sweeteners (e.g., aspartame, neotame, and sucralose), polysaccharides (e.g., carrageenan and xanthan gum), and others. Syrups are usually sugar–water solutions and are prepared as follows:

- Solubilization of a sugar ingredient into purified water
- Mixing with or without heating
- Strain, where needed
- Addition of water to the right volume

REFERENCES

Al-Farsi, M., C. Alasalvar, M. Al-Abid, K. Al-Shoaily, M. Al-Amry, and F. Al-Rawahy. 2008. Compositional and functiona characteristics of dates, syrups and their by-products. *Food Chemistry*, 104: 943–947.

Batu, A. 2005. Production of liquid and white solid pekmez in Turkey. *Journal of Food Quality, 28*: 417–427

Baumann, M. 2005. Haines birch syrups attract gourmet following. *Alaska Journal of Commerce*, May 29, 2005.

Camire, M. E., and M. P. Dougherty. 2003. Raisin dietary fiber composition and *in vitro* bile acid binding. *Journal of Agriculture and Food Chemistry*, *51*, 834–838.

Cargill-Cerestar BVBA. 2003. Application for the Approval of Isomaltulose Regulation (EC) No. 258/97 of the European Parliament and of the Council of 27 January 1997 Concerning Novel Foods and Novel Food Ingredients.

Cheetham, P. S. J., and C. Bucke. 1999. Production of isomaltulose using immobilized bacterial cells. In: *Carbohydrate Biotechnology Protocols* Series: *Methods in Biotechnology*, vol. 10, 255–260, DOI: 10.1007/978-1-59259-261-6_20.

Codex Alimentarius Commission. 2000. Draft revised standard for gluten free foods (CX/NFSDU 98/4). Codex Committee on Nutrition and Foods for Special Dietary Uses, 22nd session, Berlin, Germany

Eggleston, G., and M. Grisham. 2003. Oligosaccharides in cane and their formation on cane deterioration. *ACS Symposium Series*, *849*, 211–232.

Ermurat, Y. 2006. Carbonation of evaporated grape syrup "pekmez" formulation and production of carbonated evaporated grape syrup pekmez soft drinks: Pekmez gazoz, pekmez cola and pekmez limon. *Electronic Journal of Environmental, Agricultural and Food Chemistry, 5,* 1: 1221–1223.

European Commission (EC). Commission Regulation (EC) No. 1573/1999 of 19 July 1999 laying down detailed rules for the application of Council Regulation (EC) No. 2201/96 as regards the characteristics of dried figs qualifying for aid under the production aid scheme. *Official Journal of the European Communities* L 187/27–31.

Food Standards Australia New Zealand (FSANZ). 2007. Final Assessment Report Application A578 Isomaltulose As A Novel Food. http://www.foodstandards.gov.au/scienceandeducation/factsheets/factsheets/isomaltulose.cfm.

General Chemical State Laboratory. 1976. Greek code of food and drinks. Athens, General State Laboratory.

Gutierrez, D. 2008. New low-glycemic sweetener isomaltulose (Palatinose) approved by FDA. http://www.naturalnews.com/022601.html. Accessed July, 2011.

Irwin, W. E., and P. J. Sträter. 1991. Isomaltulose. In: L. O'Brien-Nabors and R. C. Gelardi (eds.), *Alternative Sweeteners, 2nd Rev. Expanded Ed.* 299–307. New York: Marcel Dekker.

Karathanos, V. T., A. F. Kostaropoulos, and G. D. Saravacos. 1994. Viscoelastic properties of raisins. *Journal of Food Engineering*, *23*, 481–490.

Kislev, M. E., A. Hartmann, and O. Bar-Yosef. 2006a. Early domesticated fig in the Jordan Valley. *Science*, *312: 5778*, 1372. doi:10.1126/science.1125910.

Kislev, M. E., A. Hartmann, and O. Bar-Yosef. 2006b. Response to comment on "Early domesticated fig in the Jordan Valley." *Science*, *314: 5806*, 1683b. doi:10.1126/science.1133748.

Kochilas, D. 2009. Petimezi: The modern revival of an ancient sweetener. http://www.tanea.gr/potitismos/article/?aid=4536889. Accessed July 2011.

König, D., W. Luther, V. Poland, S. Theis, G. Kozianowski, and A. Berg. 2007. Metabolic effects of low-glycemic Palatinose™ during long-lasting endurance exercise. *Annals of Nutrition Metabolism, 51*: 1, 69.

Lev-Yadun, S., G. Neeman, S. Abbo, and M. A. Flaishman. 2006. Comment on "Early domesticated fig in the Jordan Valley." *Science*, *314: 5806*, 1683a. doi:10.1126/science.1132636.

Lina, B., D. Jonker, and G. Kozianowski. 2002. Isomaltulose (Palatinose): A review of biological and toxicological studies. *Food and Chemical Toxicology*, *40: 10*, 1375–1381. doi:10.1016/S0278-6915(02)00105-9. PMID12387299.

Matz, S. A. (ed.). 1989. Sweeteners and malt syrup. In: *Technology of the materials of baking* (96). New York: Elsevier Science Publishers.

Matz, S. A. 1996. *Ingredients for bakers*. McAllen, TX: Pan-Tech International, Inc.

Millares, Y. 2011. De la savia de las palmeras, un jarabe llamado "miel." http://www.pellagofio.com/?q=node/370. Accessed July 2011.

Nakajima, Y. 1984. Palatinose production by immobilized α-glucosyl-transferase, *Proc. Res. Soc.* Japan Sugar Refineries Tech. *33*: 54–63.

Ofiyeva, M. 2011. Pure concentrate. http://www.dermaharmony.com/ingredients/prune.aspx. Accessed October 2011.

Okuda, Y., K. Kawai, Y. Chiba, Y. Koide, and K. Yamashita. 1986. Effects of parenteral palatinose on glucose metabolism in normal and streptozotocin diabetic rats. *Hormone and Metabolic Research, 18*, 361–364.

Papadakis, S. E., C. Gardeli, and C. Tzia. 2006. Spray drying of raisin juice concentrate. *Drying Technology, 24: 2*, 173–180.

Payne, T. J. 1994. California Raisin Report. California Raisin Advisory Board Food Technology Newsletter.

Porter, M. C., M. H. M. Kuijpers, G. D. Mercer, R. E. Hartnagel, Jr., and H. B. W. M. Koeter. 1991. Safety evaluation of Protaminobacter rubrum: Intravenous pathogenicity and toxigenicity study in rabbits and mice. *Food and Chemical Toxicology, 29:* 10, 685–688.

Puoci, F., F. Iemma, U. G. Spizzirri, D. Restuccia, V. Pezzi, R. Sirianni, L. Manganaro, M. Curcio, O. I. Parisi, G. Cirillo, and N. Picci. 2011. Antioxidant activity of a Mediterranean food product: Fig syrup. *Nutrients, 3*, 317–329. doi:10.3390/nu3030317.

Purdue University. 2011. Horticulture and Landscape Architecture: Fig, *Ficus carica*. Retrieved from http://www.hort.purdue.edu/newcrop/morton/fig.html#Toxicity. Accessed July 2011.

Riedel, H. R. 1976. Seedless Californian raisins for chocolate and confectionary products. *Confectionary Production, 8:* 42–44.

Sabanis, D., C. Tzia, and A. Papadakis. 2007. Effect of different raisin juice preparations on selected properties of gluten-free bread. *Food and Bioprocess Technology*. doi:10.1007/s11947-007-0027-9.

Sanders, L. M. 2009. Disorders of carbohydrate metabolism. http://www.merckmanuals.com/home/childrens_health_issues/hereditary_metabolic_disorders/disorders_of_carbohydrate_metabolism.html. Accessed July 2011.

Sanders, S. W. 1991. Using prune juice concentrate in whole wheat bread and other bakery products. *Cereal Foods World, 36*: 280–283.

Scientific Committee on Food (SCF). 1984. Report of the Scientific Committee on Food on Sweeteners (Opinion expressed in 1984), 16th Series, 1985. Commission of the European Communities. Food-science and techniques http://ec.europa.eu/food/fs/sc/scf/reports/scf_reports_16.pdf

Scientific Committee on Food (SCF). 1997. Minutes of the 107th Meeting of the Scientific Committee on Food. 12–13 June.

Shin Mitsui Sugar Company. 2003. Palatinose Catalogue.

Siddiqua, I. R., and B. Furhala. 1967. Isolation and characterization of oligosaccharides from honey—Part I: Disaccharides. *Journal of Apicultural Research, 6*, 139–145.

Simsek, A., N. Artik, and E. Raspinar. 2004. Detection of raisin concentrate (pekmez) adulteration by regression analysis method. *J. Food Composition and Analysis, 17*, 155–163.

Solomon, A., S. Golubowicz, Z. Yablowicz, S. Grossman, M. Bergman, H. E. Gottlieb, A. Altman, Z. Kerem, and M. A. Flaishman. 2006. Antioxidant activities and anthocyanin content of fresh fruits of common fig (*Ficus carica* L.). *Journal of Agricultural and Food Chemistry, 54*, 7717–7722.

Südzucker AG. 1996. Amendment of GRAS Petition 6G0321. 29 January 1996. In: Cerestar, 2003.

Takazoe, I. 1985. New trends on sweeteners in Japan. *International Dental Journal, 35: 2*, 58–65.

USDA Nutrient Database. 2009. Dried figs nutrition facts-dried figs calories. http://www.lose-weight-with-us.com/dried-figs-nutrition.html. Accessed October 2011.

Van Can, J. G. P., T. H. Ijzerman, L. J. C. Van Loon, F. Brouns, and E. E. Blaak. 2009. Reduced glycaemic and insulinaemic responses following trehalose ingestion: Implication for postprandial substrate use. *The British Journal of Nutrition, 102:* 10, 1408–1413.

Van Can, J. G. P., L. J. C. Van Loon, F. Brouns, and E. E Blaak. 2011. Reduced glycaemic and insulinaemic responses following trehalose ingestion: Implications for postprandial substrate use. First view article. *The British Journal of Nutrition, 102: 10*, 1–8.

Veberic, R., M. Colaric, and F. Stampar. 2008. Phenolic acids and flavonoids of fig fruit (*Ficus carica* L.) in the northern Mediterranean region. *Food Chemistry, 106: 11*, 153–157. doi:10.1016/j.foodchem.2007.05.061.

Vinson, J. A. 1999. Functional food properties of figs. *Cereal Foods World, 44: 2*, 82–87.

Vinson, J. A., L. Zubik, P. Bose, N. Samman, and J. Proch. 2005. Dried fruits: Excellent *in vitro* and *in vivo* antioxidants. *Journal of the American College of Nutrition, 24: 1*, 44–50.

West, D., R. Morton, J. W. Stephens, S. C. Bain, L. P. Kilduff, S. Luzio, R. Still, and R. M. Bracken. 2011. Isomaltulose improves postexercise glycemia by reducing CHO oxidation in T1DM. *Medicine and Science in Sports and Exercise, 43: 2*, 204–210. doi: 10.1249/MSS.0b013e3181eb6147.

Ziemke, W. H. 1977. Raisins and raisin products for the baking industry. *Baker's Digest, 4: 51*.

INTERNET SOURCES

Innova Market Insights for FoodIngredientsFirst.com, June 2010. Global Sugars and Sweeteners Product Trends.

http://www.petimezi.gr/petimezi.html. Accessed July 2011.

Prune and senna for chronic constipation. eHow.com. http://www.ehow.com/about_6712505_prune-senna-chronic-constipation.html. Accessed July 2011.

http://www.californiadriedplums.org/industrial/products/juice-concentrate. Accessed October 2011.

Grape Must in Greek Cooking. http://dianekochilas.com/889/grape-must-in-the-greek-cooking. Accessed December, 2011.

Health Directories. http://www.health-directories.com/constipation-prunejuice.html.

Cretan Petimezi. http://www.petimezi.gr/petimezi.html.

CHAPTER 8

Application of Sweeteners in Food and Drinks (Bakery, Confectionery, Dairy Products, Puddings, Fruit Products, Vegetables, Beverages, Sports Drinks, Hard Candies, Loukoumia, Marmalades, Jams, Jellies, Baked Goods, Sorbet)

Theodoros Varzakas and Barbaros Özer

CONTENTS

8.1 INTRODUCTION

Sweeteners or sweetener blends are being used to either enhance existing products or create new ones. In applications ranging from energy drinks to functional beverages, sweeteners can give mild sweetness and simultaneously enhance mouthfeel. Moreover, sweeteners in combination with bulk sweeteners optimize flavor delivery in beverages. In confectionery products, including fondants, crèmes, fudge, hard candies, caramels, and toffees, the incorporation of sweeteners can reduce sweetness intensity, enhance the flavor profile, reduce any off flavors resulting from browned or intense notes, and aid in crystallization control. In chewing gums, sweeteners can be used to modify sweetness and extend the overall flavor release. Moreover, the acid stability, nonreducing character, and low sweetness of sweeteners, as well as the protein protection effect from freezing and desiccation, help in the extension of the shelf life of processed foods and maintenance of their texture, color, and flavor. Finally, the same effect can be seen in dairy products, where the nonhygroscopic nature of sweeteners reduces moisture sensitivity and product clumping while giving a pleasant, mild, sweet flavor to pasteurized dessert toppings, puddings, or yogurts.

8.2 PRODUCTION OF HARD CANDIES

Candies are a confectionery product that consist of sugar (56%), corn syrup (28%), and water (16%), i.e., a mass of sweeteners with a glassy or not glassy appearance. With these ingredients, aromatic compounds, coloring agents, citric acid, or other additives could be incorporated, depending on the product (e.g., mint, orange, or lemon). Processing of candies includes the following steps:

1. Boiling of sugar solution
2. Processing under vacuum (humidity removal)
3. Addition of aromatic compounds, coloring agents, and other additives
4. Kneading and cooling
5. Shaping in matrices
6. Cooling
7. Packaging

The basic quality points in the production of hard candies are described as follows:

- *Heat intensity*—Temperatures reaching 145°C–148°C in 12–15 min. Any deviation from the stated processing parameters results in product browning. The maximum limit of temperature should not exceed 155°C–160°C (browning of caramel mass). The higher the temperature is, the harder the mass becomes, and the faster the glassiness appears.
- *Cooling of candy mass*—This is carried out using fans.
- *Packaging*—Packaging should be carried out below 20°C–30°C in sealed containers and rooms with low humidity.
- *Residues*—These residues are incorporated into the mass up to 10%–20%, after grinding in small granules.
- *Hygroscopicity*—Viscosity problems are a result of the high relative humidity of the atmosphere.

In order to produce an extrasaturated sugar solution in which sugar is not seen crystallized, the appropriate quantity of water should be used in relation to temperature, as seen in Table 8.1.

Color formation in glucose syrups during the manufacture of boiled candies can be a serious problem for the confectionery industry, as it may lead to the loss of acceptable color and to the development of off flavors (Kearsley and Birch 1985).

Table 8.1 Correlation between Temperature and Solids for Saturated Sugar Solution

Water Temperature (°C)	Sugar Solids (%)
20	67
40	70
60	75
80	78
121	90
160	98

There is a lack of information about reducing the end color of syrups by effective methods, such as the adsorption process.

Adsorption of Maillard reaction products (MRP) onto a styrene–divinylbenzene copolymer-based adsorbent resin (AR) was studied by Serpen, Atac, and Gokmen (2007). The effects of various parameters, such as temperature, AR concentration, contact time, and sugar concentration, were investigated.

Among various isotherms, the equilibrium data fitted best to the Guggenheim–Anderson–De Boer (GAB) isotherm, which indicated a multilayer adsorption of MRP onto AR. The temperature strongly influenced the adsorption process. The thermodynamic parameters Ea, DH, DG, and DS indicated the adsorption process was exothermic and spontaneous, which is favored at lower temperatures, and also occurred at both physical adsorption and weak chemical interactions. The adsorption of MRP onto AR obeyed the pseudo-second-order kinetic model and occurred at both surface sorption and intraparticle diffusion. The pore diffusion controlled the adsorption process at low sugar concentrations, and the film diffusion governed the adsorption mechanism at high sugar concentrations. It was concluded that the adsorption process should be performed at sugar concentrations lower than 50% to overcome film resistance.

When the temperature reaches 80°C, the concentration of sugar solids increases to 78%, i.e., an increase of 12% is evident (Table 8.1).

8.3 CHEWING GUM FORMULATION AND PROCESS

Gum base is heated at 71°C, and liquid sorbitol is added to the base. Then, one-third dry sorbitol is added, followed by another one-third dry sorbitol. Glycerine/flavor is mixed in, and the other one-third of dry sorbitol is added. Then, the sweetener is added, and the mixture is cooled to 43°C and extruded to tablets, balls, or strips. During the initial steps until cooling, kneading for 5 min is carried out after each step (Figure 8.1).

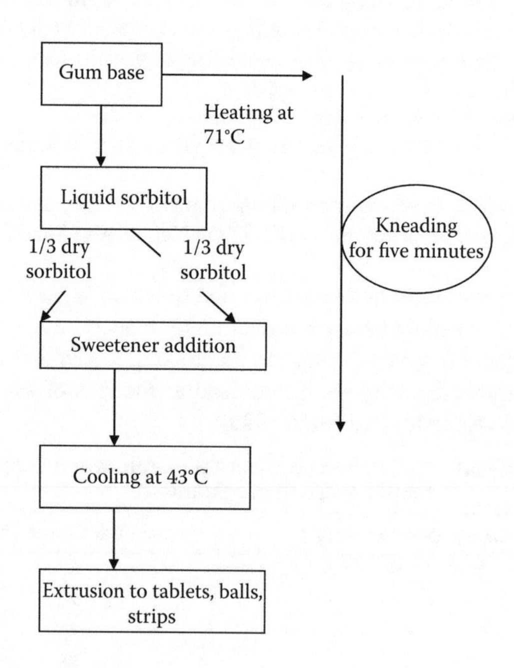

Figure 8.1 Chewing gum manufacture.

Trehalose, because of its lower solubility, can be used in chewing gum to modify sweetness and extend flavor release. Its sweet flavor can persist slightly longer than sucrose; hence, the longer the extension of sweetness, the longer the perception of the flavor system (Pszczola 2003).

8.4 LOUKOUMIA PRODUCTION

Loukoumi is a Greek product defined as a sweet starchy gel enriched with aromatic compounds. The major ingredients added are starch, corn syrup, and sugar or sweeteners. The speed and the strength of the produced gel depends on the following:

1. *Type and quantity of starch.* Starches, predominantly the linear polymer amylase, have a difficult gelatinization and only dissolve in water under high temperatures. The absence of amylose results in easy and fast gelatinization. Moreover, the texture of a potato starch gel is different from that of cornstarch gel.
2. *Percentage of added humidity.* Water added should reflect the hydration of starch and other substances and should not be very high.
3. *Type and quantity of additives.* These are very important, because they can cause undesired hydrolysis. The addition of preservatives and aromatic and coloring agents should occur at the end of processing.
4. *Type and quantity of added sweeteners.* Sweeteners compete with starch in the absorption of humidity, increasing the total solids and creating a strong gel. However, the type and the quantity of sweeteners used is of great importance, because a very high sweetness could result.
5. *Speed and way of heating.* The degree and the rate of gelatinization, as well as the final product quality, depend on the rate and way of heating.

Loukoumia production is detailed as follows:

1. Boiling of starch and water with intense agitation
2. Addition of water, corn syrup, and sugar under heating with continuous agitation
3. Cessation of heating and addition of honey, citric acid, and aromatic compounds
4. Placement of gel in shallow discs containing small amounts of scattered fine sugar
5. Shaping and cutting in cubes
6. Fine sugar coating

8.5 GRANITA OR SORBET PRODUCTION

Good frozen sorbet has the look and texture of a stack of snow. The production is easy compared to other frozen desserts and is carried out quickly. It is based on saccharose, water, and juice or fruit pulp. These three ingredients are mixed in the same machine as used for ice cream production, and then additives, flavoring agents, taste enhancers, or coloring agents are added. The dose of sugar is important for the successful preparation of sorbet because of the following reasons:

1. If less saccharose is added, the sorbet produced will be tasteless and will be cooled more than necessary, so it will form like dry snow.
2. If extra saccharose is added, this might delay freezing, and the sorbet might end up like a dense syrup. It should be noted that, when fruits are used with a sweet flavor for the production of sorbet, the addition of lemon juice is required.

8.6 MARMALADE PRODUCTION

Marmalade, confitures, gels, and confectionery products, in general, are derived from the thermal processing of fruits and juices at high sugar concentrations, approximately 65%–75%. In most cases, these products arise from the boiling of fruits or juices with sugar. The final product has a semisolid shape, and it may contain the fruit pulp and is named marmalade, the whole fruit or the fruit in pieces and is called jam, or the fruit juice only and is called jelly. Marmalade is derived from the Portuguese word *marmello*, meaning "quince," which was first used in the past for marmalade production. The basic ingredients are the following:

- **Fruits.** They contribute to the final product with their sugars, part or all of their pectin, their acids required to form the gel, their coloring agents, their cellulose, their vitamins, and their inorganic compounds. For the manufacture of marmalade, ripe fruits are used and not overripe ones, because the aromatic substances and pectins of the latter have deteriorated.
- **Sugars.** The final product for marmalades, jams, and jellies should contain approximately 65%–70% sugars. Hence, sugar addition during fruit preparation is essential in such quantity to achieve the correct final concentration at the end of processing. Hence, in order to prepare 100 g of marmalade from strawberries (9% sugars), an additional 61 g of sugar should be added. Instead of adding sugar, sweeteners or glucose syrup in proportion to the sugar can be added to avoid crystallization of the marmalade during preservation.
- **Acids.** They enhance taste and aroma and contribute to gel formation. The pulp or fruit juice pH should range between 3.0–3.3. Low-acidity fruits need the addition of citric acid or lemon juice to correct the acidity.

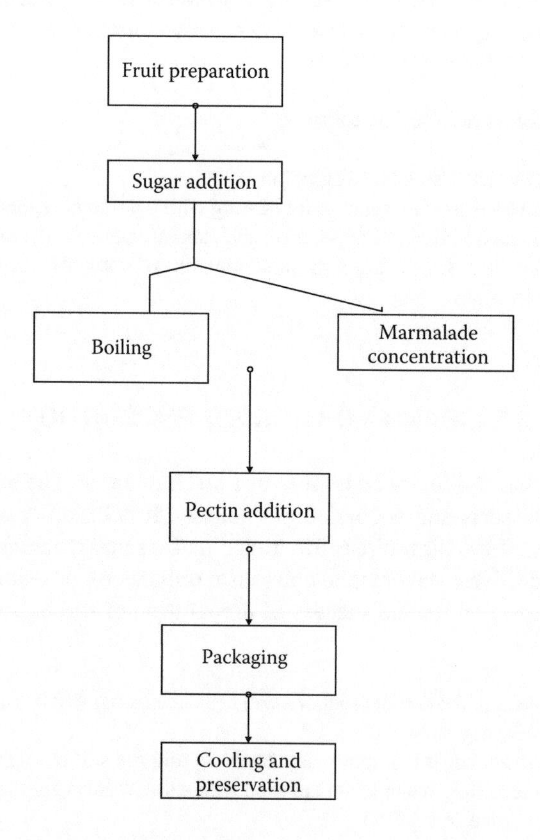

Figure 8.2 Flow diagram of marmalade preparation.

- **Pectin.** Natural fruit pectin is low in concentration; hence, commercial pectin is usually added. Pectin is a polysaccharide consisting of small molecules of galacturonic acid. Pectin power differs in fruits. Hence, in quince marmalades, pectin should not be added, because its pectins have good thickening strength. Pectins are distinguished as slow and fast thickening.

Marmalade preparation is shown in Figure 8.2 and is described as follows:

- Fruit preparation depends on the type of fruit and the type of product; for example, marmalade requires mashed fruits, whereas confiture requires thick pulp with small fruit pieces.
- Saccharose to be added is determined by the sugar concentration in the fruit. The final product should have total sugars of approximately 65%–70%.
- Boiling is carried out under vacuum. The purpose of this step is to remove water evaporating the product to the desired sugar level.
- Pectin is added just before the end of boiling.
- The product is packaged in glass jars while still warm (80%–85%). Following lid insertion, the jars are reversed to sterilize the product. Rapid cooling follows. Additives may be added to the mixture for better preservation.

8.7 CANNING FRUIT

The process of canning fruit is described as follows:

1. Fill a water-bath canner with enough hot water to cover the tops of the jars. Place the canner on heat.
2. Discard unsuitable jars with cracks or chips in their sealing surface.
3. Scald the jars and leave them in hot water.
4. Prepare an ascorbic acid solution and make a suitable syrup for fruit.
5. Wash and sort the fruit, selecting firm ripe fruit.
6. Pack the fruit into hot jars. A hot pack may be used for all fruits. Precook the fruit in syrup. Pack boil hot fruit loosely into hot jars, leaving appropriate headspace. Cover the fruit with boiling syrup, leaving the same amount of headspace.
7. A cold pack may be used for all fruits, except for apples, apple sauce, and rhubarb. Pack the fruit firmly into jars, leaving appropriate headspace. Pour boiling syrup into the jar, leaving appropriate headspace. Packing jars too full, without adequate headspace, may result in the loss of syrup from the jar during processing.
8. Remove air bubbles from filled jars by inserting a narrow metal spatula down the sides of each jar. This helps in keeping the syrup above the fruit.
9. Lower the jars into the hot water-bath canner. Cover the canner and heat.
10. Count the processing time when the water comes to the boil. Keep the water boiling gently during the processing time.
11. Remove the jars from the canner when the processing time is up. Cool jars upright on a folded cloth.
12. Test seals when the jars are cold. If the jar is not sealed, the lid will bounce up and down when pressed at the center.
13. Label jars with contents, date, and lot number if canning of more than one batch took place during 1 day.
14. Store in a cool and dry place.

The fruit to be canned is placed in a syrup of greater sugar concentration than that in the fruit itself. The dissolved sugar in the syrup diffuses into the fruit (osmosis) and improves its flavor. As the fruit cooks in the syrup, the cell walls become more permeable, the fruit texture grows more tender, and the retention of sugar renders the fruit plump and attractive. Whole fruits with tough skins, such as Kieffer pears and kumquats, are impermeable to the sugar syrup, unless they are precooked or the skins are pierced.

8.8 FREEZING FRUIT

Fruits to be frozen benefit from either a dry sugar pack or freezing in sugar syrup. For a dry sugar pack, the fruit is gently mixed with sugar in a given proportion so that each piece is coated. The choice of dry or syrup pack generally depends on the use to which the frozen fruit is to be put. Fruits packed in syrup are usually chosen for dessert, and fruits packed in dry sugar are preferred for cooking purposes.

Some fruits, such as blueberries, cranberries, raspberries, and rhubarb, may be frozen in a dry pack without sugar; however, these and all other fruits benefit greatly from a sugar pack regardless of the type used (dry or syrup).

Sugar helps protect the surfaces of frozen fresh fruit from contact with air, which produces enzymatic browning discoloration resulting from oxidation. In some fruits, such as peaches, nectarines, and apricots frozen in a syrup pack, ascorbic acid is also added to prevent darkening. The presence of sugar also lessens flavor change by retarding possible fermentation. In addition, texture, fresh fruit aroma, and normal size are retained upon thawing when sugar is used in freezing fruit (Pintauro 1990).

1. Check the variety of fruit to give the highest quality when frozen.
2. Make and chill syrup. If ascorbic acid or any other antibrowning agent is used, this should be added to the syrup before use (only when freezing light-colored fruits in dry sugar to avoid browning). A medium syrup (40%) is preferred for most fruits.
3. Use fully ripe but not overripe fruits of fine flavor and color. Wash them in ice cold water, then drain, hull, and pit or peel. Fruits should be handled gently to avoid bruising.
4. Freeze fruits in syrup or sugar, depending on the fruit being frozen and the intended use. Most fruits have better color, texture, and flavor when frozen in syrup and are usually best for dessert.
5. Pack fruits while leaving an appropriate amount of headspace.
6. Place containers into the freezer as soon as they are packed. Freezer should be at –18°C during freezing and storing.

8.9 SUGARS AND SWEETENERS IN BAKERY PRODUCTS

Cakes, cookies, breads, and croissants require sugar for flavor, color, texture, improved shelf life, yeast fermentation, and moisture retention.

What are the functional roles sugar plays in baked products?

8.9.1 Gluten Development

During dough mixing, flour proteins are hydrated, forming gluten strands. Gluten forms thousands of small, balloon-like pockets that trap gases produced during leavening. These gluten strands are highly elastic and allow the dough to stretch under gas expansion. However, if too much gluten develops, the dough becomes rigid and tough (Gates 1981; Ponte 1990). Gluten results from the mixing of the wheat proteins gliadin and glutenin in the presence of water.

Sugar competes with these gluten-forming proteins for water in the dough and prevents the full hydration of the proteins during mixing. If sugar is added correctly in the recipe, gluten maintains optimum elasticity, allowing for gases to be held within the dough. Sugar prevents gluten development and gives the final baked product good texture and volume (Beranbaum 1985; Ponte 1990).

8.9.2 Leavening

Under moisture and growth, sugar is broken down by yeast cells, and carbon dioxide gas is released at a faster rate. Sugar increases the effectiveness of yeast. The leavening process is hastened and dough rises at a faster rate (Knecht 1990).

8.9.3 Creaming

In cakes and cookies, sugar and sweeteners promote lightness by incorporating air in the form of small air cells into the shortening during mixing. During baking, these air cells expand when filled with gases from the leavening agents (Ponte 1990).

8.9.4 Egg Foams

Sugar serves as a whipping agent for the stabilization of egg foams. In foam-type cakes, sugars interact with egg proteins to stabilize the whipped foam structure, making it more elastic, so air cells can expand (Ponte 1990; Gates 1981).

8.9.5 Gelatinization

During baking, sugar absorbs liquid and delays gelatinization. In cakes, the heat of baking causes the starch in flour to absorb liquid and swell. This is called gelatinization. As more liquid is absorbed by the starch, the batter goes from a fluid to a solid state, setting the cake. Sugar competes with starch for liquid and slows gelatinization. The temperature at which the cake turns from fluid to solid is attained only when the optimum amount of gases are produced by leavening agents.

The same mechanism occurs in breads. A bread with a more tender crumb texture results (Ponte 1990; The Sugar Association 1992).

Starch gelatinization is a physicochemical phenomenon involving the disaggregation of starch granules within an aqueous environment at a suitable temperature. At 50°C–70°C, depending on the type of starch and its origin, water swells the starch granules, and a more or less homogeneous gel is formed.

Starch gelatinization has been studied by many researchers, who have advanced hypotheses about the molecular mechanism and studied the influence of some components, such as water, lipids, sugar, and salts, on the phenomenon (Horton, Lauer, and White 1990; White and Lauer 1990; Slade and Levine 1991; Liu and Lelievre 1992).

It should be noted that most of these studies were based on the application of differential scanning calorimetry (DSC). DSC studies have shown that starch gelatinization appears as a single endothermic peak starting at about 60°C at high moisture levels (>70%; Zanoni, Schiraldi, and Simonetta 1995).

Zanoni, Schiraldi, and Simonetta constructed a naive model of starch gelatinization kinetics to determine a control parameter of the baking process (i.e., a baking index). Bread dough samples were instantaneously heated at different temperatures (60°C–90°C) for varying times and then subjected to DSC to evaluate the extent of starch gelatinization. Calorimetric traces, after smoothing and standardization, were deconvoluted into one or two Gaussian curves, depending on the treatment temperature and time. This suggests the system is a mixture of two components, the second of which was found to have a lower gelatinization rate.

8.9.6 Caramelization

Sugar caramelizes when it is heated above its melting point, adding flavor and leading to surface browning, which improves moisture retention in baked products. At approximately 175°C, melted dry sugar takes on an amber color and develops an appealing flavor and aroma. Caramel is that amorphous substance resulting from the breakdown of sugar. Caramelization takes place when heat is applied in the oven when baking dough containing sugar and represents one of the ways in which surface browning occurs. The golden-brown, flavorful, and slightly crisp surface of breads, cakes, and cookies gives a good taste and helps retain moisture in the baked product, thus prolonging freshness (Ponte 1990).

None of the known alternative sweeteners participate in caramelization.

8.9.7 Maillard Reactions

At oven temperatures, sugar chemically reacts with proteins in the baking product, contributing to the food's browned surface.

These Maillard reactions are the second way in which bread crusts, cakes, and cookies get their familiar brown surfaces. During the baking of breads, cakes, and cookies, Maillard reactions occur among the sugar and amino acids, peptides, or proteins from other ingredients in the baked product, causing browning. These reactions also result in the aroma associated with the baked good. The higher the sugar content of the baked good is, the darker golden brown the surface appears (Ponte 1990; The Sugar Association 1992).

A large number of reactions, such as the Maillard reaction, caramelization, and chemical oxidation of phenols, may contribute to browning (Manzocco et al. 2001). In the case of sugar syrups, browning is a consequence of the Maillard reaction, which results from an initial reaction of a reducing sugar with an amino compound, followed by a cascade of consecutive and parallel reactions to form a variety of colored and colorless products (Hodge 1953; Maillard 1912; Martins and Van Boekel 2005).

Aspartame, alitame, thaumatin, and glycyrrhizin participate in Maillard browning reactions but, at typical usage levels, do not impart a brown color, and sweetness is lost after the reaction (Nelson 2000a).

8.9.8 Surface Cracking

Sugar helps produce the desirable surface cracking of some cookies. Because of the relatively high sugar concentration and the low water content in cookies, sugar crystallizes on the surface. As sugar crystallizes, it gives off heat that evaporates the water that it absorbed during baking and mixing. At the same time, leavening gases expand and cause cracking of the dry surface (Godshall 1990).

8.9.9 Heat Stability

Most high-intensity sweeteners can withstand heat and provide sweetness after baking. Alitame has shown some losses during baking, but higher than 80% is retained in most baked goods (Freeman 1989).

Acesulfame K shows excellent heat stability in cookies and cakes with excellent recovery rates (Peck 1994).

Sucralose is also heat stable, and performance tests have shown 100% recovery in baked goods, such as cakes, cookies, and graham crackers (Barndt and Jackson 1990).

Aspartame is not heat stable; however, it is available in a more stable encapsulated form. Recovery rates after baking in cookies and cakes are 95% and 79%, respectively. Glycyrrhizin and stevioside are heat stable; however, they are not used in baked goods, because they impart off flavors.

8.9.10 Bakery Products: Yeast-Raised Breads, Rolls, Bagels, and Sweet Rolls

Yeast breads require sugar initially to speed up the yeast in producing carbon dioxide for the leavening of dough. During the mixing phase, sugar absorbs a high proportion of water, delaying gluten formation. The delayed gluten formation makes the bread dough's elasticity ideal for trapping gases and forming a good structure (Ponte 1990).

Sugar in Maillard reactions contributes to the brown crust and delicious aromatic odor of bread. In addition, some of the yeast fermentation by-products and proteins from the flour react with sugar, contributing to the bread's color and flavor. They are called hard wheat products, in which a gluten structure is developed by kneading the wheat flour proteins in the presence of water. These include yeast-raised breads, rolls, bagels, and sweet rolls. The typical concentration of traditional sweeteners varies between 0%–15%, depending on how sweet the product is. Nutritive sweeteners act as a fermentable carbohydrate for the yeast. Fermentation produces gas, causing the product to rise and form an air cell structure. Fermentation also creates reducing sugars, which participate in Maillard browning and caramelization, giving the product its brown crust color.

Alternative sweeteners are not used to replace traditional sweeteners in breads and rolls because nutritive sweeteners are used at low concentrations. Calories are reduced by using less fat, replacing flour with a fiber source, adding emulsifiers and vital wheat gluten, and increasing moisture (Altschul 1993).

Yeasts utilize starch but preferentially utilize glucose or sucrose. Sugar alcohols inhibit yeast activity, and the product may not fully proof or rise adequately (Nelson 2000a,b).

It may be necessary to add sucrose or enzymes to aid in starch breakdown to provide fermentable carbohydrates. This will aid the browning process.

In sweet dough products, alternative sweeteners could be added, replacing nutritive sweeteners. These products are moist and soft because of their high sugar levels, which inhibit gluten development. Hence, partial replacement of sugars can result in a drier product. Hygroscopic agents, such as sugar alcohols, can be added to increase moistness. Gums, cellulose, or starches can also be added to help bind the moisture. To reduce toughness, plasticizing agents can be added, either by increasing the fat content or by adding emulsifiers if calories are of concern.

8.9.11 Shortened and Unshortened Cakes

In shortened cakes, sugar aids creaming to incorporate air into the shortening of these cakes. During mixing, sugar tenderizes cakes by absorbing liquid and preventing complete hydration of gluten strands. During baking, sugar tenderizes shortened cakes by absorbing water and delaying gelatinization (Ponte 1990).

Unshortened cakes, such as sponge and angel food cakes, contain no fat but include a large proportion of eggs or egg whites. Much of the cellular structure of the cake is derived from egg protein. The leavening agent is the air that has been beaten into the eggs. Sugar serves as a whipping aid to stabilize the beaten foam. Part of the sugar is also combined with flour before being folded into the foam mixture (Ponte 1990).

Nutritive sweeteners provide bulk, batter viscosity, and proper setting of the crumb by delaying starch gelatinization and protein denaturation. They influence crumb characteristics (moistness and tenderness) and shelf life. Sucrose is used in cake formulations at 25%–30% and high-intensity sweeteners at 0.1%–2%.

Maltodextrins and granular sugar alcohols can be used during creaming to add bulk, and emulsifiers help in the incorporation and retention of air. Polydextrose can affect the pH of the formulation; hence, the leavening system is adjusted by changing the sodium bicarbonate levels. Polydextrose is added in cakes at 7%–15% by weight. If too much is used, a heavy cake with a syrupy mouthfeel and bitter aftertaste can result (Khan 1993). Shrinkage of the final product can also occur.

Powdered cellulose, an insoluble bulking agent, has also shown good results (Ang and Miller 1991). Sugar alcohols could also be used to provide bulk and viscosity. Frye and Setser (1992) have reported the use of polydextrose, maltodextrin, sorbitol, lactitol, isomalt, and hydrogenated starch hydrolysates (HSH) in cake formulations, showing that, when the first two are used, a puffed appearance and air pocket formation do not occur.

Finally, sugar alcohols maintain crumb tenderness and moistness. Other humectants, such as water-binding starches and gums, are used.

8.9.12 Cookies

Cookies, like cakes, are chemically leavened with baking soda or baking powder. Cookies, however, have more sugar and shortening and less water proportionately. In cookies, sugar introduces air into the batter during the creaming process. Approximately half the sugar remains undissolved at the end of mixing. When the cookie dough enters the oven, the temperature causes the shortening to melt and the dough to become more fluid. The undissolved sugar dissolves as the temperature increases and the sugar solution increases in volume. This leads to a more fluid dough, allowing the cookies to spread during baking (Godshall 1990; Ponte 1990).

Sugar also serves as a flavorant, caramelizing while the cookies are baked.

Nutritive sweeteners affect cookie texture; some produce hard, crispy products and others soft, chewy products. Sweeteners also influence the spread, surface cracking, and brown color formation of the final product. The typical nutritive sweetener content of cookies is 20%–40%; whereas, concentration of high-intensity sweeteners in cookies is 0.5%–2%.

Bulking agents are added in the creaming stage to reduce lumping in the dough and decrease the possibility of a gummy texture (Dartey and Briggs 1987). However, most bulking agents bind water, creating problems with cookies that need to be crispy and hard. This is not the case with soft, chewy cookies and bulking agents, such as polydextrose, cellulose powders, fibers, gums, and maltodextrins, are used with good results (Khan 1993). For a hard, crispy texture, some sucrose must be present in order for recrystallization to take place, resulting in the formation of a glass structure. Sugar alcohols with low hygroscopicity (mannitol, isomalt, and lactitol) aid in achieving crispness (Altschul 1993).

Reduction in spread is one of the most notable results of replacing nutritive sweeteners. Polydextrose increases the spread of certain cookie formulations; whereas, it decreases dough cohesiveness. Sugar alcohols can be used, and fat content can be increased to improve the spread.

Cracking on the cookie surface results from recrystallization of sucrose and surface drying. Reducing the moisture level can improve cracking but can also reduce dough-handling quality, inhibit spread, and result in a dry, brittle product (Nelson 2000a,b).

8.10 CANDY PROCESSING

Hard candy formulations comprise sugar, a glucose syrup (used at a 20%–50% range in the formula with a DE of 38–48 to enhance flavors and maintain good moisture control), color, and flavor.

In candy making, sugar is first dissolved in water at room temperature to the point at which no more sugar will dissolve (sugar to water ratio 1:1/2). The result is a saturated solution. This saturated solution is placed over heat and stirred continuously, allowing more sugar to dissolve into the solution. The solution is then heated to boiling (152°C–168°C), creating a supersaturated solution. The supersaturated sugar solution is then heated to above BP, forcing more and more water to evaporate and the solution to become even more concentrated (Gates 1981). If crystals do not completely dissolve, they act as sites for nucleation and crystal growth. If viscosity of a syrup is too low, crystallization occurs and can be increased by the addition of a low-DE glucose-containing syrup (Alexander 1998).

The three methods of boiling hard candies include open pan, vacuum, and continuous film or scraped-surface heat exchanger. The ratio of sucrose to glucose in open-pan boiling is 70:30; whereas, in the other two, it is approximately 60:40. Vacuum boilers are used in clear or colorless candies because they decrease boiling temperature and produce less browning or off colors. High-maltose syrups are used in processes requiring extremely high temperatures.

The degree of sugar concentration of the supersaturated solution can determine the candy's final consistency. By monitoring the stages of the supersaturated solution with a candy thermometer and by testing a small sample of the sugar syrup in cold water, one can determine the specific concentration of the sugar syrup.

These concentrated supersaturated solutions are very unstable because the sugar molecules are prone to prematurely recrystallizing as the solution becomes increasingly concentrated. During heating of the solution, agitation should not occur because it can cause premature recrystallization.

Crystallization can be a major determinant of quality in sugar-based products. In products, such as hard candies, the formation of sugar crystals is inhibited during formation of the glassy state; whereas, in products, such as fondants, the presence of crystals is necessary for the desired texture.

Crystallization of an amorphous sugar matrix is affected by numerous factors, including water content, ingredients or additives, and environmental conditions of relative humidity and temperature. Both additives and water content can affect the glass transition temperature (T_g) of a product. T_g is a determining factor in the stability and, thus, crystallization of a sugar glass (Levenson and Hartel 2005).

In hard candies, corn syrup prevents unwanted crystallization or graining.

Levenson and Hartel (2005) made sugar glasses with a mixture of corn syrups (ADM, Decatur, Illinois) and analytical-grade sucrose (Fisher Scientific, Chicago, Illinois).

Each formulation had a 30:70 sucrose to corn syrup ratio on a dry weight basis. The mixtures, with excess water to ensure sugar dissolution, were cooked over an open flame to about 150°C to ultimately reduce the water content to approximately 3.5%.

The mixtures were cooled into the glassy state in a constant volume in closed-system inside metal washers (2.36-mm height and 9.47-mm inner diameter).

Colors are usually added during boiling. The solution is then cooled and poured onto an oiled table where flavors are added and kneaded into the molten mass. The warm and plastic mass is then either molded, rolled, or machined into the final form.

Candy types can be divided into categories: candies in which sugar is present in the form of crystals and candies in which the sugar is present in an uncrystallized form.

Usual defects are a result of sugars that can invert because of acid addition (citric, malic, or tartaric acid) usually used for flavor. This causes the candy to be hygroscopic and sticky. Vacuum boiling reduced inversion because it operates at lower temperatures and time. Crystallization is another already mentioned problem caused by incomplete boiling, when the sucrose to glucose ratio is too high, and if final moisture is higher than 1%–2%. Reworked candy could also cause crystallization as could humid storage conditions.

8.10.1 Crystalline Candies

Crystalline candies can be subdivided into two groups: candies with perceptible crystals, such as rock candy, and cream candies in which crystals are too small to be detected by the tongue, such as fondant and fudge.

Rock candy is prepared simply by immersing a string in a supersaturated sugar solution, heating the solution to the hard ball stage, and then allowing it to cool. Left to cool, sugar from the solution will recrystallize on the string. With no stirring or other interfering agents, sugar molecules will continue to clump, and the crystals will increase in size as long as the mass is immersed. The resulting product is pure sugar because only chemically pure sucrose will recrystallize (McGee 1984).

Inverted sugar helps prevent recrystallization. Inverted sugar is the result of the breakdown, or the inversion, of the sucrose into fructose and glucose. This process takes place when sucrose is heated with moist heat or, as in candy making, when a water and sugar solution is heated. The amount of water used and the length and intensity of the cooking of the supersaturated solution both control how much of the sucrose is inverted. The process may be accelerated by added acid from candy ingredients, such as fruit, brown sugar, molasses, honey, or chocolate (Beranbaum 1985).

8.10.2 Noncrystalline Candies

Noncrystalline or amorphous candies are much simpler to make. The sugar solution must simply contain sufficient interfering agents or cook to a high enough temperature to prevent recrystallization. In taffies, butterscotch, brittles, and caramels, either inverted sugar in the form of molasses or acid that will produce inverted sugar or corn syrup are added to the mixture to prevent the formation of crystals in the candy. These candies are cooked to a higher temperature than crystallized candies so as to reduce the water content to 2% or less, which also prevents recrystallization (Gates 1981).

Noncrystalline candy can be cooked by dry heat as well as moist heat. Some peanut brittles, for example, are made by melting dry sugar. The brittle does not recrystallize because the lack of water during the cooling period causes it to take the form of a noncrystalline, glassy solid.

8.10.3 Fudge and Fondants

Fudge and fondant are a chewy candy grained or crystallized. They are manufactured using a caramel base to which a fondant (a grained confection used to cause the crystallization of the fudge's caramel base) is added (Alexander 1998).

Higher boiling temperatures, longer boiling times, more fondant, and higher sucrose ratios lead to faster crystallization of sugars in the fudge and firmer, shorter textures. Texture is coarse and gritty if the temperature is higher than 52°C.

Fudge can be produced by batch or continuous processes (higher ratios of sucrose to glucose 10:1).

Fondants are made by mixing a supersaturated sugar solution (12% water) that is usually sucrose with inverted sugar or glucose-containing syrup. A combination of lactose and honey can be used instead. The solution is boiled via an open-fire kettle, steam-jacketed kettle, or vacuum boiler. Then, in the creaming step, it is cooled with intense mixing or beating. The sucrose to glucose ratio is 80:20. Cooled fondants are heated to 57°C–82°C and a syrup bob is added to the reheated mixture. The bob syrup contains a higher proportion of a glucose-containing syrup (Alexander 1998).

8.11 CARAMELS

These belong to the chewy candy group because they incorporate proteins from dairy-based ingredients, such as condensed milk, whey powder, cream, or skim milk. Proteins react with the reducing sugars present, forming the pigments and distinctive flavor associated with caramels. Lower temperatures are used, up to 121°C, for cooking except toffee, which is boiled at higher temperatures, up to 149°C, with a lower amount of dairy ingredients added and less fat (Alexander 1998).

Lower-DE syrups are added to chewy candies to increase the viscosity of the formula and produce a less sweet product. Fats from butter and cream help soften caramels and act as release agents that keep the candy from sticking to the teeth.

Gummed candies include jelly beans, fruit slices, and gummy bears made with gums (pectins or agar) or with starch used to set the shape and texture of the products. Some gummed candies are made with 40% honey. Regular corn syrups may be used because of their control of crystallization and hygroscopicity.

8.12 ICINGS

Sugar's role in icings is similar to that in candies. Sugar is the most important ingredient in icings, providing sweetness, flavor, bulk, and structure.

8.13 GELATION

Sugar is essential in the gelling process of jams, preserves, and jellies to obtain the desired consistency and firmness. This gel-forming process is called gelation, and the fruit juices are enmeshed in a network of fibers. Pectin, a natural component of fruits, has the ability to form this gel only in the presence of sugar and acid. Sugar is essential because it attracts and holds water during the gelling process. In addition, acid must be present in the proper proportions. The optimum acidity is a pH between 3.0 and 3.5. Some recipes include lemon juice or citric acid to achieve this proper acidity (Meschter 1990).

The amount of gel-forming pectin in a fruit varies with ripeness (less ripe fruit has more pectin) and the variety (apples, cranberries, and grapes are considerably richer in pectin than cherries and strawberries). In the case of a fruit too low in pectin, some commercial pectin may be added to produce gelling, especially in jellies. In recipes that use commercial pectin, the proportion of sugar may be slightly higher or lower than the one part fruit to one part sugar ratio (Beranbaum 1985).

8.14 PRESERVING

Preserves are made with crushed or whole fruit.

Sugar prevents spoilage of jams, jellies, and preserves after the jar is opened. Properly prepared and packaged preserves and jellies are free from bacteria and yeast cells until the lid is opened and exposed to air. Once the jar is opened, sugar incapacitates any microorganisms by its ability to attract water. This is accomplished through osmosis (the process whereby water will flow from a weaker solution to a more concentrated solution when they are separated by a semipermeable membrane). In the case of jellies and preserves, the water is drawn from these microorganisms toward the concentrated sugar syrup. The microorganisms become dehydrated and are unable to multiply and cause food spoilage. In jellies, jams, and preserves, a concentrated sugar solution of at least 65% is necessary to perform this function. Because the sugar content naturally present in fruits and their juices is less than 65%, it is essential to add sugar to raise it to this concentration in jellies and preserves (Meschter 1990; Beranbaum 1985).

8.15 COLOR RETENTION

Sugar helps retain the color of the fruit through its capacity to attract and hold water. Sugar absorbs water more readily than other components, such as fruit, in preserves and jellies. Thus, sugar prevents the fruit from absorbing water, which would cause its color to fade through dilution (Gates 1981).

8.16 HIGH-FRUCTOSE CORN SYRUP

HFCS is used in many commercial jellies and is comparable in sweetness to sugar. The major disadvantage of HFCS is that it is a liquid and may contain as much as 29% water. The extra water may be evaporated in the final stage of production, a process that causes part of the volatile fruit flavors to be lost.

Concentrated fruit juices could also be used as a sweetening ingredient because they are similar in composition to sugar syrups (Meschter 1990).

The addition of corn syrup to sugar confections both inhibits sucrose nucleation and increases the induction time before the onset of nuclei formation.

Various grades of HFCS can be produced by the process outlined in Figure 8.3.

High-DE (95+), i.e., a measure of the extent of starch hydrolysis, dextrose syrup is made as follows: A two-stage process is used for the production of corn syrups. The first step involves the acid hydrolysis of starch to about 42 DE. Starch makeup goes through a heat exchanger and a retention cell. If a higher DE is required, starch is kept in a holding tank, and a second hydrolysis with enzymes is used until the desired DE is reached. Finished corn syrups and dextrose syrups are produced by standard refining (filtration, carbon bleaching, filtration again) and evaporation processes (Howling 1992). Most common syrups have a DE of 42 and 62 also ranging from 24 to 82 DE. Corn syrup solids with a DE of 20–30 are made by a process similar to that of maltodextrin production (corn syrups spray dried and crystallized).

Hydrolysate is then passed over immobilized glucose isomerase, and the glucose is partially converted to fructose by a process called isomerization. Then it is clarified through a vacuum filter;

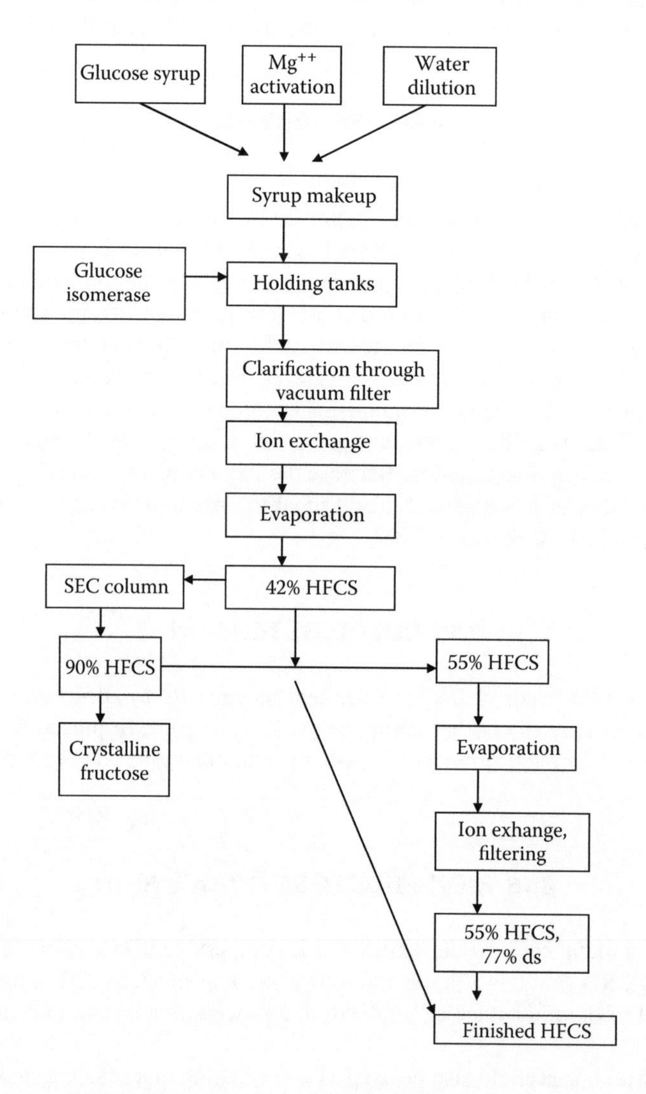

Figure 8.3 Manufacture of HFCS from glucose syrup. (Adapted from Alexander, J., *Sweeteners Nutritive*, Eagan Press, American Association of Cereal Chemists, St. Paul, MN, 1998. Howling, D., in F.W. Schenck and R.E. Hebeda (eds.), *Starch Hydrolysis Products*, pp. 277–317, VCH Publishers, New York, 1992. With permission.)

passed through an ion-exchange column; and evaporated to give a syrup containing 42% fructose, 53% glucose, and 5% more complex sugars. This 42% HFCS is converted to 90% HFCS by size exclusion chromatographic columns. The 90% syrup is blended with 42% syrup to give a 55% HFCS (1.2 times as sweet as sucrose) (Alexander 1998). All syrups are then refined by evaporation, ion exchange, and vacuum filters to give finished HFCSs. Part of the 42% syrup is concentrated and sold as 42% HFCS (71%–80% dry substance). This is similar for 55% HFCS (77% dry substance). Some of the 90% HFCS is blended with alcohol and evaporated to give 90% HFCS (90% dry substance). Then it is crystallized with fructose crystals, centrifuged, dried, and cooled to give crystalline fructose.

8.17 BISCUITS

The effect of sugar on dough behavior is an important factor in biscuit making. In excess, sugar causes a softening of the dough, a result in part of competition between the added sugar and the availability of water in the system (Bure 1980). Using a Farinograph, Olewnik and Kulp (1984) observed that an increase in the sugar concentration in a cookie dough reduces its consistency and cohesion. Mizukoshi (1985) studied the effect of varying sugar content on shear modulus measured during cake baking while keeping the proportion of other ingredients constant. He showed that below 20% sugar has no effect on shear modulus; whereas, an increase from 30% to 40% reduces it appreciably, revealing the existence of a threshold value associated with the variation of sugar content in the formula. Vettern (1984) studied the effects of sugar quantity and its grain size on biscuit spreading. The conclusion was that a fine grain size and a high concentration of sugar contributed to a significant spreading of the biscuit. Sucrose acts as a hardening agent by crystallizing as the cookie cools, which makes the product crisp. However, at moderate amounts, it acts as a softener, because of the ability of sucrose to retain water (Schanot 1981). Sugar makes the cooked product fragile because it controls hydration and tends to disperse the protein and starch molecules, thereby preventing the formation of a continuous mass (Bean and Setser 1992).

The addition of sugar to the formula decreases dough viscosity and relaxation time. It promotes biscuit length and reduces thickness and weight. Biscuits that are rich in sugar are characterized by a highly cohesive structure and a crisp texture (Maache-Rezzoug et al. 1998).

8.18 FROZEN DESSERTS

Frozen desserts are made by freezing a liquid mixture of sugar with cream, milk, fruit juices, or purees. In the liquid mixture, the dissolved sugar's ability to attract and hold water diminishes the water available for water crystallization during freezing. As a result, the FP of the liquid mixture is lowered. Because less "free" water is available, the ice crystals that form tend to be smaller (Smith 1990).

As part of the liquid mixture begins to freeze, the sugar in the remaining unfrozen solution becomes more concentrated, further lowering the FP of the remaining unfrozen solution. Therefore, a temperature much lower than the FP of the liquid mixture is used to ensure rapid, consistent cooling. This combination of lower FP provided by the dissolved sugar and a colder than freezing temperature produces a frozen product with tiny ice crystals. Tiny ice crystals give the frozen dessert its smooth, creamy texture. Large crystals are undesirable because they impart a gritty or sandy texture in the frozen dessert. Hardening and storage of frozen desserts are the final steps to achieving a high-quality frozen dessert (Smith 1990).

HFCS, if used as an alternative sweetener, lowers the FP twice as much as sugar does, which produces an icy texture.

In frozen desserts, sugar also functions to balance flavors and mouthfeel. Sugar acts to enhance flavors, thereby eliminating the need for additional flavor ingredients. Sugar also increases the viscosity (thickness) of frozen desserts, which helps impart a thick, creamy mouthfeel. It provides a clean, sweet taste preferable to the syrupy taste produced by corn-derived sweeteners. About 16% sugar, by weight, is recommended for ice cream. Somewhat higher proportions of sugar are used for lower-fat desserts, such as ice milk and sherbet, in order to counterbalance the reduced amount of butterfat. When cream is replaced with lower-fat ingredients, such as milk or fruit puree, additional sugar is necessary to ensure a smooth, creamy mouthfeel and balanced flavor (Smith 1990).

8.19 MOLASSES

Molasses, a by-product of sugar cane processing, defined as the concentrated liquid extract of the sugar refining process, is one of the most basic, natural sweeteners available to bakers. Although not realized by many bakers, molasses is available in many different forms for use in baked goods. There are light or dark molasses, refined syrups, and final molasses (blackstrap).

Molasses can be derived from either sugar beets or sugar cane. However, sugar beet molasses has a very astringent off flavor and aroma and is not used in food applications.

Various types of sugar cane molasses are grouped relative to the general processing methods used.

8.19.1 Imported Unsulphured Molasses

This type originates in the Caribbean area and is derived from sun-ripened sugar cane, which has been grown for 12–15 months. The sugar cane juice is extracted from the cane, usually by pressing, and then clarified and evaporated to about 79.5% solids. Because it contains all of the sugar of the original cane, it is sweeter than other grades of molasses, which have undergone further processing. Also, it is subjected to controlled curing and maturing to develop the desirable rum-like flavor. Because of its naturally light color, it is not necessary to bleach with sulfur dioxide. This type of molasses, sometimes called cane juice molasses, is considered premium grade molasses in terms of its light color, mellow flavor, mild aroma, high level of sweetness, and clarity (Hickenbottom 1996).

8.19.2 Mill (Domestic or New Orleans) Molasses

Mill molasses originates from sugar cane in Louisiana, Florida, and Texas. There are five grades of molasses in this category, which are produced in raw sugar mills.

8.19.2.1 Whole Juice Molasses

This grade consists of the whole juice, which has been clarified and evaporated to about 79.5% solids. The basic differences between this material and the imported types are in flavor and color. Whole juice molasses (WJM) tends to be just a bit harsher in flavor and slightly darker in color than its imported counterpart. These differences are the result of a shorter, weather-related cane ripening period.

8.19.2.2 First Molasses

This grade is the product obtained after one extraction of sugar from the mother liquor. It is not as delicate in flavor as WJM and sometimes is slightly darker in color (Hickenbottom 1996).

8.19.2.3 Second Molasses

After another sugar extraction, the resulting liquor is termed second molasses. It is again a shade darker and stronger in flavor than first molasses.

8.19.2.4 Third Molasses

The liquor is again extracted, and this results in a very dark, strong, and fairly bitter-flavored molasses.

8.19.2.5 Final Molasses

This is the final residual liquor from which no additional sugar can be extracted economically, and it is the darkest and strongest in flavor and aroma.

8.19.3 Refiner's Syrup or Refiner's Molasses

This is the product extracted from raw sugar during the sugar refining process—thus, the name, refiner's syrup. It has a definite flavor, which is quite different from both imported and mill molasses. Its flavor is more of a caramelized sugar flavor than a true cane flavor. This flavor difference results from removal of almost all the original cane flavor constituents during the raw sugar milling process.

8.19.4 Dry Molasses

This type is essentially a dry form of the final molasses discussed above. Not all the various liquid counterparts are, however, available in dry form nor are the dry products 100% molasses. Drying agents, such as dextrins, wheat flour, wheat starch, and/or flow agents must be added in order to dry molasses in either drum dryers or spray dryers. Most dry molasses products available today are made from liquid molasses types at the lower end of extraction because inverted sugar is more difficult to dry than sucrose. Thus, when dry molasses is rehydrated, a liquid product with a darker color and a fairly strong flavor will result. Most dry molasses is hygroscopic as well and must be kept away from moisture prior to use (Hickenbottom 1996).

8.19.5 Characteristics and Composition of Molasses

Attributes associated with molasses are

- Natural sweeteners, ideal for use in bakery products and many other foods, nutrients (high levels of vitamins and minerals), and unique flavor characteristics.
- All molasses must be carefully filtered, pasteurized, and tested for physical grading characteristics as well. Very stringent quality control measures must be maintained in order to produce uniform, food-grade, batch-to-batch, year-round, standard molasses types from the wide variations in raw materials. Climate, soil conditions, cane growing period, cane maturity at harvest, degree of sugar extraction, and the process used all affect the physical characteristics of the base raw molasses.
- The higher the invert (glucose + fructose) content of a high total sugar solids product in conjunction with low ash, the lighter and sweeter the molasses type. Conversely, the higher the sucrose of a low total sugar solids product with higher ash, the darker and more bitter the molasses type.
- Final molasses ends up with the most micronutrients. Calcium, potassium, and iron, the nutritionally significant nutrients, are present in appreciable amounts in all types of molasses. Molasses also

contains B vitamins. Molasses does not contain fat or fiber. Being a food component of plant origin, it also does not contain cholesterol.

- Molasses (imported, dry, mill first, and refiner's) can be used in most confectionery products (Hickenbottom 1996).

8.19.6 Molasses for Color and Sweetness

The wide color ranges available among types of molasses provide the baker with a wide choice and source for natural, nonammoniated coloring agents. Molasses is an effective colorant in whole wheat, cracked wheat, and bran items to mask their unappetizing colors with its own golden brown color. Where color is the main criteria, second or third molasses, or even final molasses, are selected, keeping sweetness and flavor in mind as well.

If a product is to have a a light crust and crumb with a sweet, mild flavor, a refiner's syrup, imported molasses, or perhaps a molasses having a similar sucrose-to-invert ratio are better choices. To achieve a finished product with a darker color and a definite molasses flavor, choose a molasses from the lower spectrum of sucrose-to-invert ratio, perhaps a third molasses, for example. In calculating the quantity of sugar to be replaced by molasses, the total sugar content as well as the ratio of sucrose to inverted sugar in the molasses must be considered. Sugar (sucrose) is not quite as sweet, pound for pound, as inverted sugar. Thus, if imported or first molasses is used in place of sucrose, a sweeter baked product will result, assuming replacement was made on a solids basis. To achieve a more equal replacement, as far as sweetness goes, a second molasses or a blend could have been chosen (Hickenbottom 1996).

8.19.7 Molasses for Flavor

The flavor-masking ability of molasses, particularly in high-fiber products, is unsurpassed. When used as a flavor enhancer, molasses is especially compatible with caramel, chocolate, coffee, cocoa, maple, butterscotch, peanut butter, vanilla, and mocha.

8.19.8 Molasses as a Leavening Agent

Molasses has natural acids present, which makes it a valuable leavening agent in cookies using baking soda. The reaction between the acids and the soda releases gas (carbon dioxide), which raises the dough. The acids also affect the spread and spring of cookies. Because of these reasons, it is very important that the molasses of choice has a uniform acidity or pH. Most commercially available food-grade molasses products are processed to have consistent pH levels.

8.19.9 Molasses as a Buffering Agent

Molasses contains between 2% and 9% organic and inorganic salts, giving the molasses buffering capacity. Buffering contributes to the control and stabilization of pH in fermented systems, the protection of acid-sensitive flavors, and the prevention of texture deterioration by controlling available moisture.

8.19.10 Molasses and Increase in Shelf Life

Humectant, antioxidant, and water activity properties of molasses are very important to the shelf life of products. Humectancy allows a product to absorb and retain moisture, which relates directly to softness, moistness, freshness, and the staling rate of baked items. Molasses retards oxidation because of its natural antioxidant property. This property is unique in that it is polar or, more simply, very easily used in complex food systems, particularly those high in fat content, such as wheat germ items. For best effect, molasses levels used should be about 3%, based on total fat content of the final product.

Water activity of molasses is at least equal to that of sugar when compared on an equal solids basis. Because this activity is probably the most important factor affecting the shelf life of intermediate moisture baked items (granola cookies, chewy brownies, fruit cakes, etc.), it follows that this characteristic alone would warrant use of molasses. The water in a food item is controlled by its being rendered unavailable for chemical or biochemical reactions. In effect, the water is bound by the molasses itself and, thus, is not available to participate in other reactions detrimental to the product's freshness (Hickenbottom 1996).

8.19.11 Addition of Molasses to Produce Brown Sugar

Granulated sugar can be "painted" with the appropriate type of molasses to produce brown (light, medium, or dark) sugar. This painting helps bakers to eliminate the problem of brown sugar lumping during storage, create their own brown sugar color and flavor standards, maintain uniform brown sugar products, and save on ingredient costs.

All that is required to produce in-house brown sugar is for the user to choose the optimum color liquid molasses and add it to granulated sugar at about one part molasses to nine parts sugar (Hickenbottom 1996).

8.20 FRUIT PUNCH BEVERAGE DRY MIX PROCESS

The following ingredients are mixed well until uniformly blended: malic acid, maltodextrin, sweeteners, salt, sodium citrate, ascorbic acid, and flavoring and coloring agents.

A quantity of the above mix is diluted with water to make the finished beverage.

Sweeteners, such as aspartame, should be mixed according to the following instructions:

1. Aspartame should be dissolved in water at a 1:12 ratio. A small mixing tank should be used with a strong mixing vortex. Water should be neutral pH treated water.
2. The sweetener should be sprinkled into water while under agitation; otherwise, the formation of lumps might be possible.
3. Approximately 15 min is sufficient to fully dissolve the sweetener; however, this depends on the temperature and mixing speed.
4. Solution should be immediately transferred to a syrup tank.

8.21 FROZEN DESSERT PROCESS

The following ingredients should be added: milk fat, nonfat dry matter, polydextrose, whey solids, maltodextrin, sweeteners, stabilizers, flavoring agents, emulsifiers, gelatin, and sorbitol. All ingredients are mixed except coloring and flavoring agents.

Heating to 45°C–50°C follows with agitation. Pasteurization of the mixture is then carried out by HTST (high temperature short time) at 75°C for 25–30 s. Homogenization at 2000–2500 PSI/500 PSI, and cooling then follows. Aging for 1 day should be carried out followed by the addition of color, flavor, and freezing.

Trehalose can extend the shelf life of frozen products and help maintain their texture, flavor, and color because it can protect proteins and other substances from freezing and desiccation.

Moreover, off flavors can be minimized in pasteurized, dairy-based desserts, pasteurized yogurts or pasteurized and aseptic puddings. Trehalose gives a mild, sweet flavor, minimizing moisture sensitivity and product clumping because of its hygroscopic nature.

Neotame can also be used in frozen desserts. It is rapidly metabolized, completely eliminated, and does not accumulate in the body. The major metabolic pathway is hydrolysis of the methyl ester by esterases present throughout the body, yielding deesterified neotame and methanol. Very small amounts of neotame are needed to sweeten foods; hence, the amount of methanol derived from neotame is small relative to that derived from fruit and vegetable juices (Pszczola 2003).

8.21.1 Ice Cream

Ice cream is the largest frozen dessert category, and its principles apply to the making of related products, such as frozen yogurt, mousses, ice milks, sherbets, and ices.

Ice cream production starts with the making of an ice cream mix comprised of nonfat milk solids, fat, sweeteners, stabilizers, and emulsifiers. Blending of dry ingredients and mixing with liquid ingredients follows.

Sweeteners can be either sucrose as a sole ingredient or corn syrups (42 or 62 DE), molasses, honey, fructose, brown sugar, or maple sugar as additional sweeteners.

Total sweeteners range from 14%–16% for most products with ices and sherbets having a higher percentage of corn syrups, about 4%.

Pasteurization and homogenization of the mix then follows. Freezing finally occurs with simultaneous whipping to incorporate air. The factor determining the product's final body and texture is the overrun (the volume of ice cream obtained in excess of the initial volume of mix). Fluffy texture means too high an overrun; whereas, products with too low an overrun are soggy and too dense (Alexander 1998). Sweetener choice and level play an important role in determining the overrun.

After the whipping and freezing of the mix, the product is molded or packaged and then further hardened by storage at a cold temperature.

Final product defects, such as weak body or sandiness (a result of lactose crystallization) can be related to the sweetening system of the formulation. High ice cream quality is related to sucrose only as a sweetener.

Sucrose alone is not a good sweetening system for sherbets and ices because of crystallization at the product surface; hence, dextrose is added in the form of corn syrup solids or corn syrup at approximately one-third of the amount of sucrose in the formula.

Sugar alcohols sorbitol, lactitol, xylitol, and isomalt can be used to lower the freezing point of ice creams and frozen desserts and inhibit crystallization of other sugars (Khan 1993). Maltitol, mannitol, and HSH can also be used.

Lactitol does not cause a sandy texture in ice cream and frozen desserts. Xylitol cannot be used as the sole sweetener because it makes the product too soft, requiring the addition of thickening agents (Hyvonen, Koivistoinen, and Voirol 1982). Isomalt can crystallize because of its low solubility when used at levels higher than 15%. HSH improves the freeze–thaw stability of ice creams.

Soukoulis, Rontogianni, and Tzia (2010) investigated the role of rheological, thermal, and physical properties to the establishment of specific sensory attributes critical for the quantification of the sensorial quality of vanilla ice cream made with bulk sweeteners.

The bulk sweeteners used included (a) polyols, e.g., sorbitol (Neosorb 70/70, Roquette Pharma, Italy), xylitol (X3335, Sigma–Aldrich, U.S.A.), maltitol (Maltisorb P200, Roquette Pharma, Italy), mannitol (Pearlitol 160C, Roquette Pharma, Italy), and fructose (Zografos, Athens, Greece); (b) disaccharides, e.g., trehalose (Merck, Darmstadt, Germany) and maltose (Merck, Darmstadt, Germany); (c) cornstarch hydrolysates, e.g., glucose 39DE (Glucidex 39, Roquette Pharma, Italy), glucose 22DE (Glucidex 22, Roquette Pharma, Italy), maltodextrin 17DE (MD1210QS, Syral, France), and maltodextrin 12DE (MD1210QS, Syral, France); and (d) oligosaccharides, e.g., inulin (Fibruline XL, MW > 3.300, DP > 20, Cosucra, Belgium), oligofructose (Fibrulose F97, MW > 1000, DP < 20, Cosucra, Belgium), and soluble dietary fiber (Nutriose FB06, Roquette Pharma, Italy).

The ice cream mix's composition was 8% fat (provided as fresh cream), 11% milk solids nonfat (MSNF), 16% total sugar solids (provided as crystalline sucrose and partially as bulk sweeteners), 0.2% stabilizer (an 8:2 blend of xanthan gum and microcrystalline cellulose), and 0.2% emulsifier (mono-diglycerides of fatty acids, 60% monoester content). Bulk sweeteners were used as partial substitutes of sucrose (control sample) at ratios of 10:90 and 30:70. The ice cream mixes were prepared by dispersing under agitation (at 1000 rpm) the stabilizer/sweeteners/emulsifier/MSNF dry blend into the liquid materials at 50°C for 10 min using a mechanical stirrer. The mixes were then batch pasteurized at 76°C for 20 min using a water bath and, consequently, two-stage (200 and 30 bars, respectively) twice homogenized using a laboratory single piston homogenizer. Then, they were rapidly cooled to 4°C and aged at the same temperature for 18 h. The aged mixes were frozen using a batch freezer at a set draw temperature of −5.5°C, packaged into 300-mL high-density polyethylene (HDPE) containers, hardened, and stored in a deep freezer under quiescent freezing conditions at −25°C (Soukoulis, Rontogianni, and Tzia 2010).

They also reported that creaminess, apart from thermal properties (glass transition temperature-Tg, ice crystal uniformity-DT_{mcurve}, and unfrozen water content or UFW), was poorly related with rheological and physical properties. On the contrary, coarseness, wateriness, greasiness, and gumminess perception were found to be well correlated with rheological and thermal properties, overrun, air cell mean size, and melting rate. Flavor and taste characteristics were interrelated particularly with melting and thermal characteristics, overrun, and air cell mean size.

The structural elements of ice cream contribute significantly to the perception pattern of texture and flavor (Kokini and van Aken 2006). Texture quantification can be accomplished by the determination of the properties that are related to the colloidal aspects of ice cream, such as microstructure, serum viscoelasticity, emulsion characteristics, and thermal properties (Herrera et al. 2007; Regand and Goff 2003; Muse and Hartel 2004).

The physical structure of ice cream is composed of air, ice, fat, and an unfrozen serum phase, which consists of proteins, emulsifiers, stabilizers, milk salts, lactose, and sweeteners within the unfrozen water.

Ice crystal size plays an important role in determining the quality of ice cream; small ice crystal sizes are desired to enhance consumer acceptance. During freezing, nuclei formation must be promoted and ice crystal growth and recrystallization minimized to create many small ice crystals. The effects of sweetener type, draw temperature, dasher speed, and throughput rate on ice crystal size distributions during freezing of ice cream were investigated by Drewett and Hartel (2007). These operating variables affect the mechanisms by which ice crystals form and ripen into the disc-shaped crystals observed exiting the freezer. Increasing the throughput rate, reducing the residence time in the freezer allows less time at warmer temperatures when recrystallization occurs rapidly, which leads to smaller ice crystals at the freezer exit. Smaller ice crystals were also found at lower draw temperatures because coolant temperatures were reduced, increasing the driving force for nucleation.

Increased ice crystal size at draw occurred when faster dasher speeds were utilized because of the addition of frictional heat into the system, causing melting of small crystals. Residence time was found to have the most pronounced effect on mean ice crystal size followed by draw temperature and dasher speed; whereas, type of sweetener had no significant effect.

An ice cream mix with 12% milk fat, 17% sweetener, 11.5% MSNF, 0.1% Polmo emulsifier, and 0.28% gelatin stabilizer was pasteurized for 19 s at 85°C, homogenized at 6895 and 20,684 kPa, and placed into a storage tank at 1°C–2°C for overnight aging.

Sweeteners, 28 dextrose equivalent (DE) corn syrup, sucrose, and HFCS, were chosen to obtain a wide range of FP depressions (Drewett and Hartel 2007).

8.22 YOGURT FORMULATION AND PROCESS

The fruit phase usually consists of the addition of thawed fruit or fresh fruit, sweeteners instead of sugar, water, glucose, pectin, citric acid, flavor, potassium sorbate, and antifoaming agents. The yogurt phase consists of skim milk, milk solids, and culture.

The finished yogurt should contain a ratio of 80 to 20, yogurt to fruit phase, respectively.

The process is based on heating skim milk to 43°C followed by the addition of MSNF-gelatin or other stabilizer. Heating to 82°C for 30 min is then carried out to pasteurize the product followed by cooling to 45°C held constant until pH is 4.3–4.5.

Cooling at 35°C is then carried out and sweeteners added along with coloring, flavoring agents, and fruit.

Filling into cups finishes the process with subsequent cooling to 4°C.

The U.S. FDA approved the use of aspartame in yogurts in 1988 (Van der Ven 1988; Keller et al. 1991). When used in yogurt making, a 52% reduction in the calories was achieved per cup of product.

Keller et al. (1991) evaluated the degradation of aspartame in yogurt in relation to microbial growth in the product. Three commercial yogurt starter cultures (i.e., single strains and mixed) containing the *Lactobacillus delbrueckii* ssp. *bulgaricus* and *Streptococcus thermophilus* were used. The proportion of degraded aspartame was related to the growth and activity of the starter culture with no significant differences between single or mixed strains. The results showed that only 10% of the aspartame was degraded, and it remained stable in yogurt over 6 weeks of storage at refrigeration temperature (Pinheiro et al. 2005).

However, Fellows, Chang and Shazer (1991b) evaluated the stability of aspartame during milk fermentation to produce a sundae-style yogurt, containing fruit and using different incubation periods and temperatures. Aspartame's stability, determined by HPLC, was 95% in yogurt made in 6 h and incubated at 43°C, and a parallel product that was fermented at 32°C for 13 h showed a 90% stability of the sweetener. The study demonstrated that aspartame has excellent stability during the preparation of yogurts with fruit.

Brandão, Silva, and Reis (1995) reported the effect of adding different amounts of Dairy-Lo™ (Pfizer Inc., New York, U.S.A.) (0, 1, 2, 3, and 4 g/100 g; this is a microparticulated whey protein used as a fat substitute) on the quality of yogurt made from skim milk and aspartame. It was recommended that smaller amounts of aspartame should be used during the manufacture of a yogurt containing Dairy-Lo™. However, Monneuse, Bellisle, and Louis-Sylvestre (1991) concluded that aspartame used at a rate of 0.028 g/100 g of yogurt was preferred by a taste panel.

Decourcelle et al. (2004) concluded that the presence of pectin and starch in low-fat stirred yogurt reduced the release of aroma compounds in the headspace of the sample, and yogurt prepared with locust bean gum significantly increased the flavor release. Furthermore, yogurts made with guar gum and sweeteners (fructose, fructooligosaccharides, aspartame, or acesulfame) appeared to have no effect on flavor release.

Hyvonen and Slotte (1983) studied the effect of different sweeteners (e.g., xylitol, sorbitol, fructose, cyclamate, and saccharin—sucrose was used as a reference) on the quality of yogurt, and they were added to the milk base before the incubation period or to the fermentate. All the sweeteners used were satisfactory to sweeten the yogurt, especially after incubation, except saccharin. Nevertheless, saccharin could be used when mixed with xylitol to mask its bitter aftertaste. In addition, sorbitol reduced the growth rate of the starter culture when added before fermenting the milk, and its effect was only satisfactory when used in conjunction with sucrose. They also reported the following blends of sweeteners should be used: 8% xylitol, 7% fructose, and 0.07% cyclamate; or 4% xylitol and 0.007% saccharin as an alternative replacement of sucrose equivalent to 8 g/100 g concentration.

Regarding cyclamate, a 10:1 ratio of cyclamate to saccharin is commonly used and provides the intense sweetness of saccharin combined with the ability of cyclamate to lessen the bitter aftertaste of saccharin (Wilkes 1992).

As sucralose does not interact with food molecules, it can be used in a great number of foods, such as carbonated and noncarbonated beverages, chewing gums, dairy products, fruit compotes, frozen desserts, and salad dressings, among others (Grice and Goldsmith 2000). The quality of dietetic yogurts made with skim milk, graviola pulp, starter cultures, and three sweeteners (e.g., aspartame, sucralose, and fructose) was evaluated by Fonseca and Neves (1998). The calorific values and sensory profiles of all the products were compared with a parallel yogurt sweetened with 10 g/100 g sucrose, and the results showed calorific values of all the experimental yogurts lower than the product sweetened with sucrose with preferred ratings of yogurts to be aspartame > fructose > sucralose.

Pinheiro et al. (2002) studied the effect of sucrose (6 g/100 g) or a mixture of dextrose (2 g/100 g) and sweeteners (aspartame, aspartame + saccharin, and sucralose) on the physicochemical properties, microbiological quality, and sensory profiling of probiotic yogurts.

8.23 PRODUCTION OF GELS AND FRUIT-GLAZED PRODUCTS

Fruit gel production is carried out at a high sugar concentration of approximately 75% inside the fruit pulp. This high concentration is achieved through successive soaking of used fruits in high-concentrated syrups. Following that, fruits or gels in fruit-glazed products are exposed to an external glazing attained by an instant soaking in a very dense sugar syrup. Fruit gel production follows similar principles to those of marmalades and gels. The basic steps of the sugar-based product are presented in Figure 8.4.

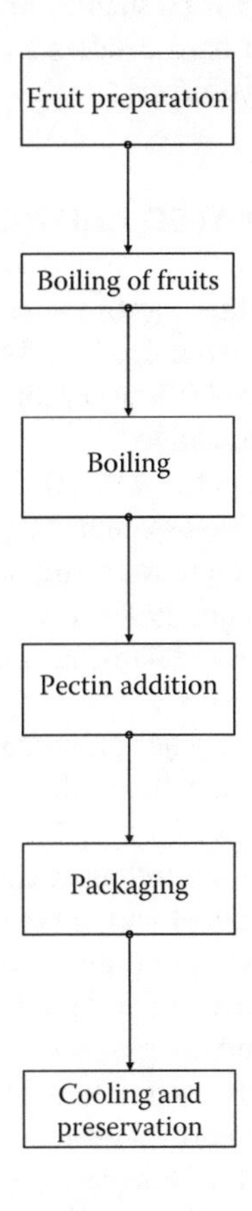

Figure 8.4 Flow diagram of jelly production or sugar-based production.

Quality of these jelly products depends on the following:

- Fruit preparation involves cleaning of the fruit, washing, kernel removal where necessary, and chopping.
- First boiling softens the fruit flesh to make the absorption of syrup easier.
- Second boiling is carried out after a 24-h stay of the fruits in syrup. Fruits are strained. Then they are mixed in hot water with an equal weight of glucose syrup and sugar until final dilution. This mixing will give a concentration of 35–40°Bx when fruits are soaked once more. Boiling follows for 2–3 min and another stay for 24 h in the syrup.
- The above procedure is carried out every day for a number of days. Syrup is produced and added every day until its concentration reaches 75°Bx. At every syrup addition, boiling of 2–3 min follows with a stay for 24 h.
- Sugar fruits are strained from the final syrup and are then soaked instantly in boiling water. Filtration and drying then follow. Drying is carried out by heating at 50°C–55°C in an oven. Sugar fruits are then glazed or packaged. Glazing is done by instant soaking in a dense syrup of 90°Bx. These sugar jelly products are named fruit-glazed products.

8.24 CONFECTIONERY PRODUCTS

The incorporation of trehalose to reduce sweetness intensity can enhance the flavor profile of different confectionery products, such as chocolates, caramels, toffees, fondants and crèmes, fudge, and hard candies. It can also reduce off flavors, such as intense caramel or browned notes, because of its high thermal stability. It is used in hard candies along with the addition of citric, malic, or tartaric acid because it is stable to acid hydrolysis.

8.25 SUGAR DRINKS: CARBONATED AND NONCARBONATED BEVERAGES

Most beverages are sweetened with either sucrose or HFCS (usually 88% fructose). Maltose and lactose are naturally present in alcoholic and dairy-based beverages, respectively.

Carbonated beverage formulations are 90% water and 10% sweetener. Color, carbon dioxide, flavor, acidulants, or preservatives are often added.

Today, many carbonated drinks contain the 42% HFCS although the 55% product is also used. The latter should be used because in sucrose-sweetened carbonated beverages part of the sucrose is converted to inverted sugar in the low-pH environment, and the mixture is sweeter than would be predicted from the sucrose alone (Alexander 1998). Inversion of sucrose may also be a deliberate processing step and may be induced by an acidulant, such as citric acid, with or without added heat. Invertase may also be used.

Noncarbonated beverages include those made from fruits, vegetables, and dairy-based ingredients. Sucrose or HFCS are the sweeteners mostly used. Corn syrups or maltodextrins are also used to build body and aid stability and emulsification.

In functional beverages and energy drinks, trehalose can contribute mild sweetness and enhance mouthfeel and body. It could also be combined with other bulk sweeteners to optimize flavor delivery in beverages. Cargill has launched various products based on trehalose, such as a fitness water and two lemon-lime energy drinks. Moreover, in 2003, it launched a raspberry tea product based on isoflavones for healthy bones, inulin, a fructooligosaccharide for calcium absorption, and calcium.

Sugar alcohols, however, are not widely used in liquid beverage formulations because they have a laxative effect at the levels needed to achieve sweetness.

Polydextrose used at levels of 3%–5% in beverage formulations provides mouthfeel (Altschul 1993). HSHs provide mouthfeel and inhibit microbial activity. Factors considered in choosing an alternative sweetener for a beverage formulation include quality of taste, ease of incorporation into processing, stability at various pH levels (2–8), heat stability (pasteurization and UHT) and solubility.

Blends of alternative sweeteners are used, such as saccharin, acesulfame potassium, stevioside, and cyclamates.

Saccharin is rarely used alone in these formulations because of its bitter and metallic aftertastes. Cyclamate masks the aftertaste of saccharin. Sucrose, acesulfame K, and aspartame also mask saccharin's aftertaste. Blends of saccharin and aspartame are often used in the United States in fountain applications. Formulations designed for this purpose are concentrated and then are mixed with water and carbonated (Holleran 1996). Saccharin maintains sweetness and extends the product shelf life.

Acesulfame K has a metallic aftertaste and is often not used alone. It is blended with aspartame which masks its off flavor. It forms synergistic blends with aspartame and sucrose, and its sweetening ability is greater at pH below neutral (Grenby 1989).

Aspartame is widely used in soft drink formulations in the United States (Nelson 2000a,b). It is not stable at low pH and degrades in fountain formulas. It is blended with saccharin to provide stability.

Alitame is also stable in many beverage formulations. It is used in carbonated and still beverages in countries with approved use; however, off flavors have been noticed after storage (Grenby 1989).

Sucralose is very stable, provides a clean taste, and can be used in beverages preserved with heat (light juice drinks).

Thaumatin has a licorice-type flavor, enhancing coffee flavors, used to mask the bitterness of saccharin. Its sweetening effect can be reduced in formulas containing high concentrations of gums, such as locust bean, xanthan, carrageenan, alginate, or carboxymethyl cellulose (CMC) (Anonymous 1996).

Soft drinks are consumed by all ages. The average consumption in Greece is 0.5 L/day; whereas, in the United States, it is 1 L/day. They cause thirst and hypothermia because of the sugar concentration they contain. They are also cariogenic and cause infection in the gums. Carbon dioxide or carbonic acid aids the digestion or accelerates the sugar absorption, which, however, causes flatulence. They have a lower nutritional value compared to juices.

8.25.1 Soft Drink Types: Cola

Sugar is the main ingredient, as well as water and cola extracts. Coca-Cola's concentrated solution or extract has been a secret for many years; the only data that is known regards acidity regulators (carbonic acid, phosphoric acid, citric acid, caffeine), color (caramel color, a by-product from heating of sugar), preservative (sodium benzoate), aromatic substances, and antifoaming agents.

Concentrate and beverage bases are produced by special plants of The Coca-Cola Company. Concentrate is liquid and contains the flavor materials for the brand Coca-Cola.

Acidifiers give beverages their distinctive tart taste. The Coca-Cola Company has approved many acidifiers for their products. For instance, phosphoric acid is the one used in Coca-Cola, and citric acid is used in many other products.

Colorants include the caramel in Coca-Cola, as well as the natural and artificial colorants used in other products.

Preservatives are used in very small amounts to prevent microbiological spoilage. Those approved by The Coca-Cola Company include sodium, potassium benzoate and sodium, or potassium sorbate. They may be used alone or in combination with each other.

8.25.2 Soft Drinks: Light

Composition of these products is similar to normal soft drinks; however, natural sugars have been replaced with synthetic high-intensity sweeteners, such as aspartame, cyclamates, acesulfame potassium, and sucralose. Cola with vanilla flavor, as well as other flavors, are the new cola-type products.

Beverage bases may be either liquid or dry and contain the flavor materials for all other beverages, including Diet Coke and Coca-Cola Light. Concentrates and beverage bases are packaged in precisely measured units. This helps in the determination of the units of syrup needed to make the required cases of product, the quantities of each ingredient, quantity of liquid sweetener, and amount of additional water.

Concentrates and beverage bases come in bulk packages, intermediate bulk containers, 5- and 10-gal pails, and 2- and 3-gal jugs. Package sizes can range from 4 L in one unit to a package of bulk concentrate that is 20,000 L (The Coca-Cola Company 2011).

There are two methods for manufacturing syrup—batch and continuous blend. In continuous blend, the ingredients are added in continuous streams that are combined in a metered flow to produce the final syrup.

Screens of various mesh sizes are required—1 in for dry materials, 10 mesh screen for liquid ingredients, and 40 mesh screen for the syrup. Moreover, agitators should be working effectively.

Batch processing includes preparation of the syrup containing the sweetener and the water. After the syrup is mixed, the degrees Brix is checked to make sure it is correct, as well as the level in the tank using a sight glass. Second, the addition of concentrates or beverage bases and any other ingredients occurs at the mixing tank. If beverage bases are frozen, these should be thawed before use. Agitation then follows. Mixing continues with the addition of additives, e.g., benzoates. Care should be taken because they are not soluble in the presence of acidic mixes. Then the concentrate or beverage base is added to the tank and and followed by rinsing with water. Finally, adjustment of the degrees Brix of the final syrup should be carried out. The degrees Brix test measures the percentage by weight of the sweetener in the final syrup.

The primary application areas for FRUCTOPURE™ fructose include dry mix beverages, enhanced or flavored waters, still and carbonated beverages, sports and energy drinks, fruit spreads, breakfast cereals, baked goods and baking mixes, dairy products and ice cream, and reduced-calorie foods and beverages. Because of its high level of sweetness and label friendliness, FRUCTOPURE™ fructose is also an excellent choice for sweetening organic foods and drinks subject to certain criteria (Tate and Lyle 2010).

FRUCTOPURE™ fructose is made from non-GMO corn and is available in crystalline and various liquid forms. FRUCTOPURE™ fructose is one of the most effective monosaccharides for binding moisture. Its humectancy, high solubility, and low tendency to recrystallize provide new opportunities in the formulation of intermediate-moisture foods, such as granola bars, icings, fruit pieces for cereal, and fruit bars.

FRUCTOPURE™ fructose has a low GI (~11), resulting in a moderate release of insulin to the bloodstream, relative to other major carbohydrates, such as glucose or sucrose. One does not get an insulin spike after consumption.

This is particularly important for athletes as it avoids hypoglycemia caused by the insulin peak.

Isoglucose syrups—the ISOSUGAR™ family—are a standard raw material for soft drinks in many parts of the world. ISOSUGAR™ 041, a syrup with 55% fructose in its dry substance is an important sweetener in this product category.

ISOSUGAR™ 111, a syrup with 42% fructose, is another important member of this family, especially in the formulation and development of fruit-flavored drinks.

Other syrups with variable fructose content (10%–55%) have been developed and adapted for different applications and product categories. The key ingredient of this product category is fructose because of its ability to enhance fruity flavors. ISOSUGAR™ is an alternative to inverted and other liquid sugar types.

Many other properties of these syrups are attractive to the beverage industry. Primary ones include

- Extraordinary purity and high color stability
- Microbiological safety

- No inversion reactions and high stability
- Cost effectiveness and ease of handling

The combination of different sugars generating synergistic effects in terms of relative sweetness and the adaptation of the sugar spectrum to meet the specific needs of each product are the biggest advantages to the ISOSUGAR™ family.

ISOSUGAR™ 200, 300, 400, 500, 600, and 700 offer real advantages to low degrees Brix beverages, that is, combinations of bulk sweeteners with intensive sweeteners.

They combine the clean sweetness of fructose with the mouthfeel properties of high sugars, putting lost body back into a drink. In the case of nectars and concentrated beverages (syrups, cordials, squashes), fructose containing glucose syrups with higher concentrations of long-chain saccharides are the product of choice. They improve mouthfeel and body and can be found in the GLUCAMYL™ F and GLUCAMYL™ H families.

For drinks, such as iced teas, juices, milk drinks, and others, the flexibility of fructose-containing glucose syrups offers substantial advantages.

Tate and Lyle's product range also includes dry sweeteners that are particularly useful in beverages.

8.26 CHOCOLATE MANUFACTURE

Chocolate manufacturing processes generally involve mixing, refining, and conching of chocolate paste. Finally, tempering and final product preparation, which may include panning, enrobing (covering a base food material with a melted coating that hardens to form a solid surrounding layer), and molding (Figure 8.5) occurs. The final step is cooling.

Chocolates contain cocoa liquor, sugar, cocoa butter, milk fat, and milk powder (depending on product category). A mix of sugar, milk solids, and cocoa liquor at an overall fat content of

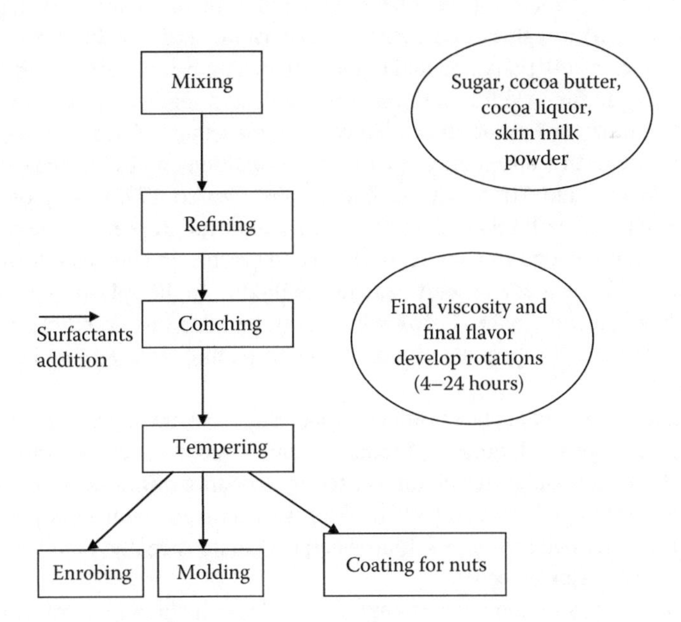

Figure 8.5 Flow diagram of chocolate manufacture. (Adapted from Alexander, J., *Sweeteners Nutritive*, Eagan Press, American Association of Cereal Chemists, St. Paul, MN, 1998. Chandran, R., *Dairy-Based Ingredients*, American Association of Cereal Chemists, St. Paul, MN, 1997. With permission.)

8%–24% is refined to a particle size <30 mm, normally using a combination of two- and five-roll refiners (Beckett 2000). Final particle size critically influences the rheological and sensory properties. A five-roll refiner consists of a vertical array of four hollow cylinders, temperature controlled by internal water flow, held together by hydraulic pressure. A thin film of chocolate is attracted to increasingly faster rollers. Roller shearing fragments solid particles, coating new surfaces with lipid, so these become active, absorbing volatile flavor compounds from cocoa components (Afoakwa, Paterson, and Fowler 2007).

Texture in milk chocolate appears improved by a bimodal distribution of particles with a small proportion having sizes up to 65 mm. Optimum particle size for dark chocolate is lower at <35 mm although values are influenced by product and composition (Awua 2002). Refiners, in summary, not only affect particle size reduction and agglomerate breakdown but also distribute particles through the continuous phase, coating each with lipid.

Refined mixtures then move into conching (slow mixing of heated chocolate paste to reduce particle size and increase thickness), a process that contributes to development of viscosity and final texture and flavor (Figure 8.5). This is the endpoint for manufacture of bulk chocolate.

Fine crystalline sucrose, the main sweetener in chocolate products, compound coatings, and cocoa drinks, is utilized at up to 50% of the formula on a weight basis in chocolate confectionery (Kruger 1999). It has a low hygroscopicity and is easy to blend. Sucrose must be dry to eliminate caking and should not contain inverted sugar because this would decrease viscosity.

Lactose, in milk solids, is present at lower levels in an amorphous form, and in its glassy state, it holds a proportion of milk fat (Beckett 2000), influencing chocolate flavor and flow properties, and is present in milk chocolate. Lactose enhances the browning by participating in Maillard reactions (Bolenz, Amtsberg, and Schape 2006; Kruger 1999). Monosaccharides, glucose and fructose, are rarely used in chocolate as they are difficult to dry. Consequently, the additional moisture present in chocolate increases interactions between sugar particles and increases viscosity. Dextrose and lactose can successfully replace sucrose in milk chocolate (Bolenz, Amtsberg, and Schape 2006; Muller 2003).

Recently, sucrose-free chocolates have become popular among consumers and manufacturers because of reduced calorific values and being noncariogenic and suitable for diabetics (Olinger 1994; Olinger and Pepper 2001; Sokmen and Gunes 2006; Zumbe and Grosso 1993).

Sugar alcohols, including xylitol, sorbitol, mannitol, and lactitol are used for the manufacture of lower-calorie or sugar-free products. However, replacement of sucrose with sugar alcohols affects rheological properties and, hence, processing conditions and chocolate quality (Sokmen and Gunes 2006; Wijers and Strater 2001; Zumbe and Grosso 1993). Moreover, Sokmen and Gunes (2006) reported that maltitol results in similar rheological properties of chocolate to sucrose and, thus, may be recommended as a good alternative to sucrose in chocolate formulations. These authors also observed that chocolate with isomalt resulted in higher plastic viscosity, and xylitol causes a higher flow behavior index. Polydextrose may be added as an edible carbohydrate and intense sweetener. The EU limits consumption of sugar alcohols to 20 g/day because of laxative effects (Kruger 1999).

Sugar bloom is a defect in chocolate products, appearing as white or grey flecks on the chocolate surface. There are two types of bloom: fat bloom, when the incorrect crystalline form of the fat is present in the final product, and sugar bloom caused by moisture migration rather than by crystallization changes. Sugar bloom is caused by damp storage; damp cooler air; impure, hygroscopic, or brown sugars; or improper packaging, resulting when chocolate is coated onto a center with higher ERH than the chocolate (Alexander 1998).

Sugar alcohols are used in chocolate manufacture. These include maltitol, sorbitol, mannitol, lactitol, isomalt, and xylitol.

Sorbitol provides a slight cooling effect. It is very soluble and is usually added during mixing. The γ crystalline form should be used in a fine grade (Grenby, Parker, and Lindley 1983).

Sorbitol has a low melting point, hence, refining and conching temperatures should be reduced. A wet conching procedure should be used because of sorbitol's recrystallization.

Similarly, mannitol is used because of its high melting point; however, its sweetness and solubility are much lower than those of sucrose and gives a chalky mouthfeel to the end product (Happel 1995).

Maltitol is incorporated during mixing. Conching can be carried out at higher temperatures to enhance final flavor development (Happel 1995; Rapaille, Gonze and Van der Schueren 1995).

Anhydrous lactitol can be used during refining and conching. Lactitol has good solubility and good mouthfeel (Olinger 1995).

Isomalt has a low melting point, and, hence, refining and conching temperatures should be about 45°C. Cocoa butter should be added to the formula for good melt-down characteristics (Irwin 1990; Olinger 1994).

Regarding xylitol, the most noteworthy effect in chocolate manufacture is its cooling effect in the final product, making it attractive for chocolate mint-type products, as well as its use as a sole sweetener. Conching should be carried out at temperatures up to 55°C. Because of its low viscosity, solids should be added (Pepper and Olinger 1988).

The final product can be very sandy if ERH during storage is >85%.

8.27 SWEETENERS FOR DAIRY PRODUCTS

In the manufacture of flavored milk products, the addition of various sweeteners to the product is usually desired. The prime purposes of using sweeteners in milk products are to reduce the acidic taste of the product and to make the products tastier and organoleptically more attractive to consumers (Nahon 2005). Among dairy products, fruit and flavored yogurts, low-calorie yogurts and yogurt-type products, ice cream, flavored UHT milks, and dairy-based sweets are mostly produced by incorporating nutritive or artificial sweeteners. The level of incorporation of sweetening agents to dairy products varies depending on the following criteria:

- Type of sweeteners used
- Consumer preferences
- Type of fruit used
- Possible inhibitory effects on the starter bacteria
- Legal limitations
- Economical considerations (Tamime and Robinson 2007)

Many kinds of sweetening agents are used in the manufacture of sweetened dairy products, and the list of such compounds is given in Table 8.2. Sweetening agents are usually classified into two groups: nutritive sweeteners and artificial sweeteners.

8.27.1 Nutritive Sweeteners

Sucrose is the most widely used nutritive sugar in the manufacture of flavored milk products. Sucrose can be used either in liquid form (60%–67% total solids) or in granulated form. In the case of the use of liquid sucrose, the total solids content of the initial mix should be adjusted properly to avoid the dilution of the total solids content of the end product. Fruit and flavored set and stirred yogurts (both fruit corner and sundae-style) are among the dairy products in which natural or artificial sweeteners are widely employed in their production. There are three major sources of sugar in flavored and fruit yogurt: residual milk sugar (lactose, galactose, and glucose), natural sugars present in fruit preparation (sucrose, fructose, and maltose), and sugar added externally by the

Table 8.2 List of Some Sweetening Compounds

Sweetening Compounds	Relative Sweetness (Sucrose = 1)
Lactose	0.4
Dulcitol	0.4
Maltose	0.4
Sorbitol	0.5
Mannose	0.6
Galactose	0.6
Xylose	0.7
Mannitol	0.7
Glycine	0.7
Inverted sugar	0.7–0.9
Glycerol	0.8
Sucrose	1.0
Fructose	1.1–1.5
Cyclamate	30–80
Acesulfame K	300–400
Aspartame	200
Saccharine	240–300
Sucralose	400–800
Neohesperidin DC	1500–2000
Thaumatin	3000
Neotame	7000–13,000

manufacturers (Tamime and Robinson 2007). The level of added sugar is decided by considering the perceived sweetness of fruits in fruit-flavored yogurts. In general, the concentration of added nutritive sweeteners into processed fruits for yogurt manufacture ranges from 25% to 65% (Tamime and Robinson 2007). Because of the inhibitory effect of sugars on the yogurt starter bacteria, the concentration of added sugar should not exceed the level of 10%–11% (Chandan and O'Rell 2006). The development of acidity by yogurt starter bacteria was demonstrated to be suppressed as the sugar level was increased from 6% to 12%. It is a well-established fact that thermophilic yogurt bacteria (*Str. thermophilus*) is more tolerant to sugar than lactobacilli (Dimitrov et al. 2005; Kim, Baick, and Yu 1995). In an extensive study, Keating and White (1990) compared different sweeteners, including sucrose, sorbitol, HFCS, HFCS + monoammonium glycyrrhizinate (MAG), sucrose + MAG, aspartame, Ca-saccharine, Na-saccharine, acesulfame-K, and dihydrochalcone, with regard to their effects on acid development, viscosity, and the microbiological and sensory properties of flavored yogurt. The type of sweeteners and storage time affected the growth of yogurt cultures. Independently from the type of sweeteners, the apparent viscosities of the yogurts increased over time. The yogurts supplemented with fructose syrup had the highest viscosity value. The yogurts added with sucrose + MAG and aspartame received close sensory scores to the yogurt added with sucrose. Hyvonen and Slotte (1983) found that 15% sorbitol added prior to incubation inhibited the culture growth such that no acid, aroma, or coagulation developed. Regarding the behavior of yogurt starter bacteria in the presence of nutritive sweeteners, contradictory results have been published. Tamime and Robinson (2007) reviewed most of the commercially available yogurt starter bacteria and learned they are tolerant to sugar up to 12% in a milk base. On the contrary, Song et al. (1996) reported the growth of yogurt starter cultures was inhibited by various sweetening agents at different levels (Table 8.3). The inhibitory effect of added sugar on yogurt starters stems from an adverse osmotic effect of solutes in the milk and reduced water activity (a_w), the former being more effective.

Table 8.3 Inhibitory Levels of Different Sweeteners on Yogurt Starter Bacteria

Sweeteners	Level of Inhibitory Effect (%)
Sucrose	>4.00
Fructose	>2.70
Aspartame	>0.02
Fructooligosaccharides	>7.30
Isomaltooligosaccharides	>7.70

The incorporation of nutritive sweeteners other than sucrose in yogurt production is rather limited. Fructose, which is derived mainly from the conversion of starch, may be an alternative to sucrose to sweeten yogurt. A negligible effect of fructose on the flavor of yogurt after storage was reported (Hyvonen and Slotte 1983). Grape must is another source of fructose, and the optimum rate of use of grape must in the manufacture of fruit and flavored yogurt should be 20% (Calvo, Ordonezi, and Olano 1995). Fernandez-Garcia, McGregor and Traylor (1998) investigated the chemical and microbiological properties of plain yogurt supplemented with oat fiber and various natural sweeteners (e.g., sucrose, fructose). Addition of fructose at a level of 5.5% caused an increase in the fermentation time by 60% and slowed down the production of organic acids, such as lactic, pyruvic, acetic, and propionic acids. Apparent viscosities of the products increased with the addition of sweeteners and oat fibers.

An undesirable beany aroma is the major drawback of soy milk yogurt (Favaro et al. 2001). On the other side, the high nutritional value of soy products has stimulated efforts to convert soy yogurt into a more appealing product to consumers. Adding a sweetening agent, mainly sucrose, is one of the options to mask undesirable flavor in soy yogurt. Sucrose is not utilized by *Lb. delbrueckii* subsp. *bulgaricus* while *Str. thermophilus* is capable of fermenting this sugar. Estevez et al. (2010) investigated the role of different sugar combinations (42% glucose + 58% fructose and 42% glucose + 58% sucrose) on the organoleptic acceptability of soy milk-based yogurt. The yogurt fortified with a glucose–fructose combination received the highest sensory scores. The effects of types (sucrose, fructose, and sucrose/fructose at 1:1) and concentrations (6%, 8%, and 10%) of sweeteners on the sensory properties of soy-fortified yogurt were studied by Drake, Gerard, and Chen (2001). With the increase in the concentration of sweeteners, the soy flavors and astringency decreased. The sweetener concentration was found to be more effective on the sensory quality of soy-fortified yogurt than the type of sweetener.

Another alternative nutritive sweetener that may be used in the production of sweetened dairy products is glucose–galactose syrup (Thomet et al. 2005). Corn syrup, which is another alternative sweetener to sucrose, is the product in which 20%–70% of the glycoside linkages have been hydrolyzed (Chandan and O'Rell 2006). Corn syrup (with 36 or 42 DE) is generally used in the production of low-fat and nonfat yogurt and frozen desserts. The full replacement of sucrose by corn syrup may carry the risk of flavor defects. Therefore, the use of this sweetening agent in frozen desserts is limited to 25%–30% of the total sweetener level. Corn syrup decreases the stiffness of frozen yogurt while improving the shelf life of the end product. During the last decade, HFCS has gained a relative popularity in the yogurt and ice cream industry. According to Chandan and O'Rell (2006), the most common blends of sugars and HFCS are made up in a 50:50 mix based on solids. HFCS had no negative effects on the quality of yogurt, and the acceptability of the end product increased (McGregor and White 1986). On the other hand, owing to its high OP, HFCS may show an inhibitory effect on yogurt starter bacteria.

Other nutritive sweeteners, such as maltodextrin, crystalline fructose, and inverted sugar, are far less commonly employed in the manufacture of yogurt, but they are mostly preferred for the manufacture of low-fat and nonfat frozen yogurt or frozen desserts. These sweeteners are more widely employed in the manufacture of dietetic yogurts. Maple sugar; brown sugar; and syrup made from rice, oats, and other grains have the potential to be used in the manufacture of flavored yogurt, but at present, their commercial applications are fairly limited. These sweeteners are generally incorporated into yogurt after fermentation.

Honey, which is a natural sweetener, consists of inverted sugar (74.5%), moisture (17.5%), sucrose (2%), dextrin (2%), and miscellaneous (4%) (Marshall and Arbuckle 2000). The incorporation of honey in the production of sweetened dairy products is not common. However, successful attempts were made to substitute sucrose with bee-honey to produce a naturally sweetened ice cream (Groschner 1998). The major drawback of incorporating honey in ice cream processing is that the honey flavor poorly blends with other flavors. Rizai and Ziar (2008) investigated the effect of natural honey on the growth characteristics of *Str. thermophilus* and *Lb. delbrueckii* subsp. *bulgaricus* employed in yogurt making as single or mixed cultures. The growth and acidification capacity of the *Str. thermophilus* monoculture was increased in the presence of 10% honey. However, the growth of the *Lb. delbrueckii* subsp. *bulgaricus* monoculture was limited under the same conditions. The counts and metabolic activities of mixed yogurt cultures were affected adversely, resulting in an increased curdling time by about 1 h. Varga (2006) showed the addition of acacia honey to flavored yogurt at levels varying from 1% to 5% did not influence the counts of yogurt starter bacteria. Regarding the sensory quality of the final product, the addition of acacia honey at a rate of 3% was recommended.

In a novel approach, extruded blends of whey protein concentrate 35 (WPC 35), corn starch, and sugar cane bagasse with 0%, 25%, and 50% substituted yogurt formula were developed (Verdalet-Guzman et al. 2011). Yogurt substituted with extruded blends showed higher viscosity and lower syneresis values than that without blends. Additionally, yogurts with extruded blends at different levels had sensory attributes of taste, acidity, and texture scores similar to those observed for yogurt with no extruded blends.

In another approach, Jaros et al. (2003) investigated the possibility of replacing sucrose with whey-based powders in the manufacture of flavored yogurt. The authors demonstrated that when incorporating powders from whey ultrafiltrate permeates into the fruit preparation, up to 25% of sucrose was replaced by an iso-sweet amount before sensory differences were detected. When using dried, nanofiltered whey ultrafiltrate permeate, this amount could be increased to 30%. According to the same authors, the concentration of monovalent ions was the limiting factor for the replacement of sucrose by an iso-sweet amount of whey permeate powder.

For safety reasons, the sweeteners are usually incorporated into the yogurt mix prior to heat treatment. The advantages of this application are the elimination of osmophilic yeasts and molds that might be present in the sugar ingredients, preventing postpasteurization contamination and better textural quality in the final product. In some cases, incorporation of sugar into fermented milk after incubation is required. In this situation, sugar should be added as pasteurized liquid sugar or flavored sweetened syrups, and extra care must be taken to avoid any contamination. Using liquid sugar is more widely preferred by yogurt manufacturers because of the efficiency of its handling. From a technological point of view, the use of sugar seems to be more advantageous to manufacturers. However, production of liquid sugar from dry sugar requires additional installations, such as storage tanks, pumps, heaters, strainers, etc., which bring about additional operational costs. In order to avoid microbial contamination to the pasteurized liquid sugars, the storage tanks are attached with a UV light source. Additionally, the moisture condensation, which eases the growth of microorganisms, should be prevented by a proper ventilation system (Özer 2010). The optimum storage temperature of the liquid sugar is around 30°C–32°C (Chandan and O'Rell 2006). Other specifications of liquid sugar are presented in Table 8.4.

Sucrose is the main sweetener used in ice cream making. The desired sweetness of ice cream, based on equivalence to sucrose, ranges from 13%–16% (Marshall and Arbuckle 2000). Sweeteners

Table 8.4 Some Specifications of Liquid Sugars

Specifications	
Total solids level	66%–67% (67°Bx)
	Sucrose level: 99.7%
	Inverted sugar level <0.3%
Ash content	0.04%
	Iron level: <0.5 mg kg^{-1}
pH	6.7–8.5
Viscosity	2.0 Poise
Color	<35 ICU
Storage temperature	30°C–32°C

Source: Chandan, R. C., and K. R. O'Rell, in R. Chandan (ed.), *Manufacturing Yogurt and Fermented Milks*, pp. 179–193, Blackwell Publishing, Ames, IA, 2006. With permission.

contribute to the smoothness of texture and rate of melting in ice cream. In general, the molecular weight of a sweetener is inversely proportional to its effect on FP. HFCS that is used as an alternative to sucrose in the ice cream industry can reduce the FP of the product, resulting in softer body. The effect of alternative sweeteners to sucrose, including HFCS; glucose syrup; honey; sucrose + glucose syrup; and sucrose + HFCS and honey + sucrose mixtures on the quality of ice cream was investigated by Özdemir et al. (2008). The honey + sucrose mixture yielded the highest viscosity, and honey alone increased the fat destabilization in ice cream. Glucose syrup caused the lowest melting ratio in the end product. Panelists preferred the ice cream samples sweetened with the mixture of glucose syrup and sucrose. Substituting a 50:50 blend of two corn sweeteners (HFCS-90 and high-maltose corn syrup or HMCS) for sucrose led to a smoother texture in ice cream, but the degree of sweetness decreased (Conforti 1994).

8.27.2 Nonnutritive Sweeteners

Although sweetened dairy products are more desired by Western society than acidic or sour products, the relatively large amounts of sugar in dairy products are a cause of concern (Dufseth, 2004). This concern has triggered changes in the eating habits and lifestyle of people, especially in developed countries. The trend of consuming foods with improved health benefits and reduced risk of disease has gained enormous popularity (Matsubara 2001), and the food industry has responded to this demand strongly (Tamime et al. 1994; Sivieri and Oliviera 2001). During the last two decades, great numbers of no-calorie sweeteners were discovered, and new sugar-free products have been developed for people suffering from diabetes and for those on special diets and/or for the obese (Pinheiro et al. 2005; Özdemir and Sadıkoğlu 1998).

Artificial sweeteners are chemicals that offer the sweetness of sugar without calories (Özer 2010; Khurana and Kanawjia 2007). Because the substitutes are much sweeter than regular sugar, it takes a much smaller quantity of them to create the same sweetness. Therefore, products made with artificial sweeteners have a much lower calorie count than do those made with sugar. The major nonnutritive artificial sweeteners employed in the production of various dairy products include aspartame, acesulfame K, saccharine, and sucralose. Reis et al. (2011) compared the sweetness equivalence of different sweeteners in strawberry-flavored yogurts. The authors showed the sweetness concentrations equivalent to strawberry yogurt sweetened with 11.5% sucrose in the tested sweeteners were 0.072% for aspartame, 0.042% for aspartame/acesulfame K (2:1), 0.064% for cyclamate/saccharine (1:1), 0.043% for cyclamate/saccharine (2:1)/stevia (1.8:1), and 0.30% for sucralose.

Aspartame is one of the first artificial sweeteners that was approved by the FDA in 1988 to use in yogurt-type products (Van der Ven 1988; Keller et al. 1991). The sweetness characteristics of aspartame show similarities to that of sucrose (Keller et al. 1991; NutraSweet 1996). According to Farooq and Haque (1992), the replacement of sucrose by aspartame in yogurt making results in some 50% reduction in the calories per cup. The sweetening power of aspartame is generally greater at low concentrations and in products stored at room temperature.

The stability of sweetening agents during processing or cold storage of yogurt is a decisive factor for their incorporation into the manufacturing process. Lotz, Klug, and Kreuder (1992) found that aspartame was degraded during fermentation, but acesulfame K formed a stable sweetening. An opposite finding was reported by Fellows, Chang, and Keller (1991) who demonstrated that the recovery of aspartame was 95% and 90% in sundae-style yogurts produced using a 6-h incubation at 43.3°C and 13 h at 32.2°C, respectively. The stability of aspartame in three fruit preparations stored for 6 months at three different temperatures (i.e., 32.2°C, 21.1°C, and 4.4°C) was assessed by Fellows, Chang, and Shazer (1991). The authors showed that ~60% of original aspartame remained stable after 6 months at 21.2°C. The joint effect of pH, time, and storage temperature was found to be determinative on the stability of aspartame in fruit preparations. The rate of degradation of aspartame in yogurt is closely related with the growth and metabolic activities of the starter bacteria (Keller et al. 1991). In general, the lower the metabolic activity of the starter bacteria, the lower the degradation of aspartame during processing and/or storage. In reverse, the presence of aspartame did not affect the viability of human-derived probiotic bacteria (*Lb. acidophilus*, *Lb. agilis*, and *Lb. rhamnosus*) in ice cream (Basyigit, Kuleasan, and Karahan 2006).

Aspartame can withstand a wide range of heat treatments (Table 8.5). This is particularly important for flavored milks and yogurts to which aspartame is added prior to fermentation.

The milk components may affect the sweetness perception in yogurts sweetened by artificial sweeteners. Although milk fat contributes to the sweetness perception in yogurt, the effect of aspartame on this characteristic is much greater (King, Lawler, and Adams 2000). With the increase in the level of aspartame (i.e., 400 or 600 ppm), the fruity flavor, metallic notes, and sweetness were stronger. On the contrary, this effect was not found in yogurt containing low levels of aspartame (i.e., 200 ppm) (King, Lawler, and Adams 2000). Replacement of milk fat by microencapsulated whey proteins as a fat substitute (e.g., Dairy Lo™) affected the perception of acid taste in fruit yogurt sweetened with aspartame (Brandão, Silva, and Reis 1995). Therefore, the level of aspartame may be reduced when Dairy Lo™ is used at relatively higher amounts (e.g., >2%–3%).

Yogurt gel is primarily based on the protein–protein interactions, and this association is influenced by amphipathic peptides, such as aspartame (Mozaffar and Haque 1992; Haque 1993). This, in turn, may affect the void spaces of the casein network, which also influences the level of syneresis of yogurt. Haque and Aryana (2002) compared the effects of sugar and aspartame on the microstructure of flavored yogurt. The authors concluded that when aspartame was used, the casein micelles formed double longitudinal polymers. In contrast, sugar caused casein micelles to form clusters. In the yogurt containing no sweetener, the gel network was formed predominantly by casein micelles arranged in longitudinal polymers. Aspartame stimulated the polymerization of casein micelles. Kumar (2000) incorporated aspartame (at a level of 0.08%) in the production of

Table 8.5 Stability of Aspartame toward Various Heat Treatments

Heat Treatment	pH	Loss in Aspartame (%)
UHT (121°C–135°C for 15 s)	2.5–3.5	0.5%–1.5%
Pasteurization (VAT system 80°C for 15 min)	2.6–4.2	2.0%–4.0%
HTST (72°C for 15 s)	3.5	<1.0%
Hot filling (88°C)	3.5	3.0%–6.0%
High-heat treatment (90°C for 20 min)	3.8	<5.0%

low-calorie lassi—an Indian-origin fermented food. Aspartame had no effect on the development of acidity, but the viscosity of lassi decreased remarkably after the addition of aspartame.

Aspartame contains phenylalanine, which is not metabolized by individuals with lack of phenylalanine hydroxylase (PAH). The accumulation of phenylalanine, which is further converted to phenylpyruvate, can cause serious damages in brain development, leading to mental retardation, brain damage, and seizures. Therefore, people who have such a genetic disorder should avoid consuming foods containing phenylalanine. According to EU regulations, labeling of foods containing aspartame or aspartame/acesulfame K must bear the following warning: "Aspartame contains a source of phenylalanine." The U.S. FDA recommends the consumption of <50 mg of aspartame per each kg of total body weight daily (FDA 2011). In the European Union, the European Food Safety Authority has recommended a slightly lower acceptable daily intake (ADI) for aspartame at 40 mg/kg body weight.

Acesulfame K is 5.6-dimethyl-1,2,3-oxathiazine-4(3H)-one-2,2-dioxide combined with potassium. In dairy applications, acesulfame K is used in combination with aspartame. This mixture consists of 36% acesulfame K and 64% aspartame. Acesulfame K/aspartame is slightly soluble in water and ethanol, and its sweetness intensity compared with that of sucrose is 300–400. An acesulfame K/aspartame mixture can withstand thermal treatments and shows great stability in acid and alkaline pH. The recommended level of acesulfame K/aspartame for fruit yogurt production is 0.06% (Lotz, Klug, and Kreuder 1992). According to King, Arents, and Duineveld (2003), when natural yogurt was sweetened with aspartame or a mixture of aspartame/acesulfame K, no residual sweetness or bitterness was organoleptically evident. In contrast, the sourness appeared to be masked by aspartame either alone or blended with acesulfame K. A very bitter aftertaste was noted in the product sweetened with sucrose. The use of acesulfame K/aspartame in the manufacture of dairy products is rather limited compared with aspartame alone. This mixed sweetener is more widely used in the production of carbonated beverages and as a tabletop sweetener. The recommended daily consumption amount of acesulfame K sits at 15 mg per each kg of body weight (FDA 2011).

Saccharine is the first artificial sweetener produced commercially. It can be synthesized chemically from toluene or phthalic anhydride (Candido and Campos 1996). Saccharine is well-known for its bitter taste and metallic aftertaste. Therefore, it is recommended that saccharine should be mixed with other sweeteners, such as xylitol (i.e., 4% xylitol + 0.007% saccharine to obtain sucrose equivalent to 8%) (Hyvonen and Slotte 1983), and/or bulking agents, such as alginate, pectin, or carrageenan (Wolf 1979) to mask its bitter aftertaste. Candido (1999) proposed to use saccharine plus aspartame to improve the taste and stability of sweetness in food products. Saccharine is extremely stable against thermal treatments (i.e., no change after 1 h at 150°C) and pH changes (stable between pH 2.0 and 8.0). To the best of our knowledge, saccharine has little or no use in dairy processing. Very limited data is available on the possible inhibitory effect of saccharine on the metabolic activities (e.g., acid production) of *Str. thermophilus* and *Lb. delbrueckii* subsp. *bulgaricus*. In an early work, Gautneb, Steinsholt, and Abrahamsen (1979) demonstrated that acid production by yogurt starter bacteria was suppressed when mixed with a sweetening agent containing 99.9% sorbitol and 0.1% saccharine. Therefore, it is much safer to add such agents after the fermentation of yogurt.

Cyclamate is the commercial name of cyclamic acid or cyclohexylsulfamic acid, and it is usually used in the form of a sodium or calcium salt (Nelson 2000b; Pinheiro et al. 2005). Cyclamate is currently used in food processing in many countries, including some EU countries and China, but has been banned in the United States since 1969 because of its possible toxic effect. Today, some countries are considering banning the use of cyclamate in foods. Cyclamate can diminish the bitter aftertaste of saccharine; therefore, its use in combination with saccharine at a ratio of 10:1 is recommended (Wilkes 1992). Regarding its sweetening power, stability during storage, and effect on the yogurt starter bacteria, cyclamate seems to be a suitable alternative to sucrose. The ADI value of sodium cyclamate is 10–11 mg per kg body weight (Zhu et al. 2005).

Sucralose is the generic name of a relatively new high-intensity, noncaloric sweetener derived from ordinary sugar. It is 400–800 times sweeter than sucrose (Pinheiro et al. 2005) and is easily soluble in water, methanol, and ethanol. It is useful in a wide range of foods and beverages, including frozen desserts (Marshall and Arbuckle 2000). The sweetening power of sucralose is largely affected by sucralose concentration, pH, and temperature. In general, the increase in the sucralose concentration, the pH of the food, and the decrease in temperature led to a reduction in the sweetness of sucralose. Owing to its high stability during pasteurization treatment and at a pH range of 3–7 makes this sweetener suitable for the manufacture of yogurt, ice cream, and dairy desserts (Marshall and Arbuckle 2000). According to Candido (1999), the perceived sweetness of sucralose in yogurt is fast, and compared to sucrose, it persists in yogurt for a slightly longer period. No bitter taste and/or metallic aftertaste associated with sucralose was reported. Sucralose is increasingly used in the manufacture of light and low-calorie yogurts and drinks (Chandan and O'Rell 2006). Fonseca and Neves (1998) investigated the role of sweeteners (e.g., aspartame, sucralose, and fructose) on the quality of low-calorie yogurt made from skimmed milk with added graviola pulp. The results of the sensory test panel revealed the yogurt supplemented with sucralose was preferred less often than that supplemented with fructose or aspartame. The use of aspartame, aspartame + saccharine, or sucralose with the overall sweetness rate being equivalent to 6 g of sucrose per 100 g had no significant effect on the growth and viability of both yogurt bacteria and probiotics (*Lb. acidophilus* and *Bifidobacterium* spp.). This finding indicates the production of probiotic dietetic yogurt with an acceptable sensory quality was possible. The accepted daily intake of sucralose is 5 mg per kg body weight (FDA 2011). Among nonnutritive sweeteners, sucralose is the unique one that does not have any reported disadvantages when applied to food processing.

In the United Kingdom, a new sweetener has been introduced with all the benefits of probiotics without the refrigeration requirements. It is called Nevella Low Calorie Sweetener with probiotics and has less than one calorie per tablet. The company Heartland Sweeteners indicates that a healthy digestive system supports a healthy immune system, and with Nevella, you get the health benefits of probiotics in your favorite beverages (Figure 8.6).

Neotame™, which was developed by NutraSweet, is a derivative of aspartic acid and phenylalanine and is some 7000–13,000 times sweeter than sucrose. Neotame™ generates no off flavor or metallic aftertaste in dairy products (Pinheiro et al. 2005). It can withstand the high heat treatment (e.g., UHT) and fermentation conditions of yogurt. This commercial sweetener was reported to remain stable at a level of 99% after UHT treatment and 88% after yogurt fermentation and storage for 5 weeks at 4°C (Pinheiro et al. 2005). The levels of Neotame™ needed to achieve a 6%–10% sucrose level in fruit-flavored yogurt and fruit for yogurt preparation are 11–17 ppm and 60–120 ppm, respectively (Pinheiro et al. 2005). To achieve the rate of sweetness in yogurt sweetened with aspartame, the concentrations of aspartame should be 550–800 ppm and 2500–3500 ppm, in the same order. The effect of the partial substitution of sucrose by Neotame™ on the sensory and physical characteristics of plain yogurt was investigated by Hernandez-Morales, Hernandez-Montes, and Villegas de Gante (2007). A substitution of 37.5% and 50% of sucrose with Neotame™ resulted in a sweeter taste and longer sweetness duration in yogurt. The sweetness profile of 25% substituted by Neotame™ yogurt was found to be comparable with sucrose.

Neohesperidine DC is an intense sweetener that was approved by the EU in 1994 (Official Journal of the European Union 1994a). It has a dual function in foods: sweetener and flavor-enhancer (Official Journal of the European Union 1995). Neohesperidine DC is effective on suppressing the bitter aftertaste in yogurt-type products (Montijano, Tomás-Barberan, and Borrego 1997). The stability of neohesperidine DC during yogurt manufacturing and storage was investigated by Montijano, Tomás-Barberan, and Borrego (1995). This intense sweetener is more stable at acidic pH, and no significant decomposition of neohesperidine DC was found in yogurt after 6 weeks of storage at 3°C (Montijano, Tomás-Barberan, and Borrego 1995). The same work demonstrated that this sweetening agent did not influence the acid production by yogurt bacteria. It was also showed

Figure 8.6 United Kingdom: Nevella low-calorie sweetener with probiotics, less than 1 calorie per tablet. (Courtesy of Innova Market Insights for FoodIngredientsFirst.com, June 2010; http://www.nevella.com/press-area/downloads/NEVELLAPRO_BLISTER_300TABS_CMYK.png.)

that neohesperidine DC withstood heat treatment ranging from 70°C to 100°C without significant degradation (Montijano, Tomás-Barberan, and Borrego 1997).

It is well known that thickening agents can modify the flavor release in dairy products (Dubois, Lubbers, and Voilley 1995). In an early study, a negative correlation between pectin concentration and aroma release was reported by Pangborn and Szczesniak (1974). Decourcelle et al. (2004) made fruit preparations by incorporating thickeners (starch, pectin, locust bean gum, and guar) and sweeteners (fructose, fructooligosaccharide, aspartame, and acesulfame K), which were then introduced in fat-free stirred yogurt. The release of flavor was reduced in the presence of pectin and starch and tended to increase in the presence of locust bean gum. The combination of sweetening agents and guar did not affect the aroma release mechanism in yogurt. The same authors concluded that, under shear conditions, the composition of fruit preparations was more effective on the release of aroma than the rheological properties of yogurt. The incorporation of aspartame at a level of 200 ppm along with starch as a stabilizer at a level of 500 ppm improved the perceived sweetness and overall quality of skimmed milk yogurt (Farooq and Haque 1992). The interaction between sweeteners and stabilizers is also important in ice recrystallization in ice cream. Miller-Livney and Hartel (1997) investigated the role of interactions of various sweeteners (20- and 42-DE corn syrup solids, sucrose, and HFCS) and stabilizers (gelatin, locust bean gum, xanthan gum, and carrageenan). Overall, carrageenan and locust bean gum were found to be the most effective in retarding ice growth in ice cream.

Large numbers of sweeteners are available for the production of milk and milk-derived products. Artificial (or no-calorie) sweeteners have proved to be suitable alternatives to sucrose and other nutritive sweeteners, as they have little, if any, adverse effect on the quality of dairy products.

Table 8.6 Maximum Permitted Limits of Artificial Sweeteners in Milk and Dairy Products in the EU

Sweeteners	Milk and Milk Derivative–Based Drinks Energy Reduced[a] or with No Added Sugar[b] (mg/L)	Milk and Milk Derivative–Based Preparations Energy Reduced[a] or with No Added Sugar[b] (mg/L)
Acesulfame K (E 950)	350	350
Aspartame (E 951)	600	1000
Cyclamic acid (E 952)	400	250
Saccharine and its Ca, Na, and K salts (E 954)	80	100
Neohesperidine DC (E 959)	50	50

Source: Official Journal of the European Union, 1994; No. L 237/5. With permission.
[a] Without any added monosaccharides, disaccharides, or any other foodstuff used for its sweetening properties.
[b] With an energy value reduced by at least 30% compared with the original foodstuff or a similar product.

The use of sweeteners in combination offers some advantages over their single applications. These include better stability, improved sweetness, and reduced costs (Reis et al. 2011; Portman and Kilcast 1998). The technological and legal limitations must be considered in deciding the upper limits of artificial sweeteners. Maximum permitted limits of no-calorie sweeteners in milk and milk-derived products by the EU are presented in Table 8.6. Additionally, sweeteners authorized by EU regulations must comply with the specific purity criteria defined by Directive 2008/60/EC.

8.28 SPORTS AND ENERGY DRINKS

The category of sports and energy drinks occupies an important place in the beverage industry. It has been a growing market over the last two decades that attracts mainly young consumers but not necessarily athletes or exercising individuals. The main natural and artificial sweeteners used in these drinks are presented in Table 8.7. It should be emphasized that, although the use of these sweeteners is similar for both sports and energy drinks, these products have some distinct differences.

Energy drinks, except for carbohydrates, contain caffeine in various quantities (70–200 mg per 16 oz. of serving), glucuronolactone, amino acids such as taurine and L-carnitine, several herb extracts such as guarana, ginseng, and ginkgo biloba, and various B-complex vitamins (Higgins, Tuttle, and Higgins 2010). Mainly because of their caffeine and herb extract contents, energy drinks have been criticized, and certain concerns have been raised concerning safety (Clauson et al. 2008;

Table 8.7 Natural and Artificial Sweeteners That Are Used in Sports and Energy Drinks

Glucose (Dextrose)
Sucrose
Fructose
Maltodextrin
Glucose syrup
High-fructose corn syrup
Saccharin
Sucralose
Trehalose
Aspartame
Acesulfame potassium (ACE K)
Truvia
Rebiana

Reissig, Strain, and Griffiths 2009; Kaminer 2010; Ballard, Wellborn-Kim, and Clauson 2010). Although the amounts of guarana, taurine, and ginseng found in most popular energy drinks are far below the amounts expected either to cause adverse effects or to provide therapeutic benefits (Clauson et al. 2008; Higgins, Tuttle, and Higgins 2010), the increasingly common practice of consuming such energy beverages with alcohol may elevate the incidence of caffeine-related disorders, including caffeine intoxication, dependence, and withdrawal (Reissig, Strain, and Griffiths 2009; Ballard, Wellborn-Kim, and Clauson 2010). In addition, safety concerns have been expressed regarding driving safety, because it is a common false perception that the consumption of energy drinks can reverse alcohol-related impairment, including motor coordination and visual reaction time (Kaminer 2010). In terms of improvements in exercise performance after the ingestion of energy drinks, conflicting results have been reported during cycling exercise (Candow et al. 2009; Ivy et al. 2009), whereas such beverages are not recommended for rehydration because of the diuretic properties of caffeine (Higgins, Tuttle, and Higgins 2010).

On the other hand, sports drinks usually contain 5%–8% carbohydrates in the form of dextrose, fructose, sucrose, and maltodextrin and some electrolytes, which are mainly sodium and potassium (Coombes and Hamilton 2000). Some formulations may contain vitamins or even proteins and amino acids, depending on their use. Glucose polymers, such as maltodextrin, are frequently used instead of glucose to provide more energy with less sweetness than normal glucose and to keep osmolality close to that of human plasma (280 mOsm/kg). This way, sports beverages maintain a hypotonic or isotonic nature that will promote water absorption (Hunt, Smith, and Jiang 1985) compared to hypertonic solutions, which stimulate less water absorption and more secretion into the gastrointestinal lumen, resulting in the potential for dehydration, especially during prolonged (more than 1 h) exercise in a hot and humid environment. Sports drinks may also be hypertonic if the priority is not maintaining fluid balance but a higher delivery of energy in the form of blood glucose for replacing muscle and liver glycogen stores that have been used during exercise (Maughan and Noakes 1991). In terms of effectiveness, research has shown that consumption of sports drinks containing electrolytes and carbohydrates have positive effects on fluid-electrolyte balance and exercise performance and are considered superior to water (American College of Sports Medicine 2007). A more detailed discussion regarding the mechanisms by which sports drinks can assist human performance will be presented in the section of this book dealing with the nutritional and health aspects of sweeteners.

REFERENCES

Afoakwa, E. O., A. Paterson, and M. Fowler. (2007). Factors influencing rheological and textural qualities in chocolate: A review. *Trends in Food Science and Technology, 18*, 290–298.

Alexander, J. (1998). *Sweeteners Nutritive*. St. Paul, MN: Eagan Press, American Association of Cereal Chemists.

Altschul, A. M. (1993). *Low-Calorie Foods Handbook*. New York: Marcel Dekker.

American College of Sports Medicine. (2007). Exercise and fluid replacement: Position stand. *Med. Sci. Sports Exerc., 39*, 377–390.

Ang, J. F., and W. B. Miller. (1991). Multiple functions of powdered cellulose as a food ingredient. *Cereal Foods World, 36*, 558–564.

Anonymous. (1996). Thaumatin: The sweetest substance known to man has a wide range of food applications. *Food Technol., 50*, 74–75.

Awua, P. K. (2002). *Cocoa Processing and Chocolate Manufacture in Ghana*. Essex, UK: David Jamieson and Associates Press, Inc.

Ballard, S. L., J. J. Wellborn-Kim, and K. A. Clauson. (2010). Effects of commercial energy drink consumption on athletic performance and body composition. *Phys. Sportsmed., 38*, 107–117.

Barndt, R. L., and G. Jackson. (1990). Stability of sucralose in baked goods. *Food Tech., 44*, 62–66.

Basyigit, G., H. Kuleasan, and A. G. Karahan. (2006). Viability of human-derived probiotic *lactobacilli* in ice cream produced with sucrose and aspartame. *J. Ind. Microbio. Biot., 33*(9), 796–800.

Bean, M. M., and C. S. Setser. (1992). Polysaccharide, sugars, and sweeteners. In: J. Bowers (Ed.), *Food Theory and Applications* (69–198). New York: Macmillan.

Beckett, S. T. (2000). *The Science of Chocolate*. London: Royal Society of Chemistry Paperbacks.

Beranbaum, R. L. (1985). *Sugar in Baking and Cooking: The Research Report*. Atlanta, GA: International Association of Culinary Professionals.

Bolenz, S., K. Amtsberg, and R. Schape. (2006). The broader usage of sugars and fillers in milk chocolate made possible by the new EC cocoa directive. *Int. J. Food Sci. Technol., 41*, 45–55.

Brandão, S. C. C., R. C. Silva, and J. S. Reis. (1995). Produtos de laticinios light: Uma nova opção para o consumidor. *Leite e Derivados, 22*, 22–24.

Bure, J. (1980). *La pate de farine de ble: la chimie du ble*. Massy, France: Collection PROMO-ENSIA. Edit. S.E.P.A.I.C.

Calvo, M. M., J. A. Ordonezi, and A. Olano. (1995). The use of grape must in the elaboration of yogurt: Changes of carbohydrate composition during manufacture. *Milchwissenschaft, 50*(9), 506–508.

Candido, L. M. B. (1999). Edulcorantes em leites fermentados. In: *Seminario Internacional De Leite Fermentados* (46–59). Campinas, Brazil: Chr. Hansen and Tetra Pak. (extracted from Pinheiro, M. V. S., Oliveira, M. N., Penna, A. L. B., and Tamime, A. Y. (2005). The effect of different sweeteners in low-calorie yogurts: A review. *Int. J. Dairy Technol., 58*(4), 193–199.

Candido, L. M. B., and A. M. Campos. (2009). Edulcorantes não nutritivos (181–208). In: D. G. Candow, A. K. Kleisinger, S. Grenier, and K. D. Dorsch (Eds.), Effect of sugar-free Red Bull energy drink on high-intensity run time-to-exhaustion in young adults. *J. Strength Cond. Res., 23*, 1271–1275.

Candow, D. G., A. K. Kleisinger, S. Grenier, and K. D. Dorsch. (2009). Effect of sugar-free Red Bull energy drink on high-intensity run time-to-exhaustion in young adults. *J. Strength Cond. Res., 23*: 1271–1275.

Chandan, R. C., and K. R. O'Rell. (2006). Ingredients for yogurt manufacture. In: R. Chandan (ed.), *Manufacturing Yogurt and Fermented Milks* (179–193). Ames, IA: Blackwell Publishing.

Chandran, R. (1997). *Dairy-Based Ingredients*. St. Paul, MN: American Association of Cereal Chemists.

Clauson, K. A., K. M. Shields, C. E. McQueen, and N. Persad. (2008). Safety issues associated with commercially available energy drinks. *J. Am. Pharm. Assoc., 48*, e55–e63.

Conforti, F. D. (1994). Effect of fat content and corn sweeteners on selected sensory attributes and shelf stability of vanilla ice cream. *J. Soc. Dairy Technol., 47*(2), 69–75.

Coombes, J. S., and K. L. Hamilton. (2000). The effectiveness of commercially available sports drinks. *Sports Med., 29*, 181–209.

Dartey, C. K., and R. H. Briggs. (1987). Reduced-calorie baked goods and methods for producing same. U.S. Patent 4, 668, 519.

Decourcelle, N., S. Lubbers, N. Vallet, P. Rondeau, and E. Guichard. (2004). Effect of thickeners and sweeteners on the release of blended aroma compounds in fat-free stirred yoghurt during shear conditions. *Int. Dairy J., 14*, 783–789.

Dimitrov, T., S. Boicheva, T. Iliev, and N. Jeleva. (2005). Production of yoghurt with different additives. *Dairy Science Abstracts, 67*, 348.

Drake, M. A., P. D. Gerard, and X. Q. Chen. (2001). Effects of sweeteners, sweetener concentration, and fruit flavor on sensory properties of soy-fortified yogurt. *J. Sensory Studies, 16*, 393–405.

Drewett, E. M., and R. W. Hartel. (2007). Ice crystallization in a scraped surface freezer. *J. Food Eng., 78*, 1060–1066.

Dubois, C., S. Lubbers, and A. Voilley. (1995). Revue bibliographique: Interactions aromes autres constituents—Applications á l'aromatisation. *Industries Alimentaries et Agricoles, 4*, 186–193.

Dufseth, H. (2004). Yoghurt or not yoghurt. *Dairy Science Abstracts, 66*, 902.

Estevez, A. M., J. Mejia, F. Figuerola, and B. Escobar. (2010). Effect of solids content and sugar combinations on the quality of soymilk-based yogurt. *J. Food Processing and Preservation, 34*, 87–97.

Farooq, K., and Z. U. Haque. (1992). Effect of sugar esters on the textural properties of nonfat low-calorie yogurt. *J. Dairy Sci., 75*, 2676–2680.

Favaro, C., S. Terzi, L. Trugo, R. Dellamodesta, and S. Couri. (2001). Development and sensory evaluation of soymilk-based yoghurt. *Archivos Latinoamericanos de Nutrición, 51*(1), 100–104.

Fellows, J. W., S. W. Chang, and S. E. Keller. (1991). Effect of sundae-style yogurt fermentation on the aspartame stability in fruit preparation. *J. Dairy Sci., 74*(10), 3345–3347.

Fellows, J. W., S. W. Chang, and W. H. Shazer. (1991). Stability of aspartame in fruit preparations used in yogurt. *J. Food Sci., 56*(3), 689–691.

Fernandez-Garcia, E., J. U. McGregor, and S. Traylor. (1998). The addition of oat fiber and natural alternative sweeteners in the manufacture of plain yogurt. *J. Dairy Sci., 81*, 655–663.

Fonseca, R. R., and R. C. A. Neves. (1998). Elaboração de iogurte dietéticode graviola. In: 16th Congresso Brasilero de Ciencia e Technologia de Alimentos. Rio de Janeiro, Brazil. (extracted from Pinheiro, M. V. S., Oliveira, M. N., Penna, A. L. B., and Tamime, A. Y. (2005). The effect of different sweeteners in low-calorie yogurts: A review. *Int. J. Dairy Technol., 58*(4), 193–199.

Freeman, T. M. (1989). Sweetening cakes and cake mixes with alitame. *Cereal Foods World, 34*, 1013–1015.

Frye, A. M., and C. S. Setser. (1992). Optimizing texture of reduced-calorie yellow layer cakes. *Cereal Chem., 69*, 338–343.

Gates, J. C. (1981). *Basic Foods, 2nd edition*. New York: Holt, Rinehart and Winston.

Gautneb, T., K. Steinsholt, and R. K. Abrahamsen. (1979). Effect of different sweeteners and additives on acidification of frozen yogurt. *Dairy Science Abstracts, 41*(8), 4232.

Godshall, M. A. (1990). Use of sucrose as a sweetener in foods. *Cereal Foods World, 35*, 384–389.

Grenby, T. H. (1989). *Progress in Sweeteners: Applied Food Science Series*. New York: Elsevier.

Grenby, T. H., K. J. Parker, and M. G. Lindley. (1983). *Developments in Sweeteners, vol. 2*. New York: Applied Science Publishers.

Grice, H. C., and L. A. Goldsmith. (2000). Sucralose: An overview of the toxicity data. *Food Chem. Toxicol., 38*, S1–S6.

Groschner, P. (1998). Production of ice cream by using honey instead of sugar. *Deutsche Lebensmittel-Rundschau, 94*(7), 214–217.

Happel, B. L. (1995). Crystalline maltitol in the manufacture of chocolate. *Manufacture Confectionery, 75*(11), 96–99.

Haque, Z., and K. J. Aryana. (2002). Effect of sweeteners on the microstructure of yogurt. *Food Sci. Technol. Res., 8*(1), 21–23.

Haque, Z. U. (1993). Influence of milk peptides in determining the functionality of milk proteins: A review. *J. Dairy Sci., 76*, 311–320.

Hernandez-Morales, C., A. Hernandez-Montes, and A. Villegas de Gante. (2007). Effect of the partial substitution of sucrose by neotame on the sensory and consistency characteristics of plain yogurt. *Revista Mexicana de Ingenieria Quimica, 6*(2), 203–209.

Herrera, M. L., J. I. M'Cann, C. Ferrero, N. E. Zaritzky, and R. W. Hartel. (2007). Thermal, mechanical and molecular relaxation properties of frozen sucrose and fructose solutions containing hydrocolloids. *Food Biophysics, 2*, 20–28.

Hickenbottom, J. (1996). Use of molasses in bakery products. *AIB Technical Bulletin*, G. Ranhotra (Ed.), Vol. XVIII (6), 1–6.

Higgins, J. P., T. D. Tuttle, and C. L. Higgins. (2010). Energy beverages: Content and safety. *Mayo Clin. Proc., 85*, 1033–1041.

Hodge, J. E. (1953). Dehydrated foods: Chemistry of browning reactions in model systems. *J. Agric. Food Chem., 1*, 928–943.

Holleran, J. (1996). Sweet success: Use of high-intensity sweeteners in soft drinks. *Beverage Ind., 87*(4), 50–52.

Horton, S. D., G. N. Lauer, and J. S. White. (1990). Predicting gelatinization temperatures of starch/sweetener systems for cake formulation by differential scanning calorimetry—Part II: Evaluation and application of a model. *Cereal Food World, 35*, 734–740.

Howling, D. (1992). Glucose syrup: Production, properties and applications. In: F.W. Schenck and R.E. Hebeda (eds.), *Starch Hydrolysis Products* (277–317). New York: VCH Publishers.

Hunt, J. N., J. L. Smith, and C. L. Jiang. (1985). Effect of meal volume and energy density on the gastric emptying of carbohydrates. *Gastroenterology, 89*, 1326–1330.

Hyvonen, L., P. Koivistoinen, and F. Voirol. (1982). Food technological evaluation of xylitol. *Adv. Food Res., 28*, 373–403.

Hyvonen, L., and M. Slotte. (1983). Alternative sweetening of yogurt. *J. Food Technol., 18*, 97–112.

Innova Market Insights for FoodIngredientsFirst.com, June 2010. Global Sugars and Sweeteners Product Trends, powerpoint presentation.

Irwin, W. E. (1990). Reduced-calorie bulk ingredients: Isomalt. *Manufact. Confect., 70*(11), 55–60.

Ivy, J. L., L. Kammer, Z. Ding et al. (2009). Improved cycling-time-trial performance after ingestion of a caffeine energy drink. *Int. J. Sport Nutr. Exerc. Metab., 19*, 61–78.

Jaros, D., C. Spieler, T. Kleinschmidt, and H. Rohm. (2003). Using whey permeate powders for partial sucrose substitution in flavored yogurt. *Milchwissenschaft, 63*(2), 174–178.

Kaminer, Y. (2010). Problematic use of energy drinks by adolescents. *Child Adolesc. Psychiatr. Clin. N. Am., 19*, 643–650.

Kearsley, M. W., and G. G. Birch. (1985). The chemistry and metabolism of the starch-based sweeteners. *Food Chemistry, 16*, 191–207.

Keating, K. R., and C. H. White. (1990). Effect of alternative sweeteners in plain and fruit flavored yogurts. *J. Dairy Sci., 73*, 54–62.

Keller, S. E., S. S. Newberg, T. M. Kreiger, and W. H. Shazer. (1991). Degradation of aspartame in yogurt related to microbial growth. *J. Food Sci., 56*(1), 21–23.

Khan, R. (1993). *Low-Calorie Foods and Food Ingredients*. New York: Chapman and Hall.

Khurana, H. K., and S. K. Kanawjia. (2007). Recent trends in development of fermented milks. *Curr. Nutrition and Food Sci., 3*, 91–108.

Kim, Y. J., S. C. Baick, and J. H. Yu. (1995). Effect of oligosaccharides and bifidobacterium growth promoting materials in yogurt base made by *Bifidobacterium infantis* 420. *Korean J. Dairy Sci., 17*, 167–173.

King, B. M., P. Arents, and C. A. A. Duineveld. (2003). A comparison of aspartame and sucrose with respect to carryover effects in yogurt. *Food Quality and Preferences, 14*(1), 75–81.

King, S. C., P. J. Lawler, and J. K. Adams. (2000). Effect of aspartame and fat on sweetness perception in yogurt. *J. Food Sci., 65*(6), 1056–1059.

Knecht, R. L. (1990). Properties of sugar. In: N. L. Pennington and C. W. Baker (eds.), *Sugar: A User's Guide to Sucrose*. New York: Van Nostrand Reinhold.

Kokini, J., and G. van Aken. (2006). Discussion of food emulsions and foams. *Food Hydrocolloids, 20*, 438–445.

Kruger, C. (1999). Sugar and bulk sweetener. In: S. T. Beckett (ed.), *Industrial Chocolate Manufacture and Use, 3rd ed.* (36–56). Oxford: Blackwell Science.

Kumar, M. (2000). Physicochemical characteristics of low-calorie lassi and flavored dairy drink using fat replacer and artificial sweeteners. M.Sc. Thesis, National Dairy Research Institute, Karnal, India.

Levenson, D. A., and R. W. Hartel. (2005). Nucleation of amorphous sucrose–corn syrup mixtures. *J. Food Eng., 69*, 9–15.

Liu, H., and J. Lelievre. (1992). A differential scanning calorimetry study of melting transitions in aqueous suspensions containing blends of wheat and rice starch. *Carbohydr. Polym., 17*, 145–149.

Lotz, A., C. Klug, and K. Kreuder. (1992). Sweetener stability. *Dairy Industries Int., 57*, 27–28.

Maache-Rezzoug, Z., J.-M. Bouvier, K. Alla, and C. Patras. (1998). Effect of principal ingredients on rheological behaviour of biscuit dough and on quality of biscuits. *J. Food Eng., 35*, 23–42.

Maillard, L.-C. (1912). Action des acides amines sur les sucres formation des melanoidines par voie methodique. *Council of Royal Academy Science Series 2, 154*, 66–68.

Manzocco, L., S. Calligaris, D. Mastrocola, M. C. Nicoli, and C. R. Lerici. (2001). Review of nonenzymatic browning and antioxidant capacity in processed foods. *Trends in Food Sci. Technol., 11*, 340–346.

Marshall, R. T., and W. S. Arbuckle. (2000). *Ice Cream*. Gaithersburg, MD: Aspen Publishers Inc.

Martins, S. I. F. S., and M. A. J. S. Van Boekel. (2005). A kinetic model for the glucose/glycine Maillard reaction pathways. *Food Chemistry, 90*, 257–269.

Matsubara, S. (2001). Alimentos funcionais: Uma tendĕncia que abre perspectivas aos laticinios. *Industria de Laticinios, 6*, 10–18.

Maughan, R. J., and T. D. Noakes. (1991). Fluid replacement and exercise stress: A brief review of studies on fluid replacement and some guidelines for the athlete. *Sports Med., 12*, 16–31.

McGee, H. (1984). *On Food and Cooking*. New York: Charles Scribner's Sons.

McGregor, J. U., and H. C. White. (1986). Effect of sweeteners on the quality and acceptability of yogurt. *J. Dairy Sci., 69*, 698–703.

Meschter, E. E. (1990). Sugar in preserves and jellies. In: N. L. Pennington and C. W. Baker (Eds.), *Sugar: A User's Guide to Sucrose*. New York: Van Nostrand Reinhold.

Miller-Livney, T., and R. W. Hartel. (1997). Ice recrystallization in ice cream: Interactions between sweeteners and stabilizers. *J. Dairy Sci., 80*(3), 447–456.

Mizukoshi, M. (1985). Model studies of cake baking—Part VI: Effects of cake ingredients and cake formula on shear modulus of cake. *Cereal Chemistry, 62*, 247–251.

Monneuse, M. O., F. Bellisle, and J. Louis-Sylvestre. (1991). Impact of sex and age on sensory evaluation of sugar and fat in dairy products. *Physiol. Behav., 50*, 1111–1117.

Montijano, H., F. A. Tomás-Barberan, and F. Borrego. 1995. Stability of the intense sweetener neohesperidine DC during yogurt manufacture and storage. *Z. Lebensm. Unters. Fosch A., 201*, 541–543.
Montijano, H., F. A. Tomás-Barberan, and F. Borrego. (1997). Accelerated kinetics study of neohesperidine DC hydrolysis under conditions relevant to high-temperature-processed dairy products. *Z. Lebensm. Unters. Fosch A., 204*, 180–192.
Mozaffar, Z., and Z. U. Haque. (1992). Casein hydrolysate—Part 3: Some functional properties of hydrophobic peptides synthesized from casein hydrolysate. *Food Hydrocolloids, 5*, 573–579.
Muller, T. (2003). Schokolade-neue wege gehen. *Susswaren, 11*, 13–15.
Muse, M. R., and R. W. Hartel. (2004). Ice cream structural elements that affect melting rate and hardness. *J. Dairy Sci., 87*, 1–10.
Nahon, D. (2005). Sweetening dairy products. *European Dairy Magazine, 3*, 44–45.
Nelson, A. L. (2000a). Properties of high-intensity sweeteners (17–30). In: *Sweeteners: Alternative—Practical Guide for the Food Industry*. New York: Marcel Dekker.
Nelson, A. L. (2000b). *Sweeteners alternative*. Eagan Press, American Association of Cereal Chemists. *Nutr., 30*, 115–359.
NutraSweet. (1996). Ingredient overview. *NutraSweet Kelco Company Bulletin*, 5200, 1–4.
Official Journal of the European Union. (1994a). 10- No: L 237/3.
Official Journal of the European Union. (1994b). No. L 237/5.
Official Journal of the European Union. (1995). 18- No: 61/1.
Olewnik, M. C., and K. Kulp. (1984). The effect of mixing time and ingredient variation on farinograms of cookie doughs. *Cereal Chemistry, 61*(6), 532–537.
Olinger, P. M. (1994). New options for sucrose-free chocolate. *Manufact. Confect., 74*(5), 77–84.
Olinger, P. M. (1995). Lactitol: Its use in chocolate—Xylitol as a sanding medium. *Manufact. Confect., 75*(11), 92–95.
Olinger, P. M., and T. Pepper. (2001). Xylitol. In: O. L. Nabors (ed.), *Alternative Sweeteners* (335–365). New York: Marcel Dekker.
Özdemir, C., E. Dagdemir, S. Özdemir, and O. Sagdic. (2008). The effects of using alternative sweeteners to sucrose on ice cream quality. *J. Food Quality, 31*, 415–428.
Özdemir, M., and H. Sadıkoğlu. (1998). Characterization of rheological properties of systems containing sugar substitutes and carrageenan. *Int. J. Food Sci. Technol., 33*, 439–444.
Özer, B. (2010). Strategies for yogurt manufacturing. In: F. Yildiz (ed.), *Development and Manufacture of Yogurt and Other Functional Dairy Products*, (47–96). Boca Raton, FL: CRC Press.
Pangborn, R. M., and A. S. Szczesniak. (1974). Effect of hydrocolloids and viscosity on flavor and odor intensities of aromatic flavor compounds. *J. Texture Studies, 4*, 467–482.
Peck, A. (1994). Use of acesulfame K in light and sugar-free baked goods. *Cereal Foods World, 39*, 743–745.
Pepper, T., and P. Olinger. (1988). Xylitol in sugar-free confections. *Food Tech., 42*(10), 98–106.
Pinheiro, M. V. S., L. P. Castro, F. L. Hoffmann, and A. L. B. Penna. (2002). Estudo comparativo de edulcorantes em iogurtes probióticos. *Revista do Instituto de Laticínios Cândido, Tostes, 57*, 142–145.
Pinheiro, M. V. S., M. N. Oliveira, A. L. B. Penna, and A. Y. Tamime. (2005). The effect of different sweeteners in low-calorie yogurts: A review. *Int. J. Dairy Technol., 58*(4), 193–199.
Pintauro, N. D. (1990). Sugar in processed foods. In: N. L. Pennington and C. W. Baker (eds.), *Sugar: A User's Guide to Sucrose*. New York: Van Nostrand Reinhold.
Ponte, J. G. (1990). Sugar in bakery. In: N. L. Pennington and C. W. Baker (eds.), *Sugar: A User's Guide to Sucrose*. New York: Van Nostrand Reinhold.
Portman, M. O., and D. Kilcast. (1998). Descriptive profiles of synergistic mixtures of bulk and intense sweeteners. *Food Quality and Preferences, 9*(4), 221–229.
Pszczola, D. (2003). Sweetener and sweetener enhances the equation. *Food Technol., 57*(11), 48–61.
Rapaille, A., M. Gonze, and F. Van der Schueren. (1995). Formulating sugar-free chocolate products with maltitol. *Manufact. Confect., 49*(7), 51–54.
Regand, A., and H. D. Goff. (2003). Structure and ice recrystallization in frozen stabilized ice cream model systems. *Food Hydrocolloids, 17*, 95–102.
Reis, R. C., V. P. R. Minim, H. M. A. Bolini, B. R. P. Dias, L. A. Minim, and E. B. Ceresino. (2011). Sweetness equivalence of different sweeteners in strawberry-flavored yogurt. *J. Food Quality, 34*, 163–170.
Reissig, C., E. C. Strain, and R. R. Griffiths. (2009). Caffeinated energy drinks: A growing problem. *Drug Alcohol Depend., 99*, 1–10.

Rizai, A., and H. Ziar. (2008). Growth and viability of yogurt starter organisms in honey-sweetened yogurt. *African J. Biotech., 7*(12), 2055–2063.

Schanot, M. S. (1981). Sweeteners: functionality in cookies and crackers. *AIB Technology Bulletin, 3*, 4.

Serpen, A., B. Atac, and V. Gokmen. (2007). Adsorption of Maillard reaction products from aqueous solutions and sugar syrups using adsorbent resin. *J. Food Eng., 82*, 342–350.

Sivieri, K., and M. N. Oliviera. (2001). Avaliação da vida-de-prateleira de bebidas lacteas preparadas com "fat replacers" (Litesse e Dairy Lo™). *Ciencia e Technologia de Alimentos, 22*, 24–31.

Slade, L., and H. Levine. (1991). Beyond water activity: Recent advances based on an alternative approach to the assessment of food quality and safety. *Crit. Rev. Food Sci. 30*(2–3): 115–360.

Smith, D. (1990). Sugar in dairy products. In: N. L. Pennington and C. W. Baker (Eds.), *Sugar: A user's guide to sucrose*. New York: Van Nostrand Reinhold.

Sokmen, A., and G. Gunes. (2006). Influence of some bulk sweeteners on rheological properties of chocolate. *LWT-Food Sci. Technol., 39*, 1053–1058.

Song, T. B., Y. H. Kim, J. H. Shin, Y. K. Lee, K. J. Cha, J. H. Wang, and J. H. Yu. (1996). Effects of types and sweetness intensity of low-calorie sweeteners on growth and lactic acid production by *Lactobacillus bulgaricus* and *Streptococcus thermophilus*. *Dairy Science Abstracts, 58*, 243.

Soukoulis, C., E. Rontogianni, and C. Tzia. (2010). Contribution of thermal, rheological and physical measurements to the determination of sensorially perceived quality of ice cream containing bulk sweeteners. *J. Food Eng., 100*, 634–641.

Tamime, A. Y., M. N. I. Barclay, G. Davies, and E. Barrantes. (1994). Production of low-calorie yogurt using skim milk powder and fat substitute: A review. *Milchwissenschaft, 49*, 85–88.

Tamime, A.Y., and R. K. Robinson. (2007). *Yogurt Science and Technology, 3rd edition*. Cambridge: CRC Press.

Tate and Lyle. (2010). Sweeteners: Tastier, fruitier, lighter, sweeter. www.tateandlyle.com/.../SWEETENERS BrochureTateLyleNov09.pdf. Accessed October 14, 2011.

The Coca Cola Company (2011). www.thecoca-colacompany.com/.../Coca-Cola_10-K-Ownership&Ops.pdf-Forward-looking statements report. Accessed October 14, 2011.

The Sugar Association. (1992). Sugar's functional roles in cooking and food preparation. Washington, DC.

Thomet, A., B. Rehberger, B. Wyss, and W. Bisig. (2005). Sugar syrup from dairies for the food industry. *10th Aachen Membrane Colloquium, March*, 451–457.

United States Food and Drug Administration (FDA). (2011). Official web site of the United States Food and Drug Administration. Retrieved from http://www.fda.gov. Accessed October 14, 2011.

Van der Ven, A. A. (1988). Aspartame: Propriedades e aplicações. *Alimentaria, 61*, 61–64.

Varga, L. (2006). Effect of acacia honey (*Robina pseudo-acacia L.*) on the characteristic microflora of yogurt during refrigerated storage. *Int. J. Food Microbiol., 108*(2), 272–275.

Verdalet-Guzman, I., R. Viveros-Contreras, S. L. Amaya-Llano, and F. Martinez-Bustos. (2011). Effects of extruded sugar bagasse blend on yogurt quality. *Food and Bioprocess Technol., 4*(1), 155–160.

Vettern, J. L. (1984). Technical Bulletin VI. American Institute of Baking, Manhattan, KS.

White, D. C., and G. N. Lauer. (1990). Predicting gelatinization temperatures of starch/sweetener systems for cake formulation by differential scanning calorimetry—Part I: Development of a model. *Cereal Foods World, 35*, 728–733.

Wijers, M. C., and P. J. Strater. (2001). Isomalt. In: O. L. Nabors (ed.), *Alternative sweeteners* (265–281). New York: Marcel Dekker.

Wilkes, A. P. (1992). Expanded options for sweetening low calorie foods. Food Product Design, July 1992. Retrieved from http://www.foodproductdesign.com/archive/1992/0792DE.html.

Wolf, E. (1979). Practical problems of cyclamate and saccharine incorporation in foodstuffs. In: Guggenhei, B. (ed.), *Health and sugar substitutes* (153–158). Cologne, Federal Republic of Germany: European Reseach Group of Oral Biology 20th Symposium.

Zanoni, B., A. Schiraldi, and R. Simonetta. (1995). Naive model of starch gelatinization kinetics. *J. Food Eng., 24*, 25–33.

Zhu, Y., Y. Guo, M. Ye, and F. S. James. (2005). Separation and simultaneous determination of four artificial sweeteners in food and beverages by ion chromatography. *J. Chromatography A, 1085*, 143–146.

Zumbe, A., and C. Grosso. (1993). Product and process for producing milk chocolate. U.S. Patent 5238698/ EP Patent 0575070 A2.

CHAPTER 9

Quality Control of Sweeteners: Production, Handling, and Storage

Molisch Test, Feligion Test, Barfoed Test, Resorkin Test, Quality Control of Sugars and Inverted Sugar, Color Determination, Corn Syrup Determination, and Artificial Sweeteners Quality Control

Theodoros Varzakas

CONTENTS

9.1 INTRODUCTION

Qualitative and quantitative determination of the physicochemical and organoleptic properties of sweeteners will be described here. Some basic tests will be outlined. Identification is easy for pure sugar solutions, i.e., solutions containing only one sugar.

9.2 MOLISCH TEST

This is a general test for the determination of sugars or carbohydrates in unknown solutions. Phenol a-naphthol is the substrate used for the Molisch test as are other phenols, such as thymol, resorcinol, or phloroglucinol. In this test, furfural or hydroxymethylfurfural is formed by the reaction of carbohydrates with sulfuric acid and is then condensed with phenol to give the characteristic color.

9.3 FELIGION TEST

This is based on using reducing sugars to reduce an alkaline solution of copper tartrate and form a red-yellow copper oxide. A precipitate will not be formed in solutions that do not have reducing sugars. This is carried out after the Molisch test.

9.4 BARFOED TEST

This is used to distinguish reducing sugars between monosaccharides and disaccharides and is applied to solutions that are positive in the Feligion test. The Barfoed test is used to detect lactose or maltose in dairy products.

9.5 RESORKIN TEST

The Resorkin test is based on ketoses producing a red color in the presence of resorcinol and hydrochloric acid. It is used to distinguish ketoses and aldoses and, most specifically, to detect fructose and distinguish it from the presence of other monosaccharides such as glucose.

9.6 PHYSICAL METHODS

Physical methods are indirect methods and include refractometry, densitometry, and polarimetry for sugar determination.

9.6.1 Densitometry

Densitometry is a determination method for the concentration of sugar solutions that is not as exact as refractometry. Densitometers have a glass tube that is thicker on its base and full of spheres (depending on the concentration range) with a thinner tube on the upper part, where there is a scale for density measurement. Densitometers measure the total solids of a solution in degrees Baumé, Twaddell, or Quevenne. Degrees Baumé is the most used. Measurement is carried out through a direct read on the level of solution placed on a volumetric cylinder.

9.6.2 Refractometry

Light refraction is the change in direction when air comes into water and is expressed by the ratio (sin a/sin b) of angles a and b and is called the refraction index.

If a sugar solution replaces water, then the refractive index depends on the concentration and temperature of the solution. The concentration of the solution can be calculated from the refractive index.

Refractometers measure the refractive index and the concentration of soluble solids in solutions. There are tables correlating the refractive index, saccharose (in percent), Brix, degrees Baumé, etc.

Saccharose solutions differ in the refractive index from the dextrose or fructose solutions of the same concentration; hence, there are coefficients of conversion of saccharose into dextrose and fructose as described below:

1. Saccharose (%) × 1.020 = dextrose (%).
2. Saccharose (%) × 1.022 = fructose (%).

Refractometers are based on the principle of increase in refractive index with an increase in density. There are two detection systems: (1) the reflection (digital) system and (2) the transparent system (mobile and Abbe refractometers).

9.6.3 Brix (%)

Brix (%) measures the percentage of soluble solids in a water solution. This includes sugars, salts, proteins, and acids.

9.6.4 Polarimetry

Sugar solutions are optically active, i.e., they turn the polarized light level. This is called the specific rotational ability (SRA) and is directly proportional to the concentration and type of sugar, as calculated in Equation 9.1:

$$SRA = [\alpha]_D^{20} = \frac{100 * a}{L * C}, \tag{9.1}$$

where

$+[\alpha]_D^{20}$ = SRA 10% sugar solution at 20°C

a = turn

L = length of polarimeter tube (in centimeters)

C = solution concentration (in grams per 10 ml)

Optical balance is achieved through

1. Adding some drops of ammonia to the solution
2. Heating the solution
3. Applying ultrasound to the solution

Calculating the turn of the polarized light α and taking the value of SRA from a table, we can calculate the concentration in the solution from Equation 9.2:

$$C = \frac{100 * a}{L * [a]_D^{20}} \text{ gr/100 ml.} \tag{9.2}$$

9.7 LANE–EYNON DETERMINATION OF REDUCING SUGARS

The Lane–Eynon method determines only the reducing sugars (monosaccharides, lactose, and maltose), whereas nonreducing sugars, e.g., saccharose, need to be inverted so that they can be determined. This is a volumetric determination of sugars in a solution and is based on the fact that, when basic copper solutions are heated with reducing sugar solutions, copper is reduced to copper oxide, giving a red color. According to this method, the volume of the sugar solution is determined to reduce or neutralize 10- or 25-ml feligian liquid using methylene blue as an indicator. Based on the used milliliters of sugar solution with the coefficients of sugars from specific tables, the sugar concentration is calculated according to Equation 9.3:

$$\text{Sugar concentration (\%)} = \frac{\Sigma * 100}{A}, \tag{9.3}$$

where A refers to sugars in milligrams per 100 ml solution, and Σ refers to a coefficient from tables.

9.7.1 Determination of Reducing and Total Sugars Using the Schoorl–Regenbogen Method in Raw Materials and Final Products

9.7.1.1 Principle

The Schoorl–Regenbogen method is based on the titration of the remaining quantity of copper arising from boiling of the sugar solution with Feligian liquid. It is an indirect method with a great reproducibility and the advantage of being carried out under the same conditions for sugars, including pentoses and reducing disaccharides.

This method requires the following equipment for reagents:

- Flask, 400 ml
- High-speed mixer
- Erlenmeyer flask, 250 ml
- Bunsen burner
- Water bath at 100°C

The following reagents are also needed:

- Carrez I solution: 36.0 g of ($K_4[Fe(CN)_6] \cdot 3H_2O$) per liter
- Carrez II solution: 72.0 g of zinc sulphate ($ZnSO_4 \cdot 7H_2O$) per liter
- NaOH solution, 0.1 N

A sample pretreatment is described as follows:

- A 5.0-g well-ground sample is weighed in a 400-ml flask.
- Deionized water, 175 g, is added.
- Mixing is carried out for 70–80 s in a high-speed mixer to dissolve the soluble solids.
- Then, the following solutions are added under stirring:
  - 5-ml Carrez I
  - 5-ml Carrez II
  - 10-ml NaOH 0.1 N
- Stirring is continued for another 30 s.
- Filtration with a paper towel follows. Approximately 70 ml of filtrate is enough for the determinations.

9.7.1.2 Determination of Reducing Sugars

The following reagents are needed for determination of reducing sugars:

- Fehling A solution: 69.278 g of ($CuSO_4 \cdot 5H_2O$) is dissolved in water up to a volume of 1 l.
- Fehling B solution: 346 g of ($C_4H_4KNaO_6 \cdot 4H_2O$) is dissolved in water up to a volume of 1 l.
- Potassium iodide: 30% v/v (30 g dissolved in 100 ml of H_2O).
- Sulfuric acid: 25% (1 + 6).
- $Na_2S_2O_3 \cdot 5H_2O$ solution, 0.1N: 25 g of $Na_2S_2O_3 \cdot 5H_2O$ is dissolved in 1000 ml of distilled water and 0.1 g of Na_2CO_3 is added as a preservative.
- Starch solution (indicator): 2 g of soluble starch is added to 25 ml of H_2O, and the formed paste is added slowly and under continuous stirring in 250 ml of boiling water. Boiling continues for 1–min, and the solution is cooled and kept in glass flasks with lids.
- HCI: 0.5 N.
- NaOH: 0.25 N.

The procedure is described as follows:

- In a 250-ml conical flask, a 25-ml sample is taken for examination, and 10-ml Fehling A, 10-ml Fehling B, and H_2O are added to reach a final volume of 50 ml.
- The mixture is heated in a Bunsen burner for 5 min.
- Cooling follows after 2 min, adding 10-ml KI 30% and 10-ml H_2S0_4 25%, and titration with 0.1 N $Na_2S_2O_3 \cdot 5H_2O$ follows.
- At the end of titration, starch solution is added, and titration continues until the solution changes color.
- Blank determination is then carried out. The difference of the two titrations gives the reduced quantity of copper, and through the corresponding table, the quantity of sugar as inverted sugar is calculated.

Results are expressed as follows:

$$\text{Reducing sugars (\%), as inverted sugar} = \frac{(\text{Inverted sugar from table}) \times 8 \times 100}{(\text{Weight of sample in grams}) \times 100}.$$

A 5-g sample is calculated as follows:

$$\text{Reducing sugars \%, as inverted sugar} = (\text{inverted sugar from table}) \times 0.16.$$

9.7.1.3 Determination of Total Sugars as Inverted Sugar

For the determination of total sugars, saccharose contained in the solution should be inverted by heating it in an acid environment. A 25-ml sugar sample is transferred to a 250-ml conical flask, and 2.5-ml HCI 0.5 N is added. Heating at 100°C in a water bath for 30 min, followed by cooling and neutralization with 5-ml NaOH 0.25 N, is then carried out. Then, the procedure described above follows.

Results are expressed as follows:

$$\text{Saccharose (\%)} = \frac{(\text{inverted sugar from table}) \times 0.95 \times 8 \times 100}{(\text{Weight of sample in grams}) \times 100}.$$

A 5-g sample is calculated as follows:

$$\text{Saccharose (\%)} = (\text{inverted sugar from table}) \times 0.152.$$

In Table 9.1, sugars are estimated in milligrams from the used N/10 $Na_2S_2O_3 \cdot 5H_2O$ solution in milliliters according to Schoorl and Regenbogen. Each milliliter of N/10 $N\alpha_2S_2O_3$ corresponds to 6.36-mg copper.

Table 9.1 Sugars Estimation (in mg) from the Used N/10 $Na_2S_2O_3 \cdot 5H_2O$ Solution (in ml) according to Schoorl

N/10 Solution ($N\alpha_2S_2O_3$)	Copper (Cu)	Glucose ($C_6H_{12}O_6$)	Fructose ($C_6H_{12}O_6$)	Inverted Sugar ($C_6H_{12}O_6$)	Galactose ($C_6H_{12}O_6$)	Mannose ($C_6H_{12}O_6$)	Lactose ($C_{12}H_{22}O_{11}$+H_2O)	Maltose ($C_{12}H_{22}O_{11}$)	Arabinose ($C_5H_{10}O_5$)	Xylose ($C_5H_{10}O_5$)	Ramnose ($C_6H_{12}O_5$)
1	6.4	3.2	3.2	3.2	3.3	3.1	4.6	5.0	3.0	3.1	3.2
2	12.7	6.3	6.4	6.4	7.0	.3	9.2	10.5	6.0	6.3	6.5
3	19.1	9.4	9.7	9.7	10.4	9.5	13.9	16.0	9.2	9.5	9.9
4	25.4	12.6	13.0	13.0	14.0	12.8	18.6	21.5	1.3	12.8	13.3
5	31.8	15.9	16.4	16.4	17.5	16.1	23.3	27.0	15.5	16.1	16.8
6	38.2	19.2	20.0	19.8	21.1	19.4	28.1	32.5	18.7	19.4	20.2
7	44.5	22.4	23.7	23.2	24.7	22.8	33.0	38.0	21.9	22.8	23.7
8	50.9	25.6	27.4	26.5	28.3	26.2	38.0	43.5	25.2	26.2	27.2
9	57.3	28.9	31.1	29.9	32.0	29.6	43.0	49.0	28.6	29.6	30.8
10	63.6	32.3	34.9	33.4	35.7	33.0	48.0	55.0	32.0	33.0	34.4
11	70.0	35.7	38.7	36.8	39.4	36.5	53.0	60.5	35.4	36.5	38.0
12	76.3	39.0	42.4	40.3	43.1	40.0	58.0	65.0	38.8	40.0	41.6
13	82.7	42.4	46.2	43.8	46.8	43.5	63.0	72.0	42.2	43.5	45.2
14	89.1	45.8	50.0	47.3	50.5	47.0	63.0	78.0	45.6	47.0	48.8
15	95.4	49.3	53.7	50.8	54.3	50.6	73.0	83.5	49.0	50.6	52.4
16	101.8	52.8	57.5	54.3	58.1	54.2	78.0	89.0	52.4	54.2	56.0
17	108.1	56.3	61.2	58.0	61.9	57.9	83.0	95.0	55.8	57.9	59.8
18	114.4	59.8	65.0	61.8	65.7	62.2	88.0	101.0	59.3	62.6	63.5
19	120.8	63.3	68.7	65.5	69.6	65.3	93.0	107.0	62.9	65.3	67.3
20	127.2	66.9	72.4	69.4	73.4	69.2	98.0	112.5	66.5	69.2	71.0
21	133.5	70.7	76.2	73.3	77.2	73.1	103.0	118.5	70.2	73.1	74.8
22	139.8	74.5	80.1	77.2	81.2	77.0	108.0	124.5	74.0	77.0	78.6
23	146.2	78.5	84.0	81.2	85.1	81.0	113.0	130.5	77.9	81.0	82.4
24	152.6	82.6	87.8	85.2	89.0	85.0	118.0	135.5	81.8	85.0	86.2
25	159.0	86.6	91.7	89.2	93.0	89.0	123.0	142.5	85.7	89.0	90.0

9.7.2 Determination of Nonreducing Sugars

Nonreducing sugars (saccharose) can be determined using the Lane–Eynon titration if they are converted to reducing sugars. For example, saccharose can be converted into a reducing sugar through inversion when it is hydrolyzed into glucose and fructose.

There are different methods of inversion, differing in temperature conditions, hydrolysis time, and hydrolysis agents, which could be either an acid or an enzyme.

9.8 QUALITY CONTROL IN SUGARS

Quality control in sugars is necessary in many food industries using sugar as a raw material, such as confections, biscuits, ice cream, and candies. Sugar specifications are shown in Table 9.2. The International Commission for Uniform Methods of Sugar Analysis (ICUMSA) has suggested certain parameters:

- Density of crystals
- Total solids determination
- Inverted sugar concentration
- Color determination

Comparison between different sugars is shown in Table 9.3.

9.8.1 Density of Crystals

The use of sugars in many foods requires the knowledge of other parameters:

- Size of crystals
- Number of crystals per gram
- Outside surface of crystals

This knowledge is required for the production of mixtures' solubility speed. Density is affected by the size and number of crystals, the uniformity of size and shape, and the formation of incorporations. Duration after sugar production plays an important role in the quality of fine sugar. The more ground the fine sugar is, the more light it becomes, and the lower density it has. On the contrary, the older the production of sugar is, the more cohesive and clotted it becomes, and the higher the density is. The ratio of weight/volume (in grams per milliliter) gives the density of sugar crystals.

Table 9.2 Sugar Specifications

Physicochemical Parameters	Value
Color	60 ICUMSA Units (IU)
Saccharose	Min 99.7%
Inverted sugars	Max 0.04%
Ash	Max 0.04%
Moisture	Max 0.1%
SO_2	Max 20 mg/kg
Arsenic (As)	Max 1 mg/kg
Copper (Cu)	Max 2 mg/kg
Lead (Pb)	Max 2 mg/kg

Note: ICUMSA, International Commission for Uniform Methods of Sugar Analysis.

Table 9.3 Comparison between Different Sugars

Sweeteners	Melting Point (°C)	Molecular Weight (g/mole)	Relative Sweetness to Sucrose (%)	Degree of Hygroscopicity/% ERH at 20°C
Ribose	80–90	150	<10	High
Dextrose	80–85	180	75	Medium
Fructose	102–105	180	100–170	High
Maltose	120–125	342	30	Medium
Sucrose	160–186	342	100	Low
Sorbitol	99–101	182	60	Medium
Xylitol	92–95	152	100	High

9.8.2 Inverted Sugar Concentration

Inverted sugar concentration is a very important quality parameter. High percentages of inverted sugar reduce the total percentage of saccharose and cause hygroscopic and adhesive problems. For inverted sugar determination, the ethylene diamine tetra acetic (EDTA) acid method is used. The Lane–Eynon method cannot be used because of the low concentration of sugar as inverted sugar. Inverted sugar concentration can be determined according to Equation 9.4:

$$\text{Inverted sugar (\%)} = 0.0199 - (0.0015 \times \text{used ml } 0.005 \text{ N EDTA}). \quad (9.4)$$

9.8.3 Color Determination

The two mostly used methods for color determination in sugar are the following:

- The macroscopic method, which refers to the color of the crystals as seen by a naked eye compared to a control.
- Spectrophotometric determination in two wavelengths, depending on the solution. For white crystalline sugar, a 420-nm wavelength is used. For darker sugar solutions (brown sugar), a 560-nm wavelength is used.

For absorption, the absorption index α_ς is used according to Equation 9.5:

$$\left(\alpha_\varsigma\right) = \frac{A_\varsigma}{L_x C}, \quad (9.5)$$

where

A_ς = absorption (absorption of sugar solution–water absorption)
L_x = width of cuvette in spectrophotometer (in centimeters)
C = concentration of sugar solution (total solids; in grams per milliliter).

Based on the absorption index, we can calculate the color in ICUMSA units according to Equation 9.6:

$$\frac{a_s}{\beta} 1000 \text{ ICUMSA units.} \quad (9.6)$$

9.9 QUALITY CONTROL IN CORN SYRUP

Corn syrup is a sweetener manufactured with the partial hydrolysis of starch suspensions with acids, enzymes, and their combination. The hydrolysis degree determines the following basic attributes:

- Dextrose concentration
- Dextrin concentration
- Sweetness (Table 9.4)
- Dextrose equivalent (DE)

DE is a measure expressing the concentration of corn syrup or other sweeteners in reducing sugars; hence, it is a measure of the degree of hydrolysis. It is expressed as follows:

$$\text{DE} = \text{Reducing sugars of syrup (dextrose)/total solids of syrup.}$$

The degree of hydrolysis depends on conditions such as pH, temperature, and time. For the determination of DE, we should first determine the total solids of the syrup and, then, the reducing sugars of the solution expressed as dextrose. Lane–Eynon is the method described for the determination of reducing sugars; however, the official Association of Official Analytical Chemists (AOAC) method is Zerban–Sattler, which is iodometric. Corn syrup can also be determined spectrophotometrically.

Table 9.4 Sweeteners and Sweetening Intensity Relative to Sucrose

Sweeteners	Sweetness
Saccharose	1.0
Glucose	0.7
Fructose	1.7
Lactose	0.4
Maltose	0.5
Sorbitol	0.6
Mannitol	0.7
Glycine	0.7
Saccharin	300–500
Cyclamate	30–60
Aspartame	180–200
Acesulfame K	130–200
Dulcin	200
Thaumatin	2000–3000
Neohesperidin DC	400–600
Sucralose	600

Source: Adapted from Kibbe, A. H., *Handbook of Pharmaceutical Excipients, 3rd edition*, American Pharmaceutical Association and Pharmaceutical Press, London, 2000. Mullarney, M. P. et al., *Int. J. Pharmaceutics*, 257, 2003. With permission. International Sweeteners Association (ISA), *Sweeteners in All Confidence*, 2010. With permission.

9.10 QUALITY CONTROL IN ARTIFICIAL SWEETENERS

Artificial sweeteners are produced with physicochemical or biotechnological methods. Many sweeteners have been manufactured for different uses with a higher or lower sweetness than standard sugar. Saccharin can be detected with the following tests:

1. Salicylic acid test
2. Resorcinol test (saccharin presence can be detected from the green fluorescent color)
3. Taste test
4. Phenol sulphuric acid test.

9.11 GRANULATED SUCROSE SPECIFICATIONS

9.11.1 General Specifications

Granulated sucrose must meet the specifications in the United States Food Chemicals Codex (FCC) and the Joint FAO/WHO Expert Committee on Food Additives (JECFA) standards, including some specific requirements of the Coca-Cola Company (Table 9.5). Upon receipt, it should comply with the following, plus any other standards defined by local food regulations, consistent with its intended use.

Granulated sucrose must be manufactured, packaged, stored, and shipped under sanitary conditions appropriate to food products and must be in compliance with all applicable food hygiene, health, and sanitary requirements and regulations in effect at the manufacturing and receiving locations. This includes all applicable good manufacturing practices, laws and regulations (national and local), and all local, national, and international transport regulations as appropriate between manufacturing and receiving locations.

The use of chemicals for controlling microorganisms is discouraged. There must be no residue of any substance used to control microorganisms in the final product. If chemicals are used for controlling microorganisms, they must be identified by name and usage level.

The supplier must perform residual pesticide testing on a semiannual basis. Residual pesticides must be defined as any organic or synthetic material used to protect crops from pests.

9.11.2 Granulated Sucrose Storage Conditions

To prevent sugar from absorbing moisture and to minimize caking problems, it should be stored at temperatures above 10°C and a relative humidity below 60%.

9.12 LIQUID SUCROSE SPECIFICATIONS

9.12.1 Purpose

This section defines liquid sucrose specifications used during product manufacturing.

9.12.2 General Specifications

Liquid sugar (sucrose) must meet the specifications in Table 9.6 plus any other standards defined by local food regulations consistent with its intended use. It must be manufactured and stored under sanitary conditions appropriate to food products and in compliance with all applicable health and sanitary regulations.

Table 9.5 Granulated Sucrose Specifications

Physicochemical Parameters	Specifications	Reference
Appearance	White crystals or crystalline powder, with no more than 4 black specks per 500 g	
Taste	Typically sweet and free from foreign tastes	
Odor	Free from foreign odors	
Odor after acidification	Free from objectionable odor	
Assay (Purity)	Not less than 99.9% w/w: polarization or calculated as 100% sucrose minus ash, minus moisture, minus invert.	ICUMSA
	Polarization not less than 99.7°Z	Codex STAN 212-1999
Invert sugar content	Not more than 0.04% by weight	COUNCIL DIRECTIVE 2001/111/EC
	Not more than 0.1% w/w (FCC)	FCC
Loss on drying	Not more than 0.06% by weight	COUNCIL DIRECTIVE 2001/111/EC
Conductivity ash	Not more than 0.015% w/w	ICUMSA
Color	Not more than 45 Reference basis units (RBU) [or ICUMSA Units (IU) equivalents].	ICUMSA
	Not more than nine points determined in accordance to the method of the Brunswick Institute for Agricultural and Sugar Industry Technology.	COUNCIL DIRECTIVE 2001/111/EC
Floc potential	Must pass test	Floc potential tests
Copper	Not more than 1.5 mg/kg	ICUMSA
Arsenic (As)	Not more than 1.0 mg/kg (FCC)	ICUMSA/FCC
Lead	Not more than 0.5 mg/kg (FCC/JECFA)	ICUMSA/AOAC/ FCC
Moisture	Not more than 0.06% w/w	ICUMSA
Quaternary ammonium compounds	Not more than 2 mg/kg in sugar refined by any process that uses quaternary ammonium compounds (QAC)	
Screen size	Not more than 7.5% finer than 65 mesh when screened for 10 minutes. Mean aperture: 0.45–0.67 mm CV (%): 25–35	ICUMSA, or equivalent method
Sediment (insoluble or suspended matter)	Not more than 5 ppm visual gravimetric insolubles	
Sulfur dioxide	Less than 6.0 mg/kg	AOAC/ICUMSA
Turbidity	None in 50% w/w solution	
Microbiological		
Total viable count	Less than 200 cfu/10 g	ICUMSA
Yeasts	Less than 10 cfu/g	
Molds	Less than 10 cfu/g	
Staphylococcus aureus	Absent in 1 g	
Salmonella	Absent in 25 g	

Source: Adapted from Coca-Cola Manufacturing Manual, 2001. British Sugar, www.hbingredients.co.uk/specsheets/1642.pdf, 2001. Official Journal of the European Committees, Council Directive 2001/111/EC, December 20, 2001. CODEX Standard for Sugars, CODEX STAN 212–1999, 1–5.

Note: Granulated sucrose consists of white free flowing granules of a crystalline appearance. White sugar is refined from sugar beet or sugar cane in a complex continuous process, which involves cleaning, purification, decolorization, concentration, crystallization, separation, drying and conditioning.

Table 9.6 Liquid Sucrose Specifications

Physicochemical Parameters	Specifications	Reference
General description	Only first or second run bone char liquors may be used	
Taste	Typically sweet and free of foreign tastes	
Odor	Free from foreign odors	
Acidification	Free from objectionable odor	
Assay (Purity)	Not less than 99.9% w/w: polarization or, calculated as 100% sucrose minus ash	ICUMSA
Conductivity ash	Not more than 0.05% w/w (dry basis)	ICUMSA
°Brix	67.5 ± 0.3°	ICUMSA
Chloride (NaCl)	Not more than 50 mg/kg (dry basis)	ICUMSA
Color	Not more than 50 reference basis units (RBU) or ICUMSA Units (IU) equivalents. Clear to light straw	ICUMSA
Copper (Cu)	Not more than 1.5 mg/kg (dry basis)	AOAC/ICUMSA
Floc Potential	Must pass test	
Arsenic (As)	Not more than 1mg/kg (FCC) (dry basis)	ICUMSA/FCC
Lead	Not more than 0.5 mg/kg (FCC/JECFA) (dry basis)	ICUMSA/FCC/AOAC
Invert sugar	Not more than 0.5% w/w (dry basis)	AOAC/FCC
Iron (Fe)	Not more than 3.0 mg/kg (dry basis)	FCC
pH	6.5–7.0	ICUMSA
Temperature	Not more than 35°C at time of receipt	ICUMSA
Turbidity	None	
Sediment	Not more than 7 mg/kg gravimetric insolubles	
Sulfur dioxide	Less than 6 mg/kg (dry basis)	ICUMSA
Microbiological		
Mesophilic bacteria	Less than 200 cfu/g	ICUMSA
Yeasts	Less than 10 cfu/g	
Molds	Less than 10 cfu/10 g	

Source: Adapted from the Coca-Cola Manufacturing Manual, 2001. Codex Alimentarius; Food Chemicals Codex, 2011, The United States Pharmacopeial Convention, 2011. Joint FAO/WHO Expert Committee on Food Additives, Combined Compendium of Food Additive Specifications, FAO JECFA Monograph 1, ISSN 1817–7077, 2006.

With regard to ash, conductivity measures only ionized soluble salts, whereas sulfated ash or residue on ignition measures total salts and matter, both soluble and insoluble.

Liquid sucrose must be manufactured, packaged, stored, and shipped under sanitary conditions appropriate to food products and must be in compliance with all applicable food hygiene, health, and sanitary requirements and regulations in effect at the manufacturing and receiving locations. This includes all applicable good manufacturing practices, laws and regulations (national and local), and all local, national, and international transport regulations as appropriate between manufacturing and receiving locations.

Water used in the finished product must meet the standards.

9.12.3 Liquid Sucrose Storage Conditions

Storage temperature must be above 25°C. Temperature at 25°C is approximately the crystallization temperature of liquid sucrose at 67°Brix.

9.13 HIGH-FRUCTOSE STARCH-DERIVED SYRUP

9.13.1 Purpose

This section refers to specifications based on the United States Food Chemical Codex (FCC) standards and specific requirements (Table 9.7).

Table 9.7 HFSS Specifications

Physicochemical Parameters	Specifications	Reference
Percent solids	70.5%–71.5% w/w	ISBT
Fructose	42.0%–44.0% w/w of total solids	ISBT
Glucose and fructose	Not less than 93.0% w/w of total solids	ISBT
Other saccharides	Not more than 7.0% w/w of total solids	ISBT
Taste	Free from foreign tastes	—
Appearance	Free from turbidity	—
Odor	Free from objectionable odor	—
Odor after acidification	Free from objectionable odor	—
Color	Not more than 25 RBU/IU at the time of receipt	ISBT
Temperature	Not more than 35°C upon receipt nor below 16°C	—
Acetaldehyde	Not more than 0.08 mg/kg	ISBT
Chlorides	Not more than 50 mg/kg	CRA
Sulfated ash	Not more than 0.05% w/w	ISBT
Floc	Must not produce a floc in products	
pH	4.0 ± 0.5	ISBT
Arsenic (As)	Not more than 1 mg/kg	AOAC
Copper (Cu)	Not more than 1.5 mg/kg	AOAC
Heavy metals (as Pb)	Not more than 5 mg/kg (AOAC)	AOAC
Iron (Fe)	Not more than 3 mg/kg	CRA
Lead (Pb)	Not more than 0.5 mg/kg	AOAC
Sediment	Not more than 6 mg/kg gravimetric insolubles	
Silica (SIO2)	Not more than 20 mg/kg	Atomic absorption/ICP
Sulfonated polystyrene	Not more than 1.0 mg/kg through 1 cm at 720 nm	ISBT
Sulfur dioxide	Not more than 3 mg/kg	ISBT
Titratable acidity	Not more than 4.0 ml of 0.05N NaOH to raise 100 g to pH 6.0	ISBT
Turbidity	Free of turbidity in products	
Microbiological		
Mesophilic bacteria	Not more than 200 cfu/g	—
Yeast	Not more than 10 cfu/g	ISBT
Mold	Not more than 10 cfu/g	—

Source: Adapted from the Coca-Cola Manufacturing Manual, 2001. Codex Alimentarius, Food Chemical Codex, 2011. www.codexalimentarius.net/download/standards/338/CXS_212e_u.pdf. Codex Alimentarius, Food Chemical Codex, 2011, the United States Pharmacopeial Convention, 2011. Joint FAO/WHO Expert Committee on Food Additives, Combined Compendium of Food Additive Specifications, FAO JECFA Monograph 1, ISSN 1817-7077, 2006; ISBT.

9.13.2 General Specifications

High-fructose starch-derived syrup 42 (HFSS-42) upon receipt should comply with the following, plus any local food regulations consistent with its intended use.

HFSS-42 must be manufactured, packaged, stored, and shipped under sanitary conditions appropriate to food products and must be in compliance with all applicable food hygiene, health, and sanitary requirements and regulations in effect at the manufacturing and receiving locations. This includes all applicable good manufacturing practices, laws and regulations (national and local), and local, national, and international transport regulations as appropriate between manufacturing and receiving locations.

The supplier must perform residual pesticide testing on a semiannual basis. Residual pesticides must be defined as any organic or synthetic material used to protect crops from pests.

HFSS-42 must be labeled and delivered in containers approved for food use by all applicable regulatory agencies.

HFSS containers must be equipped with nontoxic, supplier-identifiable, tamper-evident seals.

9.13.3 Storage Conditions

The ideal storage temperature for HFSS is 27°C–30°C. It should not be stored above 35°C.

9.14 SEDIMENT: NUTRITIVE SWEETENERS

9.14.1 Purpose

This procedure is used to identify potential sources of sediment in syrup and beverages HFSS-42, HFSS-55, and liquid sucrose. This procedure is not a routine analysis. It should be used only if a visual check shows that sediment is present.

9.14.2 General Description

Identification of sediment requires the following equipment:

- Top-loading balance
- Drying oven
- Desiccator
- Membrane filter apparatus
- Hot plate
- Glass beaker, 600–800 ml
- Standard sediment disc
- Matched-weight, 0.8-µm membranes or 8-µm membrane for granulated sucrose
- Filter funnel
- Filter paper (Whatman No. 54)

Distilled water is used as the reagent.

For the visual method, sample preparation for the visual sediment test requires diluting or dissolving 300 g of sweetener in about 300 ml of hot, distilled water.

Test completion for the visual method is described as follows:

1. Filter the solution through a coarse filter paper (Whatman No. 54 or its equivalent).
2. Wash the filter paper with distilled water and dry in an oven at 105°C.
3. If sediment is present, compare with a standard sediment disc with a pan scale equal to 2 mg/kg. Do the gravimetric sediment test if the sediment level is more than 2 mg/kg.

For the gravimetric method, sample preparation for the gravimetric test requires diluting or dissolving 300 g of sweetener in about 300 ml of hot, distilled water.

Completion of test for the gravimetric method is described as follows:

1. Filter the solution through a stacked pair of matched-weight, 0.8-μm membrane filters.
2. Wash well with distilled water.
3. Dry both membranes in an oven at 105°C, cool in a desiccator, and weigh separately.

Results Evaluation. The difference in weight represents the weight of the sediment, which can be quantified using the following equation:

$$\text{Sediment (mg/kg)} = \text{Weight difference (g)} \times 10{,}000{,}000 \times 300 \text{ g of solids.}$$

Ensure that sediment does not exceed these amounts:

- Liquid or granular sucrose: 7 mg/kg
- HFSS-42 or HFSS-55: 6 mg/kg

9.15 NUTRITIVE SWEETENERS: TASTE, ODOR, AND APPEARANCE

9.15.1 Purpose

This procedure is used to evaluate the taste, odor, and appearance of nutritive sweeteners sucrose and HFSS.

9.15.2 General Description

Evaluation of the taste, odor, and appearance of nutritive sweeteners requires the following equipment:

- Top-loading balance
- pH meter
- High-intensity light source
- Water bath or incubator
- Turbidimeter (optional)
- Glass beakers: 250 ml, 500 ml, 1 L, and 2 L
- Graduated cylinders: Class A, 100 ml and 1000 ml
- Filter funnel
- Whatman No. 54 filter paper
- Burette (25-ml capacity): Class A
- 200-ml, wide-mouth, screw-capped bottle
- Thermometer
- Taste-test glassware
- Watch glass

Distilled water, phosphoric acid, or 75% weight/volume H_3PO_4 can be used as reagents.

9.15.3 Test Samples Preparation

Preparation of test samples for evaluation is described as follows:

Granulated sucrose (turbidity). Prepare a 50°Bx test sample by dissolving 246 g of sucrose in 246 g of distilled water.

Liquid sucrose. Adjust the liquid sucrose sample to 54°Bx with distilled water (about 740 ml of liquid sucrose and 240 ml of water).

HFSS. Dilute samples to about 50°Bx. Use 637 ml of HFSS-42 and 368 ml of distilled water, and 576 ml HFSS-55 and 431 ml of distilled water. Acidify this solution to pH 1.5 using 75% w/v phosphoric acid.

9.15.4 Test Completion and Results Evaluation for Appearance

9.15.4.1 Granulated Sucrose

Using the 50°Bx test solution, examine the sample in a glass container against a white background with a high-intensity light source. Ensure that no color, cloud, or haze (turbidity) is seen. Turbidimeter readings must be <10 NTU (Joint FAO/WHO Expert Committee on Food Additives 2006).

If there is turbidity in granular sucrose, test as follows to determine if more filtering will eliminate it:

1. Filter the 50°Bx test solution using a Whatman No. 54 filter paper.
2. Check for turbidity in the filtrate. If none is found, the granular sucrose may be acceptable for use after more filtering.

9.15.4.2 Liquid Sucrose

Use the 54°Bx sample, acidified to pH 1.5 with 75% phosphoric acid. Examine the sample in a glass container against a white background with a high-intensity light source. Ensure that no color, cloud, or haze (turbidity) is seen. Turbidimeter readings must be <10 NTU.

9.15.4.3 HFSS

Examine the sample in a glass container against a white background with a high-intensity light source. Ensure that no color, cloud, or haze (turbidity) is seen. Turbidimeter readings must be <10 NTU.

To determine if HFSS is a source of flocculation in syrup or beverage, cover the acidified HFSS samples used in the appearance test and let them stand undisturbed. Examine for flocculation with a high-intensity light source daily for 10 days.

9.15.5 Taste Testing and Evaluation of Results

Granular Sucrose

Use the 50°Bx sucrose solution. Taste the sample and note any off-flavor taste.

Liquid Sucrose

Use the 54°Bx liquid–sucrose solution. Taste the sample and note any off-flavor taste.

HFSS

Use the 50°Bx HFSS solution. Taste the sample and note any off-flavor taste.

9.15.6 Odor Testing and Results Evaluation

Granulated Sucrose

Use the 50°Bx sucrose solution, smell the headspace, and note any off odor. To increase odor sensitivity, half fill a wide-mouth glass bottle with a granulated sucrose sample. Apply a screw cap, and heat the bottle to 50°C. Open the bottle, smell the headspace, and note the nature of any off odor.

Liquid Sucrose

Use the 54°Bx liquid–sucrose solution. Smell the container's headspace and note any off odor.

HFSS

1. Half fill a wide-mouth, screw-capped bottle with 50°Bx HFSS solution acidified to pH 1.5.
2. Heat to 30°C in a water bath or incubator.
3. Check the odor every 5 min for 30 min.
4. Ensure that samples are free from off odors.

9.16 SULFONATED POLYSTYRENE IN HFSS

9.16.1 Purpose

Use this procedure to test HFSS to determine if it is likely to cause floc and neck-ring problems in a beverage (Joint FAO/WHO Expert Committee on Food Additives 2006).

Use the rhodamine B test to rapidly evaluate the sweetener. If you get a positive result, perform the quinine hydrochloride test.

9.16.2 Rhodamine B Test

Rhodamine B Test requires the following equipment:

- Balance
- 10-ml graduated cylinder
- 100-ml graduated cylinder
- 20-ml test tubes
- 250-ml beaker
- Stirring rod
- Dropper

The following reagents can be used:

- 1% weight/volume (w/v) rhodamine B solution
- Refined sucrose
- Distilled water

Sample preparation is described as follows:

1. Measure 5-ml HFSS into a 20-ml test tube, add 5-ml distilled water, mix thoroughly, and label as "sample."
2. Add two drops of 1% w/v rhodamine B solution and shake.

Control preparation is conducted as follows:

1. Dilute 47 g of refined sucrose into 100 ml of distilled water.
2. Measure 5 ml of the sucrose solution prepared in step 1 into another test tube and label as "control."
3. Add two drops of 1% w/v rhodamine B solution and shake.

For test completion, hold the sample and control side by side. View the test tubes from the top.

Rhodamine B Test Results Evaluation. When viewing the test tubes from the top, evaluate the solutions according to the following guidelines:

- A positive test gives a violet color to the sample solution. The color will concentrate mostly in the foam (produced from shaking). If you see a violet color in the sample's foam, perform the quinine hydrochloride test (below).
- A negative test gives a similar red color to the sample and the control.

9.16.3 Quinine Hydrochloride Test

The quinine hydrochloride test requires the following equipment:

- 10-ml graduated cylinder
- 100-ml graduated cylinder with stopper
- 10-ml pipette
- 20-ml test tube
- Timer
- 400-ml beaker
- Hot plate or burner
- Spectrophotometer with 1-cm cells

For reagents, 6% w/v quinine hydrochloride or distilled water can be used.

Sample preparation is described as follows:

1. Measure 50-ml HFSS in a 100-ml graduated cylinder with a stopper. Add 50-ml distilled water, stopper the cylinder, and mix.
2. Pipette 10-ml HFSS solution into a 20-ml test tube, add 2-ml 6% w/v quinine hydrochloride, and mix.
3. Allow the mixture to stand for 30 min.
4. Place the test tube in a boiling water bath for 5 min.
5. Cool the sample immediately.

Sample reading is conducted as follows:

1. Fill a 1-cm cell with sample.
2. Fill another 1-cm cell with distilled water.
3. Set the spectrophotometer at 720 nm and read the sample's transmittance against a distilled water blank.

For results interpretation, any sample reading less than 90% T is unacceptable.

9.17 SAMPLE PREPARATION

Sample preparation for the determination of high-intensity sweeteners is relatively simple. Carbonated soft drinks are degassed prior to analysis. Liquid beverages and tabletop sweeteners are diluted or dissolved in water. Sweeteners in complex foods are extracted with water or an appropriate solvent. Then, the extract can be clarified, centrifuged, or cleaned by using solid-phase extraction (SPE) techniques.

The physicochemical properties of foods vary. Variability in the composition of a given food sample can be minimized with proper sample preparation. In general, preparation of food samples is carried out in four steps: (1) homogenization; (2) extraction; (3) cleanup; and (4) preconcentration.

Analysis of liquid samples does not require the homogenization step because of their liquid state. After sample preparation, some matrix components may coextract with analytes because of similar solubility in the solvents used for extraction, interfering in the analysis and causing problems with method accuracy. Further cleaning up the sample for analysis is the only way of problem resolution. Cleanup is usually carried out by SPE, dialysis, liquid–liquid extraction, precipitation, and filtration.

Optimal sample preparation, extraction, cleanup, and/or purification can reduce analysis time, enhance sensitivity, and enable confirmation and quantification of analytes and depends on the complexity of the sample and the sensitivity and selectivity of the method used (Self 2005).

Determination of acesulfame K and aspartame in a simple matrix is much easier, less time consuming, and less complex than the determination of these two sweeteners in a more complicated food sample matrix. Bulk samples of sweeteners have a much more simple pretreatment stage than complex food matrices. The interference of other food additives in the determination of sweeteners is more common in complicated food sample matrices. In bulk samples (e.g., tabletop tablets), additives are present in much lower quantities. Usually, three types of additives are added in bulk powders: (1) glidants for a better flow in the material; (2) pure lubricants for better mixing; and (3) antiadherents for nonadherence of the tablet to the die. Determination of sweeteners in tabletop solid tablets is done by converting them into powder, weighing the powder, dissolving it into ultrapure water, and transferring it into volumetric flasks. Liquid sweeteners and beverages with a relatively simple matrix can be diluted or dissolved in deionized water or in an appropriate buffer. Carbonated drinks have to be degassed by sonication, sparging with nitrogen, or under vacuum.

Prior to analysis, all samples are filtered, and extracts are centrifuged. SPE is used to preconcentrate the analytes and/or remove the chemical interferences. The most frequently used SPE cartridges are nonpolar C_{18}. Their stationary packing material consists of nonpolar C_{18} chains. The SPE procedure consists of four steps: (1) cartridge conditioning; (2) sample load; (3) cartridge wash; and (4) elution of analytes.

The type of SPE packing material, solvents, pH, and flow rates are very important. The interfering substances should be retained very strongly, and weakly retained substances are readily removed from the cartridge during sample load and/or cartridge wash. Analytes are eluted during the elution step, and interfering substances having a strong affinity to the sorbent stay adsorbed within the cartridge. Evaporation of the final SPE extract to dryness and its reconstitution with a smaller amount of a solvent (preconcentration) might enhance the sensitivity of a final determination. SPE-based sample-preparation protocols are simple, reproducible, reasonably quick and inexpensive, and compatible with the most popular techniques used in food analysis (Tunick 2005).

Currently, the most common technique for the determination of artificial sweeteners in soft drinks is high-performance liquid chromatography (HPLC). Sample preparation for HPLC includes filtering through a 0.45-mm membrane filter, followed by ultrasonication. Nectars are centrifuged, filtered through membrane filters, and then ultrasonicated (Lino, Costa, and Pena 2008). For the determination in a cola drink, the volume of the drink is accurately weighed into a volumetric flask and then degassed in an ultrasonic bath. If determination is carried out by HPLC, the volume of the cola drink is directly diluted in a mobile phase (Demiralay, Çubuk, and Guzel-Seydim 2006).

Sweetened beverages include two types of beverages with carbon dioxide and two types of beverages without carbon dioxide. Beverages with and without carbon dioxide include light and sugar-sweetened soft drinks. In light soft drinks with carbon dioxide, a mixture of aspartame and acesulfame K is more often used as a sweetener (Leth, Fabricius, and Fagt 2007). Micellar electrokinetic capillary chromatography (MEKC) is a rapid method used in the determination of artificial sweeteners in low-calorie soft drinks. Low-calorie soft drinks are prepared by dilution with an appropriate amount of deionized water, followed by filtering through a 0.45-μm cellulose acetate filter before analysis (Thompson, Trenerry, and Kemmery 1995a).

Preparation of gum samples is done by placing them in a flask and extracting them with a mixture of glacial acetic acid, water, and chloroform. Hard or soft candy samples are shaken with water until dissolved (Biemer 1989). Milk and dairy products are homogenized prior to analysis, and an aliquot of a homogeneous sample is transferred to a flask, followed by the addition of distilled water. The mixture is thoroughly stirred, allowed to stand, and then filtered (Ni, Xiao, and Kokot 2009).

Sweeteners are determined in diet jams by mixing the jam with water, followed by mixture sonication and filtering through a 0.45-mm filter (Boyce 1999). Preserved fruits are usually ground and homogenized, weighed into a volumetric flask, extracted ultrasonically, and diluted to volume with water after cooling at room temperature. A volume of supernatant is applied to a conditioned Sep-Pak C_{18} cartridge (Chen and Wang 2001), and the subsequent SPE step follows.

Beverages, juices, strawberry sweets, and tomato sauce containing acesulfame K and/or aspartame can be analyzed with the use of a flow injection analysis (FIA) system coupled with a UV detector. An adequate amount of beverage is taken, degassed, diluted, sometimes centrifuged (as in the cases of juices), and filtered through a 0.2-mm Millipore filter. Strawberry sweets are weighed, crushed in a glass mortar, dissolved in water with the aid of an ultrasonic bath, centrifuged, and filtered. Finally, in the case of tomato sauce, a portion is diluted in water, centrifuged, and exposed to the same procedure as for beverages (Jiménez, Valencia, and Capitán-Vallvey 2009).

Moreover, the pretreatment of a sample of soy sauce, dried roast beef, and sugared fruit includes placement in a volumetric flask, water dilution, mixing thoroughly, and the addition of a portion of the solution to a Sep-Pak C_{18} cartridge. Then, water is passed through the cartridge to remove the interfering substances. Other additives are absorbed onto the packing material, eluted with acetonitrile–water (2:3 v/v), and collected as eluate, which is then diluted to volume with acetonitrile–water (2:3 v/v; Chen and Fu 1995).

Fourier transform infrared (FTIR) spectroscopy is also a quick method for the determination of acesulfame K in commercial diet food samples without the use of an extraction step prior to analysis. Samples with difficult food matrices, such as chocolate syrup, coffee drinks, coffee creamer, cranberry juices, ice cream, and instant chocolate milk, can be easily analyzed. The ice cream sample is converted to a mixture at 60°C in a water bath with continuous agitation. All other samples are used without any pretreatment. Food samples are mixed thoroughly with ultrapure water and maintained at 40°C in an incubator, followed by ultrasonication. Carrez I (3.6-g potassium hexacyanoferrate trihydrate dissolved and made up to 100-ml distilled water) and Carrez II reagent (7.2-g zinc sulphate heptahydrate dissolved in 100-ml distilled water) solutions are added to samples, followed by centrifugation at 4°C. Supernatants are separated using a syringe to avoid the fat layers and are filtered. The water-soluble extract obtained is directly poured onto the surface of attenuated total reflectance (ATR-FTIR) according to Shim et al. (2008).

9.18 PHYSICOCHEMICAL METHODS

9.18.1 Capillary Electrophoresis

Capacitively coupled contactless conductivity detection (C^4D) is a relatively new approach for detection on capillary electrophoresis (CE–C^4D; Fracassi da Silva and do Lago 1998; Fracassi da Silva, Guzman, and do Lago 2002; Zemann et al. 1998). CE–C^4D has been successfully applied for the determination of inorganic and organic compounds in food (Escarpa et al. 2008; Richter et al. 2005).

CE–C^4D was used by Bergamo, da Silva, and de Jesus (2011) for a simple, rapid, and simultaneous determination of aspartame, cyclamate, saccharin, and acesulfame K in commercial samples of soft drinks and tabletop sweetener formulations. A buffer solution containing 100 mmol L_1 tris(hydroxymethyl)aminomethane (TRIS) and 10 mmol L_1 histidine (His) was used as a background electrolyte. A complete separation of the analytes could be attained in less than 6 min. The limits of detection (LOD) and limits of quantification (LOQ) were considered better than those usually obtained by CE with photometric detection. Recoveries ranging from 94% to 108% were obtained for samples spiked with standard solutions of the sweeteners. The relative standard deviation (RSD) for the analysis of the samples with the CE–C^4D method varied in the range of 1.5%–6.5%.

Separation and quantification of sweeteners using CE have been described by other authors as a rapid and simple method (Frazier et al. 2000; Horie et al. 2007; Pesek and Matyska 1997; Schnierle, Kappes, and Hauser 1998; Thompson, Trenerry, and Kemmery 1995).

Frazier et al. (2000) employed the MEKC mode with a 20-mM carbonate buffer at pH 9.5 as the aqueous phase and 62 mM sodium dodecyl sulfate as the micellar phase.

A rapid method for the determination of artificial sweeteners in low-joule soft drinks and other foods by MEKC is described by Thompson, Trenerry, and Kemmery (1995b). The artificial sweeteners aspartame, saccharin, acesulfame K, alitame, and dulcin and the other food additives are well separated in less than 12 min using an uncoated fused-silica capillary column with a buffer consisting of 0.05-M sodium deoxycholate, 0.01-M potassium dihydrogen orthophosphate, and 0.01-M sodium borate operating at 20 kV. Dehydroacetic acid was used as the internal standard for the determinations. The levels of artificial sweeteners were in good agreement with those determined by the HPLC procedure. The MEKC (Thompson et al. 1995a) procedure has the same order of repeatability and is faster and less costly to operate than the HPLC method.

Photometric detection in the UV region is the most commonly used detection method, although for some applications, it shows inadequate limits of detection because of the low UV absorptivities of most sweeteners, especially cyclamate (Bergamo, de Silva, and de Jesus 2011). Conductivity detection is a good alternative method for compounds lacking a strong UV-absorbing moiety. Using this detection technique, an isotachophoresis method was published by Herrmannova et al. (2006) for the determination of sweeteners in chewing gums and candies.

High-performance CE has been employed by Pesek and Matyska (1997) to analyze the aspartame in food products using a bare capillary, a pH 2.14 buffer, and detection at 211 nm. The analysis time was faster than that reported for HPLC methods.

The mobile phase can be methanol–water, methanol–acetic acid, methanol–phosphate buffer, methanol–ammonium citrate, or acetonitrile–phosphate buffer when reversed-phase columns (RP-C18) are used. Methanol–phosphoric acid and Na_2CO_3 or NaOH solutions were used with amino and ion-exchange columns, respectively. Acesulfame K, alitame, aspartame, glycyrrhizin, neohesperidin dihydrochalcone (NHDC), saccharin, and stevioside can be determined by UV absorbance (192–282 nm) and by amperometric or conductometric detection in the case of ion–chromatographic procedures (Yebra-Biurrun, 2005).

9.18.2 Ion Chromatography

Chromatographic methods (Wasik, McCourt, and Buchgraber 2007; Yang and Chen 2009; Zhu et al. 2005) have been employed to detect artificial sweeteners in food.

A novel ion chromatographic method was proposed for the simultaneous determination of artificial sweeteners (sodium saccharin, aspartame, and acesulfame K) by Chen and Wang (2001). Separation was performed on an anion-exchange analytical column operated at 408°C within 45 min by an isocratic elution with 5-mM aqueous NaH_2 PO_4 (pH 8.20) solution containing 4% (v/v) acetonitrile as eluent and the determination by wavelength-switching UV absorbance detection. The detection limits (signal-to-noise ratio of 3:1) for all analytes were below the sub–milligram per milliliet level. The results also indicated ion chromatography possibly would be a beneficial alternative to conventional HPLC for the separation and determination of these compounds.

Table 9.8 Results from Sweetener Analysis

Samples	Sweetener	Concentration Provided by the Manufacturer (mg L^{-1})	Concentration Found by Ion Chromatography (mg/kg)	Concentration by HPLC (mg/kg)	Concentration Found by CE-C^{4}D mg L^{-1}
Soft drink (cola)	Aspartame	120	—	—	(94 ± 5)
	Cyclamate	240	—	—	(232 ± 15)
	Acesulfame K	150	—	—	(143 ± 9)
	Saccharin	—	133 ± 6	137 ± 2	—
Soft drink (orange)	Aspartame	—	—	—	—
	Cyclamate	701	—	—	(725 ± 12)
	Saccharin	160	—	—	(147 ± 9)
Guarana	Aspartame	—	—	—	—
	Cyclamate	640	—	—	(619 ± 30)
	Saccharin	80	—	—	(126 ± 8)
Fruit juice drink	Acesulfame K	—	82 ± 3	81 ± 1	—
Preserved fruit	Saccharin	—	4720 ± 140	4810 ± 80	—

Source: Adapted from Bergamo, A. B. et al., *Food Chem.*, 124, 2011. Chen, Q. C., and J. Wang, *J. Chromatogr. A*, 937, 2001. With permission.

Note: Table shows a reasonable agreement between the sweetener concentrations found by CE-C^{4}D and those claimed by the manufacturers.

Results from sweetener analysis comparing different techniques such as ion chromatography and HPLC are presented in Table 9.8.

9.18.3 Planar Chromatography

Planar chromatography was used to determine sucralose as a high-potency sweetener in burfi, an Indian milk-based confection produced in-house (Morlock and Prabha 2007). It consists of a reagent-free derivatization step. Sucralose was determined on high-performance thin-layer chromatography (HPTLC) amino plates whose amino groups reacted with sucralose to fluorescent zones by just heating the plate after chromatography. According to European legislation, it has been approved for use as E955 (1,6-dichloro-1,6-dideoxy-β-D-fructofuranosyl-(2-1)-4-chloro-4-deoxy-α-D-galactopyranoside), and the limits of addition range between 10 mg/kg and 3000 mg/kg for various foods. The previously mentioned authors met these limits fully. However, concern over the safety of sucralose is under debate, because there are no long-term health studies in humans for this organochloride.

9.18.4 Spectrophotometric Techniques

Spectrophotometric methods have also been used to detect artificial sweeteners in food (Cantarelli et al. 2009; Fatibello-Fihlo, Marcolino, and Pereira 1999).

Extractive spectrophotometric techniques using colorimetric reagents (oxazine dye, Sevron blue 5G for acesulfame K; p-dimethylaminobenzaldehyde, 1,4-benzoquinone, ninhydrin for aspartame;

picryl chloride, p-quinone for cyclamate; vanillin for glycyrrhizin; Nile blue, azure A, B, or C, Sevron blue 5G, brilliant cresyl blue for saccharin; and anthrone for steviosides) are used for the direct determination of high-intensity sweeteners (Yebra-Biurrun 2005).

9.18.5 Titrimetric Methods

Titrimetric assays have been developed for acesulfame K (titrated with sodium methoxide in benzene), aspartame, sodium cyclamate, sodium saccharin (titrated with perchloric acid), and saccharin (acid form), with potassium hydroxide acting as a titrant. Precipitation, chelatometric, and redox titrations are proposed for the determination of cyclamate. The oldest methods for saccharin involve its determination by means of a Kjeldahl procedure (Yebra-Biurrun, Cancela-Perez, and Moreno-Cid-Barinaga 2005; Yebra-Biurrun, Moreno-Cid, and Cancela-Perez 2005).

9.18.6 Electrochemical Detection Systems: Biosensors

Electrochemical methods have also been used for the detection of artificial sweeteners as reported in the literature (Assumpcao et al. 2008; Medeiros et al. 2008).

Another electrochemical detection system based on a coated-carbon-rod ion-selective electrode for the determination of saccharin in dietary products has also been described recently by Fatibello-Fihlo and Aniceto (1997).

Lipid films can be used for the rapid detection or continuous monitoring of a wide range of compounds in foods and in the environment. Such electrochemical detectors are simple to fabricate and can provide a fast response and high sensitivity (Nikolelis and Pantoulias 2000).

Nikolelis et al. (2001) explored the interactions of the sweeteners acesulfame K, saccharin, and cyclamate with bilayer lipid membranes (BLMs). BLMs composed of egg phosphatidylcholine can be used for the direct electrochemical sensing of these sweeteners. The interactions of these compounds with lipid membranes were found to be electrochemically transduced in the form of a transient current signal with a duration of seconds, which reproducibly appeared within 11 s after exposure of the membranes to the sweetener. The mechanism of signal generation was investigated by differential scanning colorimetry (DSC) studies and monolayer compression techniques. DSC showed that the interactions of the sweeteners with lipid vesicles stabilize the gel phase of the lipid films.

These latter studies revealed the adsorption of sweeteners caused an increase of the average molecular area occupied by the lipids and resulted in increased structural order of the membranes.

The detection limit of acesulfame K is 1 μm, and the reproducibility is on the order of ±4 to 8% in a 95% confidence level. The recovery ranged between 96% and 106% and shows no interferences from the matrix (Nikolelis et al. 2001).

The weak points of the present sensor compared with conventional analytical methods are that there is no selectivity to discriminate between artificial sweeteners, i.e., acesulfame K, saccharin, and cyclamate, and the fragility and the very limited stability of the freely suspended BLMs. However, the present technique can be used as a one-shot sensor for the rapid detection of these sweeteners but keeps prospects for potential applications for the selective determination and analysis of mixtures of acesulfame K, saccharin, and cyclamate in granulated sugar-substitute products by using filter supported BLMs (Nikolelis et al. 1999, 2001).

Commercially available biosensors for applications in the food and beverage industry are based on either an oxygen electrode or a hydrogen-peroxide electrode in conjunction with an immobilized oxidase. Targeted compounds include artificial sweeteners such as aspartame, saccharin, cyclamate, and acesulfame, which are used in soft drinks and desserts (Luong, Bouvrette, and Male 1997).

9.18.7 High-Performance Liquid Chromatography

HPLC in combination with UV detection has very often become the method of choice for the determination of individual artificial sweeteners such as saccharin, aspartame, acesulfame, and NHDC (Montijano et al. 1997).

Saccharin is determined in foodstuffs by HPLC, as described by AOAC (2000) and Sjoberg and Alanko (1987a).

The lack of any chromophore in sucralose makes a sensitive and specific detection by direct UV absorption difficult. As a result, current analytical methods for its determination in foodstuffs involve either refractive index (RI) detection (Kobayashi et al. 2001) or UV absorbance at a wavelength of 200 nm, which lacks specificity (Lawrence and Charbonneau 1988). Further methods that have been successfully applied are HPTLC (Spangenberg et al. 2003), high-performance anion exchange chromatography with pulsed amperometric detection (Dionex 2004), and CE with indirect UV absorption (McCourt, Stroka, and Anklam 2005; Stroka, Dossi, and Anklam 2003). Moreover, cyclamate does not absorb in the usable UV/visible range. This shortcoming has been solved by using indirect UV photometry (Thompson, Trenerry, and Kemmery 1995; Choi, Hsu, and Wong 2000), postcolumn ion-pair extraction (Lawrence and Charbonneau 1988), precolumn derivatization (Casals et al. 1996), and conductivity detection (Chen et al. 1997).

A HPLC method with evaporative light scattering detection (HPLC-ELSD) has been developed by Wasik, McCourt, and Buchgraber (2007) for the simultaneous determination of multiple sweeteners, i.e., acesulfame K, alitame, aspartame, cyclamic acid, dulcin, neotame, NHDC, saccharin, and sucralose in carbonated and noncarbonated soft drinks, canned or bottled fruits, and yogurt. The procedure involves an extraction of the nine sweeteners with a buffer solution, sample cleanup using SPE cartridges, followed by an HPLC-ELSD analysis.

The trueness of the method was satisfactory, with recoveries ranging from 93% to 109% for concentration levels around the maximum usable dosages for authorized sweeteners and from 100% to 112% for unauthorized compounds at concentration levels close to the LOQ. Precision measures showed mean repeatability values of <4% (expressed as RSD) for highly concentrated samples and <5% at concentration levels close to the LOQ.

Ion-pair HPLC has been used by Verzella, Bagnasco, and Mangia (1985) to determine aspartame in finished bulk and dosage forms at levels down to 0.1%.

Moreover, a reversed-phase HPLC method was developed by Di Pietra et al. (1990) for the quality control of pharmaceutical and dietary formulations containing the synthetic sweeteners aspartame and saccharin. Artificial sweeteners were analyzed by HPLC electrospray tandem mass spectrometry (HPLC-ESI-MS/MS) after SPE according to a recently published method (Scheurer, Brauch, and Lange 2009). Because artificial sweeteners were incompletely removed in wastewater treatment plants, some of these compounds end up in receiving surface waters, which are used for drinking water production. The sum of the removal efficiency of single-treatment steps in multi-barrier treatment systems affects the concentrations of these compounds in the provided drinking water. This is the first systematic study revealing the effectiveness of single-treatment steps in laboratory experiments and in waterworks (Scheurer et al. 2010). Six full-scale waterworks using surface water–influenced raw water were sampled up to 10 times to study the fate of acesulfame, saccharin, cyclamate, and sucralose. Saccharin and cyclamate proved to play a minor role for drinking water treatment plants, as they were eliminated by nearly 100% in all waterworks with biologically active treatment units, such as river bank filtration (RBF) or artificial groundwater recharge. Acesulfame and sucralose were not biodegraded during RBF, and their suitability as wastewater tracers under aerobic conditions was confirmed. Sucralose proved to be persistent against ozone, and its transformation was <20% in laboratory and field investigations. Remaining traces were completely removed by subsequent granular activated carbon (GAC) filters. Acesulfame readily reacts with ozone. However, the applied ozone concentrations and contact times under typical waterworks conditions only led to an incomplete removal (18%–60%) in the ozonation step. Acesulfame was efficiently removed by subsequent GAC filters with a low throughput of less than 30 m^3 kg^{-1}, but removal strongly depended on the GAC preload.

Zygler, Wasik, and Jacek Namiesnik (2010) provided information about the application of SPE for the isolation of nine high-intensity sweeteners (acesulfame K, alitame, aspartame, cyclamate, dulcin, neotame, saccharin, sucralose, and NHDC) from aqueous solutions. The influence of several types of LC-MS compatible buffers (different pH values and compositions) on their recovery has been studied and discussed. A number of commercially available SPE cartridges, such as Chromabond C18 ec, Strata-X RP, Bakerbond Octadecyl, Bakerbond SDB-1, Bakerbond SPE Phenyl, Oasis HLB, LiChrolut RP-18, Supelclean LC-18, Discovery DSC-18, and Zorbax C18, were tested in order to evaluate their applicability for the isolation of analytes. Very high recoveries (better than 92%) of all studied compounds were obtained using formic acid–N,N-diisopropylethylamine buffer adjusted to pH 4.5 and C18-bonded silica sorbents. The behavior of polymeric sorbents strongly depends on their structure. Strata-X RP behaves much like a C18-bonded silica sorbent. Recoveries obtained using Oasis HLB were comparable with those observed for silica-based sorbents. The only compound less efficiently (83%) retained by this sorbent was cyclamate. Bakerbond SDB-1 shows unusual selectivity toward aspartame and alitame. Recoveries of these two sweeteners were very low (26% and 42%, respectively). It was also found that aspartame and alitame can be selectively separated from the mixture of sweeteners using a formic acid–triethylamine buffer at pH 3.5.

Stevia rebaudiana Bertoni contains glycosides, which are insoluble in carbon dioxide and soluble in mixtures of carbon dioxide and a polar solvent. Yoda et al. (2003) studied the supercritical fluid extraction of these glycosides from stevia leaves using a two-step process: (1) CO_2 extraction at 200 bar and 30°C and (2) CO_2 + water extraction. The chemical compositions of the extracts were analyzed by GC-FID, GC-MS, TLC, and HPLC. The overall extraction curves for the system stevia + CO_2 had the typical shape and were successfully described by the Sovova model. Approximately 72% of the CO_2-soluble compounds were recovered, and the major compound was austroinulin. The stevia + CO_2 + water system behaved as expected at 10°C and 16°C and 120 bar and 250 bar, but its behavior was unusual at 30°C and 250 bar. The process removed approximately 50% of the original stevioside and about 72% of the rebaudioside A.

Herrmann, Damawandi, and Wagmann (1983) reported on a method for the detection of aspartame, cyclamate, dulcin, and saccharin using an ion-pair HPLC separation with indirect photometric detection. However, such varied methods with their differing derivatization protocols make the analysis of artificial sweeteners time consuming and labor intensive. Recently, some methods employing liquid chromatography mass spectrometry have been published for the analysis of sucralose (Loos et al. 2009) and other artificial sweeteners in water samples (Scheurer, Brauch, and Lange 2009).

A methodology for the chromatographic separation and analysis of three of the most popular artificial sweeteners (aspartame, saccharin, and sucralose) in water and beverage samples was developed using liquid chromatography/time-of-flight mass spectrometry (LC/TOF-MS) by Ferrer and Thurman (2010). The sweeteners were extracted from water samples using SPE cartridges. Furthermore, several beverages were analyzed by a rapid and simple method without SPE, and the presence of the sweeteners was confirmed by accurate mass measurements below 2-ppm error. The unambiguous confirmation of the compounds was based on accurate mass measurements of the protonated molecules [M+H]+, their sodium adducts, and their main fragment ions. Quantitation was carried out using matrix-matched standard calibration, and the linearity of response over two orders of magnitude was demonstrated ($r > 0.99$). A detailed fragmentation study for sucralose was carried out by TOF, and a characteristic spectrum fingerprint pattern was obtained for the presence of this compound in water samples. Finally, the analysis of several wastewater, surface water, and groundwater samples from the U.S. showed sucralose can be found in the aquatic environment at concentrations up to 2.4 μg/L, thus providing a good indication of wastewater input from beverage sources.

9.18.8 Fluorescence Spectroscopy

Because of the presence of various aromatic compounds in sap and syrup, intrinsic fluorescence was used by Clement, Lagace, and Panneton (2010) to characterize the physicochemistry and typicity of maple syrup. Two hundred samples of sap and their corresponding syrup were obtained from various farms in 2003 and 2004. They were analyzed by conventional physicochemical tests and fluorescence spectroscopy. Two major regions of fluorescence were found, which were mostly the same for sap and syrup. The first region was at 320 nm, excited at 275 nm, and the second region was at 460 nm, excited at 360 (syrup) or 370 (sap) nm.

Maple syrup is produced by the heat evaporation of maple sap collected from maple sugar trees (*Acer saccharum Marsh*) during the early spring season in North America.

Clement, Lagace, and Panneton (2010) determined the potential of fluorescence spectroscopy to assess the major physicochemical characteristics of maple sap and syrup in a rapid and nondestructive way and explored new possibilities of automated classification, including typicity and the moment of sap harvest. They found that precise farm location, rather than the broad region of production, is the major factor of typicity.

Optical spectroscopy combined with chemometrics can characterize and classify food products (Karoui and De Baerdemaeker 2007; Kulmyrzaev et al. 2007). For instance, near-infrared (NIR; Ruoff et al. 2007) and front-face fluorescence spectroscopy (Ruoff et al. 2005) have been proposed to characterize the physicochemical properties and adulteration of honey. Data available on maple syrup have shown that various spectroscopic methods (NIR, Raman, and FTIR) could determine the presence of adulterants (Paradkar, Sivakesava, Irudayaraj 2003).

Raman spectroscopy is an important tool in the quantitative analysis of complex mixtures (Strachan et al. 2007; Peica 2009). One of its benefits lies in the fact that water does not hinder the analysis of liquid samples.

It is a nondestructive technique that yields reliable results for solid and liquid multicomponent samples, especially when the analyte concentration exceeds 0.5%–1% by weight.

Quantification of aspartame in commercial sweeteners was carried out by Mazurek and Szostak (2011). They reported the treatment of Raman data with three chemometric methods: partial least squares (PLS), principal component regression (PCR), and counter-propagation artificial neural networks (CP-ANN). Four commercial preparations containing between 17% and 36% of aspartame by weight were evaluated by applying the developed models. Concentrations found from Raman data analysis agree perfectly with the results of the UV-Vis reference analysis.

9.18.9 Atomic Absorption Spectrophotometry

An atomic absorption spectrometric method for the determination of sodium saccharin in commercially available mixtures of artificial sweeteners and pharmaceuticals by continuous precipitation with silver ion in a flow manifold is proposed by Yebra, Gallego, and Valcarcel (1995), and Yebra-Biurrun (2000).

The silver saccharinate precipitate formed is retained on a filter, washed with diluted acetic acid, and dissolved in ammonia for online atomic absorption determination of silver, the amount of which in the precipitate is proportional to saccharin in the sample. The proposed method allows the determination of sodium saccharin in the range of 5–75 μg ml^{-1} with a RSD of 2.7% at a rate of ca. 20 samples h^{-1}. The method is very selective; other sweeteners do not interfere, but chloride must be absent.

Most applications of this approach involve using flame atomic absorption spectrometry (FAAS). Several continuous automatic precipitation systems coupled online to a conventional atomic absorption instrument have been used for indirect determinations of inorganic anions.

9.18.10 Nuclear Magnetic Resonance

Rebaudioside A is a natural sweetener from *S. rebaudiana* in which four β-D-glucopyranose units are attached to the aglycone steviol. Its ^{1}H and ^{13}C NMR spectra in pyridine-d_5 were assigned using one- and two-dimensional methods. Constrained molecular dynamics of solvated rebaudioside using nuclear magnetic resonance (NMR) constraints derived from ROESY cross peaks yielded the orientation of the b-D-glucopyranose units (Steinmetz and Lin 2009). Hydrogen bonding was examined using the temperature coefficients of the hydroxyl chemical shifts, ROESY and long-range COSY spectra, and proton–proton coupling constants.

Only partial ^{13}C NMR assignments are available (Dacome et al. 2005).

9.18.11 Inductively Coupled Plasma Optical Emission Spectrometry

For liquid cyclamate–saccharine sweeteners, cadmium was found in one of 15 analyzed samples, employing the inductively coupled plasma optical emission spectrometry (ICP OES) technique (Sousa, Baccan, and Cadore 2006), arsenic and lead were found in two of four samples of liquid aspartame sweeteners, analyzed by ICP-MS (Sousa et al. 2007), and selenium was found in many samples of liquid sweeteners, containing cyclamate–saccharine and stevioside, analyzed by graphite furnace atomic absorption spectrometry (Sousa et al. 2006, 2007, 2011). ICP OES has been used by Pedro, Oliveira, and Cadore (2006) for the determination of major and minor elements in foods.

Inorganic species (calcium, cadmium, copper, cobalt, iron, magnesium, manganese, sodium, nickel, lead, selenium, and zinc) were determined by Sousa, Baccan, and Cadore (2011) in solid sweeteners by ICP OES without employing a mineralization step. Robust conditions were used in the ICP equipment, and the samples were analyzed as aqueous suspensions. The analysis of different kinds of sweeteners showed different concentration profiles. Some samples presented contamination by cadmium, but in general, the samples were found to contain calcium, iron, magnesium, manganese, sodium, and selenium in tolerable amounts. The method's accuracy was evaluated, and recovery values between 90% and 110% were attained for the recovery experiments; the RSD values were lower than 5% in most cases.

9.18.12 Titrimetric and Other Methods

Titration with a copper (II) salt (Lane–Eynon titration) is the standard method for the determination of reducing sugars (glucose and fructose) in bulk raw and white sugars. Moisture in solid sugars is determined generally by oven drying; in liquid products, by Karl Fischer titration (Yebra-Biurrun, Cancela-Perez, and Moreno-Cid-Barinaga 2005; Yebra-Biurrun, Moreno-Cid, and Cancela-Perez 2005). Inorganic content is determined by either conductivity in solution or sulfated ash gravimetric procedures.

In syrups, the solids content, degrees Brix, or refractometric dry solids, is determined by RI measurement. Tables correlating RI with sucrose and invert content are published in the ICUMSA and AOAC methods books.

Physicochemical analysis of various sweeteners and analytical techniques used in the determination of artificial sweeteners in food products are described in Table 9.9.

Table 9.9 Analytical Techniques Used in the Determination of Artificial Sweeteners in Food Products

Determined Analytes	Matrix	Technique	Mobile Phase/ Electrolyte	Column/ Capillary	LOD	RSD	Recovery and Linear Range	Reference
ACS K, SAC, benzoic acid, and sorbic acid	Beverages and jams	HPLC-UV	8% MeOH in phosphate buffer (pH = 6.7)	Spherasorb C18 ODS-1 (5 µm × 250 mm × 4.6 mm I.D)	<0.1 mg/100 ml	N/A	98.1%–104.2% and 0 mg/l–100 mg/l	Hannisdal 1992
ACS K, ASP, and SAC	Soft drinks, juices, tomato sauce, and strawberry sweets	FIA with online monolithic element	Water (0.4 M in NaCl, 5×10^{-3} M $NaClO_4$) (pH = 9)	Quaternary amine ion exchanger monolithic column	0.9 µg/ml	0.09%	97.6%–103.4% and 3–600 µg/ml	Jiménez et al. 2009
SAC, ASP, CYC, ALI, ACS K, DUL, NEO, SCL, and NHDC	Soft drinks, canned and bottled fruits, yogurt	HPLC-ELSD	TEA formate buffer-methanol-acetone	Nucleodur C18 Pyramid (5 µm, 250 mm × 3 mm I.D)	15 µg/g	0.9%–4.5%	93%–109% and N/A	Wasik et al. 2007
AK, SCL, SA, CYC, ASP, DUL, GA, STV, and REB	Solid and liquid food matrices	HPLC-ESI-MS	Acetonitrile–water (8:2)	Zorbax Eclipse XDB-C18 (150 × 2.1 mm I.D)	1–5 µg/g	5%–10.9%	75.7%–109.2% and N/A	Koyama et al. 2005
DUL, ACS K, SAC, preservatives, and antioxidants	Soy sauce, sugared fruits, and dried roast beef	Ion-paired LC-UV	ACN-aqueous α-hydroxy-isobutyric acid solution (pH = 4.5) containing hexadecyl-trimethylammonium bromide	Stainless-steel Shoko 5 C_{18} column (5 µm, 250 mm × 4.6 mm I.D)	0.5–3 µg/g	0.3%–5.69%	81.9%–103.27% and N/A	Chen and Fu 1995
SAC, ASP, ACS K, sorbic acid, benzoic acid, caffeine, theobromine, and theophylline	Cola drinks, preserved fruits, tablets, fermented milk, and fruit juice	IC-UV	Aqueous NaH_2PO_4 (pH = 8.20) –4% (v/v) ACN	IC-A3 Shim-Pac (5 µm,150 mm × 4.6 mm I.D.)	4–30 mg/L	1%–5%	85%–104% and N/A	Chen and Wang 2001

(continued)

Table 9.9 (Continued) Analytical Techniques Used in the Determination of Artificial Sweeteners in Food Products

Determined Analytes	Matrix	Technique	Mobile Phase/ Electrolyte	Column/ Capillary	LOD	RSD	Recovery and Linear Range	Reference
ACS K, SAC, ASP, antioxidants, and preservatives	Cola beverages, and low-joule jam	Micellar electrokinetic chromatography (MEKC)-UV	Sodium tetraborate solution (pH = 9.3) with Na cholate, dodecyl sulfate-10% ACN or isopropanol, or MeOH	Fused silica capillary (52 cm × 75 μm I.D)	N/A	0.9%–1.5%	98.9%–100.86% and N/A	Boyce 1999
ACS K, SAC, ASP, CYC, sorbitol, mannitol, lactitol, and xylitol	Chewing gum and candy	Capillary isotachophoresis	Two electrolytes used: HCl–Tris (pH = 7.7; E1) L-histidine- Tris (pH = 8.3; E2)	Capillary ethylene propylene copolymer (90-mm length)	0.024–0.081 mM	0.8%–2.8%	98.2%–102.5% and N/A	Herrma-nova et al. 2006
Sucralose	Burfi and Indian milk delicacy	Planar chromatography, HPTLC	Acetonitrile–water (4:1 v/v)		6 ng/band for standards, 1 mg/kg for milk-based matrix	4.2%	88% ± 4.7% and N/A	Morlock and Prabha 2007
ASP	Commercial sweeteners	FT-Raman spectroscopy, HPLC	Phosphate buffer + methanol (70:30, v/v)	Lichrospher 100, 250–254 mm	—	1.8%–2.2%	97.8%–102.2% for PLS, PCR, CP-ANN models, and N/A	Mazurek and Szostak 2011
Stevioside and rebaudioside A	Soft drinks	HPLC-UV and liquid chromatography-electrospray ionization mass spectrometry (LC-ESI-MS)	Acetonitrile–water (8:2 v/v)	C_{18} HILIC column 250 × 4.6 mm	1 μg/ml LC-ESI-MS LOD 6 ng/ml	Intraday 1.1%–2.5%	95.9%–109.2% and 1–100 μg/ml	Wolwer-Rieck et al. 2010

Steviol and glycosides	Stevia leaves and commercial sweeteners	UHPLC-ESI-MS	Methanol, SPE cleaning CH_2Cl_2	UHPLC hss C18 column (150 mm × 2.1 mm I.D., 1.8 μm)	15, 50, 10, and 1 ng ml^{-1} for stevioside, rebaudioside A, steviolbioside, and steviol, respectively	3.1%–5.3% for steviol glycosides and 6.3%–9.4% for steviol	89%–103% and 40–180 mg/g	Gardana et al. 2010
SAC and CYC	Tabletop sweeteners	PLSspecstroscopy/ FT-Raman	—	—	0.2% w/w	0.8%	N/A and N/A	Armenta et al. 2004
ACS K and SAC	Beverages	Differential pulse polarography	—	—	<1 ppm	N/A	N/A and 1–200 ppm	Hannisdal et al. 1993
SAC, CYC, and ACS K	Cola drinks, fruit drinks, and milk, dairy products	Differential kinetic spectrophotometry	—	—	0.0312 μg/ml	N/A	N/A and N/A	Ni et al. 2009
SAC	Artificial sweeteners	Potentiometry	—	—	3.6×10^{-4} mol/l	0.6%	101.5%–102.5% and N/A	Giaan et al. 2009
SAC	Instant tea powders, diet soft drinks, strawberry dietetic jam	Potentiometry	—	—	3.9×10^{-7} mol/l	1.8%–2.3%	96.7%–102% and 5×10^{-7} mol/l – 1×10^{-2} mol/l	Santini et al. 2008

Note: SAC: saccharin, CYC: cyclamate, ACS K: acesulfame k, ASP: aspartame, CYC: cyclamate, DUL: dulcin, ALI: alitame, NEO: neotame, SCL: sucralose, NHDC: neohesperidine dihydrochalcone, GA: glycyrrhizic acid, STV: stevioside, REB: rebaudioside, p-HBA: p-hydroxybenzoic acid, LOD: limit of detection, and RSD: relative standard deviation.

9.19 SENSORY ANALYSIS METHODS

9.19.1 Consumer Preference Test

Individual variation in the perception of saccharin has been related to genetic sensitivity to the bitterness of 6-n-propylthiouracil (PROP). However, data on other intense sweeteners are sparse, particularly when tasted in real foods. The objectives of this study were to identify the sensory attributes of intense sweeteners that influenced the perception and acceptability of citrus-flavored model soft drinks and to investigate the influence of PROP taster status on these responses (Zhao and Tepper 2007). The sweeteners were 10% and 8% HFCS (controls), sucralose (SUC), aspartame (ASP), acesulfame K (ACE), ASP/ACE, and SUC/ACE. Twenty-nine PROP nontasters (NT) and 30 PROP supertasters (ST) rated nine attributes for intensity and liking. Data were analyzed using principal component analysis (PCA). The sweeteners were described in three dimensions. Factor 1 was a bitter–citrus contrast for which overall liking was associated with higher citrus impact and lower bitterness. Factors 2 and 3 were related to overall flavor and carbonation, respectively.

The sensory profiles of ASP, ASP/ACE, and SUC were most similar to 10% HFCS. SUC/ACE was more bitter and less acceptable than 10% HFCS; ACE was the most bitter and was the least liked. PCA also revealed that NT placed more emphasis on the perception of sweetness and citrus flavor (Factor 1, 37% variance), whereas ST tasters placed more emphasis on bitterness (Factor 1, 43% variance).

Liking was uniquely related to lower bitterness for NT. For ST, liking was negatively related to bitterness and weakly positively related to persistence of sweetness. These data suggest that ST experiences intense sweeteners different from NT, but these differences play a minor role in soft drink acceptance.

The beverage industry frequently uses sweetener blends to overcome the sensory limitations of individual sweeteners. Blending takes advantage of the phenomenon of synergy among sweeteners. Synergy is typically observed for sweeteners that exhibit different flavor profiles, particularly if one of the sweeteners is bitter. Blending tends to enhance the sweetening power beyond the sum of the individual sweeteners, creating an improved flavor profile. Typical blends include aspartame–saccharin, aspartame–acesulfame K, and sucralose–acesulfame K (Meyer and Riha 2002; O'Brien-Nabors 2001, 2002).

Forty subjects participated in each of two experiments in which both lemon–lime- and cola-flavored beverages containing one of the six sweeteners—sucrose, sodium saccharin, aspartame, acesulfame K, and two calcium cyclamate/sodium saccharin blends (10:1 and 3.5:1)—were evaluated on similarity and adjective scales, as reported by Schiffman, Crofton, and Beeker (1985). The similarity data suggest drinks containing sucrose and aspartame cannot be discriminated from one another in either a lemon–lime or cola medium in this experimental design. Sucrose and aspartame were also statistically equivalent on every adjective scale for both lemon-lime and cola drinks. On both similarity judgments– and adjective scales, acesulfame K and sodium saccharin were most different from sucrose. The calcium cyclamate/sodium saccharin blends tended to be less similar than aspartame but not as different from sucrose as the acesulfame K– or sodium saccharin–sweetened beverages.

Measurements of apparent specific volume (ASV), an index of taste modality, i.e., salty, sweet, bitter, and sour, and even of taste acceptability for a series of alternative sweeteners (cyclamates, sulfamates, saccharins, acesulfames, and anilinomethanesulfonates) have been made (Birch et al. 2004). Taste data have been obtained for many of the new compounds, unless the toxicity of the associated metals precluded this. Apparent molar volume (AMV), isentropic specific (IASC), and isentropic molar (IAMC) compressibilities were also measured. Sixteen metallic cyclamates cyc-$C_6H_{11}NHSO_3M$ and two phenylsulfamates $ArNHSO_3Na$, namely, 3,5-dimethyl- and

3,4-dimethoxyphenylsulfamates have been examined. When the ASVs for these are combined with those for 15 aliphatic, aromatic, and alicyclic sulfamates from a previous study, many of the values are seen to fall into the region that was previously identified as the "sweet area," i.e., the ASVs lay between 0.0.5 and 0.0.7 (a few sweet compounds fall below this range, and it is suggested that it could be extended slightly to accommodate these). Interestingly, the anilinomethanesulfonates, $ArNHCH_2SO_3Na$ ($Ar=C_6H_5$-, 3-MeC_6H_4- and 3-ClC_6H_4-) lie clearly in the sweet region, but only one of them shows slight sweetness, showing the molecular structural change made (compared with the parent sulfamate–$NHSO_3^-$) cannot be accommodated at the receptor site.

9.19.2 Electronic Nose, Electronic Tongue, and Array-Based Sensing

Detection of chemically diverse analytes is done by producing specificity not from any single sensor but as a unique composite response for each analyte. Such cross-reactive sensor arrays mimic the mammalian olfactory and gustatory systems and are a widely used approach in electronic nose (Gardner and Bartlett 1999) and tongue (Anand et al. 2007; Toko 2000) technologies.

A disposable, low-cost colorimetric sensor array has been described by Musto, Lim, and Suslick (2009) and created by pin-printing onto a hydrophilic membrane 16 chemically responsive nanoporous pigments that comprised indicators immobilized in an organically modified silane (ormosil). The array has been used to detect and identify 14 different natural and artificial sweeteners at millimolar concentrations, as well as commonly used individual-serving sweetener packets. The array has shown excellent reproducibility and long shelf life and has been optimized to work in the biological pH regime.

Suslick et al. (2007) have developed an alternative optoelectronic approach using simple colorimetric sensor arrays for the detection and identification of a wide range of analytes (Suslick et al. 2007; Zhang, Bailey, and Suslick 2006; Zhang and Suslick 2007). They have recently reported the use of nanoporous pigment arrays for the discrimination of several carbohydrates (Lim et al. 2008).

9.20 MULTIVARIATE ANALYSIS/CHEMOMETRICS

Honey can be adulterated with sweeteners, such as cane sugar, beet sugar, and corn syrup. Different analytical techniques have been employed to detect the adulteration of honey, such as isotopic (Padovan et al. 2003; Cabañero, Recio, and Rupérez 2006), chromatographic (Cordella et al. 2003b, 2005; Morales, Corzo, and Sanz 2008), and thermal analysis of DSC (Cordella et al. 2003a) and NMR (Cotte et al. 2007; Doner and Philips 1981). HFCS has been detected in apple juice by $^{13}C/^{12}C$ isotopic ratio. However, these methods are time consuming, destructive, and, sometimes, expensive. Moreover, vibrational spectroscopic methods such as middle-infrared (Gallardo-Velazquez et al. 2009; Kelly, Petisco, and Downey 2006a, 2006b; Sivakesava and Irudayaraj 2002; Irudayaraj, Xu, and Tewari 2003; Bertelli et al. 2007) and NIR spectroscopy (Kelly, Petisco, and Downey 2006a; Toher, Downey, and Murphy 2007) have previously been applied to authenticate honey.

Zhu et al. (2010) used NIR spectroscopy combined with chemometric methods to detect the adulteration of honey samples. The sample set contained 135 spectra of authentic ($n = 68$) and adulterated ($n = 67$) honey samples. Spectral data were compressed using wavelet transformation (WT) and PCA, respectively. They reported five classification modeling methods, including least square support vector machine (LS-SVM), support vector machine (SVM), backpropagation artificial neural network (BP-ANN), linear discriminant analysis (LDA), and K-nearest neighbors (KNN), which were adopted to correctly classify pure and adulterated honey samples. WT proved more effective than PCA as a means for variable selection. The best classification models were achieved with LS-SVM. A total accuracy of 95.1% and the area under the receiver operating characteristic curves of 0.952 for a test set were obtained by LS-SVM. The results showed that WT-LS-SVM can be used as a rapid screening technique for the detection of this type of honey adulteration with good accuracy and better generalization.

NIRS coupled with appropriate chemometric methods seems to be an efficient and rapid technique to discriminate between authentic honeys and honeys adulterated by fructose–glucose mixtures.

9.21 ROLE OF VISCOSITY AND TEXTURE IN SWEETNESS

Most of the experimental data available about the influence of viscosity on sweetness deal with sweetener solutions thickened with different hydrocolloids. Generally, in these solutions, increasing viscosity leads to a decrease in sweetness (Cook et al. 2002), although some authors have reported the opposite effect: in some cases, increasing viscosity may enhance sweetness (Kanemaru, Harada, and Kasahara 2002).

The big differences in the composition, structure, and rheological behavior of the diversity of food matrices and the well-known variations in the mastication process among individuals make the interpretation of the perceived differences in sweetness highly difficult (Wilson and Brown 1997). In gelled systems (food or model) sweetened with one compound at a certain concentration, the higher the hydrocolloid concentration, the lower the sweetness intensity (Wilson and Brown 1997; Costell, Durlan, and Peyrolon 2000).

Morris (1995) has concluded that the sensory response also depended on some textural characteristics, such as breaking strain or brittleness. Other gel characteristics, such as elasticity, cohesiveness, or melting point, may also play a role (Clark 2002).

Bayarri, Duran, and Costell (2003) analyzed the influence of the type and concentration of two hydrocolloids, k-carrageenan and gellan gum, and two sweeteners, sucrose and aspartame, on the gel resistance to compression, sweetener diffusion, and intensity of the gel sweetness and studied the relationships between the gels' physical properties and their perceived sweetness.

The addition of sucrose produced a bigger increase in gel strength at the higher hydrocolloid concentration. The main effect detected on the sweeteners' diffusion constant was the higher value observed in low concentration (3 g L^{-1}) k-carrageenan gels. Gellan gels were perceived sweeter than k-carrageenan gels. Variations in sweetener concentration, true rupture strain, and deformability modulus values explained 93% of the variability in sweetness for gels with sucrose and 94% for gels with aspartame.

9.22 ADULTERATION

The application of carbon stable isotope ratio analysis has been very useful in detecting the addition of corn and cane sugars to apple juice (Doner and Phillips 1981). When sweeteners are used to adulterate, organic acids will also have to be added to obtain the specified sugar-to-acid ratio.

The development of enzymic methods for the analysis of L-malic acid has been a breakthrough for detecting the addition of synthetic malic acid (Evans, Van Soestbergen, and Ristow 1983).

Sharkasi, Bendel, and Swanson (1982) measured deviations in the sorbitol-to-sucrose and sorbitol-to-total sugar ratios for the adulteration of apple juice.

Sugar, nonvolatile acid, minerals, and UV spectral profiles were determined by Pilando and Wrolstad (1992) for seven commercial fruit juice concentrates (hard pear, soft pear, white grape, pineapple, prune, fig, and raisin) and three sweeteners (inverted beet, inverted cane, and HFCS).

Inverted cane and inverted beet have a similar profile, consisting of 40%–50% sucrose, with the remainder being equal proportions of glucose and fructose. HFCS contains 80% fructose and 20% glucose.

The absence of a maltose peak (retention time = 7.97 min) reflects the advances in corn syrup technology, with the complete hydrolysis of starch to monosaccharides taking place and/or maltose being removed in the ion-exchange resin refining process. This is significant, as maltose levels have been used as an indicator for corn syrup addition to products such as honey.

The glucose-to-fructose ratio, along with sorbitol and sucrose content, are useful indices for distinguishing these juice concentrates and sweeteners. If apple juice were adulterated with prune juice concentrate, the high sorbitol content and high glucose-to-fructose ratios of prune (five times higher for prune than apple) should allow for its detection if the level of adulteration is substantial.

Inverted cane, inverted beet, white grape, pineapple, fig, and raisin juice concentrates all have glucose-to-fructose ratios that are two to three times greater than apple. The glucose-to-fructose ratios for HFCS, soft pear, and hard pear, however, are within the range of authentic apple juice (0.21–0.54; Pilando and Wrolstad 1992).

Sugar and nonvolatile acids were quantitated by HPLC. Sugar analyses included glucose, fructose, sucrose, and sorbitol content, and nonvolatile acid determinations included quinic, malic, citric, tartaric, shikimic, and fumaric acids.

L-Malic content was also determined by enzymic procedures. Mineral composition was measured by ICP spectroscopy. Fruit juice concentrates and sweeteners have characteristic compositional profiles that are useful for evaluating juice quality and authenticity.

The sweeteners are low in metal ions, except for the sodium content of inverted beet and cane sugar. The absence of sodium in HFCS supports White's conclusion that the sodium-to-potassium ratio is not reliable for detecting the adulteration of honey with corn syrup (White 1977).

9.23 MECHANICAL PROPERTIES

The physical, flow, and mechanical properties of four common pharmaceutical sweeteners were measured by Mullarney et al. (2003) to assess their relative manufacturability in solid dosage formulations. Sucrose, acesulfame potassium, saccharin sodium, and aspartame were evaluated to determine significant differences in particle shape, size distribution, and true density. Powder flow, cohesivity, and compact mechanical properties such as ductility, elasticity, and tensile strength were measured and found to be noticeably different. Among these sweeteners, sucrose and acesulfame potassium demonstrated excellent flowability and marginal mechanical property performance relative to more than 100 commonly used pharmaceutical excipients evaluated in the authors' laboratory. Saccharin sodium and aspartame demonstrated poor flowability and superior compact strength relative to sucrose and acesulfame, despite their noticeably higher brittleness.

The true densities of the sweeteners tested were between 1.35 g/cm^3 and 1.83 g/cm^3. These small differences in density are not likely to incite significant segregation in typical blends. Powder densification behavior was determined by performing bulk and tapped density measurements. There was no change in ranking by converting from density to solid fraction, (acesulfame > sucrose > saccharin > aspartame).

Mullarney et al. (2003) showed that sucrose and acesulfame can be tapped to a much higher solid fraction than the smaller, more irregularly shaped particles (saccharin and aspartame).

The dynamic indentation hardness (H_0) is defined as the pressure (force/area) required to plastically deform a compact during a very fast compression operation using a pendulum impact device (Hiestand 2002). The subscript on the indentation hardness symbol, e.g., Ht, indicates the approximate dwell time (t) of the indenter in minutes.

The acesulfame compacts demonstrated a relatively low H_0, and the aspartame compacts demonstrated a relatively high H_0. This indicates acesulfame compacts (which were in the lower 20% of the excipient population) deformed rather easily, suggesting interparticulate bonding surfaces formed relatively easily.

One would typically prefer to select materials with moderate H_0, for example, within the interquartile range ($H_0 \approx$ 100–250 MPa), such as sucrose and saccharin, whose particles deform well under rapid compression.

The quasistatic indentation hardness (H_{10}) of a compact is defined as the pressure required to plastically deform a compact during extremely slow indentation.

The two indentation hardness measurements for each material followed a similar trend: aspartame > saccharin ~= sucrose > acesulfame.

The elastic moduli of the sweeteners were determined by the measurement of the dent recovery during quasistatic indentation hardness testing. These materials are considered to demonstrate moderate elasticity. The experimental results showed that most of the tensile strengths of the sweetener compacts were notably different (aspartame > saccharin > sucrose ~= acesulfame). Because tensile strength is proportional to the tablet crushing strength, this has a significant effect on measurable tablet property performance. The aspartame compacts were approximately 5 times stronger than the sucrose compacts, 7 times stronger than the acesulfame compacts, and 2.5 times stronger than the saccharin compacts (Mullarney et al. 2003).

9.24 PURIFICATION TECHNIQUES

For the application of sweeteners as a dietary supplement or a drug, safe and sustainable purification techniques are needed.

The use of solvents should be avoided as much as possible. Although the sweeteners can be extracted with water, current purification techniques still make use of methanol or ethanol (Carakostas et al. 2008). Kutowy, Zhang, and Kumar (1999) were the first to propose a membrane-based purification process. After the extraction column, their process consists of a microfiltration and an ultrafiltration membrane stage to remove impurities with a higher molecular weight than rebaudioside A and, finally, a nanofiltration membrane stage to remove impurities with a lower molecular weight than stevioside. Zhang, Kumar, and Kutowy (2000) presented some interesting lab-scale results concerning this process configuration; however, product purity and yield were not mentioned. Although the purity can be measured easily based on the sweetener content and total dry weight of a sample, it is very impractical to measure for lab-scale membrane processes, especially when many different membranes and conditions are tested to optimize the process because of the presence of a large amount of permeate, so an accurate reading of the dry weight on a standard balance is taken.

Reis et al. (2009) studied different microfiltration membranes and conditions and proposed to use the UV absorption at 420 nm and 670 nm as specific indicators for the impurity content as the sweeteners only absorb at 210 nm.

Vanneste et al. (2011) prepared polyethersulfone membranes using the diffusion-induced phase inversion technique (Boussu et al. 2006). In this process, a thin layer of the polymer dissolved in an appropriate solvent is cast on a support, and phase separation is induced by a nonsolvent. The most efficient way of inducing the phase inversion is by immersing the polymer solution film in a nonsolvent bath. By changing the preparation factors, an optimized membrane for a specific purpose can be obtained. Three parameters were varied to study the effect on permeability and selectivity: the polymer concentration, composition of the nonsolvent bath (water–isopropanol, IPA, mixture), and a mixture of solvents—dimethylformamide (DMF) or N-methyl pyrrolidone (NMP).

For the plant extract, the best commercial membrane (PW010) had a selectivity and flux similar to the best lab-made membrane (27% PES), but the lab-made membrane was preferred, because it showed a slightly lower retention of the sweeteners, as desired. Starting from an extract purity of 11%, with the overall process (microfiltration, ultrafiltration, and nanofiltration), a purity of 37% and a yield of 30% could be reached (Vanneste et al. 2011).

REFERENCES

Anand, V., M. Kataria, V. Kukkar, V. Saharan, and P. K. Choudhury. 2007. *Drug Discovery Today*, *12*, 257–265.

Association of Official Analytical Chemists (AOAC). 2003. *Official Methods of Analysis*, *979.08.17th Ed.* Arlington: AOAC.

Armenta, S., S. Garrigues, and M. de la Guardia. 2004. Sweeteners determination in tabletop formulations using FT-Raman spectrometry and chemometric analysis. *Anal. Chim. Acta.*, *52*, 149–55.

Assumpcao, M. H. M. T., R. A. Medeiros, A. Madi, and O. Fatibello-Filho. 2008. Development of a biamperometric procedure for the determination of saccharin in dietary products. *Quimica Nova*, *31*, 1743–1746.

Bayarri, S., L. Duran, and E. Costell. 2003. Compression resistance, sweetener's diffusion and sweetness of hydrocolloids gels. *Int. Dairy J.*, *13*, 643–653.

Bergamo, A. B., J. A. F. da Silva, and D. P. de Jesus. 2011. Simultaneous determination of aspartame, cyclamate, saccharin and acesulfame-K in soft drinks and tabletop sweetener formulations by capillary electrophoresis with capacitively coupled contactless conductivity detection. *Food Chem.*, *124*, 1714–1717.

Bertelli, D., M. Plessi, A. G. Sabatini, M. Lolli, and F. Grillenzoni. 2007. Classification of Italian honeys by mid-infrared diffuse reflectance spectroscopy (DRIFTS). *Food Chem.*, *101*(4), 1565–1570.

Biemer, T. A. 1989. Analysis of saccharin, acesulfame-K and sodium cyclamate by high-performance ion chromatography. *J. Chromatogr.*, *463*, 463–468.

Birch, G. G., K. A. Haywood, G. G. Hanniffy, C. M. Coyle, and W. J. Spillane. 2004. Apparent specific volumes and tastes of cyclamates, other sulfamates, saccharins and acesulfame sweeteners. *Food Chem.*, *84*, 429–435.

Boussu, K., B. Van der Bruggen, A. Volodin, C. Van Haesendonck, J. A. Delcour, P. Van der Meeren, C. Vandecasteele. 2006. Characterization of commercial nanofiltration membranes and comparison with self-made polyethersulfone membranes. *Desalination* *191*(1–3), 245–253.

Boyce, M. C. 1999. Simultaneous determination of antioxidants, preservatives and sweeteners permitted as additives in food by mixed micellar electrokinetic chromatography. *J. Chromatogr. A*, *847*, 369–375.

Cabañero, A. I., J. L. Recio, and M. Rupérez. 2006. Liquid chromatography coupled to isotope ratio mass spectrometry: A new perspective on honey adulteration detection. *J. Agric., and Food Chem.*, *54*(26), 9719–9727.

Cantarelli, M. A., R. G. Pellerano, E. J. Marchevsky, and J. M. Camina. 2009. Simultaneous determination of aspartame and acesulfame-K by molecular absorption spectrophotometry using multivariate calibration and validation by high performance liquid chromatography. *Food Chem.*, *115*, 1128–1132.

Carakostas, M. C., L. L. Curry, A. C. Boilea, and D. Brusick. 2008. Overview: The history, technical function and safety of rebaudioside A, a naturally occurring steviol glycoside, for use in food and beverages. *Food and Chem. Toxicol.*, *46*(7), S1–S10.

Casals, I., M. Reixach, J. Amat, M. Fuentes, and L. Serra-Majem. 1996. Quantification of cyclamate and cyclohexylamine in urine samples using high-performance liquid chromatography with trinitrobenzenesulfonic acid pre-column derivatization. *J. Chromatogr. A*, *750*(1–2), 397–402.

Chen, B. H., and S. C. Fu. 1995. Simultaneous determination of preservatives, sweeteners and antioxidants in foods by paired-ion liquid chromatography. *Chromatographia*, *41*, 43–50.

Chen, Q. C., and J. Wang. 2001. Simultaneous determination of artificial sweeteners, preservatives, caffeine, theobromine and theophylline in food and pharmaceutical preparations by ion chromatography. *J. Chromatogr. A*, *937*, 57–64.

Chen, Q. C., S. F. Mou, K. N. Liu, Z. Y. Yang, and Z. M. Ni. 1997. Separation and determination of four artificial sweeteners and citric acid by high-performance anion-exchange chromatography. *J. Chromatogr. A*, *771*(1–2), 135–143.

Choi, M. M. F., M. Y. Hsu, and S. L. Wong. 2000. Determination of cyclamate in low-calorie foods by high-performance liquid chromatography with indirect visible photometry. *Analyst, 125*(1), 217–220.

Clark, R. 2002. Influence of hydrocolloids on flavour release and sensory-instrumental correlations. In: P. A. Williams and G. O. Phillips (eds.), *Gums and Stabilisers for the Food Industry, Vol. 11* (217–225). Cambridge, U.K.: Royal Society of Chemistry.

Clement, A., L. Lagace, and B. Panneton. 2010. Assessment of maple syrup physico-chemistry and typicity by means of fluorescence spectroscopy. *J. Food Eng., 97*, 17–23.

Coca Cola Standard methods manual. 2003. TCCQS Auditors toolkit. The Coca-Cola Quality system. Coca Cola Europe, Eurasia and Middle East. Confidential report.

Codex Alimentarius. 2011. Food Chemical Codex. www.codexalimentarius.net/download/standards/338/CXS_212e_u.pdf.

Cook, D. J., T. A. Hollowood, R. S. T. Linforth, and A. J. Taylor. 2002. Perception of taste intensity in solutions of random-coil polysaccharides above and below c*. *Food Qual. Prefer., 13*, 473–480.

Cordella, C., J. P. Faucon, D. Cabrol-Bass, and N. Sbirrazzuoli. 2003a. Application of DSC as a tool for honey floral species characterization and adulteration detection. *J. Thermal Anal. Cal., 71*(1), 279–290.

Cordella, C., J. S. L. T. Milito, M. C. Clement, P. Drajnudel, and D. Cabrol-Bass. 2005. Detection and quantification of honey adulteration via direct incorporation of sugar syrups or bee feeding: Preliminary study using high-performance anion exchange chromatography with pulsed amperometric detection (HPAEC-PAD) and chemometrics. *Analytica Chim. Acta, 531*(2), 239–248.

Cordella, C. B. Y., J. S. L. T. Milito, M. C. Clement, and D. Cabrol-Bass. 2003b. Honey characterization and adulteration detection by pattern recognition applied on HPAEC-PAD profiles—Part 1: Honey floral species characterization. *J. Agric., Food Chem., 51*(11), 3234–3242.

Costell, E., L. Durlan, and M. Peyrolon. 2000. Influence of texture and type of hydrocolloid on perception of basic tastes in carrageenan and gellan gels. *Food Sci. Technol. Int., 6*(6), 495–499.

Cotte, J. F., H. Casabianca, J. Lheritier, C. Perrucchietti, C. Sanglar, H. Waton et al. 2007. Study and validity of C-13 stable carbon isotopic ratio analysis by mass spectrometry and H-2 site-specific natural isotopic fractionation by nuclear magnetic resonance isotopic measurements to characterize and control the authenticity of honey. *Analytica Chim. Acta, 582*(1), 125–136.

Corn Refiners Association, 2011. Nutritive sweeteners from corn. http://www.corn.org/publications/technical-booklets/. Accessed October 16, 2011. CRA: Corn Refiners Association.

Dacome, A. S., C. C. da Silva, C. E. M. da Costa, J. D. Fontana, J. Adelmann, and S. C. da Costa. 2005. Sweet diterpenic glycosides balance of a new cultivar of Stevia rebaudiana (Bert.) Bertoni: Isolation and quantitative distribution by chromatographic, spectroscopic, and electrophoretic methods. *Proc. Biochem., 40*(11), 3587–3594.

Demiralay, E., Ö. G. Çubuk, and Z. Guzel-Seydim. 2006. Isocratic separation of some food additives by reversed phase liquid chromatography. *Chromatographia, 63*, 91–96.

Di Pietra, A. M., V. Cavrini, D. Bonazzi, and L. Benfenati. 1990. HPLC analysis of aspartame and saccharin in pharmaceutical and dietary formulations. *Chromatographia, 30*(3/4), 215–219.

Dionex Corporation. 2004. Dionex application note 159, 2004. Sunnyvale, CA, USA.

Doner, L. W., and J. G. Phillips. 1981. Detection of high-fructose corn syrup in apple juice by mass spectrometric $^{13}C/^{12}C$ analysis: Collaborative study. *J. Assoc. Off. Anal. Chem., 64*, 85–90.

Escarpa, A., M. C. Gonzalez, M. A. L. Gil, A. G. Crevillén, M. Hervas, and M. García. 2008. Microchips for CE: Breakthroughs in real-world food analysis. *Electrophoresis, 29*, 4852–4861.

Evans, R. H., A. W. Van Soestbergen, and K. A. Ristow. 1983. Evaluation of apple juice authenticity by organic acid analysis. *J. Assoc. Off. Anal. Chem., 66*, 1517–1519.

Fatibello-Fihlo, C., and C. Aniceto. 1997. Potentiometric determination of saccharin in dietary products using a coated-carbon rod ion-selective electrode. *Anal. Lett., 30*(9), 1653–1666.

Fatibello-Fihlo, O., L. H. Marcolino, and A. V. Pereira. 1999. Solid-phase reactor with copper (II) phosphate for flow-injection spectrophotometric determination of aspartame in tabletop sweeteners. *Analytica Chimica Acta, 384*, 167–174.

Food Chemicals Codex. 2011. The United States Pharmacopeial Convention. http://www.usp.org/USPNF/notices/sucrose.html. Accessed October 2011. FCC: United States.

Ferrer, I., and E. M. Thurman. 2010. Analysis of sucralose and other sweeteners in water and beverage samples by liquid chromatography/time-of-flight mass spectrometry. *J. Chromatogr. A, 1217*, 4127–4134.

Fracassi da Silva, J. A., and C. L. do Lago. 1998. An oscillometric detector for capillary electrophoresis. *Anal. Chem.*, *70*, 4339–4343.

Fracassi da Silva, J. A., N. Guzman, and C. L. do Lago. 2002. Contactless conductivity detection for capillary electrophoresis: Hardware improvements and optimization of the input-signal amplitude and frequency. *J. Chromatogr. A*, *942*, 249–258.

Frazier, R. A., E. L. Inns, N. Dossi, J. M. Ames, and H. E. Nursten. 2000. Development of a capillary electrophoresis method for the simultaneous analysis of artificial sweeteners, preservatives and colors in soft drink. *J. Chromatogr. A*, *876*, 213–220.

Gallardo-Velazquez, T., G. Osorio-Revilla, M. Zupiga-de Loa, and Y. Rivera-Espinoza. 2009. Application of FTIR–HATR spectroscopy and multivariate analysis to the quantification of adulterants in Mexican honeys. *Food Res. Int.*, *42*(3), 313–318.

Gardana, C., M. Scaglianti, and P. Simonetti. 2010. Evaluation of steviol and its glycosides in Stevia rebaudiana leaves and commercial sweetener by ultra-high-performance liquid chromatography-mass spectrometry. *J. Chromatogr. A*, *1217*, 1463–1470.

Gardner, J. W., and P. N. Bartlett. 1999. *Electronic Noses: Principles and Applications*. New York: Oxford University Press.

Hannisdal, A. 1992. Analysis of acesulfame-K, saccharin and preservatives in beverages and jams by HPLC. *Z. Lebensm.Unters Forsch.*, *194*, 517–519.

Hannisdal, A., and H. Schreder Knut. 1993. Differential pulse polarographic determination of the artificial sweeteners acesulfame-K and saccharin in beverages. *Electroanal.*, *5*,183–185.

Herrmann, A., E. M. J. Damawandi, and J. Wagmann. 1983. Determination of cyclamate by high-performance liquid chromatography with indirect photometry. *J. Chromatogr. C*, *280*, 85–90.

Herrmannova, M., L. Krivankova, M. Bartos, and K. Vytras. 2006. Direct simultaneous determination of eight sweeteners in foods by capillary isotachophoresis. *J. Sep. Sci.*, *29*, 1132–1137.

Hiestand, E. N. 2002. Mechanics and Physical Principles for Powders and Compacts. West Lafayette, IN: SSCI, Inc.

Horie, M., F. Ishikawa, M. Oishi, T. Shindo, A. Yasui, and K. Ito. 2007. Rapid determination of cyclamate in foods by solid-phase extraction and capillary electrophoresis. *J. Chromatogr. A*, *1154*, 423–428.

ICUMSA: International Commission for the Uniform Methods of Sugar Analysis. (1994). *ICUMSA Methods Book*. Port Talbot, UK: ICUMSA Publications.

International Commission for Uniform Methods of Sugar Analysis (ICUMSA). (1998). *ICUMSA Methods Book Supplement*. Port Talbot, UK: ICUMSA Publications.

International Sweeteners Association (ISA). 2011. Sweeteners in all confidence. Sweeteners_confidence_ISArelativesweetnessADIEN.pdf. Accessed October 2011.

Irudayaraj, J., F. Xu, and J. Tewari. 2003. Rapid determination of invert cane sugar adulteration in honey using FTIR spectroscopy and multivariate analysis. *J. Food Sci.*, *68*(6), 2040–2045.

ISBT: International Society of Beverage Technologists.

Joint FAO/WHO Expert Committee on Food Additives. 2006. Combined compedium of food additive specifications. Analytical methods, test procedures and laboratory solutions used by and referenced in the food additive specifications. FAO JECFA Monograph 1. ISSN 1817-7077.

Jiménez, G. J. F., M. C. Valencia, and L. F. Capitán-Vallvey. 2009. Intense sweetener mixture resolution by flow injection method with on-line monolithic element. *J. Liq. Chromatogr. R.T.*, *32*, 1152–1168.

Jiménez, G. J. F., M. C. Valencia, and L. F. Capitán-Vallvey. 2006. Improved multianalyte determination of the intense sweeteners aspartame and acesulfame-K with a solid sensing zone implemented in an FIA scheme. *Anal. Lett.*, *39*, 1333–1347.

Kanemaru, N., S. Harada, and Y. Kasahara. 2002. Enhancement of sucrose sweetness with soluble starch in humans. *Chem. Sens.*, *27*, 67–72.

Karoui, R., and J. De Baerdemaeker. 2007. A review of the analytical methods coupled with chemometric tools for the determination of the quality and identity of dairy products. *Food Chem.*, *102*, 621–640.

Kelly, J. D., C. Petisco, and G. Downey. 2006a. Application of Fourier transform midinfrared spectroscopy to the discrimination between Irish artisanal honey and such honey adulterated with various sugar syrups. *J. Agric. Food Chem.*, *54*(17), 6166–6171.

Kelly, J. D., C. Petisco, and G. Downey. 2006b. Potential of near infrared transflectance spectroscopy to detect adulteration of Irish honey by beet invert syrup and high fructose corn syrup. *J. Near Infrared Spectrosc.*, *14*(2), 139–146.

Kibbe, A. H. 2000. *Handbook of Pharmaceutical Excipients, 3rd ed.* London: American Pharmaceutical Association and Pharmaceutical Press.

Kobayashi, C., M. Nakazato, Y. Yamajima, I. Ohno, M. Kawano, and K. Yasuda. 2001. Determination of sucralose in foods by HPLC. *J. Food Hyg. Soc. Jpn.*, *42*, 139–143.

Koyama, M., K. Yoshida, N. Uchibori et al. 2005. Analysis of nine kinds of sweeteners in foods by LC/MS. *J. Food Hyg. Soc. Jpn.*, *46*, 72–78.

Kulmyrzaev, A. A., R. Karoui, J. De Baerdemaeker, and E. Dufour. 2007. Infrared and fluorescence spectroscopic techniques for the determination of nutritional constituents in foods. *Int. J. Food Prop.*, *10*, 299–320.

Kutowy, O., S. Q. Zhang, and A. Kumar. 1999. Extraction of sweet compounds from *Stevia rebaudiana* Bertoni. *United States Patent*, Patent number: 5972,120.

Lawrence, J. P., and C. F. Charbonneau. 1988. Determination of seven artificial sweeteners in diet food preparations by reverse-phase liquid chromatography with absorbance detection. *J. Assoc. Off. Anal. Chem.*, *71*(5), 934–937.

Leth, T., N. Fabricius, and S. Fagt. 2007. Estimated intake of intense sweeteners from nonalcoholic beverages in Denmark. *Food Additiv. Contam. A*, *24*(3), 227–235.

Lim, S. H., C. J. Musto, E. Park, W. Zhong, and K. S. Suslick. 2008. *Org. Lett.*, *10*, 4405–4408.

Lino, C. M., I. M. Costa, and A. Pena. 2008. Estimated intake of the sweeteners, acesulfame-K and aspartame, from soft drinks, soft drinks based on mineral waters and nectars for a group of Portuguese teenage students. *Food Additiv. Contam. A*, *25*(11), 1291–1296.

Liquid Sucrose, Quality Guidelines and Analytical Procedures, High Fructose Syrup 42 and 55, Quality Guidelines and Analytical Procedures, Granular Sucrose, Quality Guidelines and Analytical Procedures.

Loos, R., B. M. Gawlik, K. Boettcher, G. Locoro, S. Contini, and G. Bidoglio. 2009. Sucralose screening in European surface waters using a solid-phase extraction-liquid chromatography-triple quadrupole mass spectrometry method. *J. Chromatogr. A*, *1216*(7), 1126–1131.

Luong, J. H. T., P. Bouvrette, and K. B. Male. 1997. Development and applications of biosensors in food analysis. *Trends Biotech.*, *15*, 369–377.

Mazurek, S., and R. Szostak. 2011. Quantification of aspartame in commercial sweeteners by FT-Raman spectroscopy. *Food Chem.*, *125*, 1051–1057.

McCourt, J., J. Stroka, and E. Anklam. 2005. Experimental design–based development and single laboratory validation of a capillary zone electrophoresis method for the determination of the artificial sweetener sucralose in food matrices. *Anal. Bioanal. Chem.*, *382*(5), 1269–1278.

Medeiros, R. A., A. E. de Carvalho, R. C. Rocha-Filho, and O. Fatibello-Filho. 2008. Simultaneous square-wave voltametric determination of aspartame and cyclamate using a boron-doped diamond electrode. *Talanta*, *76*, 685–689.

Meyer, S., and W. E. III Riha. 2002. Optimizing sweetener blends for low calorie beverages. *Food Technol.*, *56*, 42–45.

Montijano, H., F. Borrego, I. Canales, and F. A. Tomas-Barberan. 1997. Validated high-performance liquid chromatographic method for quantitation of neohesperidine dihydrochalcone in foodstuffs. *J. Chromatogr. A*, *758*(1), 163–166.

Morales, V., N. Corzo, and M. L. Sanz. 2008. HPAEC-PAD oligosaccharide analysis to detect adulterations of honey with sugar syrups. *Food Chem.*, *107*(2), 922–928.

Morlock, G. E., and S. Prabha. 2007. Analysis and stability of sucralose in a milk-based confection by a simple planar chromatographic method. *J. Agric. Food Chem.*, *55*, 7217–7223.

Morris, E. R. 1995. Polysaccharide rheology and in-mouth perception. In: A. M. Stephen (ed.), *Food Polysaccharides and Their Applications* (517–546). New York: Marcel Dekker, Inc.

Mullarney, M. P., B. C. Hancock, G.T. Carlson, D. D. Ladipo, and B. A. Langdon. 2003. The powder flow and compact mechanical properties of sucrose and three high-intensity sweeteners used in chewable tablets. *Int. J. Pharmaceutics*, *257*, 227–236.

Musto, C. J., S. H. Lim, and K. S. Suslick. 2009. Colorimetric detection and identification of natural and artificial sweeteners. *Anal. Chem.*, *81*, 6526–6533.

Ni, Y., W. Xiao., and S. Kokot. 2009. Differential kinetic spectrophotometric method for determination of three sulphanilamide artificial sweeteners with the aid of chemometrics. *Food Chem.*, *113*, 1339–1345.

Nikolelis, D. P., T. Hianik, and U. J. Krull. 1999. Biosensors based on thin lipid films and liposomes. *Electroanalysis*, *11*(1), 7–15.

Nikolelis, D. P., and S. Pantoulias. 2000. A minisensor for the rapid screening of acesulfame-K, cyclamate, and saccharin based on surface stabilized bilayer lipid membranes. *Electroanalysis*, *12*(10), 786–790.

Nikolelis, D. P., S. Pantoulias, U. J. Krull, and J. Zeng. 2001. Electrochemical transduction of the interactions of the sweeteners acesulfame-K, saccharin and cyclamate with bilayer lipid membranes (BLMs). *Electrochim. Acta 46,* 1025–1031.

O'Brien-Nabors, L. 2001. *Alternative Sweeteners, 3rd Ed.* New York: Dekker.

O'Brien-Nabors, L. 2002. Sweet choices: Sugar replacements for foods and beverages. *Food Technol.*, *56*, 28–45.

Official Journal of the European Communities. 2002. COUNCIL DIRECTIVE 2001/111/EC relating to certain sugars intended for human consumption. December 20, 2011, L10/53–57.

Official Methods of Analysis, Food Compositions; Additives, Natural Contaminants, 15th ed; AOAC: Arlington, VA, 1990, Vol. 2.; AOAC Official Method 979.08: Benzoate, caffeine, saccharine in carbonated beverages.

Padovan, G. J., D. De Jong, L. P. Rodrigues, and J. S. Marchin. 2003. Detection of adulteration of commercial honey samples by the 13C/12C isotopic ratio. *Food Chem.*, *82*(4), 633–636.

Paradkar, M. M., S. Sivakesava, and J. Irudayaraj. 2003. Discrimination and classification of adulterants in maple syrup with the use of infrared spectroscopic techniques. *J. Sci. Food Agric.*, *83*, 714–721.

Pedro, N. A. R., E. Oliveira, and S. Cadore. 2006. Study of mineral content of chocolate flavoured beverages. *Food Chem.*, *95*, 94–100.

Peica, N. 2009. Identification and characterization of the E951 artificial food sweetener by vibrational spectroscopy and theoretical modeling. *J. Raman Spectrosc.*, *40*, 2144–2154.

Pesek, J. J., and M. T. Matyska. 1997. Determination of aspartame by high-performance capillary electrophoresis. *J. Chromatogr. A*, *781*, 423–428.

Pilando, L. S., and R. E. Wrolstad. 1992. Compositional profiles of fruit juice concentrates and sweeteners. *Food Chem.*, *44*, 19–27.

Reis, M. H. M., F. V. Da Silva, C. M. G. Andrade, S. L. Rezende, M. R. W. Maciel, and R. Bergamasco. 2009. Clarification and purification of aqueous stevia extract using membrane separation process. *J. Food Process Eng.*, *32*(3), 338–354.

Richter, E. M., D. P. de Jesus, R. A. A. Munoz, C. L. do Lago, and L. Angnes. 2005. Determination of anions, cations, and sugars in coconut water by capillary electrophoresis. *J. Brazilian Chem. Soc.*, *16*, 1134–1139.

Ruoff, K., R. Karoui, E. Dufour, W. Luginbuhl, J. O. Bosset, S. Bogdanov, and R. Amado. 2005. Authentication of the botanical origin of honey by front-face fluorescence spectroscopy: A preliminary study. *J. Agric. Food Chem.*, *53*, 1343–1347.

Ruoff, K., W. Luginbuehl, S. Bogdanov, J. O. Bosset, B. Estermann, T. Ziolko, S. Kheradmandan, and R. Amado. 2007. Quantitative determination of physical and chemical measurands in honey by near-infrared spectrometry. *Eur. Food Res. Technol.*, *225*, 415–423.

Santini, A. O., S. C. Lemos, H. R. Pezza et al. 2008. Development of a potentiometric sensor for the determination of saccharin in instant tea powders, diet soft drinks and strawberry dietetic jam. *Microchem. J.*, *90*, 124–128.

Scheurer, M., H. J. Brauch, and F. T. Lange. 2009. Analysis and occurrence of seven artificial sweeteners in German waste water and surface water and in soil aquifer treatment (SAT). *Anal. Bioanal. Chem. 394* (6): 1585-1594.

Scheurer, M., F. R. Storck, H.-J. Brauch, and F. T. Lange. 2010. Performance of conventional multibarrier drinking water treatment plants for the removal of four artificial sweeteners. *Water Res.*, *44*, 3573–3584.

Schiffman, S. S., V. A. Crofton, and T. G. Beeker. 1985. Sensory evaluation of soft drinks with various sweeteners. *Physiol. Behav.*, *34*(3), 369–377.

Schnierle, P., T. Kappes, and P. C. Hauser. 1998. Capillary electrophoresis determination of different classes of organic ions by potentiometric detection with coated-wire ion-selective electrodes. *Anal. Chem.*, *70*, 3585–3589.

Self, R. 2005. *Extraction of Organic Analytes from Foods: A Manual of Methods*. London: The Royal Society of Chemistry.

Sharkasi, T. Y., R. B. Bendel, and B. G. Swanson. 1982. Dilution and solids adulteration of apple juice. *J. Food Qual.*, *5*, 59–72.

Shim, J. Y., I. K. Cho, H. K. Khurana et al. 2008. Attenuated total reflectance–Fourier transform infrared spectroscopy coupled with multivariate analysis for measurement of acesulfame-K in diet foods. *J. Food Sci.*, *73*(5), C426–C431.

Sivakesava, S., and J. Irudayaraj. 2002. Classification of simple and complex sugar adulterants in honey by mid-infrared spectroscopy. *Int. J. Food Sci. Technol.*, *37*(4), 351–360.

Sjoberg, A. M., and T. A. Alanko. 1987a. Spectrophotometric determination of cyclamate in foods: NMKL collaborative study. *J. Assoc. Off. Anal. Chem.*, *70*(3), 588–590.

Sjoberg, A. M., and T. A. Alanko. 1987b. Liquid chromatographic determination of saccharin in beverages and sweets: NMKL Collaborative Study. *J. Assoc. Off. Anal. Chem.*, *70*(1), 58–60.

Sousa, R. A., N. Baccan, and S. Cadore. 2006. Analysis of liquid stevioside and cyclamate–saccharine dietetic sweeteners by inductively coupled plasma optical emission spectrometry without sample treatment. *J. Braz. Chem. Soc.*, *17*(7), 1393–1399.

Sousa, R. A., N. Baccan, and S. Cadore. 2011. Determination of elemental content in solid sweeteners by slurry sampling and ICP OES. *Food Chem.*, *124*, 1264–1267.

Sousa, R. A., A. S. Ribeiro, M. A. Vieira, A. J. Curtius, N. Baccan, and S. Cadore. 2007. Determination of trace elements in liquid aspartame sweeteners by ICP OES and ICP-MS following acid digestion. *Microchim. Acta*, *159*, 241–246.

Spangenberg, B., J. Stroka, I. Arranz, and E. Anklam. 2003. A simple and reliable HPTLC method for the quantification of the intense sweetener Sucralose®. *J. Liquid Chromatogr. Rel. Technol.*, *26*(16), 2729–2739.

Steinmetz, W. E., and A. Lin. 2009. NMR studies of the conformation of the natural sweetener rebaudioside A. *Carbohydrate Res.*, *344*, 2533–2538.

Strachan, C. J., T. Rades, K. C. Gordon, and J. Rantanen. 2007. Raman spectroscopy for quantitative analysis of pharmaceutical solids. *J. Pharm. Pharmacol.*, *59*, 179–192.

Stroka, J., N. Dossi, and E. Anklam. 2003. Determination of the artificial sweetener Sucralose® by capillary electrophoresis. *Food Add. Contam.*, *20*(6), 524–527.

Suslick, K. S., D. P. Bailey, C. K. Ingison, M. Janzen, M. E. Kosal, W. B. McNamara, III, N. A. Rakow et al 2007. Seeing smells: Development of an optoelectronic nose. *Quimica Nova*, *30*(3), 677–681.

Thompson, C. O., V. C. Trenerry, and B. Kemmery. 1995a. Micellar electrokinetic capillary chromatographic determination of artificial sweeteners in low-joule soft drinks and other foods. *J. Chromatogr. A*, *694*, 507–514.

Thompson, C. O., V. C. Trenerry, and B. Kemmery. 1995b. Determination of cyclamate in low-joule foods by capillary zone electrophoresis with indirect ultraviolet detection. *J. Chromatogr. A*, *704*(1), 203–210.

Toher, D., G. Downey, and T. B. Murphy. 2007. A comparison of model-based and regression classification techniques applied to near-infrared spectroscopic data in food authentication studies. *Chemom. Intell. Lab. Syst.*, *89*(2), 102–115.

Toko, K. 2000. *Biomimetric Sensor Technology*. Cambridge, UK: Cambridge University Press.

Tunick, M. H. 2005. *Methods of Analysis of Food Components and Additives*. Semih Ötleş (Ed.), 1–14. Boca Raton, FL: CRC Press Taylor & Francis Group.

Vanneste, J., A. Sotto, C. M. Courtin, V. Van Craeyveld, K. Bernaerts, J. Van Impe, J. Vandeur, S. Taes, and B. Van der Bruggen. 2011. Application of tailor-made membranes in a multistage process for the purification of sweeteners from *Stevia rebaudiana*. *J. Food Eng.*, *103*, 285–293.

Verzella, G., G. Bagnasco, and A. Mangia. 1985. Ion-pair high-performance liquid chromatographic analysis of aspartame and related products. *J. Chromatogr.*, 349, 83–89.

Wasik, A., J. McCourt, and M. Buchgraber. 2007. Simultaneous determination of nine intense sweeteners in foodstuffs by high performance liquid chromatography and evaporative light scattering detection: Development and single-laboratory validation. *J. Chromatogr. A*, *1157*, 187–196.

White, J. W. Jr. 1977. Sodium-potassium ratios in honey and in high-fructose corn syrup. *Bee World*, *58*, 31–35.

Wilson, C. E., and W. E. Brown. 1997. Influence of food matrix structure and oral breakdown during mastication on temporal perception of flavor. *J. Sens. Stud.*, *21*, 69–86.

Wolwer-Rieck, U., W. Tomberg, and A. Wawrzun. 2010. Investigations on the stability of stevioside and rebaudioside A in soft drinks. *J. Agric. Food Chem.*, *58*, 12,216–12,220.

Yang, D.-J., and B. Chen. 2009. Simultaneous determination of nonnutritive sweeteners in foods by HPLC/ESI-MS. *J. Agric. Food Chem.*, *57*, 3022–3027.

Yebra, M. C., M. Gallego, and M. Valcarcel. 1995. Precipitation flow–injection method for the determination of saccharin in mixtures of sweeteners. *Anal. Chim. Acta*, *308*, 275–280.

Yebra-Biurrun, M. C. 2005. *Sweeteners*, 562–572.

Yebra-Biurrun, M. C., S. Cancela-Perez, and A. Moreno-Cid-Barinaga. 2005. Coupling continuous ultrasound-assisted extraction, preconcentration and flame atomic absorption spectrometric detection for the determination of cadmium and lead in mussel samples. *Anal. Chim. Acta*, *533*, 51–56.

Yebra-Biurrun, M. C., A. Moreno-Cid, and S. Cancela-Perez. 2005. Fast online ultrasound-assisted extraction coupled to a flow injection-atomic absorption spectrometric system for zinc determination in meat samples. *Talanta*, *66*, 691–695.

Yebra-Biurrun, M. C. 2000. Flow injection determination of artificial sweeteners: A review. *Food Additiv. Contam.*, *17*, 733–738.

Yoda, S. K., M. O. M. Marques, A. J. Petenate, and M. A. A. Meireles. 2003. Supercritical fluid extraction from *Stevia rebaudiana* Bertoni using CO_2 and CO_2 + water: extraction kinetics and identification of extracted components. *J. Food Eng.*, *57*, 125–134.

Zemann, A. J., E. Schnell, D. Volgger, and G. K. Bonn. 1998. Contactless conductivity detection for capillary electrophoresis. *Anal. Chem.*, *70*, 563–567.

Zhang, C., D. P. Bailey, and K. S. Suslick. 2006. Colorimetric sensor arrays for the analysis of beers: A feasibility study. *J. Agric. Food Chem.*, *54*(14), 4925–4931.

Zhang, C., and K. S. Suslick. 2007. Colorimetric sensor array for soft drink analysis. *J. Agric. Food Chem.*, *55*(2), 237–242.

Zhang, S. Q., A. Kumar, and O. Kutowy. 2000. Membrane-based separation scheme for processing sweeteners from stevia leaves. *Food Res. Int.*, *33*(7), 617–620.

Zhao, L., and B. J. Tepper. 2007. Perception and acceptance of selected high-intensity sweeteners and blends in model soft drinks by propylthiouracil (PROP) non-tasters and super-tasters. *Food Qual. Prefer.*, *18*, 531–540.

Zhu, X., S. Li, Y. Shan, Z. Zhang, G. Li, D. Su, and F. Liu. 2010. Detection of adulterants such as sweeteners materials in honey using near-infrared spectroscopy and chemometrics. *J. Food Eng.*, *101*, 92–97.

Zhu, Y., Y. Y. Guo, M. L. Ye, and F. S. James. 2005. Separation and simultaneous determination of four artificial sweeteners in food and beverages by ion chromatography. *J. Chromatogr. A*, *1058*, 143–146.

Zygler, A., A. Wasik, and J. Jacek Namiesnik. 2010. Retention behavior of some high-intensity sweeteners on different SPE sorbents. *Talanta*, *82*, 1742–1748.

CHAPTER 10

EU, U.S., and Third World Country Regulations and Japanese Legislation

Theodoros Varzakas

CONTENTS

10.1 INTRODUCTION

Sweeteners form an important class of food additives that are used in an increasingly wide range of food products and beverages.

Directive 94/35/EC (European Commission 1994), as amended by Directives 96/83/EC (European Commission 1997), 2003/115/EC (European Commission 2004b), and 2006/52/EC (European Commission 2006b), specifically deals with food additives used to impart a sweet taste to foodstuffs.

These directives stipulate which sweeteners may be placed on the market for consumer sale or for use in the production of foodstuffs. Prior to their authorization, sweeteners are evaluated for their safety by the European Food Safety Authority (EFSA). Sweeteners can then be authorized to the *quantum satis* level or for a maximum usable dose or may remain unauthorized.

The list of authorized sweeteners is revised regularly by the European Commission in line with the opinion of EFSA, which takes into account the latest scientific advances in the field.

At present, eight high-intensity (nonnutritive) sweeteners are included in European Union (EU) legislation for use in foods: acesulfame K (ACS K), aspartame (ASP), ASP–ACS salt, cyclamate (CYC), saccharin (SAC), sucralose (SCL), neohesperidine dihydrochalcone (NHDC), and thaumatin. Some of them are either synthetic (ACS K, ASP, ASP–ACS salt, CYC, SAC, and SCL) or semi-synthetic (NHDC), and thaumatin occurs naturally (Yebra-Biurrun 2000).

10.2 LEGISLATION

Directive 2006/52/EC of the European Parliament and the Council of 5 July 2006, amending Directive 95/2/EC (European Commission 2004a) on food additives other than colors and sweeteners and Directive 94/35/EC on sweeteners for use in foodstuffs, cover the following.

On the basis of an opinion of the EFSA, expressed on 26 November 2003, changes were made to current authorizations in order to keep the level of nitrosamines as low as possible by bringing down the levels of nitrites and nitrates added to food while maintaining the microbiological safety of food products. The EFSA recommends the levels of nitrites and nitrates are set in the legislation as an "added amount." The EFSA is of the opinion that the added amount of nitrite rather than the residual amount contributes to the inhibitory activity against *C. botulinum*. The current provisions should be amended such that the maximum levels permitted, as mentioned by the EFSA, in nonheat-treated or heat-treated meat products in cheese and in fish are expressed as added amounts. Exceptionally, however, for certain traditionally manufactured meat products, maximum residual levels should be set on the condition that the products are adequately specified and identified. The levels set should ensure that the acceptable daily intake (ADI) established by the Scientific Committee on Food in 1990 is not exceeded.

Directive 2003/114/EC amending Directive 95/2/EC required the European Commission and the EFSA to review the conditions for the use of E214–E219 p-hydroxybenzoates and their sodium salts before 1 July 2004. The EFSA assessed the information on the safety of p-hydroxybenzoates and expressed its opinion on 13 July 2004. The EFSA established a full-group ADI of 0–10 mg/kg body weight for the sum of methyl and ethyl p-hydroxybenzoic acid esters and their sodium salts. The EFSA considered that propyl paraben should not be included in this group's ADI because propyl paraben, contrary to methyl and ethyl paraben, had effects on sex hormones and the male reproductive organs in juvenile rats. Therefore, the EFSA was unable to recommend an ADI for propyl paraben because of the lack of a clear no-observed adverse effect level (NOAEL). It is necessary to withdraw E216 propyl p-hydroxybenzoate and E217 sodium propyl p-hydroxybenzoate from Directive 95/2/EC. In addition, it is necessary to withdraw the use of p-hydroxybenzoates in liquid dietary food supplements.

Commission Decision 2004/374/EC suspended the placing on the marketing and import of jelly minicups containing gel-forming food additives derived from seaweed and certain gums because of the risk of choking on these products. Following a review of that decision, it is necessary to exclude the use of certain gel-forming food additives in jelly minicups.

The Scientific Committee on Food (SCF) assessed information on the safety of soybean hemicellulose and expressed its opinion on 4 April 2003. The committee concluded that the use of soybean hemicellulose is acceptable in certain foods with respect to which the request was made and at certain inclusion levels. It is therefore appropriate to permit such use for certain purposes. In order to facilitate matters for allergy sufferers, however, such use should not be permitted for unprocessed foods in which soybean is not expected to be found. In any event, consumers should be informed when products contain hemicellulose derived from soybean in accordance with the provisions of Directive 2000/13/EC of the European Parliament and of the Council of 20 March 2000, on the approximation of the laws of the Member States relating to the labeling, presentation, and advertising of foodstuffs [amended by Directive 2003/89, European Commission (2003)].

The EFSA assessed information on the safety of pullulan and expressed its opinion on 13 July 2004. The EFSA found the use of pullulan acceptable in the coating of food supplements that are in the form of capsules and tablets, as well as in breath fresheners in the form of films. It is therefore appropriate to permit these uses.

The EFSA assessed information on the safety of tertiary butyl hydroquinone and expressed its opinion on 12 July 2004. The EFSA established an ADI of 0–0.7 mg/kg body weight for this antioxidant and found that its use would be acceptable in certain foodstuffs at certain inclusion levels. It is therefore appropriate to permit this additive.

The SCF assessed information on the safety of starch aluminum octenyl succinate and expressed its opinion on 21 March 1997. The committee found the use of this additive as a component of microencapsulated vitamins and carotenoids may be regarded as acceptable. It is therefore appropriate to permit this use.

During the manufacture of sour-milk cheese, E500ii sodium hydrogen carbonate is added to pasteurized milk in order to buffer the acidity caused by the lactic acid to an appropriate pH value, thereby creating the necessary growth conditions for the ripening cultures. It is therefore appropriate to permit the use of sodium hydrogen carbonate in sour-milk cheese.

Currently, the use of a mixture of sorbates (E200, E202, and E203) and benzoates (E210–E213) is authorized in cooked shrimp for preservation. It is appropriate to extend that authorization to its use in all cooked crustaceans and mollusks.

E551 silicon dioxide is permitted as a carrier for food colors at a maximum level of 5%. The use of silicon dioxide as a carrier for food colors E171 titanium dioxide and E172 iron oxides and hydroxides should also be permitted at a maximum level of 90% relative to the pigment.

Directive 95/2/EC limits the use of additives listed in Annex I to that directive in the traditional French bread *pain courant francais*. The same limitation should apply to a similar traditional Hungarian bread. It is also appropriate to authorize the use of ascorbic acid (E300), sodium ascorbate (E301), and calcium disodium EDTA (E385) in Hungarian liver patties.

It is necessary to update the current provisions regarding the use of sulfites (E220–E228) in cooked crustaceans, table grapes, and lychees.

In accordance with a request from a member state and the opinion of the SCF expressed on 5 March 2003, 4-hexylresorcinol, which was authorized at the national level under Directive 89/107/EEC, should be authorized at the community level.

10.2.1 European Parliament and Council Directive 94/35/EC of 30 June 1994 on Sweeteners Intended for Use in Foodstuffs

This directive is specific, arising from the framework Directive 89/107/EEC on food additives. It applies to food additives used to impart a sweet taste to foodstuffs, together with those for a particular nutritional use or for use as tabletop sweeteners. It does not apply to foodstuffs with sweetening properties, such as honey (Table 10.1).

Sweeteners are used to replace sugar in the production of energy-reduced foodstuffs, noncariogenic foodstuffs, or food without added sugars for the extension of shelf life, as well as for the preparation of dietetic products.

Different sweeteners that may be placed on the market, as well as their conditions for use in foodstuffs, are specified in the annex of Directive 94/35. The doses specified refer to ready-to-eat (RTE) foodstuffs only.

Sweeteners cannot be used in foods for infants and young children mentioned in Directive 89/398/EEC (European Commission 1989), including foods for infants and young children who are not in good health, except if provided otherwise.

Member states are required to establish a system of regular surveys to monitor sweetener consumption. On the basis of this information, conditions of the use of sweeteners as laid down in this directive may, if necessary, be amended.

Labeling of table sweeteners containing polyols and/or ASP must bear the following warnings:

- Polyols—"Overconsumption may have laxative effects"
- ASP—"Contains a source of phenylalanine"
- ASP–ACS salt—"Contains a source of phenylalanine"

Table 10.1 Acts, Amending Acts, Entry into Force, Deadlines for Transposition, and Related Official Journal

Act	Entry into Force	Deadline for Transposition in the Member States	Official Journal
Directive 94/35/EC	31 December 1995	31 December 1995: authorization of trade in and use of products conforming to the Directive 30.6.1996: prohibition of trade in and use of products not conforming to the directive	*OJ L 237* of 10 September 1994
Amending act(s)			
Directive 96/83/EC	26 February 1997	19 December 1997: authorization of trade in products conforming to the Directive 19.6.1998: prohibition of trade in products not conforming to the directive	*OJ L* 48 of 19 February 1997
Regulation (EC) No. 1882/2003	20 November 2003	—	*OJ L* 284 of 31 October 2003
Directive 2003/115/EC	29 January 2004	29 July 2005 29 January 2006 (products marketed before the directive came into force)	*OJ L* 24 of 29 January 2004, p. 65
Directive 2003/114/EC	29 January 2004	29 July 2005 29 January 2006 (products marketed before the directive came into force)	*OJ L* 24 p. 58
Directive 2006/52/EC	15 August 2006	15 February 2008	*OJ L* 204 of 26 July 2006

Note: For related acts, see http://europa.eu/legislation_summaries/other/l21069_en.htm.

These are as last amended by Directive 2003/115/EC (European Commission 2004b).

Sweeteners authorized by community regulations must comply with the specific purity criteria defined by Directive 2008/60/EC (European Commission 2008b).

10.2.2 Regulation (EC) No 1333/2008 of the European Parliament and of the Council of 16 December 2008 on Food Additives

This regulation replaces previous directives and decisions concerning food additives permitted for use in foods, with a view of ensuring the effective functioning of the internal market while ensuring a high level of protection of human health and a high level of consumer protection, including the protection of consumer interests via comprehensive and streamlined procedures (European Commission 2008c)

This regulation harmonizes the use of food additives in foods in the community. This includes the use of food additives in foods covered by Council Directive 89/398/EEC (European Commission 1989) and the use of certain food colors for the health marking of meat and the decoration and stamping of eggs. It also harmonizes the use of food additives and food enzymes, thus ensuring their safety and quality and facilitating their storage and use. This has not been regulated at the community level.

Food additives should be approved and used only if they fulfill the criteria laid down in this regulation. Food additives must be safe when used, there must be a technological need for their use, and their use must not mislead the consumer and must be of benefit to the consumer. Misleading the consumer includes, but is not limited to, issues related to the nature, freshness, and quality of ingredients used; the naturalness of a product or of the production process; or the nutritional quality of the product, including its fruit and vegetable content. The approval of food additives should also take into account other factors relevant to the matter under consideration, including societal, economic, traditional, ethical, and environmental factors and the precautionary principle and the feasibility of controls. The use and maximum levels of a food additive should take into account the intake of the food additive from other sources and the exposure to the food additive by special groups of consumers (e.g., consumers with allergies).

Sweeteners authorized under this regulation may be used in tabletop sweeteners sold directly to consumers. Manufacturers of such products should make information available to consumers by appropriate means to allow them to use the product in a safe manner. Such information could be made available in a number of ways, including on-product labels, Internet websites, consumer information lines, or at the point of sale. In order to adopt a uniform approach to the implementation of this requirement, guidance drawn up at the community level may be necessary.

Following the adoption of this regulation, the commission, assisted by the Standing Committee on the Food Chain and Animal Health, should review all the existing authorizations for criteria, other than safety, such as intake, technological need, and the potential of misleading the consumer.

All food additives that are to continue to be authorized in the community should be transferred to the community lists in Annexes II and III to this regulation.

Annex III to this regulation should be completed with the other food additives used in food additives and food enzymes, as well as carriers for nutrients and their conditions of use in accordance with Regulation (EC) No 1331/2008 (establishing a common authorization procedure for food additives, food enzymes, and food flavorings).

To allow a suitable transition period, the provisions in Annex III, other than the provisions concerning carriers for food additives and food additives in flavorings, should not apply until 1 January 2011.

10.2.3 Specific Purity Criteria

Directives for specific criteria of purity are listed as follows.

Directive 2008/60/EC (OJ L 158 of 18 June 2008)—Specific criteria of purity concerning sweeteners for use in foodstuffs. Purity includes water content, sulphated ash, reducing sugars, chlorides, sulphates, nickel, lead, and arsenic as heavy metals.
Directive 98/66 (OJ L257, 19 September 1998)—Purity criteria for isomaltitol E953.
Directive 2000/51 (OJ L198, 04 August 2000)—Purity criteria for mannitol E421 and maltitol syrup E965ii.
Directive 2001/52 (OJ L190, 12 July 2011)—Purity criteria for mannitol and acesulfame-K E950.
Directive 2006/128/EC (European Commission 2006a)—It is necessary to adopt specific criteria for E968 erythritol, a new food additive authorized by Directive 2006/52/EC of the European Parliament and of the Council of 5 July 2006, amending Directive 95/2/EC on food additives other than colors and sweeteners and Directive 94/35/EC on sweeteners for use in foodstuffs.

A number of language versions of Directive 95/31/EC contain some errors regarding the following substances: E954 SAC and its sodium, potassium, and calcium salts; E955 SCL; E962 salt of ASP–ACS; E965(i) maltitol; and E966 lactitol. Those errors need to be corrected. In addition, it is necessary to take into account the specifications and analytical techniques for additives as set out in the Codex Alimentarius as drafted by the Joint FAO/WHO Expert Committee on Food Additives (JECFA). In particular, where appropriate, the specific purity criteria have been adapted to reflect the limits for individual heavy metals of interest to the EFSA in its scientific opinion expressed on 19 April 2006, which concluded that the composition of maltitol syrup based on a new production method will be similar to the existing product and will be in accordance with the existing specification. It is therefore necessary to amend the definition of E965(ii) maltitol syrup set out in Directive 95/31/EC for E965 by including that new production method. Maltitol syrup is a mixture consisting of mainly maltitol with sorbitol and hydrogenated oligosaccharides and polysaccharides. It is manufactured by the catalytic hydrogenation of high maltose-content glucose syrup or by the hydrogenation of its individual components followed by blending. The article of commerce is supplied both as a syrup and a solid product.

10.2.4 Labeling

10.2.4.1 Directive 96/21/EC [OJ L 88 of 05 April 1996]

The Council Directive of 29 March 1996 amends Commission Directive 94/54/EC concerning the compulsory indication on the labeling of certain foodstuffs of particulars other than those provided for in Council Directive 79/112/EEC.

This directive requires food labels to include relevant information on the sweeteners contained in foodstuffs. It also stipulates that warnings should appear on the labels of foodstuffs containing ASP or polyols.

10.2.4.2 Directive 2008/5/EC [OJ L 27 of 31 January 2008]

Other compulsory indications, apart from those provided for in Directive 2000/13/EC, must be shown on the labels of certain foodstuffs (for example, "packaged in a protective atmosphere" or "with sweeteners") in order to better inform the consumer (European Commission 2008a). For foodstuffs containing ASP, labeling should indicate that it "contains a source of phenylalanine." Confectionery containing glycyrrhizinic acid or its ammonium salt as a result of the addition of the substance(s) as such or the licorice plant *Glycyrrhiza glabra* at concentrations of 4 g/kg or above and beverages containing glycyrrhizinic acid or its ammonium salt as a result of the addition of the substance(s) as such or the licorice plant *Glycyrrhiza glabra* at concentrations of 50 mg/l or above or of 300 mg/l or above in the case of beverages containing more than 1.2% by the volume of alcohol should be labeled as follows: "Contains licorice. People suffering from hypertension should avoid excessive consumption." For foodstuffs that may contain more than 10% added polyols, labeling should indicate that excessive consumption may produce laxative effects, etc.

10.2.5 Nutrition and Health Claims Regulation

Regulation (EC) N 1924/2006 of the European Parliament and of the Council on Nutrition and Health Claims made on foods was adopted on 20 December 2006. This regulation lays down harmonized rules for the use of nutrition and health claims and contributes to a high level of consumer protection. It ensures that any claim made on a food label in the EU is clear, accurate, and substantiated, enabling consumers to make informed and meaningful choices. The regulation also aims at ensuring fair competition and promoting and protecting innovation in the area of food.

10.2.5.1 Comparative Claims

A claim is a health claim if, in the naming of the substance or category of substances, there is a description or indication of functionality or an implied effect on health.

It should be noted that all claims are subject to the general principles laid down in Articles 3 and 5. In the case of the claim "contains," this means the substance subject to the claim is present in a significant quantity and has been shown to have a beneficial nutritional or physiological effect. In addition, the use of health and nutrition claims triggers an obligation to provide nutritional information pursuant to Directive 90/496/EEC in accordance with Article 7 of the regulation.

10.2.5.2 Reduced [Name of Nutrient]

A claim stating that the content of one or more nutrients has been reduced and any claim likely to have the same meaning for the consumer may only be made where the reduction in content is at least 30% compared to a similar product, except for micronutrients, where a 10% difference in the reference values as set in Directive 90/496/EEC would be acceptable, and for sodium or the equivalent value for salt, where a 25% difference would be acceptable.

10.2.5.3 Energy-Reduced

A claim that a food is energy reduced and any claim likely to have the same meaning for the consumer may only be made where the energy value is reduced by at least 30%, with an indication of the characteristic(s) that make(s) the food reduced in its total energy value.

10.2.5.4 Light/Lite

A claim stating that a product is "light" or "lite" and any claim likely to have the same meaning for the consumer shall follow the same conditions as those set for the term "reduced." The claim shall also be accompanied by an indication of the characteristic(s) that make(s) the food light or lite (SCF 2007).

For example, for a label stating "light—50% less sugars," when the nutrient is removed from the composition of the product, this indication can be provided by a claim referring to this absence of nutrient, for example, "light—no sugars."

10.3 STEVIOL GLYCOSIDE SWEETENERS

In 2007, the JECFA established specifications for steviol glycoside sweeteners, calling for them to consist of at least 95% of the seven named steviol glycoside sweeteners (Carakostas et al. 2008).

10.3.1 Toxicology

The toxicological properties of artificial sweeteners are well studied and show that microgram per liter quantities are harmless to humans (Weihrauch and Diehl 2004). However, data on ecotoxicological properties of artificial sweeteners are scarce. For example, in the EU, there is no obligatory environmental risk assessment for artificial sweeteners, according to the European Parliament and Council directive on sweeteners for use in foodstuffs (1994), which is based on the framework directive for food additives (European Commission 1988). Therefore, it is yet unknown what the occurrence of these trace pollutants means to aquatic biocenoses.

10.4 SUGAR ALCOHOLS

All these sweeteners, except for mannitol, do not represent a hazard to health, and the JECFA deemed it not necessary to assign a numerical value for ADI but instead assigned the most favorable term "not specified." Furthermore, in the United States, they are considered generally recognized as safe (GRAS). Nevertheless, an excess consumption of mannitol may have a laxative effect, and for this reason, the JECFA has allocated a temporary ADI of 50 mg per kg (Yebra-Biurrun 2005).

Following the adoption of the EC Sweeteners Directive in 1994 (Commission of the European Communities 1994) and its implementation into the national laws, member states are required to establish a system of consumer surveys to monitor additive intake (European Commission 1994). For this reason, intake data are required through developed and validated methods for the determination of these two artificial sweeteners.

The ADI has been defined by the World Health Organization (WHO) as "an estimate by the JECFA of the amount of a food additive, expressed on a body weight basis, that can be ingested daily over a lifetime without appreciable health risk" and is based on an evaluation of available toxicological data. For example, in Europe, the ADI is set at 9 mg/kg of body weight/day for ACS K (Wilson et al. 1999). For ASP, there is a safety margin even in high-consuming diabetics (Ilbäck et al. 2003). The FDA has set the ADI for ASP at 50 mg/kg of body weight/day. An ADI of 40 mg/kg body weight per day set by the committee of experts of the FAO and the WHO is not likely to be exceeded, even by children and diabetics. A European Commission report gives a theoretical maximum estimate for adults' consumption of 21.3 mg/kg body weight per day of ASP. However, the actual consumption is likely to be lower, even for high consumers of ASP. The report also gives refined estimates for children, which show they consume 1%–40% of the ADI. People with diabetes are high consumers of foods containing ASP; their highest reported intake varies between 7.8 and 10.1 mg/kg body weight per day. As for health at the international level (JECFA), as well as for the United States Food and Drug Administration (FDA), the ADI for ACS K has been set at 15 mg/kg body weight. At the European level on 13 March 2000, the ADI has been set at 9 mg/kg body weight (SCF).

In general, in the EU, sweeteners are thoroughly assessed for safety by the EFSA before they are authorized for use. EU Directives 94/35/EC (European Commission 1994), 96/83/EC (European Commission 1996), European Commission 2003, and 2006/52/EC (European Commission 2006b) define which sweetener has been approved to be added to food products and beverages.

Currently, ACS K and ASP are used in foods, including baked goods (dry bases for mixes), beverages (dairy beverages, instant tea, instant coffee, and fruit-based beverages), soft drinks (colas, citrus-flavored drinks, and fruit-based soft drinks), sugar preserves, confections (calorie-free dustings, frostings, icings, toppings, fillings, and syrups), alcoholic drinks (beer), vinegar, pickles, sauces (sandwich spreads, and salad dressings), dairy products (yogurt and yogurt-type products, puddings, desserts, dairy analogues, and sugar-free ice-cream), fruit, vegetables, nut products, sugar-free jams and marmalades, low-calorie preserves, and other food products (i.e., chewing gums, liquid concentrates, and frozen and refrigerated desserts). Hard-boiled candies can be manufactured using ACS K as the intense sweetener. ACS K rounds the sweetness and brings the taste close to standard, sugar-containing products. In chocolate and related products, ACS K can be added at the beginning of the production process (e.g., before rolling). It withstands all treatments, including conching, without detectable decomposition (Baron and Hanger 1998). In reduced-calorie baked goods, bulking agents, such as polydextrose as a substitute for sugar and flour, may help reduce the level of fats. ACS K combines well with suitable bulking ingredients and bulk sweeteners and therefore allows the production of sweet-tasting baked goods having fewer calories. In diabetic products, combinations of ACS K and sugar alcohols, such as isomalt, lactitol, maltitol, or sorbitol, can provide volume and sweetness. Texture and sweetness intensity can be similar to sucrose-containing products.

A number of analytical methods based on different principles are available for the determination of ACS K and ASP in a broad range of food matrices. The aim here is to present the available methodology for sample pretreatment and the available protocols of analysis.

10.5 SACCHARIN

The U.S. FDA delisted SAC in 1972 because of the uncertainty of the safety of SAC, and its use in foods and beverages was proposed to be banned in 1977. However, it led to a public protest to impose a moratorium on the ban, which has been extended up to the present (Pearson 1991; Kroger, Meister, and Kava 2006). In Canada, SAC was banned in 1977 for use in foods (Arnold 1984). However, it is permitted to be sold in pharmacies as a tabletop sweetener. In other countries, SAC is permitted, although its use is restricted to varying degrees. EU Directives 94/35/EC (European Commission 1994), 96/83/EC (European Commission 1997), 2003/115/EC (European Commission 2004), and 2006/52/EC (European Commission 2006) define in which food products and in what quantity SAC can be used.

The SCF (1995) reported a full ADI for sodium SAC of 0–5 mg/kg bw. This is derived by applying a 100-fold safety factor to the NOAEL of 1% in the diet (500 mg/kg bw) for bladder tumors in rats. ADI was also expressed in terms of the free acid, because sodium SAC is not the only salt used. Taking into account the molecular weight difference between sodium SAC (MW 241) and the free acid (MW 183), the ADI expressed as the free acid is 0–3.8 mg/kg bw.

It was also reported that SAC is not a direct acting genotoxin. Support for this view comes also from the fact that it has been shown to be a carcinogen at only one site in only one sex of one species of animal.

With a relative sweetness of approximately 500, the sugar equivalent is 2500 mg.

10.6 EUROPEAN FOOD SAFETY AUTHORITY

The EFSA encourages sound scientific debate in which arguments from all sides are presented fairly and objectively, so interested parties can duly weigh all evidence available. One of the EFSA's senior scientists took part in the hearing in the European Parliament on 16 March 2011 on ASP to explain the comprehensive body of work that EFSA has carried out on the substance over the years, including a review of two recent studies on artificial sweeteners.

Unfortunately, when reporting the outcomes of this meeting, the organizers of the hearing continue to repeat errors and misinformation. The EFSA reaffirms that any possible risks from ASP have been considered by scientific bodies worldwide and the current ADI ensures consumers are protected.

Notwithstanding, the EFSA is continuously monitoring scientific literature regarding the safety of sweeteners, and its top-level independent experts on EFSA's Panel on Food Additives and Nutrient Sources Added to Food (ANS) are preparing a further scientific opinion on ASP. As for all areas of its work, the EFSA will continue to liaise with risk assessors in the EU member states and with food safety agencies outside Europe in order to benefit from the broadest expertise possible.

In a statement published recently, the EFSA concludes that two recent publications on the safety of artificial sweeteners, namely, a carcinogenicity study in mice (Soffritti et al. 2010) and an epidemiological study on the association between intakes of artificially sweetened soft drinks and increased incidence of preterm delivery (Halldorsson et al. 2010), do not give reason to reconsider previous safety assessments of ASP or of other sweeteners currently authorized in the European Union. The EFSA's review of these studies has been carried out in cooperation with the French Agency for Food, Environment, and Occupational Health Safety (http://www.FRENCHAAAT2011sa0015EN.pdf).

At its plenary meeting on 1–2 March 2011, the EFSA ANS Panel will consider the authority's statement and the possible need for further work in relation to these studies. The EFSA will continue monitoring the scientific literature in order to identify new scientific evidence for sweeteners that may indicate a possible risk for human health or which may otherwise affect the safety assessment of these food additives.

In response to a request for technical assistance from the European Commission, the EFSA reviewed the publication of Soffritti et al. (2010) on a long-term carcinogenicity study in mice exposed to the artificial sweetener ASP through feed. EFSA scientists concluded that, on the basis of the information available in the publication, the validity of the study and its statistical approach cannot be assessed and its results cannot be interpreted. Regarding the design of the study, the EFSA advised that experimental studies carried out over animals' lifetimes can lead to erroneous conclusions. Older animals, for instance, are more susceptible to illness, and when a carcinogenicity study in mice is extended beyond the recommended 104 weeks, age-related pathological changes (such as spontaneous tumors) can appear, which may confound the interpretation of any compound-related effects.

The EFSA noted that Swiss mice (used in this study) are known to have a high incidence of spontaneous hepatic and pulmonary tumors and the increased incidence of these tumors reported in the study fall within the historical control range recorded in this laboratory for these tumors in these mice. Furthermore, these hepatic tumors in mice are not regarded by toxicologists as being relevant for human risk assessment when they are induced by nongenotoxic substances such as ASP. Overall, the EFSA concluded that the findings presented by Soffritti et al. (2010) do not provide sufficient scientific evidence to reconsider the previous evaluations by the EFSA on ASP that concluded on the lack of genotoxicity and carcinogenicity of ASP.

The EFSA also assessed the publication of Halldorsson et al. (2010), which reports findings suggesting the daily intake of artificially sweetened soft drinks may be associated with an increased risk of preterm delivery. EFSA concluded there is no evidence available in this study to support a causal relationship between the consumption of artificially sweetened soft drinks and preterm delivery and additional studies would be required either to confirm or reject such an association, as indicated by the authors. Given that the association found by the authors appears to be primarily related to medically induced (rather than spontaneous) preterm deliveries, the EFSA stressed that medical history and criteria on which medical decisions to induce delivery were based, are factors that should be investigated further. The EFSA recommended future studies should also investigate other important confounding factors such as exposure to other substances in the diet, which might have an effect on pregnancy (http://www.efsa.europa.eu/en/press/news/ans110228.htm).

The EFSA keeps the safety of sweeteners and ASP under regular review. In March 2009, the ANS Panel concluded, on the basis of all the evidence currently available, including the European Ramazzini Foundation (ERF) study published in 2007, there is no indication of any genotoxic or carcinogenic potential of ASP and no reason to revise the previously established ADI for ASP of 40 mg/kg body weight. An earlier opinion, following the first study on ASP by the ERF, was adopted by the former Scientific Panel on Food Additives, Flavorings, Processing Aids and Materials in Contact with Food (AFC) in 2006.

10.7 ASPARTAME

Prior to its marketing, the safety of the high-intensity sweetener ASP for its intended uses as a sweetener and flavor enhancer was demonstrated by the results of more than 100 scientific studies in animals and humans. In the postmarketing period, the safety of ASP was further evaluated through extensive monitoring of intake, postmarketing surveillance of anecdotal reports of alleged health effects, and additional research to evaluate these anecdotal reports and other scientific issues (Butchko and Stargel 2001). The results of the extensive intake evaluation in the United States, which was done over an 8-year period, and the results of studies done in other countries demonstrated intakes that were well below the ADIs set by the FDA and regulatory bodies in other countries, as well as the JECFA. Evaluation of the anecdotal reports of adverse health effects, the first such system for a food additive, revealed the reported effects were generally mild and also common in the general population and there was no consistent or unique pattern of symptoms that could be causally linked to consumption of ASP. Finally, the results of the extensive scientific research done to evaluate these allegations did not show a causal relationship between ASP and adverse effects. Focused clinical studies would be the best way to thoroughly address the issues raised by the anecdotal reports. Thus, the weight of scientific evidence confirms that, even in amounts that people most typically consume, ASP is safe for its intended uses as a sweetener and flavor enhancer (Butchko and Stargel 2001).

The safety of ASP and its metabolic constituents was established through extensive toxicology studies in laboratory animals using much greater doses than people could possibly consume. Its safety was further confirmed through studies in several human subpopulations, including healthy infants, children, adolescents, and adults; obese individuals; diabetics; lactating women; and individuals heterozygous (PKUH) for the genetic disease phenylketonuria (PKU) who have a decreased ability to metabolize the essential amino acid phenylalanine (Meyer 2007). Several scientific issues continued to be raised after approval, largely as a concern for theoretical toxicity from its metabolic components—the amino acids, aspartate and phenylalanine, and methanol—even though dietary exposure to these components is much greater than from ASP. Nonetheless, additional research, including evaluations of possible associations between ASP and headaches, seizures, behavior, cognition, and mood, as well as allergic-type reactions and use by potentially sensitive subpopulations, has continued after approval. These findings are reviewed here. The safety testing of ASP has gone well beyond that required to evaluate the safety of a food additive. When all the research on ASP, including evaluations in both the premarketing and postmarketing periods, is examined as a whole, it is clear that ASP is safe and there are no unresolved questions regarding its safety under conditions of intended use (Butchko et al. 2002).

Magnuson et al. (2007) reviewed the scientific literature on the absorption and metabolism, current consumption levels worldwide, toxicology, and recent epidemiological studies on ASP. Current use levels of ASP, even by high users in special subgroups, remains well below the U.S. FDA and EFSA established ADI levels of 50 and 40 mg/kg bw/day, respectively. Consumption of large doses of ASP in a single bolus dose will have an effect on some biochemical parameters, including plasma amino acid levels and brain neurotransmitter levels. The rise in plasma levels of phenylalanine and aspartic acid following the administration of ASP at doses less than or equal to 50 mg/kg bw do not exceed those observed postprandially. Acute, subacute, and chronic toxicity studies with ASP and its decomposition products conducted in mice, rats, hamsters, and dogs have consistently found no adverse effect of ASP with doses up to at least 4000 mg/kg bw/day. Critical review of all carcinogenicity studies conducted on ASP found no credible evidence that ASP is carcinogenic. The data from extensive investigations into the possibility of neurotoxic effects of ASP, in general, do not support the hypothesis that ASP in the human diet will affect nervous system function, learning, or behavior. Epidemiological studies on ASP include several case-control studies and one well-conducted prospective epidemiological study with a large cohort, in which the consumption of ASP was measured. The studies provide no evidence to support an association between ASP and cancer in any tissue. The weight of existing evidence is that ASP is safe at current levels of consumption as a nonnutritive sweetener.

Butchko et al. (2002) also reported that the conversion of ASP to methanol is not sufficient to induce any toxicity from methanol or its metabolites.

The FDA evaluated the safe levels of methanol intake and concluded that the safe level of exposure to methanol is 7.1–8.4 mg/kg body weight/day (FDA 1996). This amount of methanol is about 25 times greater than the amount of methanol (0.3 mg/kg body weight) provided by ASP to the diet at the 90th percentile intake.

The Advisory Forum (AF) of the EFSA discussed the issue of ASP in 2007, and at the 24th AF meeting held on December 6 and 7 in 2007, it was agreed that a special meeting would be held on ASP. This meeting of national experts was to take place early in 2009 and, as provided in the Terms of Reference endorsed by the AF, was to review the information available on ASP, agree on the completeness of the information, ensuring any missing data were added, and identify possible data gaps and discrepancies in the available data and where such exist to consider detailed options to address the outstanding issues.

An organizing team, nominated by members of the AF, was tasked with preparing for the meetings of national experts by identifying, collating, and reviewing all published papers on ASP since the review carried out by the SCF in 2002. In addition, the organizing team also considered available nonpeer-reviewed information and anecdotal claims by individuals who attributed various symptoms and illnesses directly to ASP consumption. A meeting was held with interested parties who had submitted information in response to a call for data published by the EFSA in September 2008.

Each of the areas considered, including exposure data, brain function, satiation and appetite, allergenicity and immunotoxicity, metabolic aspects and diabetes, carcinogenicity (including cancer epidemiology), and genotoxicity were considered and reported by the organizing team. For each, consistent with the terms of reference for the national experts meeting, they made preliminary conclusions on whether there were data gaps or discrepancies that required further consideration.

This section represents the results of these deliberations.

Overall, the organizing team did not identify any major gaps in information on ASP, and the national experts agreed with their view. However, several suggestions are made within this report for additional data that would add to the available knowledge on ASP and its metabolites. To address the communications element of the terms of reference, a workshop with stakeholders and other interested parties is to be arranged to share and discuss the findings of the work.

In conclusion, the national experts have not identified any new evidence that requires a recommendation to the EFSA that the previous opinions of the EFSA and the SCF need to be reconsidered (EFSA report of the meetings on ASP with national experts EFSA Q-2009-00488).

Following a request from the European Commission, the ANS Panel was asked to deliver a scientific opinion on the results of a long-term carcinogenicity study with prenatal exposure to the artificial sweetener ASP, performed by the Cesare Maltoni Cancer Research Center of the European Ramazzini Foundation (ERF) and published in June 2007 by Soffritti et al. The authors concluded that the results of their study not only confirm but also reinforce their first experimental demonstration (published in 2005 and 2006) of ASP's multipotential carcinogenicity at a dose level close to the human ADI. Based on the results of this study, the authors further postulated that, when lifespan exposure to ASP begins during fetal life, its carcinogenic effects are increased.

The panel concluded as follows (EFSA 2009a):

- The evaluation of aggregated malignant tumor incidences as evidence of the carcinogenic potential of the test compound can only be performed based on a thorough consideration of all tumor data, including onset, and data on nonneoplastic, hyperplastic, and preneoplastic lesions, but these data were not provided by the authors.
- In accordance with the previous view of the AFC, the lymphomas and leukemias might have developed in a population of rats suffering from chronic respiratory disease.
- The increase in incidence of mammary carcinoma is not considered indicative of a carcinogenic potential of ASP, because the incidence of mammary tumors in female rats is rather high and varies considerably between carcinogenicity studies. The panel also noted that an increased incidence of mammary carcinomas was not reported in the previous ERF study with ASP, which used much higher doses of the compound.
- Overall, on the basis of all the evidence currently available from this ERF study and previous evaluations, there is no indication of any genotoxic or carcinogenic potential of ASP, and there is no reason to revise the previously established ADI for ASP of 40 mg/kg bw.

The sugar equivalent in milligrams is 8000, taking into account that the relative sweetness is 200. The term "sugar equivalent" refers to the amount of sugar that would be needed to achieve the same sweetness effect as the indicated amount of sweetener. To calculate it, we need to know the relative sweetness of low-calorie sweeteners.

ASP is stable at room temperature. The major degradation product at higher temperatures is diketopiperazine. These two compounds have been satisfactorily assessed in a large number of studies in humans, and an ADI has been established for each of them: 40 mg/kg bw for ASP and 7.5 mg/kg bw for diketopiperazine.

The metabolites of ASP in humans are compounds that are found in normal foods and are also produced by endogenous cellular metabolism. ASP is a minor source of phenylalanine, aspartic acid, and methanol compared with the standard dietary intake of these substances. They cannot, therefore, be the source of the toxic neurological effects attributed to ASP (AFSSA 2002).

Genetic toxicity assays have demonstrated that ASP and diketopiperazine are not genotoxic (AFSSA 2002).

None of the carcinogenicity tests that have been conducted on rodents indicated a relationship between treatment with ASP and the appearance of brain tumors.

The epidemiological study of Olney et al. (1996) suggested that a link between the placing on the market of ASP and a possible increase in the frequency of brain cancers in humans did not provide any scientific evidence to justify or demonstrate a basis for this suggestion; to date, it has not been confirmed.

Analysis of the scientific literature has demonstrated a lack of evidence based on the current state of knowledge, which would enable a causal link to be established between the consumption of ASP and the occurrence of epileptic seizures or anomalies on an electroencephalogram.

It has been verified in France that no anticonvulsive medication contains high quantities of ASP. The consumption of ASP in humans, even in particularly exposed populations such as diabetic children, does not exceed the ADI, notably in France.

In conclusion, the French Food Safety Agency (*Agence Francaise de Securite Sanitaire des Aliments*; AFSSA) considers that the current state of scientific knowledge does not enable a relationship to be established between the exposition to the ASP and brain tumors in humans or animals.

Diabetic children are considered an at-risk group for the consumption of sweeteners, first, because the diabetic diet partially excludes simple carbohydrates and, second, because of their lower body mass. Two studies have been conducted, one in Sweden and the other in France, using the maximum permitted values for ASP in food.

The Swedish study (Ilbäck et al. 2000) carried out a simulation of extreme consumption by taking the consumption of the 10 subjects who were the biggest consumers of sweeteners (out of 320 subjects who took part in the survey). The mean weight was 20 kg, and the estimated consumption of ASP was 46 mg/kg bw/d or 114% of the ADI. These data should be used with care as the maximum values concerned only 10 individuals, and the ASP contents used are the maximum levels permitted in food. Moreover, the dietary survey method used (frequency questionnaire) is imprecise and was not validated.

The French study (Garnier-Sagne, Leblanc, and Verger 2001), was conducted on a population of 227 young diabetics (112 girls and 115 boys), members of the association *Aide aux Jeunes Diabétiques* (Young Diabetics Support), distributed over 65 out of the 95 French departments and aged 2 to 20 years. The method used was the 5-day diary. More than 84% of the diabetic children consumed sweeteners. The mean consumption of ASP was estimated at 1.9 mg/kg bw/d (less than 5% of the ADI), and the maximum intake was 15.6 mg/kg bw/d (40% of the ADI). The results were identical for boys and girls. This study also indicated that sugar-free drinks and tabletop sweeteners were the forms in which ASP was the most commonly consumed, contributing to 56% and 16%, respectively, of the estimated consumption.

During June 2009, the U.K.-based FSA announced a new study to examine the effects of ASP because of reported symptoms of headaches and upset stomachs after consumption of ASP.

Recent developments include a method for stability improvement in aqueous solutions by the addition of a flavonoid, chlorogenic acid, or polymerized polyphenols to enhance the stability in acidic beverages.

Finally, the incorporation of sodium bicarbonate into sugar-free chewing gums along with ASP is a recent innovation. Sodium bicarbonate provides tooth whitening.

10.8 NEOTAME AND STEVIA

Neotame has recently been approved by Commission Directive 2009/163 of the European Commission and assigned a new E number, E961. The acceptable daily intake is 0–2 mg/kg body weight.

Following a request from the European Commission, the AFC was asked to deliver a scientific opinion on the safety of neotame as a sweetener and flavor enhancer (EFSA 2007).

Neotame is a dipeptide methyl ester derivate. Its chemical structure is N-[N-(3,3-dimethylbutyl)-L-α-aspartyl]-L-phenylalanine 1-methyl ester. It is intended for use in food as a sweetener and flavor enhancer. Neotame has a sweetness factor approximately 7000–13,000 times greater than that of sucrose and approximately 30–60 times greater than that of ASP, depending on the food application.

Neotame is manufactured by the reaction of ASP and 3,3-dimethylbutyraldehyde, followed by purification, drying, and milling. Neotame is generally stable under conditions of intended use as a sweetener across a wide range of food and beverage applications. Neotame degrades slowly in aqueous conditions, such as those in carbonated soft drinks.

Studies with radiolabeled neotame given orally to rats indicate no accumulation in tissues. The highest radioactivity is associated with the contents of the gastrointestinal tract and organs of metabolism and excretion (liver, kidney, and urinary bladder). In whole body autoradiography studies with pregnant rats, no radioactivity has been reported in the fetus.

The safety of neotame has been investigated in *in vitro* studies and in short- and long-term studies in mice, rats, rabbits, and dogs. The results indicate that neotame is not genotoxic, carcinogenic, teratogenic, or associated with any reproductive/developmental toxicity. The consistent findings in animal studies were reduced feed consumption, body weight, and body weight gain relative to that of controls with no clear dose response. These effects are considered not adverse or indicative of toxicity, but a consequence of reduced palatability of the neotame-containing diets. Therefore, body weight parameters were not considered appropriate end points for setting NOAELs in these studies.

The results of human studies demonstrated that neotame was well tolerated by healthy and diabetic human subjects at dose levels up to 1.5 mg/kg bw/day (the highest dose tested). The exposure to methanol, which may result from the ingestion of neotame-containing foods and beverages, is considered negligible compared to that from other dietary sources and, as such, of no concern from the safety point of view.

The panel noted that the additional phenylalanine intake expected from the ingestion of neotame as a general-purpose sweetener represents a relatively small increment in the exposure to phenylalanine of the phenylketonuric homozygous child.

After considering all the data on stability, degradation products, and toxicology, the panel concluded that neotame is not of safety concern with respect to the proposed uses as a sweetener and flavor enhancer.

The panel established an ADI of 0–2 mg/kg bw/day based on the application of a 100-fold safety factor to the NOAEL of 200 mg/kg bw from a 52-week dog study.

Conservative estimates of dietary exposure both in adults and children suggest it is very unlikely that the ADI would be exceeded at the proposed use levels.

The panel recommends that the limit for lead in the specifications should not be higher than 1 mg/kg (EFSA 2007).

The Wrigley Company has developed a method for controlling the release of neotame in chewing gum. Derivatives of ASP are treated to control their release and enhance shelf-life stability. Modified release is obtained by physical modification of sweetener properties via coating and drying (Leatherhead Food Research 2010).

Cadbury Adams described a chewing gum composition containing neotame in its free state (not encapsulated) with prolonged sweetening sensation.

Neotame and ACS K have been blended and act synergistically at a ratio of 90:10 (PepsiCo).

Stevia is a natural sweetener allowed for use in the United States and Asia, whereas in Europe, it has not been approved yet.

Following a request from the European Commission, a revised exposure assessment of steviol glycosides from their use as a food additive was carried out for children and adults based on the revised proposed uses presented in the terms of reference.

Several food consumption databases were used to conduct the revised exposure assessment. For children, data from the Individual Food Consumption Data and Exposure Assessment Studies for Children (EXPOCHI Project) and the U.K. National Diet and Nutrition Survey (NDNS) were used. Estimates for adults were based on U.K. data only (EFSA 2011).

For adults, exposure estimates give a mean dietary exposure to steviol glycosides, expressed as steviol equivalents, of 1.9–2.3 mg/kg bw/day and of 5.6–6.8 mg/kg bw/day for high-level exposures (97.5th percentile), with the main contributors being nonalcoholic flavored drinks, tabletop sweeteners, and beer and cider.

For European children (aged 1 to 14 years), exposure estimates give a mean dietary exposure to steviol glycosides, expressed as steviol equivalents, of 0.4–6.4 mg/kg bw/day; at the high level (95th/97.5th percentile), and exposure estimates range from 1.7 to 16.3 mg/kg bw/day.

Considering the limitations of consumption data, estimates can be considered conservative as in the ANS panel opinion (2010).

The EFSA comprehensive database was used to identify and assess the uncertainty resulting from the consumption data from the food group "nonalcoholic flavored drinks," the main contributor for both children and adults.

From these data, no general tendency was observed regarding the difference between the consumption levels of low-calorie nonalcoholic flavored drinks (soft drinks) and normal-calorie soft drinks.

Therefore, the consumption levels of normal-calorie soft drinks were considered an acceptable approximation of the consumption levels of low-calorie soft drinks, and thus, the exposure estimates for adults were not corrected.

For children, the data from EXPOCHI for the consumption of nonalcoholic flavored drinks (soft drinks) were found to be generally higher than the consumption of low-calorie soft drinks from the EFSA comprehensive database by a factor of 2. Consequently, the data from the EFSA comprehensive database were used to correct the consumption of this food group. The corrected exposure estimates for children who are high consumers (95th percentile) range from 1.0 to 12.7 mg/kg bw/day.

The revised mean exposure estimates differ only slightly from the exposure estimates given in the ANS panel opinion (2010). By using the EFSA comprehensive database, the upper range of high-level exposure estimates decreased from a maximum of 17.2 from the ANS panel opinion to 12.7 mg/kg bw/day for children, but high-consuming children's exposures are still above the ADI of 4 mg/kg bw/day for several countries.

The results of the new studies presented to the committee at its present meeting have shown no adverse effects of steviol glycosides when taken at doses of about 4 mg/kg bw per day, expressed as steviol, for up to 16 weeks by individuals with type 2 diabetes mellitus and individuals with normal or low-normal blood pressure for 4 weeks. The committee concluded that the new data were sufficient to allow the additional safety factor of 2 and the temporary designation to be removed and established an ADI for steviol glycosides of 0–4 mg/kg bw expressed as steviol (JECFA 2009).

The committee noted that some estimates of high-percentile dietary exposure to steviol glycosides exceeded the ADI, particularly when assuming complete replacement of caloric sweeteners with steviol glycosides, but recognized that these estimates were highly conservative and actual intakes were likely to be within the ADI range (JECFA 2009).

The committee evaluated an estimate of dietary exposure to steviol glycosides based on the replacement of all dietary sugars in the United States. Using a per capita estimate of 176 g of caloric sweetener per day, the committee calculated that the consumption of steviol glycosides would be 5.8 mg/kg bw per day. Additionally, published estimates of exposure to rebaudioside A, based on exposure to other high-intensity sweeteners and using the principle of equivalent sweetness, were evaluated by the committee. These estimates were 1.5 mg/kg bw per day for diabetic children and adults and 1.7 mg/kg bw per day for nondiabetic children consuming the sweetener at a high percentile of the exposure distribution, taken to be greater than the 90th percentile. For Japan, the per capita replacement estimate is three.

The committee used the GEMS/Food database to prepare updated international estimates of dietary exposure to steviol glycosides (as steviol). It was assumed that steviol glycosides would replace all dietary sugars at the lowest reported relative sweetness ratio for steviol glycosides and sucrose, 200:1. The dietary exposures ranged from 0.9 mg/kg bw per day (cluster J) to 5 mg/kg bw per day (clusters B and M). The committee evaluated estimates of dietary exposure per capita derived from disappearance (poundage) data supplied by Japan and China (JECFA 2009).

The AFSSA reported an opinion on a new draft order on the use of rebaudioside A, an extract of *Stevia rebaudiana*, as a food additive.

Manufacturers of breakfast cereals, cones, and wafers with no added sugar for ice cream, and dry-salted appetizers made from starch or walnuts and hazelnuts do not subject rebaudioside A to a higher temperature than 100°C to avoid its possible thermal degradation.

The maximalist exposure calculations made by AFSSA from the foods corresponding to the categories defined in the new draft order show exposure remains below the human toxicity value identified in human studies and stipulated in AFSSA's opinion expressed on 11 September 2008.

According to these calculations, the adult population would be exposed to 0.16 mg/kg bw/day on the average and up to 0.93 mg/kg bw/day at the 97.5th percentile. In adult consumers only, the average intake would be 0.36 mg/kg bw/day and 0.45 mg/kg bw/d at the 97.5th percentile. The human toxicity value used for comparison is 14.5 mg/kg bw/day for this population (AFSSA 2009).

In 3- to 17-year-old children, exposure would be 0.09 mg/kg bw/day on the average and 0.72 mg/kg bw/day at the 97.5th percentile. In these consumers only, the average intake would be 0.18 mg/kg bw/day and 0.88 mg/kg bw/d at the 97.5th percentile. The human toxicity value used for comparison is 23.5 mg/kg bw/day.

In 2- to 20-year-old diabetics, exposure would be 0.36 mg/kg bw/day on the average and 1.66 mg/kg bw/day at the 97.5th percentile, overall. In consumers only (80% of diabetic patients are consumers), the average intake would be 0.45 mg/kg bw/day and 1.70 mg/kg bw/day at the 97.5th percentile. The human toxicity value used for comparison is 23.5 mg/kg bw/day (AFSSA 2009).

The EFSA has received petitions from the European Stevia Association, Cargill, and Morita for the use of stevia sweeteners (extracts with 95% steviol glycosides). Finally, Turkey has recently approved the use of stevia extracts in foodstuffs.

Cargill and Coca-Cola have developed Truvia and Rebiana from stevia leaves, and PepsiCo and Merisant have developed Pure Via. The elimination of aftertaste is the most critical issue in the success of stevia sweeteners (Global Stevia Industry Perceptions Report 2009).

Cargill produced purified rebaudioside A compositions using solvent crystallization. Coca-Cola has also prepared rebaudioside derivatives with a more desirable flavor profile by heating a solution containing rebaudioside A with an organic acid.

PepsiCo has developed steviol glycoside isomers to be used as sweeteners in carbonated and noncarbonated beverages, oatmeal, cereals, snacks, and bakery products.

The Council of Scientific and Industrial Research in India produced steviosides using water and food-grade ion exchange resins in less time with increased yield.

Stevia extracts can be blended with maltodextrin, sucrose, fructose, and glucose and noncongruent flavor volatiles (taste-modifying composition) to produce a tabletop sweetener, as reported by Cargill.

SCL can be blended with stevia extracts including maltodextrins and erythritol to impart aftertaste and bitterness in low-calorie carbonated beverages, tea, and coffee (McNeil Nutritionals). Stevia extracts are also added to erythritol-based tabletop sweeteners, masking erythritol-induced brightening and eliminating cooling.

Finally, Cadbury used the synergistic combination of extracts from stevia leaves, *Rubus suavissimus* leaves, and *Cucurbitaceae* fruit, particularly luo han guo or *Siraitia grosvenorii*.

10.9 CYCLAMATE

The SCF reviewed the toxicity of CYC, cyclohexylamine, and dicyclohexylamine in 1985 and established a temporary ADI of 0–11 mg/kg bw, expressed as cyclamic acid, for cyclamic acid and its sodium and calcium salts (Commission of the European Communities 1985).

The new epidemiological data revealed no indications of the harmful effects on human reproduction parameters of either CYC used as a food additive or of workplace exposure to cyclohexylamine, although the latter study was considered difficult to interpret (SCF 2000).

The committee noted that no *in vitro* studies to compare the relative sensitivity of human, monkey, and rat testicular tissue to cyclohexylamine were performed but acknowledged the difficulties in performing such studies. The committee now no longer requires such studies.

After considering all the data available on the conversion of CYC to cyclohexylamine in humans, including the new data provided, the committee concluded the uncertainties with respect to the conversion rate in humans could be eliminated but the 18.9% conversion rate used for establishing the temporary ADI of 0–11 mg/kg bw is no longer appropriate. There are large interindividual variations observed in conversion rates and a lack of knowledge about the minimal time span of exposure to cyclohexylamine, which might result in testicular damage. The committee therefore concluded the maximum observed individual overall conversion of CYC to cyclohexylamine and the absorption of the latter would be 85%. This would be more appropriate for calculating an ADI. Because a maximum conversion figure would be utilized, a reduced safety factor should be applied for interindividual differences.

The committee concluded that a full ADI for CYC could now be established. Taking a NOAEL of 100 mg/kg bw for cyclohexylamine, allowing for the difference in molecular weight between cyclamic acid and cyclohexylamine, using an 85% overall conversion rate for ingested CYC, and applying a 32-fold safety factor, a full ADI of 0–7 mg/kg bw, expressed as cyclamic acid, for cyclamic acid and its sodium and calcium salts was established by the committee.

The ADI for CYC has been set at 11 mg/kg bw by the JECFA and at 7 mg/kg bw by the SCF.

With a relative sweetness of approximately 40 and the recommended ADI by the SCF, the sugar equivalent is 280.

Arcella et al. (2004) designed a food frequency questionnaire to identify adolescents who were high consumers of sugar-free soft drinks and tabletop sweeteners. A randomly extracted sample of teenagers (n = 3982) living in the District of Rome (Italy) filled it in. A consumer survey was then carried out in a randomly extracted subsample of males and females and in all females who reported a high consumption of sugar-free soft drinks and/or tabletop sweeteners. A total of 362 subjects participated in a detailed food survey by recording, at the brand level, all foods and beverages ingested over 12 days. For each sugar-free product, producers provided the concentration of intense sweeteners (SAC, ASP, ACS K, and CYC).

No intake in excess of the ADI was observed. In addition, medicines and supplements were taken into account, and these did not result in a large impact on chronic exposure to intense sweeteners. The intake levels did not exceed the ADI even under a worst-case scenario, which was performed to take into account a hypothetical future substitution of all regular food products with their sugar-free version. It can be concluded that, with the observed current consumption patterns and occurrence levels, the risk of an excessive intake of intense sweeteners by Italian teenagers is extremely low.

Ajinomoto used it as an effective agent (agonist) for appetite regulation and gastrointestinal health promotion, alleviating gastroesophageal reflux disease and functional dyspepsia.

10.10 SUCRALOSE AND ACESULFAME K

Both of these sweeteners were approved by Directive 2003/115. For SCL, we have an opinion of the SCF on SCL since 2000 (SCF 2000), following the allocation of an ADI as a sweetener in a large number of countries.

The ADI for ACS K has been set at 15 mg/kg bw by the JECFA and at 9 mg/kg bw by the SCF. With a relative sweetness of approximately 200 and with the recommended ADI of SCF, the sugar equivalent is 1800.

There is adequate evidence, both for SCL and its hydrolysis products, that there are no concerns about mutagenicity, carcinogenicity, and developmental or reproductive toxicity. Effects have been observed in some experimental animal studies on immune parameters, the gastrointestinal tract, and body weight gain. Consideration of the critical studies on these aspects have identified reduced body weight gain, when it is attributable to direct SCL toxicity rather than secondarily to reduced food intake because of impalatability of the diet, as the pivotal effect for establishing an ADI. The overall NOAEL for such reductions in body weight gain was 1500 mg/kg bw/day.

The committee concludes that SCL is acceptable as a sweetener for general food use and that a full ADI of 0–15mg/kg bw can be established based on the application of a 100-fold safety factor to the overall NOAEL of 1500 mg/kg bw/day.

The relative sweetness of SCL is 600; hence, the sugar equivalent is $600 \times 15 = 9000$.

Recent developments in SCL from Tate and Lyle Technology, Ltd., include improvements in process technology, increasing yield by multiple liquid–liquid extraction steps, multiple crystallization steps, recycling, and the use of organic solvents immiscible with water. Other improvements include purification where the partition coefficient of SCL into an organic solvent is increased.

Alternative production methods of SCL include its production from sucrose-6-acylate as proposed by Tate and Lyle Technology (Leatherhead Food Research 2010).

Mamtek International has produced SCL by the purification of SCL-6-ester using organic solvents (ethyl or petroleum ether). SCL-6-ester is used in the preparation of SCL. Another proposed method included the chlorination of sugar with bis trichloromethyl carbonate.

SCL forms include SCL crystals, stable even at high temperatures, freeze-dried SCL, and a granular form consisting of large neat or micronized SCL particles with a smoother surface.

Blends of SCL with polyols have been made by Tate and Lyle Technology for addition into beverages or sugar-free chewing gums. McNeil Nutritionals, LLC, made a blend of SCL with *S. rebaudiana Bertoni*. In that case, SCL is used to mask the licorice off flavor associated with some stevia extracts.

ACS K has been used in blends such as calorie-reduced beverages in a blend with neotame at a ratio of 90:10 (http://www.pepsicobeveragefacts.com/sweeteners.php).

ACS K and isomaltulose have been blended by Nutrinova to produce low-GI products that can promote weight control and prevent diseases.

Synergy between ASP and ACS K using solvents has been demonstrated by Nutrinova.

ACS K is also used in mixtures with HFCS 55 or 42 for calorie reduction.

Acesulfame can also be added to SCL to improve the residual sweet taste of SCL in foods. Moreover, it can also be added to thaumatin to mask any unpleasant odors and flavors from nutrients, e.g., a dietary composition useful for enhancing muscle strength or fatigue resistance (Gakic formulations).

Finally, it can act synergistically with inulin fiber and oligofructose to enhance sweetness and allow the incorporation of oligosaccharides into foods.

10.11 GLYCYRRHIZIN AND GLYCYRRHIZINIC ACID

Licorice and licorice derivatives, including ammoniated glycyrrhizin, are affirmed as GRAS for use in foods by the FDA (21 CFR 184.1408) (FDA 2011). Regulations include specifications and maximum use levels for licorice and licorice derivatives. FDA assumes that glycyrrhizin levels in foods do not pose a health hazard. Glycyrrhizinic acid was evaluated during a JECFA meeting (WHO 2005). Although a formal ADI was not established, the committee indicated consumption of 100 mg/day would not cause adverse effects in the majority of adults; however, a small set of the population may be more susceptible to its physiological effects, even at lower doses (Isbrucker and Burdock 2006). Directive 2004/77 reported that glycyrrhizinic acid and its ammonium salt should be labeled accordingly: that it contains licorice and patients suffering from hypertension should avoid excessive consumption. It is mainly found in chewing gums and drinks. Various genotoxic studies have shown that glycyrrhizin is neither teratogenic nor mutagenic. An ADI of 0.015–0.229 mg glycyrrhizin/kg body weight/day is proposed. A limit of less than 50 ppm glycyrrhizin was established by the Council of Europe and the U.K. Food Additive and Contaminants Committee. The Dutch Nutrition Information Bureau advised against daily glycyrrhizin consumption in excess of 200 mg, corresponding to 150 g of licorice confections (Fenwick, Lutomski, and Nieman 1990), although the glycyrrhizin content of confectionery products can vary as much as 30-fold. SCF (2003) did not establish an ADI for glycyrrhizinic acid; however, they reported an upper limit of consumption for the majority of the population of 100 mg/day.

10.12 LEGISLATION IN JAPAN

In Japan, a category of foods for special dietary uses exists, which includes foods for the ill. Low-calorie foods are included in this category, targeting diabetes and obesity. These foods require premarket approval. Low-calorie foods are defined as being less than 50% of the calories of comparable foods or counterparts. Low-calorie tabletop sweeteners are sold in this category. Combinations of bulking agents and intense sweeteners may also be included. Moreover, the category food for specified health uses (FOSHU) includes dietary fibers, oligosaccharides, and isoflavones. Oligosaccharides are added in table sugar to help maintain good GI condition and bowel movement (350 approved cases until the end of 2010). Xylitol with calcium monohydrogen phosphate and Fukuronori extract are added to help maintain strong and healthy teeth (73 approved cases). By August 2010, a report on health claims was released and involved FOSHU approval and regulation of health foods, when scientific evidence on targeted nutritional components for further discussion about authorization was collected (Consumer Affairs Agency 2010).

Japan is one of the leading agricultural importing nations in the world. Japan has far-reaching policies that affect caloric sweeteners (e.g., sugar and corn syrup), because it wishes to protect high-cost domestic production of sugar cane and sugar beets against foreign competition. Sugar beets are raised in the northernmost large island, Hokkaido, while sugar cane is grown in southernmost Japan, on small islands south of Kyushu, extending to Okinawa.

Sugar production is about 800,000 tons, with sugar consumption about 2.3 million tons and the total supply of sweeteners about 3.9 million tons (Fukuda, Dyck, and Stout 2002). Corn from the United States has been the main feedstock for producing HFCS in Japan, and at a level of 3 million tons a year, Japan's imports of corn for HFCS are about 20% of its total corn imports. Japan's total market for intense sweeteners is worth approximately 330 million U.S. dollars in 2009 (Leatherhead Food Research 2010), equivalent to more than 80% of the regional total (400 million U.S. dollars from Japan and the Far East). SCL is one of the most popular sweeteners used in the manufacture of soft drinks, confections, ice cream, milk, pickles, and functional foods followed by stevia, accounting for 40% of the tabletop sweeteners. ASP and ACS K are also used with the latter used in the ready-to-drink coffee drinks sector.

Guidelines for the designation of food additives and for the revision of standards for use of food additives have been described in Annex 5 (Figure 10.1).

Scientific evaluations are conducted by the Pharmaceutical Affairs and Food Sanitation Council from the view of the public health. In these evaluations, standards of the Joint FAO/WHO Codex Alimentarius Commission and conditions of Japanese food intake will be considered.

1. Safety—The safety of the targeted food additive should be proven or confirmed in the intended methods of use.
2. Effectiveness—It should be proven or confirmed that the use of the food additive comes under one or more of the purposes set out in items a–b below. However, when the manufacturing or processing method for a target food can be improved or modified at comparatively low cost and the improved or modified method does not require the food additive for the manufacture or processing of the food, the use of the food additive is not justified.
    a. To preserve the nutritional quality of the food—An intentional reduction in the nutritional quality of a food would be justified in the circumstances dealt with in item b below and also in other circumstances where the food does not constitute a significant item in a normal diet.
    b. To provide necessary ingredients or constituents for food manufactured for groups of consumers having special dietary needs, provided that the food additive is not intended to provide medical effects, such as the prevention or treatment of certain diseases.
    c. To enhance the keeping quality or stability of a food or to improve its organoleptic properties, provided that this does not change the nature, substance, or quality of the food so as to deceive the consumer.

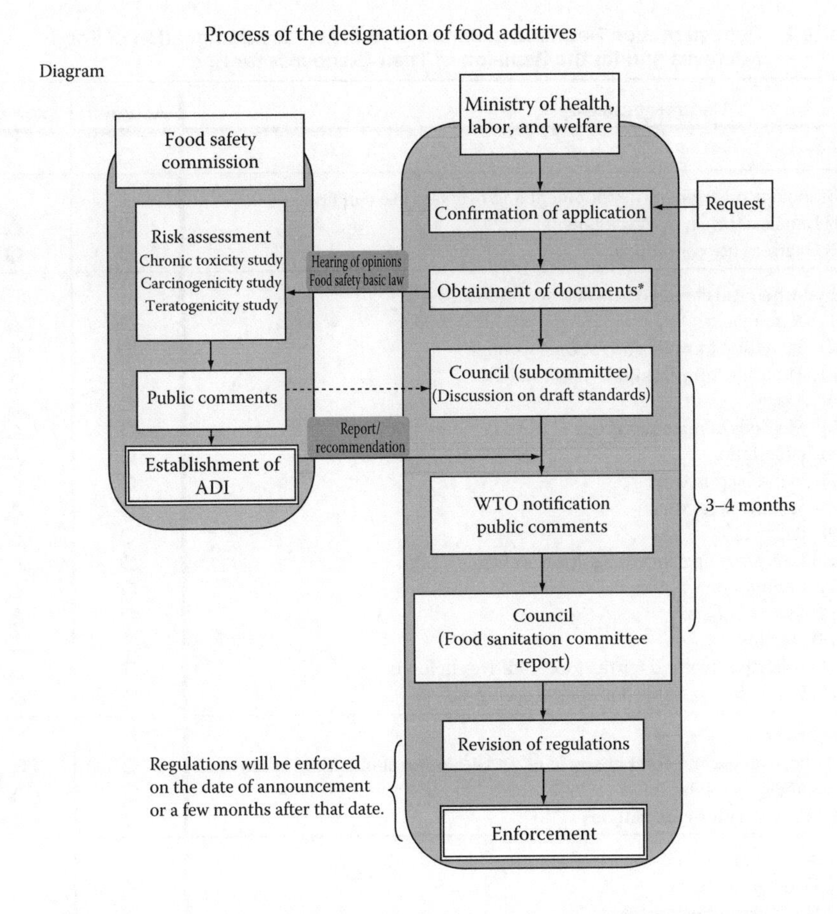

Figure 10.1 Designation of the food additives process in Japan. (Adapted from Japan Food Chemicals Research Foundation, Japanese Guidelines. Annex 5: The guidelines for designation of food additives, and for revision of standards for use of food additives (Excerpt), 1–11, 2010. With permission.)

d. To provide aids in the manufacture, processing, preparation, treatment, packing, transport, or storage of food, provided that the food additive is not used to disguise the effects of the use of faulty raw materials or of undesirable (including unhygienic) practices or techniques during the course of any of these activities.

Those who wish to apply for the designation of a food additive or to apply for the revision of standards for the use of a food additive may submit an application to the Minister of Health, Labor, and Welfare. The application should be accompanied by any required documentation on draft specifications and use standards and the safety of the food additive.

The Japan Food Chemicals Research Foundation (2010) has listed the designated additives in Table 10.2, as mentioned in Article 12 of the Enforcement Regulations under the Food Sanitation Law. These additives are listed in alphabetical order. There are 411 in total as of 13 December 2010. The number preceding the name of each additive is the sequence number given to the corresponding additive in the original Japanese table. Among those additives, a number of additives are listed, including ACS K, ASP (α-L-aspartyl-L-phenylalanine methyl ester), disodium glycyrrhizinate, mannitol, neotame, SAC, D-sorbitol, SCL (trichlorogalactosucrose), and xylitol.

Table 10.2 shows the documentation required in Japan to apply for the designation of food additives and for the revision of their standards for use.

Table 10.2 Documentation Required in Japan to Apply for the Designation of Food Additives and for the Revision of Their Standards for Use

Document category	Approval	Revision
1. Summary	O	O
2. Chronology on origin or development and overseas use conditions		
(1) Details of origin or development	O	Δ
(2) Overseas use conditions	O	O
3. Physiochemical chracteristics and specifications		
(1) Name	O	Δ
(2) Structural formula and rational formula	O	Δ
(3) Molecule formula and formula weight	O	Δ
(4) Assay	O	Δ
(5) Methods of manufacturing	O	Δ
(6) Description	O	Δ
(7) Identification tests	O	Δ
(8) Specific properties	O	Δ
(9) Purity tests	O	Δ
(10) Loss on drying, loss on ignition, or water	O	Δ
(11) Residues on ignition	O	Δ
(12) Method of assay	O	Δ
(13) Stability	O	Δ
(14) Analytical method for the food additives in foods	O	Δ
(15) Principles to establish proposed specifications	O	Δ
4. Effectiveness		
(1) Effectiveness and comparison in effect with other similar food additives	O	O
(2) Stability in foods	O	Δ
(3) Effects on nutrients of foods	O	Δ
5. Safety evaluation		
(1) Toxicity studies		
1) 28-day toxicity study	O	Δ
2) 90-day toxicity study	O	Δ
3) One-year toxicity study	O	Δ
4) Reproduction study	O	Δ
5) Teratogenicity study	O	Δ
6) Carcinogenicity study	O	Δ
7) Combined chronical toxicity/carcinogenicity study	O	Δ
8) Antigenicity study	O	Δ
9) Mutagenicity study	O	Δ
10) General pharmacological study	O	Δ
(2) Metabolism and pharmacokinetic studies	O	Δ
(3) Daily intake of food additives	O	Δ
6. Proposed sandards for use	O	O

Source: Japan Food Chemicals Research Foundation, Japanese Guidelines. Annex 5: The guidelines for designation of food additives, and for revision of standards for use of food additives (Excerpt), 1–11, 2010. With permission.

Note:

Documents marked with the symbol O are basically required.

Documents marked with the symbol Δ should be submitted, only when considered as necessary: e.g., where new information is obtained.

Table 10.3 Permitted Substances, Maximum Level of Addition, and Principal Uses in Japan

Substance Name	Permitted Food	Max. Level	Other Principal Uses
Acesulfame potassium	Substitute for sugar	Not more than 15 g/kg	Directly added to coffee and tea and used as a substitute food for sugar
	Food with nutritional function (limited to tablets)	Not more than 6 g/kg	
	Chewing gum	Not more than 5 g/kg	
	Confections and pastries	Not more than 2.5 g/kg	
	Ice cream products, tare, jam, tsukemono, ice candy, and flour paste	Not more than 1 g/kg	
	Wine, alcoholic beverages, soft drinks, milk drinks, and lactic bacteria–fermented beverage	Not more than 0.5 g/kg	
Disodium glycyrrhizinate	Soy sauce and miso (fermented soybean paste)	—	—
Saccharin	Chewing gum	Not more than 0.05 g/kg as saccharin	—
Sodium saccharin	Koji-zuke (preserved in koji, fermented rice), su-zuke (pickled in vinegar), and takuan-zuke (preserved radish in rice bran paste)	Less than 2.0 g/kg (as a residual level of sodium saccharin)	—
	Powdered nonalcoholic beverage	Less than 1.5 g/kg	—
	Kasuzuke (pickled in sake lees), misozuke (preserved in fermented soybean paste), shoyu-zuke (preserved in soy sauce), and processed fish and shellfish (excluding surimi products, tsukudani, and pickled food)	Less than 1.2 g/kg	—
	Nimame (cooked beans or peas, sweetened), processed seaweed, soy sauce, and tsukudani (preserved food boiled in soy sauce)	Less than 0.5 g/kg	—
	Edible ices (including sherbet, flavored ices), milk drinks, sauce, nonalcoholic beverages, fish-paste product, syrup, vinegar, and lactic acid bacteria drinks	Less than 0.3 g/kg (less than 1.5 g/kg) in case of materials for nonalcoholic beverages, lactic acid bacteria drinks, or fermented milk product to be diluted not less than fivefold before use, less than 0.90 g/kg in case of vinegar to be diluted not less than threefold before use	—
	An (adzuki bean paste), fermented milk product, flour paste, ice cream products, jam, and miso (fermented soybean paste)	Less than 0.2 g/kg	—
	Confectionery products	Less than 0.1 g/kg	—
	Canned or bottled food	Less than 0.2 g/kg	—

(*continued*)

Table 10.3 (Continued) Permitted Substances, Maximum Level of Addition, and Principal Uses in Japan

Substance Name	Permitted Food	Max. Level	Other Principal Uses
Sucralose	Substitute for sugar	Not more than 12 g/kg	Directly added to coffee or tea and used as a substitute food for sugar
	Chewing gum	Not more than 2.6 g/kg	
	Confections and pastry	Not more than 1.8 g/kg	
	Jam	Not more than 1.0 g/kg	
	Japanese sake (rice wine), compound sake (formulated rice wine), wine, miscellaneous alcoholic beverage, soft drinks, milk drink, and lactic acid bacteria drinks	Not more than 0.4 g/kg	
	Other foods	Not more than 0.58 g/kg	

Source: Japan External Trade Organization. Specifications and Standards for Foods and Food Additives under the Food Sanitation Act 2008, 101–103. 2009. With permission.

Permitted substances' maximum level of addition and principal uses in Japan (Japan External Trade Organization 2009) are shown in Table 10.3.

10.13 LEGISLATION IN INDIA

Nutritive sweeteners used in India include sucrose, glucose, and HFCS.

Sugar alcohols or polyols include sorbitol, mannitol, xylitol, maltitol, isomalt, and lactitol. The label declaration necessary is "Polyols may have a laxative effect."

Sorbitol is permitted in jams, jellies, fruit cheese, and fruit marmalades at a 3% maximum.

Polyols, singly or in combination, are added in the following foods (Pai 2010):

- Snacks/saoutiries (fried products)—Chivda, bhujia, dal moth, kodubale, khara boondi, spiced and fried dal, and banana chips
- Sweets (carbohydrate- and milk-based products)—Halwa, mysore pak, boondi ladoo, jalebi, khoya Burfi, peda, gulab jamun, rasgulla, and similar milk product-based sweets sold by any name
- Instant mixes—Idli mix, dosa mix, puliyogare mix, pongal mix, gulab jamun mix, jalebi mix, and vada mix
- Rice- and pulses-based papads
- Ready-to-serve beverages and tea- or coffee-based drinks
- Chewing/bubble gum
- Sugar-based or sugar-free confections
- Chocolates
- Synthetic syrup for dispensers
- Lozenges

Jaggery and khandsari are the traditional Indian sugar cane sweeteners (Rao, Das, and Das 2007).

High-intensity low-calorie sweeteners include SAC, CYC, ASP, ACS K, SCL, stevia.

Prevention of Food Adulteration (PFA) rules have been adopted since 1954. FSSA 2006 is slowly taking over and shortly will replace PFA. The Indian Food Authority has been appointed, and various scientific committees and panels are also appointed recently. Science-based regulations have been adopted using risk analysis, allowing various ingredients and additives. They are listed as follows

1. Every package of food containing permitted artificial sweetener shall carry the following:
   a. This [name of food] contains [name of artificial sweetener]
   b. Not recommended for children

c.1. Quantity of sugar added ____ g/100 g
c.2. No sugar added in the productiv. Not for phenylketonurics (if ASP is added)

2. In addition to the above declaration, the declaration "contains artificial sweetener and for calorie conscious" should also be written on packages.

Packages of artificial sweeteners marketed as tabletop sweeteners must carry following labels:

- Contains [name of artificial sweetener]
- Not recommended for children

Permitted levels of artificial sweeteners are listed as follows:

- SAC
    - Carbonated water: 100 ppm
    - Chocolates and traditional Indian sweets: 500 ppm
    - Sugar-based or sugar-free confections and chewing gum/bubble gum: 3000 ppm
- ASP
    - Carbonated water: 700 ppm
    - Biscuits, bread, and cakes: 2000 ppm
    - Indian sweets: 200 ppm
    - Jams and jellies: 1000 ppm
    - Sugar-based or sugar-free confections: 10,000 ppm
    - Ice cream: 1000 ppm
    - Flavored milk: 600 ppm
    - RTE cereal: 1000 ppm
    - Still beverages: 600 ppm
- ACS K
    - Carbonated water: 300 ppm
    - Biscuits, cakes, and pastries: 1000 ppm
    - Indian sweets: 500 ppm
    - Sugar-based or sugar-free confections: 3500 ppm
    - Still beverages: 300 ppm
- SCL
    - Carbonated water: 300 ppm
    - Biscuits and cakes: 750 ppm
    - Indian sweets: 750 ppm
    - Still beverages: 300 ppm
    - Jams and jellies: 450 ppm
    - Popsicles and candies: 800 ppm
- Neotame
    - Carbonated water: 33 ppm
    - Soft drink concentrates: 33 ppm
- Mixture of ASP and ACS K
    - Carbonated water: Allowed
    - Soft drink concentrates: Allowed
    - Synthetic syrup for dispensers: Allowed

According to the New Regulation GSR 664, the nutritional information or nutritional facts per 100 g, 100 ml, or per serving of the product shall be given on the label containing the following:

- Energy value (in kilocalories)
- Amounts of protein, carbohydrate (quantity of sugar specified), and fat (in grams)
- Amount of any other nutrient for which a nutrition or health claim is made

10.14 THE UNITED STATES

More than 4000 products containing artificial sweeteners have been launched in the United States between 2000 and 2005. However, the number of new product launches for high-intensity sweeteners worldwide has fallen by 5% in 2008 according to Tate and Lyle. For each variety of sweeteners, the regulatory status differs according to the region and country. For example, in the U.S., stevia use has increased over the last 5 years following regulatory approval, whereas CYC is banned after a poor health image (Leatherhead Food Research 2010). In 2009, the global market for intense sweeteners used in the manufacturing of food and beverages amounted to more than 1.27 billion U.S. dollars in terms of value; ASP is worth an estimated 345 million U.S. dollars (27% of the global market), followed by SCL (26%), SAC, and CYC, stevia, and neotame.

By volume, CYC is the leader, accounting for 53% of market volume in 2009, followed by SAC and ASP. These differences between value and volume are a result of the low cost of CYC and SAC (Leatherhead Food Research 2010).

In the U.S., approximately 194 million consumers eat low-calorie and low-sugar foods. Noncarbonated drinks (fruit-based beverages) and carbonated soft drinks show the highest penetration of approximately 60%, followed by ice creams and chewing gums. The North American market for intense sweeteners was worth 485 million U.S. dollars in 2009, with the leaders being ASP, ACS K, SCL, and stevia.

The Nutrition Labeling and Education Act of 1990 (21 CFR 101.9) (FDA 1990) established the principles and format for the nutrition labeling of foods. This act is complemented by a number of additional entries in the Code of Federal Regulations (CFR) relating to nutrient content claims. Some examples are listed as follows:

- "Reduced calorie" is defined as 25% fewer calories than a full-calorie equivalent product (21 CFR 101.60) (FDA 2001).
- "Low calorie" is defined as foods that contribute <40 cal/RACC or <120 cal/100 g of food.
- "Light" is defined as follows (21 CFR 101.56):
    - For foods that derive more than 50% of their calories from fat, the fat content is reduced by at least 50% compared to a full-fat equivalent product.
    - For foods that derive less than 50% of their calories from fat, the fat content is reduced by at least 50%, and the calorie content is reduced by at least one-third.

In 2004, the Calorie Control Council (www.caloriecontrol.org) entered into a dialogue with the U.S. FDA on the subject of permissible claims relating to low-calorie foods. The FDA confirmed the acceptance of the following: "Low calorie [name of food] may be useful in weight control. Obesity increases the risk of developing diabetes, heart disease, and certain cancers."

Labeling of foods with regard to sweetener ingredients and claims is a complex subject. Many countries now have their own nutrition labeling regulations, and it is important whenever contemplating the launch of a new food product to obtain local expert opinions as to labeling implications (Stowell 2006).

10.15 LEGISLATION IN EUROPE

The European market for intense sweeteners was worth an estimated 255 million U.S. dollars in 2009 (Leatherhead Food Research 2010). This has risen over the last 2 years. The U.K., France, and Germany are the largest markets in the EU. ASP accounts for the largest market in the U.K. via Canderel (tabletop sweetener). SCL is another successful sweetener (Splenda), followed by SAC. Stevia is a promising sweetener.

The U.K. accounts for the largest Western European market for diet or light carbonated beverages with volume sales of 2.2 billion L in 2009.

Intense sweeteners have also been used in noncarbonated soft drinks, e.g., Live Mate launched in France in 2010 with stevia rebaudioside A added to it.

Moreover, lactitol and maltitol are frequently used in the manufacture of low- and reduced-sugar chocolate with rising sales. Swiss company Villars made a chocolate bar with stevia in France and launched it in 2009.

Finally, ASP is used in the manufacture of sugar-free chewing gums, e.g., Wrigley, along with xylitol and sorbitol. The highest penetration of sugar-free chewing gums in 2009 occurred in Spain, followed by the U.K. and the Netherlands.

Italy has the highest penetration of sweeteners in boiled sweets, such as the Dietorelle brand, with ASP and ACS K added, competing in toffees and caramels.

Other sectors, such as yogurt, use a blend of ASP and ACS K as in the case of Activia probiotic yogurt (some of its flavors) sold in the U.K. and Mullerlight and Weight Watchers, using ASP.

Cereal bars and savory snacks are made with ASP in Europe. Walkers Snack Foods (leader in the savory snacks market) uses ASP in Walkers Sensations and Doritos products. Finally, biscuits called Boasters from McVitie's are made with a blend of ASP and ACS K.

10.16 NEW SWEETENERS

Danmark Tekniske Universitet has developed pyrrolidinone dipeptide analogues with similar structure to ASP and alitame suitable for use as taste enhancers.

PepsiCo has enzymatically produced a natural sweetener from oats, masking off flavors and reducing the use of sugar, SCL, and ACS K. The hydrolysis of oat (oat groats, oat flour, rolled oats, oatmeal, and partially milled oats) with a-amylase, b-amylase, and acid alpha-glucosidase is used to obtain modified flour and drying of the flour to obtain sweetener composition (Leatherhead Food Research 2010).

Finally, specific stereoisomers of monatin, a naturally occurring high-intensity sweetener suitable for use in tabletop sweeteners, foods, and beverages, have been employed by Cargill with superior taste characteristics and excellent stability, not requiring a phenylalanine warning for patients with phenylketonuria.

REFERENCES

AFSSA (French Food Safety Agency). (2009). Opinion of the French Food Safety Agency on a new draft order on the use of rebaudioside A, an extract of *Stevia rebaudiana*, as a food additive, 1–2.

AFSSA (French Food Safety Agency). (2002). Assessment Report: Opinion on a possible link between the exposition to aspartame and the incidence of brain tumors in humans, 1–18.

ANSES (French Agency for Food, Environment, and Occupational Health Safety). Opinion of the French Agency for Food, Environmental and Occupational Health and Safety on a publication reporting the incidence of cancer in male mice after administrattion of aspartame in their feed, and a second publication on a prospective cohort study of pregnant women reporting an association between the consumption of carbonated (fizzy) softdrinks containing sweeteners and the risk of preterm delivery. http://www.FRENCHAAAT2011sa0015EN.pdf, accessed December 2011.

Arcella, D., C. Le Donne, R. Piccinelli, and C. Leclercq. (2004). Dietary estimated intake of intense sweeteners by Italian teenagers: Present levels and projections derived from the INRAN-RM-2001 food survey. *Food and Chem. Toxicol., 42*, 677–685.

Arnold, D. L. (1984). Toxicology of saccharin. *Fundam. Appl. Toxicol., 4*, 674–685.

Baron, R. F., and L. Y. Hanger. (1998). Using acid level, acesulfame potassium aspartame blend ratio and flavor type to determine optimum flavor profiles of fruit flavored beverages. *J. of Sensory Studies, 13*, 269–283.

Butchko, H. H. et al. (2002). Aspartame—Part 2: Review of safety. *Reg. Toxicol. Pharmacol., 35*(2), S1–93.

Butchko, H. H., and W. W. Stargel. (2001). Aspartame: Scientific evaluation in the postmarketing period. *Reg. Toxicol. Pharmacol., 34*(3), 221–233.

Carakostas, M. C., L. L. Curry, A. C. Boilea, and D. Brusick. (2008). Overview: The history, technical function and safety of rebaudioside A, a naturally occurring steviol glycoside, for use in food and beverages. *Food and Chem. Toxicol., 46*(7), S1–S10.

Commission of the European Communities. (1985). Report of the Scientific Committee for Food on Sweeteners (opinion expressed on 14 September 1984). *Reports of the Scientific Committee for Food (Sixteenth Series).* CEC, Luxembourg, EUR 10210 EN.

Consumer Affairs Agency (CAA). (2010). Regulatory systems of health foods in Japan, 1–13.

European Commission. (1989). Council Directive 89/398/EEC on the approximation of the laws of the Member States relating to foodstuffs intended for particular nutritional uses. *Off. J. Eur. Un.*, L186, 27.

European Commission. (1994). Directive 94/35/EC of the European Parliament and of the Council of 30 June 1994 on sweeteners for use in foodstuffs. *Off. J. Eur. Un.*, L237, 13.

European Commission. (1997). Directive 96/83/EC of the European Parliament and of the Council of 19 December 1996 amending Directive 94/35/EC on sweeteners for use in foodstuffs. *Off. J. Eur. Un.*, L48, 16.

European Commission. (2003). Directive 2003/89/EC of the European Parliament and of the Council of 10 November 2003 amending Directive 2000/13/EC as regards indication of the ingredients present in foodstuffs. *Off. J. Eur. Un.*, L308, 15.

European Commission. (2004a). Directive 2003/114/EC of the European Parliament and of the Council of 22 December 2003 amending Directive 95/2/EC on food additives other than colours and sweeteners. *Off. J. Eur. Un.*, L24, 58–64.

European Commission. (2004b). Directive 2003/115/EC of the European Parliament and of the Council of 22 December 2003 amending Directive 94/35/EC on sweeteners for use in foodstuffs. *Off. J. Eur. Un.*, L24, 65–71.

European Commission. (2006a). Commission Directive 2006/128/EC of 8 December 2006 amending and correcting Directive 95/31/EC laying down specific criteria of purity concerning sweeteners for use in foodstuffs. *Off. J. Eur. Un.*, L346, 6–14.

European Commission. (2006b). Directive 2006/52/EC of the European Parliament and of the Council of 5 July 2006 amending directive 95/2/EC on food additives other than colors and sweeteners and Directive 94/35/EC on sweeteners for use in foodstuffs. *Off. J. Eur. Un.*, L204, 10.

European Commission. (2008a). Commission Directive 2008/5/EC of 30 January 2008 concerning the compulsory indication on the labelling of certain foodstuffs of particulars other than those provided for in Directive 2000/13/EC of the European Parliament and of the Council. *Off. J. Eur. Un.*, L27, 12.

European Commission. (2008b). Commission Directive 2008/60/EC of 17 June 2008 laying down specific purity criteria concerning sweeteners for use in foodstuffs. *Off. J. Eur. Un.*, L158, 17–40.

European Commission. (2008c). Regulation (Ec) No 1333/2008 of the European Parliament and of the Council of 16 December 2008 on food additives. *Off. J. Eur. Un.*, L354, 16.

European Food Safety Authority (EFSA). 2010. EFSA report of the meetings on aspartame with national experts. EFSA Q-2009-00488. 36th Advisory Forum Meeting, No. 1641, 1–64.

European Food Safety Authority (EFSA). (2007). Neotame as a sweetener and flavour enhancer: Scientific Opinion of the Panel on Food Additives, Flavourings, Processing Aids and Materials in Contact with Food (Question No. EFSA-Q-2003-137) on a request from European Commission on neotame as a sweetener and flavour enhancer. *The EFSA Journal, 581*, 1–3.

European Food Safety Authority (EFSA). (2009a). Scientific Opinion of the Panel on Food Additives and Nutrient Sources Added to Food on a request from the European Commission related to the 2nd ERF carcinogenicity study on aspartame. *The EFSA Journal, 945*, 1–18.

European Food Safety Authority (EFSA). (2009b). Updated opinion on a request from the European Commission related to the 2nd ERF carcinogenicity study on aspartame, taking into consideration study data submitted by the Ramazzini Foundation in February 2009. *The EFSA Journal, 1015*, 1–3.

European Food Safety Authority (EFSA). (2010). Scientific Opinion of the Panel on Food Additives and Nutrient Sources Added to Food on a request from the Commission on the safety in use of steviol glycosides as a food additive. *The EFSA Journal, 1537*, 1–87.

European Food Safety Authority (EFSA). (2011). Statement of EFSA Revised exposure assessment for steviol glycosides for the proposed uses as a food additive on request from the European Commission, Question No. EFSA-Q-2010-01214, issued on 13 January 2011. *The EFSA Journal, 9*(1), 1972, 1–19.

Fenwick, G. R., J. Lutomski, and C. Nieman. (1990). Liquorice, Glycyrrhiza glabra L.-Composition, uses and analysis. *Food Chem., 38*, 119–143.

Food and Drug Administration (FDA). 1990. Code of Federal Regulations. 21 CFR 101.9. The Nutrition Labeling and Education Act of 1990. US Department of Health and Human Services, Washington, DC. http://www.fda.gov/.../FoodLabelingNutrition/ucm063102.htm, accessed December 2011.

Food and Drug Administration (FDA). (1996). Food additives permitted for direct addition to food for human consumption. Dimethyl dicarbonate. *Fed Regist., 61*, 26,786–26,788.

Food and Drug Administration (FDA). 2001. Food Code. Food labeling regulations. 21 CFR 101.6, 101.5. US Department of Health and Human Services, Washington, DC.

Food and Drug Administration (FDA). 2011. Code of Federal Regulations. 21 CFR 184.1408. US Department of Health and Human Services, Washington, DC. http://www.accessdata.fda.gov/scripts/cdrh/cfdocs/cfcfr/CFRSearch.cfm?fr=184.1408, accessed October 2011.

Fukuda, H., J. Dyck, and J. Stout. (2002). Sweetener Policies in Japan. USDA, SSS-234-01, 1–10.

Garnier, S., J. C. Leblanc, and P. Verger. (2001). Calculation of the intake of three intense sweeteners in young insulin dependent diabetics. *Food Chem. Toxicol., 39*(7), 745–749.

Global Stevia Industry. (2009). Global Stevia Industry Perceptions Report. *Knowgenix for Steviaworld Americas.*

Halldorsson, T. I., M. Strøm, S. B. Petersen, and S. F. Olsen. (2010). Intake of artificially sweetened soft drinks and risk of preterm delivery: A prospective cohort study in 59,334 Danish pregnant women. *Am. J. Clin. Nutr., 92*, 626–633.

Ilbäck, N. G., M. Alzin, S. Jahrl, H. Enghardt Barbieri, and L. Busk. (2000). Sweetener intake and exposure. Study among Swedish diabetics. Report 2. National Food Administration, Sweden.

Ilbäck, N.-G., M. Alzin, S. Jahrl et al. (2003). Estimated intake of the artificial sweeteners acesulfame-K, aspartame, cyclamate and saccharin in a group of Swedish diabetics. *Food Add. and Contam., 20*(2), 99–114.

Isbrucker, R. A., and G. A. Burdock. (2006). Risk and safety assessment on the consumption of Licorice root (Glycyrrhiza sp.), its extract and powder as a food ingredient, with emphasis on the pharmacology and toxicology of glycyrrhizin. *Reg. Toxicol. Pharmacol., 46*, 167–192.

Japan External Trade Organization. (2009). Specifications and standards for foods, food additives, under the Food Sanitation Act 2008, 101–103.

Japan Food Chemicals Research Foundation. (2010). Japanese Guidelines. Annex 5: The guidelines for designation of food additives, and for revision of standards for use of food additives (Excerpt), 1–11.

Joint FAO/WHO Expert Group on Food Additives (JECFA). (1991). Evaluation of certain food additives and contaminants. Thirty-Seventh Report of the Joint FAO/WHO Expert Group on Food Additives. WHO Technical Report Series 806. WHO, Geneva.

Joint FAO/WHO Expert Group on Food Additives (JECFA). (2009). Joint FAO/WHO Expert Committee on Food Additives Evaluation of Certain Food Additives. 69th report WHO technical report series no. 952, 1–222.

Joint FAO/WHO Expert Committee on Food Additives (JECFA). (2001). In Toxicological evaluation of certain food additives specifications, 57th Session, *FAO Food and Nutrition Paper, 52*(9).

Kroger, M., K. Meister, and R. Kava. 2006. Low-calorie sweeteners and other sugar substitutes: A review of the safety issues. *Compr. Rev. Food Sci. Food Saf. 5*, 35–47.

Leatherhead Food Research. 2010. The global market for intense sweeteners. Market Intelligence section. www.leatherheadfood.com/market.

Magnuson, B. A., G. A. Burdock, J. Doull, R. M. Kroes, G. M. Marsh, M. W. Pariza, P. S. Spencer, W. J. Waddell, R. Walker, and G. M. Williams. (2007). Aspartame: A safety evaluation based on current use levels, regulations, and toxicological and epidemiological studies. *Crit. Rev. Toxicol., 8*, 629–727.

Meyer, H. (2007). Aspartame. In: *Sweeteners*, Wilson, R. (ed.), 3rd *edition*. London: Leatherhead Publishing and Blackwell Publishing.

Olney, J. W., N. B. Farber, E. Spitznagel, and L. N. Robins. (1996). Increasing brain tumour rates: Is there is a link to aspartame? *J. Neurpathol. Exp. Neurol., 55*(11), 1115–1123.

Pai, J. S. (2010). Food Regulations and Safety of Food Ingredients with Special Emphasis on Sweeteners. Protein Foods and Nutrition Development Association of India.

Pearson, R. L. (1991). *Alternative Sweeteners*. Lyn O'Brien-Nabors (ed.). 147–165. New York: Marcel Dekker.

Rao, P. V. K. J., M. Das, and S. K. Das. (2007). Jaggery: A traditional Indian sweetener. *Indian Journal of Traditional Knowledge, 6*(1), 95–102.

Scientific Committee on Food (SCF). (1995). Opinion on Saccharin and Its Sodium, Potassium and Calcium Salts (expressed on 2 June 1995) CS/ADD/EDUL/148-FINAL February 1997, 1–8.

Scientific Committee on Food (SCF). (2000). Opinion of the Scientific Committee on Food on Sucralose. SCF/CS/ADDS/EDUL/190 Final.

Scientific Committee on Food (SCF). (2003). Opinion of the Scientific Committee on Food on glycyrrhizinic acid and its ammonium salt. European Commission Scientific Committee on Food Report SCF/CS/ADD/EDUL/225 Final.

Scientific Committee on Food (SCF). (2007). Guidance on the implementation of Regulation No. 1924/2006 on nutrition and health claims made on foods. Conclusions of the Standing Committee on the Food Chain and Animal Health, 1–14.

Scientific Committee on Food (SCF). (2000). Revised opinion on cyclamic acid and its sodium and calcium salts. SCF/CS/EDUL/192 Final.

Soffritti, M., F. Belpoggi, M. Manservigi, E. Tibaldi, M. Lauriola, L. Falcioni, and L. Bua. (2010). Aspartame administered in feed, beginning prenatally through life span, induces cancers of the liver and lung in male Swiss mice. *Am. J. Indust. Med., 53*, 1197–1206.

Stowell, J. (2006). Calorie control and weight management. In: H. Mitchell (ed.), *Sweeteners and sugar alternatives in food technology*, 1–421. Ames, IA: Blackwell Publishing.

Weihrauch, M. R., and V. Diehl. (2004). Artificial sweeteners: Do they bear a carcinogenic risk? *Annals of Oncology, 15*(10), 1460–1465.

Wilson, L. A., K. Wilkinson, H. M. Crews, et al. (1999). Urinary monitoring of saccharin and acesulfame-K as biomarkers of exposure to these additives. *Food Add. Contam.*, 16(6), 227–238.

World Health Organization (WHO). (2005). Evaluation of certain food additives. World Health Organization Technical Report Series 928, 1–156.

Yebra-Biurrun, M. C. (2000). Flow injection determination of artificial sweeteners: A review *Food Add. Contam., 17*, 733–738.

Yebra-Biurrun, M. C. (2005). GI-1828—Grupo de quimíca analítica, automatización, simplificación, medios biolóxicos e medio ambiente. *Encyclopedia of Analytical Science*, 562–572. Elsevier.

INTERNET SOURCES

http://www.efsa.europa.eu/en/press/news/ans110228.htm, accessed October 2011.

http://www.pepsicobeveragefacts.com/sweeteners.php, accessed October 2011.

CHAPTER 11

Nutritional and Health Aspects of Sweeteners

Theodoros Varzakas and Costas Chryssanthopoulos

CONTENTS

11.1 GLYCEMIC RESPONSE TO SUGARS AND SWEETENERS

Glycemic impact, defined as "the weight of glucose that would induce a glycemic response equivalent to that induced by a given amount of food" (Miller-Jones 2007), expresses relative glycemic potential in grams of glycemic glucose equivalents (GGEs) per specified amount of food. Therefore, GGE behaves as a food component, and (relative) glycemic impact (RGI) is the GGE intake responsible for a glycemic response (Monro and Shaw 2008). RGI differs from glycemic index (GI), because it refers to food and depends on food intake, whereas GI refers to carbohydrate and is a unitless index value unresponsive to food intake. Glycemic load (GL) is the theoretical cumulative exposure to glycemia over a period of time and is derived from GI as GI × carbohydrate intake. Contracted to a single intake of food, GL approximates RGI but cannot be accurately expressed in terms of glucose equivalents, because GI is measured by using equal carbohydrate intakes with usually unequal responses. RGI, on the other hand, is based on relative food and reference quantities required to give equal glycemic responses; hence, it is accurately expressed as GGE. The properties of GGE allow it to be used as a virtual food component in food labeling and in food-composition databases linked to nutrition management systems to represent the glycemic impact of foods alongside nutrient intakes. GGE can also indicate carbohydrate quality when used to compare foods in equal carbohydrate food groupings (Monro and Shaw 2008).

Mitchell (2008) showed how the glycemic concept is being used by the food manufacturing industry, how it is perceived and understood by consumers, and how different countries rate its importance in terms of regulatory provision and consequent labeling implications. The use of GI is the most prominent form of labeling in the marketplace to date, and the use of GI symbol programs and other labeling initiatives is considered. The Australian market has been exposed to the GI phenomenon the longest, and consumer awareness in this market is very high. However, on a global scale, the picture is very different, and consumer awareness varies considerably. A broader view of how the global consumer uses nutritional labels is given. She reviewed how consumers are willing to adopt foods that offer health benefits in general and, more specifically, from the glycemic concept. She also summarized aspects to be addressed for consumers to benefit from the glycemic concept in action in the longer term.

In Table 11.1, the glycemic and insulinemic responses to bulk sweeteners and alternatives are presented for monosaccharides, disaccharides, hydrogenated monosaccharides, hydrogenated disaccharides, and other alternatives. GL should be limited to 120 g/day according to Foster-Powell et al. (2002).

Intense sweeteners are consumed in such small quantities that they have no glycemic response of their own (Livesey 2006). Aspartame and sucralose are intense sweeteners with no acute glycemic response (Rodin 1990; Abdallah et al. 1997; Mezitis et al. 1996). However, if aspartame and sucralose are compared with maltodextrins, maltose, glucose, and sucrose under controlled conditions then marked reductions in the acute glycemic response would be expected for comparable sweetness (Livesey 2006). The use of intense sweeteners in place of glycemic carbohydrates wherever bulk is necessary for technological or organoleptic reasons requires the glycemic response to bulking agents to be considered too. The glycemic (and insulinemic) response to maltodextrin, bulk sweeteners, and bulking agents varies considerably, as shown in Table 11.1.

Fructose alone is low glycemic due to both slow absorption and the need for conversion to glucose in the liver prior to appearance in the blood as glucose. In addition, the carbohydrate may be partly stored as glycogen rather than released into the circulation. Furthermore, the energy from fructose is conveyed in the circulation for oxidation partly as lactate more than is the case for glucose. A similar situation occurs for sorbitol and xylitol, although slower absorption likely gives rise to less lactate; in addition, a high proportion escapes absorption. With isomalt and lactitol, an even greater proportion escapes absorption, which gives these polyols the lowest glycemic response of all so far mentioned. Another polyol, erythritol, is almost unique; most is absorbed and is low glycemic, because it is poorly metabolized in the tissues and escapes into the urine. Mannitol behaves similarly, although it is largely (75%) unabsorbed (Livesey 2006).

Trehalose, a rearrangement of sucrose, has a glycemic response comparable to sucrose (Table 11.1) in terms of its GL, although it peaks less sharply.

Isomaltulose, also derived by the rearrangement of sucrose, has a lower glycemic response to trehalose, although most of the isomaltulose is absorbed. Other low-glycemic carbohydrates include tagatose, fructans (fructooligosaccharides and inulin), polydextrose (PDX), and resistant

Table 11.1 Glycemic and Insulinemic Responses to Bulk Sweeteners and Alternatives

Sugars or Alternatives	Glycemic Response (g GGE/100 g)	Insulin Response (g IGE/100 g)	References
		Monosaccharides	
Glucose	100	100	Livesey 2003
Fructose	19	9	Foster-Powell et al. 2002
Tagatose	3	3	Donner et al. 1999, 2010
		Disaccharides	
Maltose	105	–	Foster-Powell et al. 2002
Trehalose	72	51	Livesey 2003; van Can et al. 2009a
Sucrose	68	45	Foster-Powell et al. 2002
Lactose	46	–	Foster-Powell et al. 2002
Isomaltulose	37.32	25	Sydney University glycemic index research service, van Can et al. 2009b; Holub et al. 2010
Maltodextrin	91	90	Macdonald and Williams 1988
Maltitol syrup	48	35	Livesey 2003
Polydextrose	5	5	Foster-Powell et al. 2002
		Hydrogenated Monosaccharides	
Erythritol	0	2	Livesey 2003
Xylitol	12	11	Foster-Powell et al. 2002; Livesey 2005
Sorbitol	9	11	Livesey 2003
Mannitol	0	0	Livesey 2003
		Hydrogenated Disaccharides	
Maltitol	45	27	Livesey 2003
Isomalt	9	6	Livesey 2003
Lactitol	5	4	Livesey 2003; Foster-Powell et al. 2002

Source: Mitchell, H. L., *Am. J. Clin. Nutr.* 87 (Suppl.), 244S–246S, 2008. With permission. Livesey, G., in *Sweeteners and Sugar Alternatives in Food Technology*, Blackwell Publishing, 2006. With permission.
Note: GGE, glycemic glucose equivalent; IGE, insulin glucose equivalent.

maltodextrins. The reduction in glycemia caused by sucrose replacing high-glycemic starch is considered an advantage (Livesey 2006; Miller and Lobbezoo 1994).

Among the studies undertaken with polyols [mainly with maltitol (MTL), isomalt, and sorbitol], the glycemic response versus glucose is similar in people with normal and abnormal carbohydrate metabolism, as exemplified by type 1 and 2 diabetics, provided that insulin-dependent participants receive insulin via an artificial pancreas (Livesey 2003).

The impact of slow digestible sources of dietary carbohydrate in reducing the risk of developing obesity and related metabolic disorders is unclear, although Brand-Miller et al. (2002) hypothesized that the ingestion of slowly digestible carbohydrates attenuates the postprandial rise in glycemia and insulinemia and enhances fat oxidation rates. The latter may assist in preventing body weight gain and insulin resistance.

van Can et al. (2009b) compared the postprandial metabolic response to the ingestion of sucrose versus isomaltulose. They hypothesized that the reduced digestion and absorption rate of isomaltulose would result in lower glycemic and insulinemic responses compared with the ingestion of sucrose, leading to greater postprandial fat oxidation rates. In a randomized, single-blind, cross-over study, 10 overweight subjects ingested two different carbohydrate drinks (sucrose and isomaltulose, 75-g carbohydrate equivalents) following an overnight fast (08.40 h) and with a standardized meal (12.30 h; 25% of the total energy content was provided as either a sucrose or an isomaltulose drink). Blood samples were taken before ingestion and every 30 min thereafter for a period of 3 h, substrate use was assessed by indirect calorimetry, and breath samples were collected. Ingestion of carbohydrates with a mixed meal resulted in a lower peak glucose and insulin response and a lower change in area under the curve [difference in area under the curve (dAUC)] following isomaltulose when compared with sucrose. Together with the lower glucose and insulin responses, postprandial fat oxidation rates were higher (14%) with isomaltulose compared with sucrose when ingested with a mixed meal. The attenuated rise in glucose and insulin concentrations following isomaltulose results in reduced inhibition of postprandial fat oxidation. The metabolic response to isomaltulose coingestion suggests that this may represent an effective nutritional strategy to counteract overweight-induced metabolic disturbances. The intake of isomaltulose in combination with a mixed meal resulted in an attenuated rise in the plasma glucose and insulin responses compared with sucrose and, subsequently, less inhibition of postprandial fat oxidation. The greater postprandial fat use was accompanied by higher circulating plasma nonesterified fatty acid (NEFA) concentrations. The latter is probably attributed to a greater supply of plasma NEFA, resulting from a reduced insulin-mediated suppression of lipolysis (Wolever 2003). These data seem consistent with two other research papers that highlighted the stimulating effect of isomaltulose ingestion on fat oxidation and/or lipid deposition compared with sucrose in rats and healthy men. Sato et al. (2007) observed significant reductions in visceral fat mass, adipocyte cell size, hyperglycemia, and hyperlipidemia after 8 weeks of isomaltulose feeding compared with sucrose feeding in Zucker fatty (fa/fa) rats. Arai et al. (2007) showed that peak plasma glucose and insulin levels were lower 30 min after the ingestion of the isomaltulose-containing liquid meal compared with the control formula ingestion in healthy men. Postprandial fat oxidation rates following the ingestion of the isomaltulose meal group were higher compared with the control formula group.

The slow digestible disaccharide isomaltulose (palatinose) is available as a novel functional carbohydrate ingredient for the manufacturing of low-glycemic foods and beverages. Although basically characterized, various information on physiological effects of isomaltulose is still lacking.

Thus, Holub et al. (2010) expanded scientific knowledge of the physiological characteristics of isomaltulose by a set of three human intervention trials. Using an ileostomy model, isomaltulose was found to be essentially absorbed, irrespective of the nature of food (beverage and solid food).

The apparent digestibility of 50-g isomaltulose from two different meals was 95.5% and 98.8%; the apparent absorption was 93.6% and 96.1%, respectively. In healthy volunteers, a single-dose intake of isomaltulose resulted in lower postprandial blood glucose and insulin responses than did sucrose while showing prolonged blood glucose delivery over a 3-h test. In a 4-week trial with hyperlipidemic individuals, regular consumption of 50 g/day of isomaltulose within a Western-type diet was well tolerated and did not affect blood lipids. Fasting blood glucose and insulin resistance were lower after the 4-week isomaltulose intervention compared with baseline. This would be consistent with possible beneficial metabolic effects as a consequence of the lower and prolonged glycemic response and lower insulinemic burden. However, there was no significant difference at 4 weeks after isomaltulose compared with sucrose. In conclusion, Holub et al. (2010) showed that isomaltulose is completely available from the small intestine, irrespective of the food matrix, leading to a prolonged delivery of blood glucose. Regular isomaltulose consumption is also well tolerated in subjects with increased risk of vascular diseases. With a calculated GI of 32, isomaltulose is a low-glycemic carbohydrate, as verified by Atkinson et al. (2008).

The insulin response for the saccharides was directly proportional to their glycemic response, both in the magnitude and shape of the response curve [incremental area under the curves: 15,208 [standard deviation (SD) 8639] versus 23,347 (SD 14,451) min × pmol/L; $P = 0.005$]. Isomaltulose evoked the lowest insulin response with a maximum of 227.8 pmol/L, which is more than 50% lower compared with sucrose (470.1 pmol/L) (Holub et al. 2010).

van Can et al. (2009a) compared the postprandial metabolic response to the ingestion of glucose versus trehalose. They hypothesized that the reduced digestion and absorption rate of trehalose is accompanied by an attenuated glycemic and insulinemic response, leading to a less inhibited postprandial fat oxidation rate. In a randomized, single-blind, cross-over study, 10 overweight subjects ingested two carbohydrate drinks (75-g carbohydrate equivalents of trehalose or glucose) following an overnight fast (08:40 h) and together with a standardized mixed meal (12.30 h; 25% total energy content was provided as either glucose or trehalose). Blood samples were collected before ingestion and every 30 min thereafter for a period of 3 h; substrate use was assessed by indirect calorimetry, and expired breath samples were collected. Ingestion of carbohydrates with a mixed meal resulted in a lower peak glucose response and a lower change in area under the curve (dAUC) following trehalose compared with glucose. Differences in peak insulin response and dAUC were observed with trehalose compared with glucose in the morning and afternoon. These differences were accompanied with a reduced carbohydrate oxidation after trehalose when ingested as a drink, while no significant differences in fat oxidation between drinks were observed. Peak plasma glucose concentrations were lower after the ingestion of trehalose compared with glucose. The glycemic response (dAUC) was comparable after the intake of trehalose compared with glucose following an overnight fast ($P = 0.08$), whereas there was a lower response after trehalose compared with glucose when ingested in combination with a mixed meal ($P < 0.02$). In addition, peak insulin concentrations and the total response were lower after the ingestion of trehalose compared with glucose in the morning and afternoon.

Kurotobi et al. (2010) investigated the GI of five strawberry jams from various sugar compositions in 30 healthy adults. The jam containing the highest ratio of glucose showed a high GI, whereas that containing a high ratio of fructose, a jam made of PDX, showed a low GI. Moreover, the blood glucose level after an intake of 20 g of the high-GI jam containing the high glucose ratio was higher than that of other jams at 15 min; however, there was no significant difference after 30 min. Regardless of whether GI was high or low, differences in the jams were not observed in the postprandial blood glucose level after eating either one slice of bread or one slice of bread with less than 20 g of jam.

The only comprehensive data on honey GI are based mainly on data of different Australian honeys (Foster-Powell et al. 2002). There is a significant negative correlation between fructose content and GI, probably due to the different fructose/glucose ratios of the honey types tested. It is known that unifloral honeys have varying fructose content and fructose/glucose ratios (Persano Oddo and Piro 2004).

Some honeys with relatively high concentration of fructose, for example, acacia and yellow box, have lower GI than other honey types. There was no significant correlation between GI and the other honey sugars. The GI values of four honeys found in one study varied between 69 and 74 (Ischayek and Kern 2006), whereas in another study, the value of a honey of unidentified botanical origin was found to be 35 (Kreider et al. 2000).

Although honey is a high-carbohydrate food, its GI varies within a wide range, from 32 to 85, depending on the botanical source (Bogdanov et al. 2008). It contains small amounts of proteins, enzymes, amino acids, minerals, trace elements, vitamins, aroma compounds, and polyphenols. These covered the composition of honey, the nutritional contribution of its components, and its physiological and nutritional effects. It shows that honey has a variety of positive nutritional and health effects if consumed at higher doses of 50–80 g per intake.

Low-GI honeys might be a valuable alternative to high-GI sweeteners. In order to take into account the quantity of ingested food, a new term, GL, was introduced. It is calculated as follows. The GI value is multiplied by the carbohydrate content in a given portion and divided by 100. Values lower than 10 are considered "low," values between 10 and 20 are "intermediate," and values above 20 belong to the category "high." For an assumed honey portion of 25 g, the GL of most honey types is low, and some types are in the intermediate range (Bogdanov et al. 2008).

Renwick and Molinary (2010) explored the interactions between sweeteners and enteroendocrine cells, and the consequences for glucose absorption and insulin release. A combination of *in vitro*, *in situ*, molecular biology, and clinical studies has formed the basis of their knowledge about the taste receptor proteins in the glucose-sensing enteroendocrine cells and the secretion of incretins by these cells. Low-energy (intense) sweeteners have been used as tools to define the role of intestinal sweet-taste receptors in glucose absorption. Recent studies using animal and human cell lines and knockout mice have shown that low-energy sweeteners can stimulate intestinal enteroendocrine cells to release glucagon-like peptide-1 and glucose-dependent insulinotropic peptide. These studies have given rise to major speculations that the ingestion of food and beverages containing low-energy sweeteners may act via these intestinal mechanisms to increase obesity and the metabolic syndrome due to a loss of equilibrium between taste receptor activation, nutrient assimilation, and appetite. However, data from numerous publications on the effects of low-energy sweeteners on appetite, insulin and glucose levels, food intake, and body weight have shown that there is no consistent evidence that low-energy sweeteners increase the appetite or subsequent food intake because of insulin release or affect blood pressure in normal subjects. Thus, the data from extensive *in vivo* studies in human subjects show that low-energy sweeteners do not have any of the adverse effects predicted by *in vitro*, *in situ*, or knockout studies in animals (Renwick and Molinary 2010).

11.2 GASTROINTESTINAL TOLERANCE

Symptoms of carbohydrate malabsorption may include flatulence, bloating, increased stool frequency, diarrhea, or constipation. Delayed digestion and incomplete absorption may lead to shortened intestinal transit times due to the osmotic effects of the unabsorbed polyols. Indigestible carbohydrates reach the colon, where they feed the symbiotic colonic flora, thereby increasing the bacterial mass and formation of bacterial fermentation products such as short chain fatty acids (SCFAs) and gases such as hydrogen and methane. This enhanced metabolism of the intestinal flora may influence the colon and cause overflow diarrhea. However, gastrointestinal intolerance symptoms caused by polyols are transient and readily reversible when polyol consumption is stopped. The tolerance level may increase after continuous exposure, that is, adaptation.

Polyols have a problem of limited absorption and digestibility in the small intestine, causing malabsorption with poorly digestible carbohydrates such as raffinose (Schiweck and Ziesenitz 1986). The contribution of different polyols to the osmotic pressure is related to their molecular weight. The gastrointestinal tolerance of polyols is determined by their chemical nature, the extent of digestion and absorption, which are influenced by the total quantities and portion sizes consumed, the mode of ingestion, that is, solid or liquid, and the frequency and time of ingestion. Other dietary components such as dietary fibers might contribute to the osmotic load and, thus, the tolerance of the ingested polyols (Schiweck and Ziesenitz 1986).

When a slowly absorbed carbohydrate is occasionally ingested, the colonic flora is not adapted to this substance. This could explain why different symptoms are closely correlated with the dose of unabsorbable sugar that reaches the colon (Briet et al. 1995a,b). The tolerance of sugar alcohols varies considerably between subjects, and it has been reported to be dose dependent (Dharmaraj et al. 1987; Kruger et al. 1991; Livesey 1990).

Although glucose is easily absorbed and used for energy, the sorbitol moiety is only slowly and incompletely absorbed (Lian-Loh et al. 1982; Ziesenitz and Siebert 1987). Sorbitol is slowly absorbed by passive diffusion in the small intestine. After oral administration, it increases osmotic pressure in the bowel by drawing in water, and is thus an osmotic laxative, sometimes leading to diarrhea (Gatto-Smith et al. 1988). The bacterial fermentation of sorbitol in the bowel is associated with increased flatulence and abdominal cramping; 10 g of sorbitol can cause flatulence and bloating, and 20 g of sorbitol can cause abdominal cramps and diarrhea.

Many healthy individuals are intolerant of sorbitol and develop abdominal cramping and diarrhea with less than the usual laxative dose (Badiga et al. 1990). It has been suggested that more than 30% of healthy adults, irrespective of ethnic origin, cannot tolerate 10 g of sorbitol (Jain et al. 1987).

Certain other patients are especially sensitive to the gastrointestinal effects of sorbitol; for example, diabetics can be prone to sorbitol intolerance because of altered gastrointestinal transit time and motility. Some of them also have a higher consumption of sorbitol-containing dietary foods. Patients on chronic hemodialysis can be predisposed to sorbitol intolerance as a result of carbohydrate malabsorption (Coyne and Rodriguez 1986).

Some cases of idiopathic colonic ulcers in patients with renal failure are due to the effects of sorbitol. Five cases of extensive mucosal necrosis and transmural infarction of the colon have been reported after the use of kayexalate (sodium polystyrene sulfonate) and sorbitol enemas to treat hyperkalemia in uremic patients (Lillemoe et al. 1987). They also studied the effects of kayexalate sorbitol enemas in normal and uremic rats and concluded that sorbitol was responsible for colonic damage and the injury was potentiated in uremic rats. When sorbitol alone or kayexalate sorbitol was given, extensive transmural necrosis developed in 80% of normal rats and in all uremic rats.

Ruskoné-Fourmestraux et al. (2003) aimed at evaluating the gastrointestinal tolerance to an indigestible bulking sweetener containing sugar alcohol (MTL) using a double-blind random crossover study. MTL is a sugar alcohol produced by hydrogenation from starch hydrolyzates that have a high content of natural disaccharide maltose. After oral ingestion, MTL is slowly hydrolyzed by the enzymes of the small intestine into its constituent monomers—glucose and sorbitol. The metabolism of MTL is therefore similar to that of sorbitol. In order to simulate their usual pattern of consumption, 12 healthy volunteers ingested MTL or sucrose throughout the day, either occasionally (once a week for each sugar—first period) or regularly (every day for two 9-day periods—second period). In both patterns of consumption, daily sugar doses were increased until diarrhea and/or a grade 3 (i.e., severe) digestive symptom occurred, at which the dose level was defined as the threshold dose (TD). In the first period (occasional consumption), the mean TD was 92 ± 6 g with MTL and 106 ± 4 g with sucrose ($P = 0.059$). The mean intensity of digestive symptoms was 1.1 and 1.3, respectively (P = NS). Diarrhea appeared in six and one subjects, respectively ($P = 0.035$). In the second period (regular consumption), the mean TD was 93 ± 9 g with MTL and 113 ± 7 g with sucrose ($P = 0.008$). The mean intensity of digestive symptoms was 1.7 and 1.2, respectively (P = NS). However, diarrhea appeared in eight and three subjects, respectively ($P = 0.04$). MTL and sucrose TDs between the two periods were not different. Ruskoné-Fourmestraux et al. (2003) reported that occasional or regular consumption of MTL is not associated with severe digestive symptoms; in both patterns of MTL consumption, diarrhea frequency is higher, but it appeared only for very high doses of MTL, much greater than those currently used. MTL does not lead to intestinal flora adaptation after a 9-day period of consumption.

Other MTL tolerance studies include the work of Leroy (1982) and Koizumi et al. (1983), demonstrating that, in healthy and diabetic subjects, MTL is tolerated up to 30–50 g/day after adaptation. Higher doses cause diarrhea.

Stool excretion after the ingestion of sugar alcohols is negligible, indicating that the sugar alcohols reaching the large intestine are almost completely digested by the colonic flora (Beaugerie et al. 1990). However, this malabsorption causes certain side effects, as the fermentation of unabsorbed sugar leads to flatulence. In addition, as polyol molecules are osmotically active, diarrhea may occur when the capacity of the colonic flora to ferment these low-molecular-weight carbohydrates is exceeded and osmotic stress rises in the intestinal lumen (Hammer et al. 1989; Saunders and Wiggins 1981).

The capacity to ferment unabsorbable sugars, such as lactose and lactulose, and to reduce their laxative effects can, however, be increased by the regular ingestion of these sugars, which results in changes in the metabolic activity of the colonic flora, especially a reduced excretion of hydrogen in the breath (H_2; Florent et al. 1985; Flourié et al. 1993; Launiala 1968; Perman et al. 1981). Langkilde et al. (1994) also showed that the levels of digestion and absorption of two sugar alcohols (sorbitol and isomalt) were dose dependent, which can influence digestive tolerance. A double-blind cross-over study performed in 59 healthy volunteers has shown that an acute oral intake of 30 g of MTL in milk chocolate resulted in no significant increase in reported digestive symptoms, except for mild flatulence (Koutsou et al. 1996).

In conclusion, it seems that the gastrointestinal intolerance of sweeteners exists and symptoms and severity depend on a number of factors such as the type and amount of sweetener, the way of administration (acute or continuous), or even the mode (solid or liquid) and time of ingestion. If all these factors are considered, taking into account the individual characteristics of the person/patient, the symptoms of carbohydrate malabsorption can be kept to a minimum.

11.3 DENTAL HEALTH

Diet plays an important role in preventing oral diseases, including dental caries, dental erosion, developmental defects, oral mucosal diseases, and, to a lesser extent, periodontal disease. Moynihan (2005a) provided an overview of the evidence for an association between diet, nutrition, and oral diseases. Undernutrition increases the severity of oral mucosal and periodontal diseases and is a contributing factor to life-threatening noma (a dehumanizing oro-facial gangrene). Undernutrition is associated with developmental defects of the enamel, which increase susceptibility to dental caries. Dental erosion is perceived to be increasing. Evidence suggests that soft drinks, a major source of acids in the diet in developed countries, are a significant causative factor (Moynihan 2005a). This section will focus on dietary and behavioral factors as well as on sugar-replacing substances that influence the incidence of dental caries.

11.3.1 Dental Caries

Dental caries occurs because of the demineralization of enamel and dentine by organic acids formed by bacteria in dental plaque through the anaerobic metabolism of dietary sugars (Moynihan and Petersen 2004). Convincing evidence from experimental, animal, human observational, and human intervention studies shows that sugars are the main dietary factor associated with dental caries. Despite the indisputable role of fluoride in the prevention of caries, it has not eliminated dental caries, and many communities are not exposed to optimal quantities of fluoride. The risk of dental caries increases with the intake of nutritive sweeteners; however, this risk does not work independent of oral hygiene and fluoridation (Navia 1994). The classic evidence supporting the role of sugar in dental caries in humans includes studies that are readily recognizable by name—the Vipeholm Study, the Turku Sugar Study, World War II Food Rationing, the Hopewood House Study, Tristan da Cunha, Hereditary Fructose Intolerance, Experimental Caries in Man, and Stephan Plaque pH Response. The Vipeholm Study remains one of the most important contributions in the dental literature and definitively established that the more frequently sugar is consumed, the greater the risk becomes and that sugar consumed between meals has a much greater caries potential than when consumed during a meal (Zero 2004).

The extent of dental decay is measured using the primary/permanent-dentition decayed, missing, and filled teeth (dmft/DMFT) index. This is a count of the number of teeth in a person's mouth that are decayed, filled, or extracted. The dmft/DMFT indices are widely used for the indication of the prevalence of dental caries and the severity of dental caries experience in populations. Dental diseases—caries and periodontal disease—result in tooth loss, and therefore, the dental status of a population may also be assessed by looking at the proportion of the population who are edentulous (have no natural teeth; Moynihan and Petersen 2004). In 1982, the World Health Organization (WHO) and the Fédération Dentaire Internationale (FDI) jointly set out global goals for oral health to be achieved by the turn of the century, including that children aged 12 years on the average should have a DMFT of below 3 (WHO/FDI 1982).

Controlling the intake of sugars, therefore, remains important for caries prevention. In countries with a level of sugar consumption of less than 18 kg/person/year, caries experience is consistently low (Sreebny 1982; Woodward and Walker 1994). Ruxton et al. (1999) used the ecological data from the studies of Sreebny and of Woodward and Walker to study the relationship between changes in dental caries and in sugar intake in 67 countries between 1982 and 1994 by drawing up a simple scatter plot. This scatter plot showed that, in 18 countries where sugar supply declined, DMFT decreased. In 18 countries where caries level increased, there were increases in sugar supply.

Repeat dietary surveys of English children over three decades indicate that levels of sugar intake have remained stable while sources of sugars have changed considerably, with the contribution from soft drinks more than doubling since 1980. Dental caries eventually leads to tooth loss, which in turn impairs chewing ability, causing the avoidance of hard and fibrous foods, including fruits, vegetables, and whole grains. It has been found that edentulous subjects have a very low intake (<12 g/day) of nonstarch polysaccharides, fruits, and vegetables. In addition, the provision of prostheses alone failed to improve the diet. However, initial studies indicated that customized dietary advice at the time of denture provision resulted in increased consumption of fruits and vegetables, and positive movement through the stages of change (Moynihan 2005b).

In young children in the U.K., levels of dental caries are increasing (Pitts et al. 2004b). A recent survey of 5-year-old children conducted by the British Association for the Study of Community Dentistry has shown that 40% of 5-year-olds in England and Wales have dental caries, with an average of 1.52 teeth per child affected. The British Association for the Study of Community Dentistry survey data (Pitts et al. 2004a) shows the mean number of decayed, missing, and filled permanent teeth in 14-year-old children in England and Wales from 1990 to 2002 and indicates how the trend for a decline in caries has now stabilized, with an average of 1.5 decayed, missing, and filled permanent teeth per child. Despite a relatively low average number of decayed, missing, and filled permanent teeth, 50% of 14-year-olds in England and Wales (Pitts et al. 2004a) and 50% of 12-year-olds in the Republic of Ireland are affected by decay (Whelton et al. 2003). Dental caries is a progressive disease, and levels in European adults are very high. Even in fluoridated areas of the Republic of Ireland, the average number of decayed, missing, and filled permanent teeth for the 35- to 44-year-old age group is 18.9, and in the U.K., it is 19.0 (the WHO considers a level of ≥14.0 to be very high; World Health Organization 1996).

Over the past 20 years, the contribution of soft drinks, biscuits and cakes, and breakfast cereals to total sugar intake has risen significantly. In 1980, soft drinks contributed 15% to the total sugars intake; this percentage has overdoubled in the 20-year period to 37% in 2000. Similarly, the contribution of breakfast cereals to sugar intake has risen from 2% to 7%. The intake of sugars from confectionery, table sugar, and puddings has declined over the 20-year period. However, confectionery has remained a major source, providing 23% of the total sugars in 2000 and, together with soft drinks, provides approximately 60% of the total sugars. These findings are consistent with other surveys from industrialized countries, which indicate that children are consuming more sugars than recommended and the principal dietary sources are confectionery and soft drinks (Rugg-Gunn et al. 2007; Gregory and Lowe 2000; Guthrie and Morton 2000).

Tooth loss is associated with a reduction in both measured (Krall et al. 1998) and perceived (Rusen et al. 1993) chewing functions. The chewing function of an individual with dentures is only one-fifth of that of a dentate individual with 20 or more natural teeth (Michael et al. 1990). Early studies have reported that the loss of functional dentition results in chewing difficulties and selective food avoidance, raising concern that this situation may lead to compromised nutritional intake (Osterberg and Steen 1982; Wayner and Chauncey 1983). Foods avoided include those that are hard to chew, for example, raw vegetables and wholegrain breads, and foods containing seeds and pips such as tomatoes, grapes, and raspberries (Wayner and Chauncey 1983).

Experimental and animal studies suggest that some starch-containing foods and fruits are cariogenic, but this conclusion is not supported by epidemiological data showing that high intakes of starchy staple foods, fruits, and vegetables are associated with low levels of dental caries. Following global recommendations that encourage a diet high in starchy staple foods, fruits, and vegetables and low in free sugars and fat will protect both oral and general health (Moynihan and Petersen 2004).

The WHO (2003) recommends limiting nonmilk extrinsic sugars (NMES), that is, all monosaccharides and disaccharides added to foods by the manufacturer, cook, or consumer, plus sugars naturally present in honey, syrups, and fruit juices, and consumption to ≤10% energy to reduce the risk of unhealthy weight gain and dental caries and to restrict the frequency of intake to ≤4 times/day to reduce the risk of dental caries. Older adults, especially those from low-income backgrounds, are at increased risk of dental caries, yet there is little information on sugar intake (frequency of intake and food sources) in this age group. Bradbury et al. (2008) presented baseline data from a community-based dietary intervention study of older adults from socially deprived areas of North East England on the quantity and sources of total sugars, NMES, and intrinsic and milk sugars and on the frequency of NMES intake. Dietary intake was assessed using two 3-day estimated food diaries, completed by 201 participants (170 females and 31 males), aged 65–85 years [mean 76.7 (SD 5.5) years], recruited from sheltered housing schemes. Total sugars, NMES, and intrinsic and milk sugars represented 19.6%, 9.3%, and 10.3%, respectively, of the daily energy intake. Eighty-one participants (40.3%) exceeded the NMES intake recommendation. The mean frequency of NMES intake was 3.4 times/day. The 53 participants (26.4%) who exceeded the frequency recommendation (≤4 times/day) obtained a significantly greater percentage of energy from NMES compared with the participants who met the recommendation. The food groups "biscuits and cakes" (18.9%), "soft drinks" (13.1%), and "table sugar" (11.1%) made the greatest contributions to intakes of NMES. It was concluded that interventions to reduce NMES intake should focus on limiting the quantity and frequency of intake of these food groups (Bradbury et al. 2008).

While sugars appear to differ little in acidogenic potential, sucrose has been given special importance, being the sole substrate for the synthesis of extracellular glucans (Zero 2004). Water-insoluble glucans might enhance the accumulation of mutans streptococci on smooth tooth surfaces and appear to enhance virulence by increasing plaque porosity, resulting in greater acid production immediately adjacent to the tooth surface. Data indicating that the sugar consumption/caries relationship is now weaker have led to suggestions that recommendations to restrict sugar consumption are no longer necessary. Clearly, fluoride has raised the threshold of sugar intake at which caries will progress to cavitation, but fluoride has its limits, and caries remains a serious problem for disadvantaged individuals in many industrialized countries and is a rising problem in many developing countries. A weakening of the sugar/caries relationship may also be explained by many technical, biological, behavioral, and genetic factors (Zero 2004).

Other factors such as fluoride use and oral hygiene behavior may influence the incidence of dental caries. Moynihan (2005b) reported that the consumption of fluoridated water coupled with a reduction in NMES intake is an effective means of caries prevention. However, studies on the fluoride concentration of bottled water suggest that the increased consumption of bottled water, in preference to fluoridated tap water, would lead to a marked decrease in caries protection. Concerns have been raised about the bioavailability of fluoride from artificially fluoridated water compared with naturally fluoridated water. This issue has been addressed in a human experimental study that has indicated that any differences in fluoride bioavailability are small compared with the naturally occurring variability in fluoride absorption (Moynihan 2005b). Recently, Tseveenjav et al. (2011) assessed the effect of certain oral health-related behaviors on adults' dental health. As part of the Finnish nationwide Health 2000 Survey, dentate subjects, 30–64 years of age, reported their frequency of consumption of eight sugar- and xylitol-containing products, together with tooth brushing frequency and the use of fluoride toothpaste, and underwent clinical oral examination (n = 4361).

The mean number of teeth present (NoT) was 24.2, and the mean numbers of sound teeth (ST), filled teeth (FT), and decayed teeth (DT) were 10.8, 12.1, and 1.1 for men and 9.6, 13.8, and 0.5 for women, respectively. The consumption of sugar-sweetened beverages (SSBs) was more frequent than other sugar-containing products and greater in men than in women. Daily use of xylitol chewing gum was reported by 13% of the men and by 22% of the women. Tooth brushing at least twice daily was reported by 47% of the men and by 79% of the women; 86% and 96%, respectively, reported daily use of fluoride toothpaste. The frequency of consumption of sugar- and xylitol-containing products and of tooth brushing, as well as the use of fluoride toothpaste, play a role in the dental health of dentate adults, with the impact being weak on NoT, ST, and FT but stronger on DT, especially concerning tooth brushing frequency (relative risk = 1.5) and the use of fluoride toothpaste (relative risk = 1.8). The authors concluded that understanding the impact of certain oral health-related behaviors on dental health in adults may lead to better targeting of oral self-care messages (Tseveenjav et al. 2011).

In an attempt to replace sugar in the diet, certain substances that may possess noncariogenic or anticariogenic properties have been used. Grenby (1991) analyzed the factors that have to be considered when developing low-calorie sweeteners to replace sugar in the diet, for the benefit of dental health. He explained that the choice of low-calorie sweeteners with improved characteristics is expanding, with particular attention being paid to calorie control, dental health improvement, and developing an appealing range of foods and drinks. The use of polyol-based gum (i.e., xylitol based) can reduce the risk of dental caries compared to sucrose-sweetened gum (Makinen et al. 1995). Sugar alcohols (e.g., sorbitol, mannitol, and xylitol) do not promote tooth decay according to an authorized health claim in food labeling by the Food and Drug Administration (FDA; 2006). The Turku Study, which was a controlled dietary intervention in Finnish adults, showed that almost total substitution of sucrose in the diet with xylitol (a noncariogenic sweetener) resulted in an 85% reduction in dental caries over a 2-year period (Scheinin et al. 1976). Similarly, nonnutritive sweeteners do not promote dental caries (American Dietetic Association 1998). Moreover, sucralose does not promote dental caries according to the FDA (1996). Hence, the agency has authorized a health claim regarding noncariogenic carbohydrate sweeteners to include sucralose.

D-Xylitol is found in low content as a natural constituent of many fruits and vegetables. It is a five-carbon sugar polyol and has been used as a food additive and sweetening agent to replace sucrose, especially for noninsulin-dependent diabetics (Chen et al. 2010). It has multiple beneficial health effects, such as the prevention of dental caries and acute otitis media (AOM). In the industry, it has been produced by the chemical reduction of D-xylose mainly from photosynthetic biomass hydrolyzates. As an alternative method of chemical reduction, the biosynthesis of D-xylitol has been focused on the metabolically engineered *Saccharomyces cerevisiae* and *Candida* strains. In order to detect D-xylitol in the production processes, several detection methods have been established, such as gas chromatography-based methods, high performance liquid chromatography-based methods, liquid chromatography–mass spectrometry methods, and capillary electrophoresis methods. The advantages and disadvantages of these methods are compared in the review of Chen et al. (2010).

As an effective and safe tooth decay–preventive agent, D-xylitol is used in chewing gums, mouth rinse (Hildebrandt et al. 2010), and toothpaste (Lif Holgerson et al. 2006; Sano et al. 2007). *Streptococcus mutans* is most notably associated with human dental decay, by attachment to the acquired enamel pellicle and direct interaction with the salivary components.

However, *S. mutans* cannot utilize D-xylitol. After people take D-xylitol-containing products, the lactic acid production from fermentation by these strains will be decreased. Saliva with D-xylitol is more alkaline than that containing other sugar products.

A 40-month double-blind cohort study on the relationship between the use of chewing gum and dental caries was performed from 1989 to 1993 in Belize, Central America. The results showed that the D-xylitol-containing gum was effective in reducing caries rates, and the most effective agent was a 100% D-xylitol pellet gum (Makinen et al. 1995). Makinen et al. (2001) reported the effect of a 2-month usage of saliva-stimulating pastils containing erythritol or D-xylitol. In the D-xylitol-group, the mean weight of the total plaque mass was reduced significantly; the plaque and salivary levels of *S. mutans* and the plaque levels of the total *Streptococcus* were reduced significantly as well (Makinen et al. 2001). Milgrom et al. (2006) suggested the effective dose of D-xylitol to be between 6.44 and 10.32 g/day.

Regular use of D-xylitol in chewing gums or syrup prevented the incidence of acute otitis media (AOM) in children (Uhari et al. 1996). D-Xylitol had the ability to reduce the growth of the major otopathogen of AOM, *S. pneumoniae*, which caused 30% or more of such attacks, and also suppressed *Haemophilus influenza*, another important pathogen implicated in AOM (Kontiokari et al. 1998). The researchers reported that the exposure of epithelial cells, pneumococci, or both to 5% D-xylitol reduced the adherence of pneumococci. Some researchers implied that the inhibition of *pneumonia* growth induced by D-xylitol was mediated via the fructose phosphotransferase system. However, the mechanism remains a matter of speculation (Tapiainen et al. 2001).

Szöke et al. (2001) explored the effect of after-meal sucrose-free gum chewing on clinical caries. In this 2-year study, they investigated whether chewing sugar-free gum reduced the development of dental caries in schoolchildren. The children were split into two groups: the gum group, who were instructed to chew sugar-free gum for 20 min after eating three times daily, and the control group, who were not provided with chewing gum. The study showed that chewing gum after meals significantly reduced the incidence of dental caries. Similarly, Beiswanger et al. (1998) conducted a study to determine the effect of chewing sugar-free gum on the incidence of dental caries in children. The children were put into either a control group or a chewing gum group. Those in the gum group had to chew sugar-free gum for 20 min after each meal. The study concluded that chewing sugar-free gum after meals greatly reduces the incidence of dental caries.

He et al. (2006) discovered a novel compound (Glycyrrhizol A), from the extraction of licorice roots, with strong antimicrobial activity against cariogenic bacteria. In a current study, Hu et al. (2011) developed a method to produce these specific herbal extracts in large quantities and then used these extracts to develop a sugar-free lollipop that effectively kills cariogenic bacteria like *S. mutans*. Further studies showed that these sugar-free lollipops are safe and their antimicrobial activity is stable. Two pilot human studies indicate that a brief application of these lollipops (twice a day for 10 days) led to a marked reduction of cariogenic bacteria in oral cavity among most human subjects tested. The authors argued that this actual herbal lollipop could be a novel tool to promote oral health through functional foods (Hu et al. 2011).

In conclusion, one may argue that a combination of measures on the basis of educating the public about the dangers of frequent sugar consumption, proper oral hygiene, and fluoride use, as well as public awareness regarding anticariogenic substances, may well contribute to the prevention and reduction of dental caries.

11.4 PREBIOTICS AND DIGESTIVE HEALTH

Bulk sweeteners are sugar substitutes used most frequently by the confectionery industry. This group of sweeteners contains primarily sugar alcohols, which are not broken down in the stomach or small intestine, and nondigestible carbohydrates, which can be used in foods at levels similar to that of sucrose (Mela 1997). The term nondigestible refers to the food ingredients being undigested in the upper gut, with a large portion remaining for fermentation by the indigenous microbiota of the large intestine. MTL is an example of a bulk sweetener, whereas PDX and resistant starch (RS) are bulking agents that can substitute for the texture properties of sucrose in confectionery products. All have been found to be fermented by the indigenous bacteria of the colon (Probert et al. 2004; Ghoddusi et al. 2007; Arrigoni et al. 2005; Tsukamura et al. 1998).

Certain bacteria, namely, bifidobacteria and lactobacilli, are regarded as health promoting, interacting with the host immune system, cells of the intestinal mucosa, and other members of the gut microbiota, and play a role in immune homoeostasis and the ability to fight off infections, mucosal integrity, the production of vitamins, beneficial fats, and other metabolites used by the host.

Dietary prebiotics, recently defined as "selectively fermented ingredient(s) that result(s) in specific changes in the composition and/or activity of the gastrointestinal microbiota, thus, conferring benefit(s) upon host health" (Gibson et al. 2004), have been repeatedly shown to bring about elevated numbers of fecal bifidobacteria in human feeding studies (Kolida and Gibson 2007; Kolida et al. 2002; Gibson 1999).

Many of the sugar replacements used in the confectionery industry are nondigestible in the upper gut and therefore have the potential to be prebiotics. However, the problem for consumers is that the overconsumption of certain prebiotics has been reported to result in unwanted intestinal side effects such as increased flatulence or intestinal bloating or pain. Doses of about 15 g/day of fructooligosaccharides and inulin have been shown to induce laxation effects, increased stool frequency, and gas production (Cummings et al. 2001).

Beards et al. (2010) have measured the microbiota modulatory potential and intestinal tolerance of chocolate-containing blends of sugar replacers (in place of sucrose) likely to be used in manufacturing low-energy confectionery and compared that to traditional sucrose chocolate. To date, there are no data on the effects that these sugar replacers may exert on the gut microbiota *in vivo.* The aim of this study was therefore to assess the potential prebiotic supplementation of chocolate to selectively increase numbers of beneficial fecal bacteria and to measure the tolerability of high-level (45.6 g of sugar replacer/day) consumption of this low-energy, highly nondigestible carbohydrate chocolate. For this, a placebo-controlled, double-blinded, dose–response human feeding study was conducted using 40 healthy human volunteers in a parallel manner. Forty volunteers consumed a test chocolate (low-energy or experimental chocolate) containing 22.8 g of MTL, MTL and PDX, or MTL and RS for 14 consecutive days. The dose of the test chocolate was doubled every 2 weeks over a 6-week period. Numbers of fecal bifidobacteria significantly increased with all the three test treatments. Chocolate containing the PDX blend also significantly increased fecal lactobacilli ($P = 0.00001$) after 6 weeks. The PDX blend also showed significant increases in fecal propionate and butyrate ($P = 0.002$ and 0.006, respectively). All the test chocolates were well tolerated, with no significant change in bowel habit or intestinal symptoms even at a daily dose of 45.6 g of nondigestible carbohydrate sweetener. This is of importance not only for giving manufacturers a sugar replacement that can reduce energetic content but also for providing a well-tolerated means of delivering high levels of nondigestible carbohydrates into the colon, bringing about improvements in the biomarkers of gut health.

Worldwide interest in oligosaccharides has been increasing ever since they were accorded the prebiotic status (Patel and Goyal 2011). Oligosaccharides of various origin such as bacteria, algae, fungi, and higher plants have been used extensively both as food ingredients and pharmacological supplements. Two novel oligosaccharides—β-D-fructopyranosyl-(2 → 1)-β-D-glucopyranosyl-(2 ↔ 1)-α-D-glucopyranose and β-D-fructopyranosyl-(2 → 6)-α-D-glucopyranosyl-(1 ↔ 2)-β-D-glucopyranose—have also been isolated from a fermented beverage of 50 kinds of fruits and vegetables (Okada et al. 2010). Nondigestible oligosaccharides have been implicated as a dietary fiber, sweetener, weight-controlling agent, and humectant in confectioneries, bakeries, and breweries. Functional oligosaccharides have been found effective in gastrointestinal normal flora proliferation and pathogen suppression, dental caries prevention, the enhancement of immunity, and the facilitation of mineral absorption and as a source of antioxidant, an antibiotic alternative, and regulators of blood glucose in diabetics and serum lipids in hyperlipidemics. Apart from the pharmacological applications, oligosaccharides have found use in drug delivery, cosmetics, animal and fishery feed, agriculture, etc. Keeping in view the importance of the functional oligosaccharides, they presented an overview of their natural sources, types, structures, and physiological properties.

Functional oligosaccharides of various origin have been used extensively as food ingredients, prebiotic supplements, drug delivery agents, immunostimulators, cosmetic ingredients, animal feed, and agrochemicals (Qiang et al. 2009). Because functional oligosaccharides are attributed with multiple beneficial health effects, they are used widely in food products as anticariogenic agents and low-sweetness humectants.

Qualified prebiotics as fructooligosaccharides (FOS) and glucooligosaccharides avoid the urogenital infections by promoting the proliferation of lactobacilli (Sanchez et al. 2008). FOS improve the gut absorption of Ca and Mg, prevent urogenital infections, serve as a sweetener in beverages, improve acariogenic quality, has a positive effect on lipid metabolism, and reduce the risk of colon cancer. Galactooligosaccharides (GOS) lactulose derived from *Bifidobacterium bifidum* NCIMB 41171, *Kluyveromyces lactis*, and *Sulfolobus solfataricus* can be used as prebiotics. Lactulose is used in the treatment of hyperammonemia and portosystemic encephalopathy and as a laxative, infant formula, prebiotic, and low-calorie sweetener, as reported by Goulas et al. (2007) and Kim et al. (2006).

Prebiotics have successfully been incorporated into a wide variety of human food products such as baked goods, sweeteners, yogurts, nutrition bars, and meal replacement shakes. For instance, the introduction of GOS in baby foods has been very successful. GOS, which are identical to the human milk oligosaccharides, have emerged with strong clinical support for both digestive and immune health. Various aspects related to GOS, such as types and functions of functional food constituents with special reference to GOS, their role as prebiotics, and enhanced industrial production through microbial intervention, are dealt with in the review of Sangwan et al. (2011). GOS, also known as oligogalactosyllactose, oligogalactose, oligolactose, or transgalactooligosaccharides, because of their indigestible nature, belong to the group of prebiotics.

GOS provide their health benefits by two main mechanisms. One mechanism is by the selective proliferation of beneficial bacteria, especially bifidobacteria and lactobacilli in the gut, which provide resistance to the colonization of pathogens, thereby reducing exogenous and endogenous intestinal infections. These beneficial organisms modulate the immune system and suppress inflammatory bowel disease inflammation. The other mechanism is by the production of SCFAs. The metabolism of GOS leads to the production of SCFAs. These SCFAs show various beneficial effects, including the reduction of cancer risk, increase in mineral absorption, and improvement in bowel habit (Sangwan et al. 2011). GOS can easily be incorporated into beverages such as fruit juices, fruit drinks, breakfast drinks, and soft drinks because of their acid stability and property of forming clear solutions. GOS can easily be added together with other ingredients such as concentrated fruit juices, compounds, or sugar syrup. GOS are heat and acid stable. No decrease of GOS is measured under low-pH conditions and high temperatures.

A GOS is the perfect ingredient for use in acid drinks such as soft drinks or fruit-based drinks and juices. Because GOS are very neutral and somewhat sweet in taste, the taste of beverages will not be influenced when a GOS is added. GOS can also be used in the development of bread and baked goods that are high in fiber and have low sugar content and low calories. Furthermore, GOS have an ideal combination of functional properties such as low calorie and high-moisture retention capacity, making them an ideal component for baked products. Additionally, GOS can provide several health benefits such as the growth of bifidobacteria, relief of constipation, support of natural defenses, and improved mineral absorption (Sangwan et al. 2011).

11.5 CALORIE CONTROL AND WEIGHT MANAGEMENT

There has been much debate on the influence and mechanisms of sweetness and low-calorie sweeteners on energy intake and weight control. Nutritive sweeteners by themselves do not cause an increase in weight, and similarly, low-calorie sweeteners themselves promote neither weight gain nor weight loss. They achieve an uncoupling of sensory and caloric characteristics and can sweeten food without adding calories. Consumers can use this saving in calories to reduce or control weight or as an excuse to ingest calories in other forms (Maffeis 2009).

Most studies investigating the role of low-calorie sweeteners in weight control have shown that replacing foods in the diet with low-calorie versions containing low-calorie sweeteners reduces the overall caloric intake. Caloric intake in studies using the covert replacement of sugar with low-calorie sweeteners has shown that compensation is incomplete, although sometimes an initial rapid fall that tended to revert toward normal later was found, indicating that people are unlikely to lose weight by using low-calorie sweeteners without intentional control of their total caloric intake. Studies on subjects in weight control programs have shown that low-calorie sweeteners can be helpful in making the regime more acceptable and successful (Bellisle and Drewnowski 2007). Several studies on aspartame have suggested that it may facilitate the control of body weight and enhances weight maintenance over the long term. In 2007, Bellisle and Drewnowski examined and challenged the hypothesis that low-calorie sweeteners and the products that contain them may cause weight gain. Although their review of a variety of studies indicated that intense sweeteners may assist weight loss efforts, it did stress that low-calorie sweeteners are not a "silver bullet" solution to weight management, are not appetite suppressants, and will not result in automatic weight loss. They concluded that the ultimate effect of low-calorie sweeteners on weight loss is dependent on their integration into a whole lifestyle approach.

In earlier reviews by Barbara Rolls, the effects of low-calorie sweeteners and low-calorie products on hunger, appetite, and food intake were examined. In 1991, she published a review in which she evaluated low-calorie sweeteners' role on hunger, appetite, and food intake. In this comprehensive review, Rolls concluded that, "If low-calorie sweeteners are part of a weight control program, they could aid calorie control by providing palatable foods with reduced energy." It needs to be stressed that there are no data suggesting that the consumption of foods and drinks with low-calorie sweeteners promotes food intake and weight gain in dieters (Rolls 1991).

A meta-analysis of studies by de la Hunty et al. (2006) demonstrated that "using foods and drinks sweetened with aspartame instead of sucrose (sugar) results in a significant reduction in both energy intakes and body weight. Also, the meta-analyses both of energy intake and of weight loss produced an estimated rate of weight loss of about 0.2 kg/week. This corresponds to a weight loss of 10 kg over a one-year period" (de la Hunty et al. 2006).

A recent review paper by Mattes and Popkin (2009) analyzed findings from 224 studies on the effects of intense sweeteners on appetite, food intake, and weight. They concluded that short-term trials provide mixed evidence of reduced energy intake with intense sweetener use but that "longer-term trials—arguably the more nutritionally relevant studies—consistently indicate that the use of intense sweeteners results in slightly lower energy intakes." With regard to the impact on body mass index (BMI), the study noted that "reverse causality remains a likely explanation" for at least a portion of recent epidemiological findings linking intense sweetener use to weight gain. The researchers stated that, "Taken together, the evidence summarized by us and others suggests that if intense sweeteners are used as substitutes for higher-energy-yielding sweeteners, they have the potential to aid in weight management." Furthermore, it should be pointed out that low-calorie sweeteners offer people with diabetes the pleasure of a sweet taste without negative side effects. On the one hand, sweeteners have no impact on insulin and blood sugar levels, and on the other hand, they do not provide calories. Moreover, low-calorie foods help this group of people in the important task of controlling their weight (Maffeis 2009).

11.5.1 Artificial Sweetener Intake in Children

Beverages have been identified as a major source of artificial sweeteners in the diet (Morgan et al. 1982; Ilback et al. 2003); hence, estimates of artificial sweetener consumption are typically based on artificially sweetened drinks or sodas. Nationally representative surveys from the 1990s estimated that artificially sweetened sodas accounted for approximately 4%–18% of the total carbonated beverage intake in children (French et al. 2003; Striegel-Moore et al. 2006). Artificially sweetened soft drink consumption appears to be increasing in children, both with age and over time (Kral et al. 2008; Blum et al. 2005).

11.5.2 Observational Studies of Artificial Sweeteners and Weight Gain in Children

Brown et al. (2010) reviewed the current literature on artificial sweetener consumption in children and its health effects and identified 18 studies. Data from large, epidemiologic studies support the existence of an association between artificially sweetened beverage consumption and weight gain in children. Randomized controlled trials in children are very limited and do not clearly demonstrate either beneficial or adverse metabolic effects of artificial sweeteners. They reported that it is important to examine possible contributions of these common food additives to the global rise in pediatric obesity and diabetes.

The majority of pediatric epidemiologic studies have found a positive correlation between weight gain and artificially sweetened beverage intake. Blum et al. (2005) examined beverage consumption and BMI *Z*-scores in 164 elementary school-aged children in a longitudinal study, where increased diet soda consumption was positively correlated with follow-up BMI *Z*-score after 2 years. Similar results were reported by Berkey et al. (2004), who examined the relationship between BMI and diet soda consumption in over 10,000 children (aged 9–14 years) of Nurses' Health Study II participants over the course of 1 year. Artificially sweetened beverage intake was significantly correlated with weight gain in boys but not in girls during the study period. A long-term prospective study of 1203 children in England found that artificially sweetened beverage consumption at ages 5 and 7 was correlated both with baseline BMI and fat mass at age 9 (Johnson et al. 2007). Another longitudinal study of 2371 girls (aged 9 and 10) participating in the National Heart, Lung and Blood Institute Growth and Health Study showed that diet and regular soda consumption was significantly associated with increase in daily energy intake but not with BMI (Striegel-Moore et al. 2006; Yang 2010). The correlation between diet soda and BMI was not significant.

A much smaller study of 177 children aged 3–6 years by Kral et al. (2008) showed no association between diet soda consumption and the risk of obesity.

Forshee and Storey (2003) analyzed data from a nationally representative sample of U.S. children between 6 and 19 years of age (a cross-sectional study looking at 3111 children) and found that BMI was positively correlated with diet soda consumption. These results were consistent with the findings of Giammattei et al. (2003) in 385 sixth and seventh graders, which showed that both diet and sugar-sweetened soda intake were positively correlated with BMI *Z*-score and percentage body fat.

However, a study of 2- to 5-year-old children by O'Connor et al. (2006) using National Health and Nutrition Examination Survey (NHANES) data did not show an association between artificially sweetened beverage consumption and BMI in this age group. However, increased beverage consumption was associated with an increase in the total energy intake of the children. This noted difference between the total energy intake and mean BMI might have multiple explanations. First, the prevalence of overweight in this age group ($N = 124$; 10.7%) may be too low to detect an association between increased energy intake and increased BMI. Second, they may be capturing children who are too young to see an effect of increased total energy intake on BMI.

Prospectively studying preschool children beyond 2–5 years of age, through their adiposity rebound (approximately 5.5–6 years) to determine whether there is a trajectory increase in their BMI, may help clarify the role of beverage consumption in the total energy intake and weight status. Because the mean adiposity rebound occurs at approximately 5.5–6 years (Whitaker et al. 1998), it is possible that if preschool children were followed through their adiposity rebound, then it might be found that the increased energy intake may translate into an increase in BMI after age 6.

11.5.3 Interventional Studies of Artificial Sweeteners and Weight Gain in Children

Three small interventional studies that manipulated artificial sweetener intake have been conducted in children and did not show any metabolic effects. Shortly after the approval of aspartame by the FDA, its effects during active weight reduction and its role in glucoregulatory hormone changes were studied in 55 overweight children and young adults, aged 10–21, during a 13-week 1000-kcal/day diet (Knopp et al. 1976). There were no differences in weight loss for subjects receiving 2.7 g/day of encapsulated aspartame versus placebo.

A randomized, controlled pilot study of 103 adolescents, aged 13–18 years, examined the effect of replacing sugar-sweetened drinks with artificially sweetened beverages or water during a 25-week period (Ebbeling et al. 2006). Changes in BMI for intervention versus control (no replacement of sugar-sweetened drinks) were not significant for the entire group, although an exploratory *post hoc* analysis showed that the intervention made the greatest difference in the heaviest subjects, whose BMIs declined by 0.63 ± 0.23 kg/m^2, compared with a 0.12 ± 0.26 kg/m^2 gain in the control group. However, the authors of this study could not isolate the effect of artificial sweeteners.

In a randomized, controlled trial by Williams et al. (2007), girls aged 11–15 years consumed a 1500-kcal/day diet for 12 weeks. In one group, sugar-sweetened soda was permitted as a snack, whereas in the other group, only diet sodas were permitted. There were no differences between groups for BMI change, and it was reported that the intake of either sugar-sweetened or artificially sweetened soda did not affect BMI change.

Rodearmel et al. (2007) worked on a family-intervention study where the America on the Move small-changes approach for weight gain prevention was evaluated in families with at least one child (7–14 years old) who was overweight or at risk for overweight. These children were the primary target of the intervention, and parents were the secondary target. Families were randomly assigned to either the America on the Move group ($n = 100$) or the self-monitor-only group ($n = 92$). Families who were assigned to the America on the Move group were asked to make two small lifestyle changes: (1) to walk an additional 2000 steps per day above baseline as measured by pedometers and (2) to eliminate 420 kJ/day (100 kcal/day) from their typical diet by replacing dietary sugar with a noncaloric sweetener. Families who were assigned to the self-monitor group were asked to use pedometers to record physical activity but were not asked to change their diet or physical activity level. The results showed that, during a 6-month period, both groups of children showed significant decreases in BMI for age. However, the America on the Move group, compared with the self-monitor group, had a significantly higher percentage of target children who maintained or reduced their BMI for age and, consistently, a significantly lower percentage who increased their BMI for age. There was no significant weight gain during the 6-month intervention in parents of either group. Rodearmel et al. (2007) suggested that the small-changes approach advocated by America on the Move could be useful for addressing childhood obesity by preventing excess weight gain in families.

11.5.4 Do Artificial Sweeteners Help Reduce Weight?

When sugar was covertly switched to aspartame in a metabolic ward, a 25% immediate reduction in energy intake was achieved, promoting weight loss in rats (Porikos et al. 1977; Porikos and Koopmans 1988). Conversely, knowingly ingesting aspartame was associated with a high total energy intake, suggesting overcompensation for the expected caloric reduction (Mattes 1990). Vigilant monitoring, caloric restriction, and exercise were most likely involved in the weight loss seen in different programs associating artificial sweeteners (Rodearmel et al. 2007; Blackburn et al. 1997).

The San Antonio Heart Study examined 3682 adults over a 7- to 8-year period in the 1980s and found a positive correlation between artificial sweetener use and weight gain (Fowler et al. 2008). When matched for initial BMI, gender, ethnicity, and diet, drinkers of artificially sweetened beverages consistently had higher BMIs at the follow-up, with dose dependence on the amount of consumption.

The American Cancer Society Study conducted in the early 1980s included 78,694 women who were highly homogeneous with regard to age, ethnicity, socioeconomic status, and lack of preexisting conditions (Stellman and Garfinkel 1986, 1988). At 1-year follow-up, 2.7%–7.1% more regular artificial sweetener users gained weight compared to nonusers matched by initial weight. The difference in the amount gained between the two groups was less than 2 lb., albeit statistically significant.

Saccharin use was also associated with 8-year weight gain in 31,940 women from the Nurses' Health Study conducted in the 1970s, as described by Colditz et al. (1990).

Phelan et al. (2009) compared the dietary strategies and the use of fat- and sugar-modified foods and beverages in a weight loss maintainer group (WLM) and an always-normal weight group (NW). WLM reported consuming a diet that was lower in fat (28.7% versus 32.6%; $P < 0.0001$) and used more fat modification strategies than NW. WLM also consumed a significantly greater percentage of modified dairy (60% versus 49%; $P = 0.002$) and modified dressings and sauces (55% vs. 44%; $P = 0.006$) than NW. WLM reported consuming three times more daily servings of artificially sweetened soft drinks (0.91% versus 0.37%; $P = 0.003$), significantly fewer daily servings of sugar-sweetened soft drinks (0.07% versus 0.16%; $P = 0.03$), and more daily servings of water (4.72% versus 3.48%; $P = 0.002$) than NW. These findings suggest that WLM used more dietary strategies to accomplish their weight loss maintenance, including greater restriction on fat intake, use of fat- and sugar-modified foods, reduced consumption of SSBs, and increased consumption of artificially sweetened beverages. Ways of promoting the use of fat-modified foods and artificial sweeteners merits further research in both prevention- and treatment-controlled trials.

Vermunt et al. (2003) have evaluated whether the replacement of dietary (added) sugar by low-energy sweeteners or complex carbohydrates contributes to weight reduction. In two experimental studies, no short-term differences in weight loss were observed after the use of aspartame compared to sugar in obese subjects following a controlled energy-restricted diet. However, the consumption of aspartame was associated with improved weight maintenance after a year. In two short-term studies in which energy intake was not restricted, the substitution of sucrose by artificial sweeteners, investigated mostly in beverages, resulted in lower energy intake and lower body weight. Similarly, two short-term studies comparing the effect of sucrose and starch on weight loss in obese subjects did not find differences when the total energy intake was equal and reduced. An *ad libitum* diet with complex carbohydrates resulted in lower energy intake compared to high-sugar diets. In two out of three studies, this was reflected in lower body weight in subjects consuming the complex carbohydrate diet. In conclusion, a limited number of relatively short-term studies suggest that replacing (added) sugar by low-energy sweeteners or by complex carbohydrates in an *ad libitum* diet might result in lower energy intake and reduced body weight. In the long term, this might be beneficial for weight maintenance. However, the number of studies is small, and overall conclusions, in particular for the long term, cannot be drawn (Vermunt et al. 2003).

Donner et al. (2010) explored the metabolic effects of oral D-tagatose (D-tag) given daily to eight human subjects with type 2 diabetes mellitus (DM) for 1 year. Oral D-tag attenuates the rise in plasma glucose during an oral glucose tolerance test in subjects with type 2 DM and reduces food intake in healthy human subjects. In addition, a reduction in food consumption and less weight gain has been observed in rats fed on tagatose. Donner et al. (2010) hypothesized that the treatment period would lead to weight loss and improvements in glycated hemoglobin and the lipid profile. A 2-month run-in period was followed by a 12-month treatment period when 15 g of oral D-tag was taken three times daily with food. No serious adverse effects were seen during the 12-month treatment period. Ten of the initially 12 recruited subjects experienced gastrointestinal side effects that tended to be mild and transient. When three subjects who had oral diabetes, medications added, and/or dosages increased during the study were excluded, the mean (SD) body weight declined from 108.4 (9.0) to 103.3 (7.3) kg ($P = 0.001$). Glycated hemoglobin fell nonsignificantly from 10.6% ± 1.9% to 9.6% ± 2.3% ($P = 0.08$). High-density lipoprotein cholesterol progressively rose from a baseline level of 30.5 ± 15.8 to 41.7 ± 12.1 mg/dL at month 12 in the six subjects who did not have lipid-modifying medications added during the study ($P < 0.001$). Significant improvements in body weight and high-density lipoprotein cholesterol in this pilot study suggest that D-tag may be a potentially useful adjunct in the management of patients with type 2 DM (Donner et al. 2010).

Strategies to reverse the upward trend in obesity rates need to focus on both reducing energy intake and increasing energy expenditure. The provision of low- or reduced-energy-dense foods is one way of helping people to reduce their energy intake and therefore enable weight maintenance or weight loss to occur. The use of intense sweeteners as a substitute for sucrose potentially offers one way of helping people to reduce the energy density of their diet without any loss of palatability. de la Hunty et al. (2006) reviewed the evidence for the effect of aspartame on weight loss, weight maintenance, and energy intake in adults and addresses the question of how much energy is compensated for and whether the use of aspartame-sweetened foods and drinks is an effective way of losing weight. They identified all studies that examined the effect of substituting sugar with either aspartame alone or aspartame in combination with other intense sweeteners on energy intake or body weight. Studies that were not randomized controlled trials in healthy adults, and studies that did not measure energy intake for at least 24 h (for those with energy intake as an outcome measure) were excluded from the analysis. A minimum of 24-h energy intake data was set as the cutoff to ensure that the full extent of any compensatory effects was seen. A total of 16 studies were included in the analysis. Of these 16 studies, 15 had energy intake as an outcome measure. The studies that used soft drinks as the vehicle for aspartame used between 500 and about 2000 mL, which is equivalent to about two to six cans or bottles of soft drinks every day. A significant reduction in energy intake was seen with aspartame compared with all types of control, except when aspartame was compared with nonsucrose controls such as water. In addition, de la Hunty et al. (2006) reported that using foods and drinks sweetened with aspartame instead of those sweetened with sucrose is an effective way of maintaining and losing weight without reducing the palatability of the diet. The decrease in energy intake and the rate of weight loss that can reasonably be achieved is low but meaningful and, on a population basis, more than sufficient to counteract the current average rate of weight gain of around 0.007 kg/week. On an individual basis, it provides a useful adjunct to other weight loss regimes. Some compensation for the substituted energy does occur, but this is only about one-third of the energy replaced and is probably less when using soft drinks sweetened with aspartame. Nevertheless, these compensation values are derived from short-term studies. More data is needed over the longer term to determine whether a tolerance to the effects is acquired. To achieve the average rate of weight loss seen in these studies of 0.2 kg/week will require around a 220-kcal (0.93-MJ) deficit per day based on an energy value for obese tissue of 7500 kcal/kg. Assuming the higher rate of compensation (32%), this would require the substitution of around 330 kcal/day (1.4 MJ/day) from sucrose with aspartame (which is equivalent to around 88 g of sucrose). Using the lower estimated rate of compensation for soft drinks alone (15.5%) would require the substitution of about 260 kcal/day (1.1 MJ/day) from sucrose with aspartame. This is equivalent to 70 g of sucrose or about two cans of soft drinks every day.

11.5.5 Artificial Sweeteners and Energy

Aspartame increased subjective hunger ratings compared to glucose or water (Blundell and Hill 1986). Glucose preload reduced the perceived pleasantness of sucrose, but aspartame did not, according to the same authors. In another study, aspartame, acesulfame potassium (Ace K), and saccharin were all associated with a heightened motivation to eat and more items selected on a food preference list (Rogers et al. 1988). Aspartame had the most pronounced effect, possibly due to its nonbitter aftertaste. Artificial sweetener preloads either had no effect (Black et al. 1993; Rogers et al. 1988) or increased subsequent energy intake (Lavin et al. 1997; King et al. 1999) unlike glucose or sucrose, which decreased the energy intake at the test meal. Moreover, Rolls (1991) reported on the sweet taste of aspartame, saccharin, and Ace K, and they found it to increase ratings of hunger and, after saccharin consumption, to increase food intake. However, most investigators have found that aspartame consumption is associated with decreased or unchanged ratings of hunger. Even if aspartame consumption increases ratings of hunger in some situations, it apparently has little impact on the controls of food intake and body weight. Furthermore, aspartame does not seem to increase food intake; indeed, both short- and long-term studies have demonstrated that the consumption of aspartame-sweetened foods or drinks is associated with either no change or a reduced food intake. In addition, preliminary clinical trials suggest that aspartame may be a useful aid in a complete diet-and-exercise program or in weight maintenance. Finally, it should be mentioned that intense sweeteners have never been found to cause weight gain in humans (Rolls 1991). Those findings suggest that the calorie contained in natural sweeteners may trigger a response to keep the overall energy consumption constant.

Rodent models helped elucidate how artificial sweeteners contribute to energy balance. Rats conditioned with saccharin supplement had significantly increased total energy intake and gained more weight with increased body adiposity compared to controls conditioned with glucose according to Swithers and Davidson (2008).

Recent epidemiological evidence by Swithers et al. (2010) pointed to a link between a variety of negative health outcomes (e.g., metabolic syndrome, type II diabetes, and cardiovascular disease) and the consumption of both calorically sweetened beverages and beverages sweetened with high-intensity, noncaloric sweeteners. Research on the possibility that nonnutritive sweeteners promote food intake, body weight gain, and metabolic disorders has been hindered by the lack of a physiologically relevant model that describes the mechanistic basis for these outcomes. The researchers suggested that, based on Pavlovian conditioning principles, the consumption of nonnutritive sweeteners could result in sweet tastes no longer serving as consistent predictors of nutritive postingestive consequences. This dissociation between the sweet taste cues and the caloric consequences could lead to a decrease in the ability of sweet tastes to evoke physiological responses that serve to regulate energy balance. Using a rodent model, they have found that the intake of foods or fluids containing nonnutritive sweeteners was accompanied by increased food intake, body weight gain, accumulation of body fat, and weaker caloric compensation compared to the consumption of foods and fluids containing glucose. They also provided evidence consistent with the hypothesis that these effects of consuming saccharin may be associated with a decrement in the ability of sweet taste to evoke thermic responses and perhaps other physiological, cephalic-phase reflexes that are thought to help maintain energy balance (Swithers et al. 2010).

Moreover, saccharin-conditioned rats also failed to curb their chow intake following a sweet premeal. When a flavor was arbitrarily associated with high or low caloric content, rats ate more chow following a premeal, with the flavor predictive of low caloric content (Pierce et al. 2007). These studies pose a hypothesis: inconsistent coupling between sweet taste and caloric content can lead to compensatory overeating and positive energy balance (Yang 2010).

Using both cross-sectional and longitudinal data from the Framingham Heart Study, Dhingra et al. (2007) reported positive relationships between the consumption of diet soda and the prevalence of the metabolic syndrome that were larger than that obtained with the consumption of regular soda. Similar results have also been reported as part of independent studies of Lutsey et al. (2008) and Nettleton et al. (2009). Furthermore, Fowler et al. (2008) reported that, for humans who were normal weight or nonobese (BMI < 30 kg/m^2) at baseline, the intake of >21 nonnutritively sweetened beverages per week (diet sodas and artificially sweetened coffee and tea) was associated with about double the risk of obesity compared to nonusers at follow-up 7 to 8 years later.

11.5.6 Can Consumption of High-Intensity Sweeteners Disrupt Energy Balance?

Roy et al. (2007) have suggested that rats exposed to saccharin showed an increase in body weight and caloric intake compared to rats exposed to the caloric sweetener glucose. The authors suggested that animals, including humans, use orosensory cues from foods to signal the nutritive consequences when eating and a sweet taste that does not signal the arrival of these consequences may contribute to deficits in the regulation of energy balance. They also examined whether these effects generalized to additional caloric and noncaloric sweeteners. Rats were divided into four groups (glucose, sucrose, saccharin, or Ace K) and received fixed volumes of unsweetened or sweetened low-fat yogurt daily for 14 days in addition to *ad libitum* chow and water. During the 2-week diet exposure, animals that received high-intensity sweeteners gained significantly more weight than animals that received caloric sweeteners. Additionally, the high-intensity sweetener groups consumed significantly more total calories than the caloric sweetener groups. This data showed that the effects on food intake and body weight gain are not directly related to the consumption of a particular high-intensity sweetener but that the consumption of high-intensity sweeteners, in general, may disrupt energy balance (Roy et al. 2007).

11.5.7 High-Fructose Corn Syrup, Energy Intake, and Body Weight

High-fructose corn syrup (HFCS) is produced from the isomerization of some of the glucose in corn syrup to fructose. HFCS-55, consisting of 55% fructose and 42% glucose, is used in many sweetened beverages, whereas HFCS-42 (42% fructose; 53% glucose) is used to sweeten confections.

The GI of HFCS has not been published, but the GI of cola sweetened with HFCS is 63 ± 5 (Foster-Powell et al. 2002), a figure close to that of sucrose (68 ± 5), which might be expected because of the similarities between the sweeteners. Low-GI foods have been associated with greater satiety than high-GI foods (Brand-Miller et al. 2002). Low-GI foods may prolong satiety between meals, whereas high-GI foods may signal immediate satiety.

HFCS has been implicated in excess weight gain through mechanisms seen in some acute feeding studies and by virtue of its abundance in the food supply during years of increasing obesity. Compared with pure glucose, fructose is thought to be associated with insufficient secretion of insulin and leptin and suppression of ghrelin (Melanson et al. 2008). However, when HFCS is compared with sucrose, the more commonly consumed sweetener, such differences are not apparent, and appetite and energy intake do not differ in the short term. Longer term studies on connections between HFCS, potential mechanisms, and body weight have not been conducted. The authors examined collective data on associations between the consumption of HFCS and energy balance, with particular focus on energy intake and its regulation.

Some experts have implicated HFCS as a possible contributing factor to energy overconsumption, weight gain, and, thus, the rise in the prevalence of obesity over the past decades (Ludwig et al. 2001; Bray et al. 2004; Elliott et al. 2002).

Melanson et al. (2007) reported that HFCS results in increased plasma glucose and insulin, most likely as a result of the glucose moiety. They conducted a similar study design with two 2-day visits in 30 healthy-weight young women to compare hormonal and appetitive responses to beverages sweetened by HFCS or sucrose. The beverages were served with three meals during the day and provided 30% of the energy intake. Energy intake was controlled on the first day when the test beverages were served and appetite was rated, and food intake was *ad libitum* on the second day of each visit. Blood glucose, insulin, leptin, and ghrelin did not differ significantly between the two sweeteners. HFCS- and sucrose-sweetened beverages produced similar ghrelin suppression after each meal of approximately 200 pg/mL after both sucrose and HFCS trials. Finally, no significant differences were seen between HFCS and sucrose in *ad libitum* energy or macronutrient intakes. Appetite ratings were also similar (the one exception was a slightly greater desire to eat after sucrose consumption). The lack of differences between HFCS and sucrose in energy intake and appetite ratings are not surprising because of similar responses in plasma glucose, insulin, leptin, and ghrelin, all of which have been postulated as biomarkers of energy intake regulation.

Zuckley et al. (2007) have recently repeated the same study design to compare hormonal and appetitive responses to HFCS and sucrose in obese and overweight women. Preliminary findings showed that these responses to HFCS and sucrose do not differ significantly in persons carrying excess body weight. Similar blood glucose and hormones, as well as appetite ratings and *ad libitum* energy intake, were seen with the consumption of HFCS and sucrose.

Monsivais et al. (2007) worked on a preliminary study that compared cola sweetened with sucrose, HFCS-55, HFCS-42, or aspartame; 1%-fat milk; and a no-beverage control in 37 adults in a randomized paired design. Hunger and satiety ratings did not differ significantly among the beverage treatments. Relative to the two no-energy treatments, energy intake compensation was similar among the four energy-containing drinks at the meal 140 min later. This study examined typical HFCS loads and found similar appetite responses compared with isocaloric beverages.

Prospective epidemiologic data in adults have associated increases in SSBs with weight gain (Schulze et al. 2004). Together, these studies imply that increased energy intake by sweetened beverages is not compensated for in subsequent intake, which may lead to overconsumption.

However, these studies do not determine whether HFCS may be more of a factor in weight gain than other caloric sweeteners nor do they specifically address the implications of total dietary HFCS from all sources on energy intake and body weight.

Stanhope and Havel (2008) have investigated two hypotheses regarding the effects of fructose consumption: (1) the endocrine effects of fructose consumption favor a positive energy balance and (2) fructose consumption promotes the development of an atherogenic lipid profile. In previous short- and long-term studies, they showed that the consumption of fructose-sweetened beverages with three meals results in lower 24-h plasma concentrations of glucose, insulin, and leptin in humans than does the consumption of glucose-sweetened beverages. They have also tested whether the prolonged consumption of high-fructose diets leads to increased caloric intake or decreased energy expenditure, thereby contributing to weight gain and obesity. Results from a study conducted in rhesus monkeys produced equivocal results. Carefully controlled and adequately powered long-term studies are required to address these hypotheses. In both short- and long-term studies, they showed that the consumption of fructose-sweetened beverages substantially increases postprandial triacylglycerol concentrations compared with glucose-sweetened beverages. In the long-term studies, apolipoprotein B concentrations were also increased in subjects consuming fructose but not in those consuming glucose. Data from a short-term study comparing the consumption of beverages sweetened with fructose, glucose, HFCS, and sucrose suggest that HFCS and sucrose increase postprandial triacylglycerol to an extent comparable with that induced by 100% fructose alone. Increased consumption of fructose-sweetened beverages along with increased prevalence of obesity, metabolic syndrome, and type 2 diabetes underscores the importance of investigating the metabolic consequences of fructose consumption in carefully controlled experiments.

Similarly, Stanhope et al. (2008) carried out a short-term study suggesting that consuming HFCS- and sucrose-sweetened beverages increases postprandial triacylglycerol concentrations to the same degree as fructose alone. Similar results in both short- and long-term studies have shown that fructose consumption substantially increases postprandial triacylglycerol concentrations (Teff et al. 2004; Swarbrick et al. 2008).

11.5.8 Honey and Weight Gain

Honey is a naturally occurring sweetener that contains simple and complex sugars, as well as vitamins, minerals, acids, and enzymes. Limited clinical studies have shown that honey has a lower GI than sucrose (Shambaugh et al. 1990) and that honey, sucrose, and fructose do have differential effects on blood glucose levels (Samanta et al. 1985; Al-Waili 2003, 2004).

Chepulis and Starkey (2008) investigated whether honey and sucrose would have differential effects on weight gain during long-term feeding. Forty-five 2-month-old Sprague Dawley rats were fed a powdered diet that was either sugar-free or contained 7.9% sucrose or 10% honey *ad libitum* for 52 weeks (honey is 21% water). Weight gain was assessed every 1–2 weeks, and food intake was measured every 2 months. At the completion of the study, blood samples were removed for the measurement of blood sugar (HbA1c) and a fasting lipid profile. Dual X-ray absorptiometry analyses were then performed to determine body composition and bone mineral densities. Overall weight gain and body fat levels were significantly higher in sucrose-fed rats and similar for those fed honey or a sugar-free diet. HbA1c levels were significantly reduced, and high-density lipoprotein (HDL) cholesterol significantly increased in honey-fed rats compared with rats fed on sucrose or a sugar-free diet, but no other differences in lipid profiles were found. No differences in bone mineral density were observed between honey- and sucrose-fed rats, although it was significantly increased in honey-fed rats compared with those fed the sugar-free diet.

Certainly, the aforementioned results agree with the literature that low-GI foods can improve weight regulation compared with their higher GI counterparts (Agus et al. 2000; Spieth et al. 2000; Dumesnil et al. 2001; Brand-Miller et al. 2002; Bahrami et al. 2009), and the World Health Organization has even issued an extensive report detailing the use of low-GI foods as an appropriate way of preventing obesity (FAO/WHO Expert Consultation 1998).

As reported by Chepulis (2007), the reduced weight gain seen in honey-fed rats may be due to the insulin-mimetic effects of hydrogen peroxide produced by the honey. No studies have been undertaken to assess whether hydrogen peroxide could reach sufficient levels *in vivo* to elicit such a response, although it warrants further investigation.

Nemoseck et al. (2011) hypothesized that, in comparison with sucrose, a honey-based diet would promote lower weight gain, adiposity, and related biomarkers (leptin, insulin, and adiponectin) as well as a better blood lipid profile. Thirty-six male Sprague Dawley rats (228.1 ± 12.5 g) were equally divided by weight into two groups ($n = 18$) and provided free access to one of the two diets of equal energy densities, differing only in a portion of the carbohydrate. Diets contained 20% carbohydrate (by the weight of the total diet) from either clover honey or sucrose. After 33 days, epididymal fat pads were excised and weighed, and blood was collected for analyses of serum concentrations of lipids, glucose, and markers of adiposity and inflammation. The body weight gain was 14.7% lower ($P \leq 0.05$) for rats fed on honey, corresponding to a 13.3% lower ($P \leq 0.05$) consumption of food/energy, whereas food efficiency ratios were nearly identical. The epididymal fat weight was 20.1% lower ($P \leq 0.05$) for rats fed on honey. Serum concentrations of triglycerides and leptin were lower ($P \leq 0.05$) by 29.6% and 21.6%, respectively, and non-HDL cholesterol was higher ($P \leq 0.05$) by 16.8% for honey-fed rats. No significant differences in serum total cholesterol, HDL cholesterol, adiponectin, C-reactive protein, monocyte chemoattractant protein-1, glucose, or insulin were detected. These results suggest that, in comparison with sucrose, honey may reduce weight gain and adiposity, presumably due to lower food intake, and promote lower serum triglycerides but higher non–HDL cholesterol concentrations.

11.5.9 Sugar-Sweetened Beverages and Weight Gain

The consumption of SSBs has been linked to obesity and type 2 diabetes. Over the past decade, U.S. adult SSB consumption has increased. SSB comprises a considerable source of total daily intake and is the largest source of beverage calories. SSB consumption is highest among subgroups also at greatest risk of obesity and type 2 diabetes.

Bleich et al. (2009) examined national trends in SSB consumption among U.S. adults by sociodemographic characteristics, body weight status, and weight loss intention. They analyzed 24-h dietary recall data to estimate beverage consumption among adults (aged older than 20 years old) obtained from the NHANES III (1988–1994; n = 15,979) and NHANES 1999–2004 (n = 13,431). They reported that, from 1988–1994 to 1999–2004 on the survey day, the percentage of adult SSB drinkers increased from 58% to 63% ($P < 0.001$), the per capita consumption of SSB increased by 46 kcal/day ($P < 0.001$), and the daily SSB consumption among drinkers increased by 6 oz ($P < 0.001$). In both survey periods, per capita SSB consumption was highest among young adults (231–289 kcal/day) and lowest among the elderly (68–83 kcal/day). More adults are drinking SSBs (primarily soda; approximately 60%), and among SSB drinkers, the average caloric consumption and quantity consumed had increased, changes that parallel the rising prevalence of adult obesity and type 2 diabetes. Young blacks had the highest percentage of SSB drinkers and the highest per capita consumption compared with white and Mexican American adults ($P < 0.05$). Overweight–obese adults with weight loss intention (compared with those without) were significantly less likely to drink SSB, but they still consumed a considerable amount in 1999–2004 (278 kcal/day). Among young adults, 20% of SSB calories were consumed at work (Bleich et al. 2009).

Bremer et al. (2010) evaluated current SSB consumption trends and their association with insulin resistance-related metabolic parameters and anthropometric measurements by performing a cross-sectional analysis of the NHANES data between the years 1988 to 1994 and 1999 to 2004. Main outcome measures included SSB consumption trends, a homeostasis model assessment of insulin resistance, blood pressure, waist circumference, BMI, and fasting concentrations of total cholesterol, HDL cholesterol, low-density lipoprotein cholesterol, and triglycerides. Although the overall SSB consumption has increased, their data suggested that this increase was primarily due to an increase in the amount of SSBs consumed by males in the high-SSB intake group alone. Multivariate linear regression analyses also showed that increased SSB consumption was independently associated with many adverse health parameters. Factors other than SSB consumption must therefore be contributing to the increasing prevalence of obesity and metabolic syndrome in the majority of U.S. children.

Schulze et al. (2004) examined the association between the consumption of SSBs and weight change and the risk of type 2 diabetes in women. Prospective cohort analyses were conducted from 1991 to 1999 among women in the Nurses' Health Study II. The diabetes analysis included 91,249 women free of diabetes and other major chronic diseases at baseline in 1991. The weight change analysis included 51,603 women for whom complete dietary information and body weight were ascertained in 1991, 1995, and 1999. The researchers identified 741 incident cases of confirmed type 2 diabetes during 716,300 person–years of follow-up. Those with stable consumption patterns had no difference in weight gain, but weight gain over a 4-year period was highest among women who increased their sugar-sweetened soft drink consumption from one or fewer drinks per week to one or more drinks per day (multivariate-adjusted means, 4.69 kg for 1991–1995 and 4.20 kg for 1995–1999) and was smallest among women who decreased their intake (1.34 and 0.15 kg for the two periods, respectively) after adjusting for lifestyle and dietary confounders. Increased consumption of fruit punch was also associated with greater weight gain compared with decreased consumption. After adjustment for potential confounders, women consuming one or more sugar-sweetened soft drinks per day had a relative risk (RR) of type 2 diabetes of 1.83 [95% confidence interval (CI), 1.42–2.36; $P < 0.001$ for trend] compared with those who consumed less than one of these beverages per month. Similarly, the consumption of fruit punch was associated with increased diabetes risk (RR for ≥1 drink per day compared with <1 drink per month, 2.00; 95% CI, 1.33–3.03; $P = 0.001$). Hence, higher consumption of SSBs is associated with a greater magnitude of weight gain and an increased risk for development of type 2 diabetes in women, possibly by providing excessive calories and large amounts of rapidly absorbable sugars.

Undoubtfully, excess of sugar intake may contribute to weight gain. Furthermore, no consistent evidence exists to support the view that low-energy sweeteners increase appetite or food intake. On the other hand, low-calorie sweeteners do not lead to an automatic weight loss. Calorie control and weight management seem to be multidimensional problems that require proper lifestyle behavior.

11.6 HUMAN PERFORMANCE APPLICATIONS

The classical study of Christensen and Hansen in 1939 demonstrated the importance of carbohydrates in the diet of athletes (Christensen and Hansen 1939). In the 1960s, when the biopsy technique of obtaining muscle samples was made easier (Bergstrom 1962), several studies showed that exercise-induced reductions in the body's limited muscle glycogen stores correlated well with the development of fatigue (Ahlborg et al. 1967; Hermansen et al. 1967; Saltin and Karlsson 1971; Gollnick et al. 1973). As a consequence, "carbohydrate-loading" strategies were developed (Astrand 1967; Sherman et al. 1981), and the ingestion of carbohydrate–electrolyte drinks (sports drinks) before, during, and after exercise became a common practice for athletes. It was also the beginning of a new challenge for the food industry to develop sports-specific food products and drinks. Such sports drinks should be designed to provide fluid and energy in the form of carbohydrates to the exercising individual without causing any gastrointestinal discomfort, which is sometimes a problem during exercise (Brouns and Beckers 1993). Currently, many products have been launched in the market in liquid, semiliquid (gel), solid (bars), or powder form. Specific products with different formulations according to the intended use are offered to the athlete—products for optimizing fuel (body carbohydrate) levels before exercise, maintaining blood glucose levels and postponing the onset of fatigue during exercise, and speeding up recovery and enhancing muscle development after exercise.

Various natural and artificial sweeteners are used in all these products. Glucose (dextrose), sucrose, fructose, glucose syrup, HFCS, and glucose polymers such as maltodextrin are some common natural sweeteners used in sports drinks for providing energy. Low-calorie sweeteners like Ace K, sucralose, aspartame, and truvia (rebiana) are also used in some "low-calorie sports drinks," for example, Powerade Option by the Coca-Cola Company and All Sport Naturally Zero by PepsiCo. In this section, the claims "performance enhancement," "maintenance of euhydration," "improved recovery," and "increased muscle mass" made by manufacturers of sports beverages will be briefly examined using a nutrient timing approach. Extensive analysis of these issues is beyond the scope of this book and cán be found elsewhere (Austin and Seebohar 2011; Burke 2007; Jeukendrup and Gleeson 2010).

11.6.1 Ingestion of Carbohydrates before Exercise

The reasons for consuming carbohydrate before exercise are to enhance the availability of glycogen in the muscle and liver and of glucose in blood. However, in an early study conducted by (Costill et al. 1977), it was found that the ingestion of 75 g of glucose in 300 mL of water 45 min before the start of submaximal treadmill running caused a greater rate of muscle glycogen utilization than when exercise was performed after drinking water. This, of course, is not favorable for the athlete, because it is well known that body glycogen stores are limited and an elevation in the rate of glycogenolysis during prolonged exercise could lead to an early onset of fatigue. In fact, researchers from the same laboratory confirmed this hypothesis in a study where cyclists fatigued sooner when they ingested a similar glucose solution 30 min before exercise compared to water ingestion (Foster et al. 1979).

The findings of these two studies were explained by the hyperglycemia and hyperinsulinemia caused as a consequence of the preexercise ingestion of high-GI carbohydrates such as glucose. This hyperinsulinemia before exercise depressed fatty acids during exercise, denying in this way this important substrate to exercising muscle. As a consequence, the body relied more on the limited glycogen stores, which in turn caused premature fatigue (Costill et al. 1977; Foster et al. 1979; Hargreaves et al. 1985; Horowitz et al. 1997). However, since the early study of Foster et al. in 1979, many studies have reported either improvements in endurance capacity and performance or no effect as a result of ingesting various types of carbohydrates such as glucose, fructose, and maltodextrin in liquid or even in solid form (Coombes and Hamilton 2000). Furthermore, several attempts have been made to reduce the unfavorable metabolic perturbations caused when carbohydrates that have a high GI are ingested before exercise. These attempts have focused on the use of low-GI foods, which produce a lower postprandial hyperglycemia and hyperinsulinemia compared to high-GI carbohydrates. Such a response may elevate free fatty acid oxidation during exercise, leading to a better maintenance of carbohydrate availability (O'Reilly et al. 2010). However, although there is a potential benefit regarding exercise performance and substrate utilization when low-GI carbohydrates are ingested, variations in research methodology on the GI of meals consumed before exercise have led to inconclusive findings (American College of Sports Medicine 2009).

Regardless of controversies in the literature and because prolonged fasting is harmful to health and performance (Maughan 2010), athletes are advised to ingest a preexercise meal or snack that will provide sufficient fluid to maintain hydration, be it high in carbohydrates and low in fat and fiber to facilitate gastric emptying and minimize gastrointestinal distress and composed of familiar foods. In terms of timing and quantity for preexercise euhydration, the exercising individual should drink about 5–7 mL per kg body mass of a sports beverage at least 4 h before exercise. This would allow enough time to optimize hydration levels and for the excretion of any excess fluid as urine (American College of Sports Medicine 2009). In addition, consuming sports drinks with sodium and/or salted snacks instead of plain water alone can help stimulate thirst and retain needed fluids (American College of Sports Medicine 2007).

11.6.2 Ingestion of Carbohydrate Beverages during Exercise

The effects of carbohydrate beverages consumption on metabolism and performance during exercise is the most researched topic in sports nutrition. The main aims of sports beverages ingested during exercise are to replace sweat loss and to reduce the problems associated with dehydration and to provide a source of energy in the form of carbohydrates, which can supplement the limited stores of muscle and liver glycogen (Coombes and Hamilton 2000).

In order to thermoregulate during exercise, the human body loses substantial amounts of water and electrolytes through sweat. If these losses are not replaced, excessive dehydration occurs, which not only degrades performance but can also increase the risk of heat illness (Casa et al. 2005). The goal of drinking sports beverages during exercise is to prevent water loss that exceeds 2% of the body weight (American College of Sports Medicine 2007). However, this is not always an easy task, because sweat rate is influenced by many factors such as the duration and intensity of exercise, the environmental conditions, the type of clothing or equipment worn, and the characteristics of the individual (body weight, genetic predisposition, acclimatization state, and metabolic efficiency). Therefore, it is not surprising that sweat rates may vary from 0.3 to 2.6 L h^{-1} (American College of Sports Medicine 2007). Under these circumstances, general guidelines are difficult to give. Athletes are advised to record their body weights preexercise and postexercise in order to estimate individual sweat loss, which will help in designing individualized fluid replacement programs.

Many studies have reported an ergogenic effect when sports drinks are ingested during short term (up to 1 h; Below et al. 1995; Jeukendrup et al. 1997; Millard-Stafford et al. 1997), prolonged intermittent (1–4 h; Hargreaves et al. 1984; Murray et al. 1989, 1991; Davis et al. 1997), prolonged continuous (Maughan et al. 1989; Wright et al. 1991; Tsintzas et al. 1993; Febbraio et al. 1996; McConell et al. 1996; Tsintzas et al. 1996), or ultraendurance (>4 h; Brouns et al. 1989) exercise. These studies are some of the many investigations that have shown an improvement in performance. Several reviews offer a more detailed presentation of the topic (Coombes and Hamilton 2000; Kerksick et al. 2008; American College of Sports Medicine 2009).

In terms of quantity and timing, ingesting 200–450 mL of sports beverages (depending on body weight) every 15–20 min should deliver about 30–60 g of carbohydrates per hour (American College of Sports Medicine 2009).

As was mentioned earlier, the carbohydrates in sports drinks should provide an alternative to liver and muscle glycogen energy source for oxidation, which will result in improved maintenance of blood glucose. Several attempts of mixing glucose and sucrose, glucose and fructose, or a combination of maltodextrin and fructose in beverages that increase exogenous carbohydrate oxidation to a level greater than 1.2 g min^{-1} have been made (Jentjens et al. 2004, 2005; Jentjens and Jeukendrup 2005; Wallis et al. 2005). However, fructose ingestion on its own is not recommended, because it is not so effective in improving performance and may also cause gastrointestinal problems (Murray et al. 1989; Bjorkman et al. 1984).

Despite the fact that the majority of studies have reported positive results in exercise performance as a result of ingesting carbohydrate–electrolyte drinks during exercise, the exact mechanism by which this is achieved is not clear. A better maintenance of blood glucose and carbohydrate oxidation late in exercise (Coggan and Coyle 1987; Coyle et al. 1986) with a reduced liver glucose output (McConell et al. 1994) has been observed during cycling; whereas in running, a reduced muscle glycogen use in type I fibers has been reported (Tsintzas et al. 1995). Another mechanism, although without possessing such strong evidence as the previous ones, might be the reduced plasma-free tryptophan to branched chain amino acid ratio when carbohydrates are ingested. This leads to a reduction in the synthesis of the neurotransmitter serotonin (5-hydroxytryptamine) in the brain. Serotonin has the potential of reducing mental and physical performance during prolonged exercise (Davis et al. 2000). Recently, a number of studies have shown that sports drinks may provide a benefit to the athlete by simply mouth-rinsing these solutions and not ingesting them during short (1-h) endurance exercise. Researchers explain this amazing finding by data that show that carbohydrates in the mouth stimulate reward centers in the brain, lower the perception of effort during exercise, and increase corticomotor excitability (Rollo and Williams 2011).

11.6.3 Ingestion of Carbohydrates during Recovery

After the completion of exercise, body fluids and glycogen stores must be restored. This should be done as soon as possible if a subsequent exercise bout is to take place. In terms of restoring euhydration, athletes need to drink about 1.5 L of fluid for each kilogram of body weight lost. Consuming sports drinks and snacks with sodium will also help recovery by stimulating thirst and fluid retention (American College of Sports Medicine 2007). As far as glycogen resynthesis is concerned, the consumption of 1.0–1.5 g of carbohydrate per kilogram of body mass within 30 min after exercise and at 2-h intervals up to 6 h is recommended (American College of Sports Medicine 2009). Furthermore, carbohydrates with a high GI may result in higher muscle glycogen levels compared with the same amount of carbohydrates with a low GI (Burke et al. 1993).

In conclusion, sports drinks containing carbohydrates and electrolytes may be consumed before, during, and after exercise to help maintain blood glucose concentration, provide fuel for muscles, improve performance, and decrease the risk of dehydration.

REFERENCES

Abdallah, L., M. Chabert, and J. Louis-Sylvestre. 1997. Cephalic phase responses to sweet taste. *Am. J. Clin. Nutr.* 65: 737–743.

Agus, M. S. D., J. F. Swain, C. L. Larson, E. A. Eckert, and D. S. Ludwig. 2000. Dietary composition and physiological adaptations to energy restriction. *Am. J. Clin. Nutr.* 71: 901–907.

Ahlborg, B., J. Bergstrom, L.G. Ekelund, and E. Hultman. 1967. Muscle glycogen and muscle electrolytes during prolonged physical activity. *Acta Physiol. Scand.* 70: 129–142.

Al-Waili, N. S. 2003. Effects of daily consumption of honey solution on hematological indices and normal levels of minerals and enzymes in normal individuals. *J. Med. Food* 6 (2): 135–140.

Al-Waili, N. S. 2004. Intravenous and intrapulmonary administration of honey solution to healthy sheep: Effects on blood sugar, renal and liver function tests, bone marrow function, lipid profile and carbon tetrachloride-induced liver injury. *J. Med. Food* 6 (3): 231–247.

American College of Sports Medicine. 2007. Exercise and fluid replacement: Position statement. *Med. Sci. Sports Exercise* 39: 377–390.

American College of Sports Medicine. 2009. Nutrition and athletic performance: Position statement. *Med. Sci. Sports Exercise* 41: 709–731.

American Dietetic Association. 1998. Position of the American Dietetic Association: Use of nutritive and non-nutritive sweeteners. *J. Am. Diet. Assoc.* 98 (5): 580–587.

Arai, H., A. Mizuno, M. Sakuma, M. Fukaya, K. Matsuo, K. Muto, H. Sasaki, M. Matsuura, H. Okumura, H. Yamamoto, Y. Taketani, T. Doi, and E. Takeda. 2007. Effects of a palatinose-based liquid diet (Inslow) on glycemic control and the second-meal effect in healthy men. *Metabolism* 56: 115–121.

Arrigoni, E., F. Brouns, and R. Amado. 2005. Human gut microbiota does not ferment erythritol. *Br. J. Nutr.* 94: 643–646.

Astrand, P. O. 1967. Diet and athletic performance. *Fed. Proc.* 26: 1772–1777.

Atkinson, F. S., K. Foster-Powell, and J. C. Brand-Miller. 2008. International tables of glycemic index and glycemic load values. *Diabetes Care* 31: 2281–2283.

Austin, K. and B. Seebohar. 2011. *Performance Nutrition. Applying the Science of Nutrient Timing*. Champaign, IL: Human Kinetics.

Badiga, M. S., N. K. Jain, C. Casanova, and C. S. Pitchumoni. 1990. Diarrhea in diabetics: The role of sorbitol. *J. Am. Coll. Nutr.* 9 (6): 578–582.

Bahrami, M., A. Ataie-Jafari, S. Hosseini, H. Foruzanfar, M. Rahmani, and M. Pajouhi. 2009. Effects of natural honey consumption in diabetic patients: An 8-week randomized clinical trial. *Int. J. Food Sci. Nutr.* 60: 618–626.

Beards, E., K. Tuohy, and G. Gibson. 2010. A human volunteer study to assess the impact of confectionery sweeteners on the gut microbiota composition. *Br. J. Nutr.* 104: 701–708.

Beaugerie, L., B. Flourié, P. H. Marteau, P. Pellier, C. Franchisseur, and J. C. Rambaud. 1990. Digestion and absorption in human intestine of three sugar alcohols. *Gastroenterology* 99: 717–723.

Beiswanger, B. B., A. E. Boneta, M. S. Mau, B. P. Katz, H. M. Proskin, and G. K. Stookey. 1998. The effect of chewing sugar-free gum after meals on clinical caries incidence. *J. Am. Dent. Assoc.* 129 (11): 1623–1626.

Bellisle, F. and A. Drewnowski. 2007. Intense sweeteners, energy intake and the control of body weight. *Eur. J. Clin. Nutr.* 61: 691–700.

Below, P. R., R. Mora-Rodriguez, J. Gonzalez-Alonso, and E. F. Coyle. 1995. Fluid and carbohydrate ingestion independently improve performance during 1 h of intense exercise. *Med. Sci. Sports Exercise* 27: 200–210.

Bergstrom, J. 1962. Muscle electrolytes in man determined by neutron activation analysis on needle biopsy specimens: A study on normal subjects, kidney patients, and patients with chronic diarrhea. *Scand. J. Clin. Lab. Invest., Suppl.* 68: 1–110.

Berkey, C. S., H. R. Rockett, A. E. Field, M. W. Gillman, and G. A. Colditz. 2004. Sugar-added beverages and adolescent weight change. *Obes. Res.* 12 (5): 778–788.

Bjorkman, O., K. Shalin, L. Hagenfeldt, and J. Wahren. 1984. Influence of glucose and fructose ingestion on the capacity for long-term exercise in well-trained men. *Clin. Physiol.* 4: 483–494.

Black, R. M., L. A. Leiter, and G. H. Anderson. 1993. Consuming aspartame with and without taste: Differential effects on appetite and food intake of young adult males. *Physiol. Behav.* 53: 459–466.

Blackburn, G. L., B. S. Kanders, P. T. Lavin, S. D. Keller, and J. Whatley. 1997. The effect of aspartame as part of a multidisciplinary weight-control program on short- and long-term control of body weight. *Am. J. Clin. Nutr.* 65: 409–418.

Bleich, S. N., Y. C. Wang, Y. Wang, and S. L. Gortmaker. 2009. Increasing consumption of sugar-sweetened beverages among U.S. adults: 1988–1994 to 1999–2004. *Am. J. Clin. Nutr.* 89: 372–381.

Blum, J. W., D. J. Jacobsen, and J. E. Donnelly. 2005. Beverage consumption patterns in elementary school aged children across a two-year period. *J. Am. Coll. Nutr.* 24 (2): 93–98.

Blundell, J. E. and A. J. Hill. 1986. Paradoxical effects of an intense sweetener (aspartame) on appetite. *Lancet* 1: 1092–1093.

Bogdanov, S., T. Jurendic, R. Sieber, and P. Gallmann. 2008. Honey for nutrition and health: A review. *J. Am. Coll. Nutr.* 27 (6): 677–689.

Bradbury, J., C. E. Mulvaney, A. J. Adamson, C. J. Seal, J. C. Mathers, and P. J. Moynihan. 2008. Sources of total, nonmilk extrinsic, and intrinsic and milk sugars in the diets of older adults living in sheltered accommodation. *Br. J. Nutr.* 99: 649–652.

Brand-Miller, J. C., S. H. A. Holt, D. B. Pawlak, and J. McMillian. 2002. Glycemic index and obesity. *Am. J. Clin. Nutr.* 76 (Suppl.): 281S–285S.

Bray, G. A., S. J. Nielsen, and B. M. Popkin. 2004. Consumption of high-fructose corn syrup in beverages may play a role in the epidemic of obesity. *Am. J. Clin. Nutr.* 79: 537–543.

Bremer, A. A., P. Auinger, and R. S. Byrd. 2010. Sugar-sweetened beverage intake trends in U.S. adolescents and their association with insulin-resistance-related parameters. *J. Nutr. Metab.*: 1–8.

Briet, F., L. Achour, B. Flourié, L. Beaugerie, P. Pellier, C. Franchisseur, F. Bornet, and J. C. Rambaud. 1995a. Symptomatic response to varying levels of fructo-oligosaccharides consumed occasionally or regularly. *Eur. J. Clin. Nutr.* 49: 501–507.

Briet, F., B. Flourié, L. Achour, M. Maurel, J. C. Rambaud, and B. Messing. 1995b. Bacterial adaptation in patients with short bowel and colon in continuity. *Gastroenterology* 109: 1446–1453.

Brouns, F. and E. Beckers. 1993. Is the gut an athletic organ? Digestion, absorption and exercise. *Sports Med.* 15: 242–257.

Brouns, F., W. H. M. Saris, E. Beckers, H. Adlerceutz, G. J. van der Vusse, H. A. Keizer, H. Kuipers, P. Menheere, A. J. Wagenmakers, and F. ten Hoor. 1989. Metabolic changes induced by sustained exhaustive cycling and diet manipulation. *Int. J. Sports Med.* 10: S49–S62.

Brown, R. J., M. A. De Banate, and K. I. Rother. 2010. Artificial sweeteners: A systematic review of metabolic effects in youth. *Int. J. Pediatr. Obes.* 5 (4): 305–312.

Burke, L. 2007. *Practical Sports Nutrition*. Champaign, IL: Human Kinetics.

Burke, L. M., G. R. Collier, and M. Hargreaves. 1993. Muscle glycogen storage after prolonged exercise: Effect of the glycemic index of carbohydrate feedings. *J. Appl. Physiol.* 75: 1019–1023.

Casa, D. J., P. M. Clarkson, and W. O. Roberts. 2005. American College of Sports Medicine roundtable on hydration and physical activity: Consensus statements. *Curr. Sports Med. Rep.* 4: 115–127.

Chen, X., Z.-X. Jiang, S. Chen, and W. Qin. 2010. Microbial and bioconversion production of D-xylitol and its detection and application. *Int. J. Biol. Sci.* 6: 834–844.

Chepulis, L. 2007. The effect of honey compared to sucrose, mixed sugars and a sugar-free diet on weight gain in young rats. *J. Food Sci.* 7 (3): S224–S229.

Chepulis, L. and N. Starkey. 2008. The long-term effects of feeding honey compared with sucrose and a sugar-free diet on weight gain, lipid profiles, and DEXA measurements in rats. *J. Food Sci.* 73 (1): H1–H7.

Christensen, E. H. and O. Hansen. 1939. Arbeitsfahigkeit und Ehrnahrung. *Skand. Arch. Physiol.* 81: 160–171.

Coggan, A. R. and E. F. Coyle. 1987. Reversal of fatigue during prolonged exercise by carbohydrate infusion or ingestion. *J. Appl. Physiol.* 63: 2388–2395.

Colditz, G. A., W. C. Willett, M. J. Stampfer, S. J. London, M. R. Segal, and F. E. Speizer. 1990. Patterns of weight change and their relation to diet in a cohort of healthy women. *Am. J. Clin. Nutr.* 51: 1100–1105.

Coombes, J. S. and K. Hamilton. 2000. The effectiveness of commercially available sports drinks. *Sports Med.* 29: 181–209.

Costill, D. L., E. Coyle, G. Dalsky, W. Evans, W. Fink, and D. Hoopes. 1977. Effects of elevated plasma FFA and insulin on muscle glycogen usage during exercise. *J. Appl. Physiol.* 43: 695–699.

Coyle, E. F., A. R. Coggan, M. K. Hammert, and J. L. Ivy. 1986. Muscle glycogen utilization during prolonged strenuous exercise when fed carbohydrates. *J. Appl. Physiol.* 61: 165–172.

Coyne, M. J. and H. Rodriguez. 1986. Carbohydrate malabsorption in black and Hispanic dialysis patients. *Am. J. Gastroenterol.* 81 (8): 662–665.

Cummings, J. H., G. T. Macfarlane, and H. N. Englyst. 2001. Prebiotic digestion and fermentation. *Am. J. Clin. Nutr.* 73: S415–S420.

Davis, J. M., N. L. Alderson, and R. S. Welsh. 2000. Serotonin and central nervous system fatigue: Nutritional considerations. *Am. J. Clin. Nutr.* 72: 573S–578S.

Davis, J. M., D. A. Jackson, M. S. Broadwell, J. L. Queary, and C. L. Lambert. 1997. Carbohydrate drinks delay fatigue during intermittent, high-intensity cycling in active men and women. *Int. J. Sport Nutr.* 7: 261–273.

de la Hunty, A., S. Gibson, and M. Ashwell. 2006. A review of the effectiveness of aspartame in helping with weight control. *BNF Nutr. Bull.* 31: 115–128.

Dharmaraj, H., H. Patil, G. K. Grimble, and D. B. A. Silk. 1987. Lactitol, a new hydrogenated lactose derivative: Intestinal absorption and laxative threshold in normal human subjects. *J. Hum. Nutr.* 35: 165–172.

Dhingra, R., L. Sullivan, P. F. Jacques, T. J. Wang, C. S. Fox, and J. B. Meigs. 2007. Soft drink consumption and risk of developing cardiometabolic risk factors and the metabolic syndrome in middle-aged adults in the community. *Circulation* 116 (5): 480–488.

Donner, T. W., L. S. Magder, and K. Zarbalian. 2010. Dietary supplementation with D-tagatose in subjects with type 2 diabetes leads to weight loss and raises high-density lipoprotein cholesterol. *Nutr. Res.* 30: 801–806.

Donner, T. W., J. F. Wilber, and D. Ostrowski. 1999. D-Tagatose, a novel hexose: Acute effects on carbohydrate tolerance in subjects with and without type 2 diabetes. *Diabetes, Obes. Metab.* 1: 285–293.

Dumesnil, J. G., J. Turgeon, A. Tremblay, P. Poirier, M. Gilbert, L. Gagnon, S. St-Pierre, C. Garneau, I. Lemieux, A. Pascot, J. Bergeron, and J.-P. Despres. 2001. Effect of a low glycemic index, low fat, high protein diet on the atherogenic metabolic risk profile of abnormally obese men. *Br. J. Nutr.* 86: 557–568.

Ebbeling, C. B., H. A. Feldman, S. K. Osganian, V. R. Chomitz, S. J. Ellenbogen, and D. S. Ludwig. 2006. Effects of decreasing sugar-sweetened beverage consumption on body weight in adolescents: A randomized, controlled pilot study. *Pediatrics* 117 (3): 673–680.

Elliott, S. S., N. L. Keim, J. S. Stern, K. Teff, and P. J. Havel. 2002. Fructose, weight gain, and the insulin resistance syndrome. *Am. J. Clin. Nutr.* 76: 911–922.

Febbraio, M. A., P. Murton, S. E. Selig, S. A. Clark, D. L. Lambert, D. J. Angus, and M. F. Carey. 1996. Effect of CHO ingestion on exercise metabolism and performance in different ambient temperatures. *Med. Sci. Sports Exercise* 28: 1380–1387.

Florent, C., B. Flourié, A. Leblond, M. Rautureau, J. J. Bernier, and J. C. Rambaud. 1985. Influence of chronic lactulose ingestion on the colonic metabolism of lactulose in man (an *in vivo* study). *J. Clin. Invest.* 75: 608–613.

Flourié, B., F. Briet, C. Florent, P. Pellier, M. Maurel, and J. C. Rambaud. 1993. Can diarrhoea induced by lactulose be reduced by prolonged ingestion of lactulose? *Am. J. Clin. Nutr.* 58: 369–375.

Food and Agriculture Organization of the United Nations/World Health Organization (FAO/WHO) Expert Consultation. 1998. Carbohydrates in human nutrition. FAO Food and Nutrition Paper 66. Rome, Italy: FAO.

Food and Drug Administration. 1996. Health claims: Dietary sugar alcohols and dental caries. *Fed. Regist.* 61: 43433–43445.

Food and Drug Administration. 2006. 21 CFR Part 101 (Docket No. 2004P–0294). Food Labeling: Health Claims; Dietary Noncariogenic Carbohydrate Sweeteners and Dental Caries. Federal Register, 71 (60): Rules and Regulations.

Forshee, R. A. and M. L. Storey. 2003. Total beverage consumption and beverage choices among children and adolescents. *Int. J. Food Sci. Nutr.* 54 (4): 297–307.

Foster, C., D. L. Costill, and W. J. Fink. 1979. Effects of preexercise feedings on endurance performance. *Med. Sci. Sports Exercise* 11: 1–5.

Foster-Powell, K., S. H. Holt, and J. C. Brand-Miller. 2002. International table of glycemic index and glycemic load values. *Am. J. Clin. Nutr.* 76: 5–56.

Fowler, S. P., K. Williams, R. G. Resendez, K. J. Hunt, H. P. Hazuda, and M. P. Stern. 2008. Fueling the obesity epidemic? Artificially sweetened beverage use and long-term weight gain. *Obesity (Silver Spring)* 16: 1894–1900.

French, S. A., B. H. Lin, and J. F. Guthrie. 2003. National trends in soft drink consumption among children and adolescents aged 6 to 17 years: Prevalence, amounts, and sources, 1977/1978 to 1994/1998. *J. Am. Diet. Assoc.* 103 (10): 1326–1331.

Gatto-Smith, A. G., R. B. Scott, H. M. Machida, and D. G. Gall. 1988. Sorbitol as a cryptic cause of diarrhea. *Can. J. Gastroenterol.* 2: 140–142.

Ghoddusi, H. B., M. A. Grandison, A. S. Grandison, and K. M. Tuohy. 2007. *In vitro* study on gas generation and prebiotic effects of some carbohydrates and their mixtures. *Anaerobe* 13: 193–199.

Giammattei, J., G. Blix, H. H. Marshak, A. O. Wollitzer, and D. J. Pettitt. 2003. Television watching and soft drink consumption: Associations with obesity in 11- to 13-year-old schoolchildren. *Arch. Pediatr. Adolesc. Med.* 157 (9): 882–886.

Gibson, G. R. 1999. Dietary modulation of the human gut microflora using prebiotics oligofructose and inulin. *J. Nutr.* 129: 1438S–1441S.

Gibson, G. R., H. M. Probert, J. A. E. Van Loo, R. A. Rastall, and M. B. Roberfroid. 2004. Dietary modulation of the human colonic microbiota: Updating the concept of prebiotics. *Nutr. Res. Rev.* 17: 257–259.

Gollnick, P. D., R. B. Armstrong, C. W. Saubert, W. L. Sembrowich, R. E. Shepherd, and B. Saltin. 1973. Glycogen depletion patterns in human skeletal muscle fibres during prolonged work. *Pflügers Arch.* 344: 1–12.

Goulas, A., G. Tzortzis, and G. R. Gibson. 2007. Development of a process for the production and purification of α- and β-galactooligosaccharides from *Bifidobacterium bifidum* NCIMB 41171. *Int. Dairy J.* 17: 648–656.

Gregory, J. R. and S. Lowe. 2000. *National Diet and Nutrition Survey: Young People Aged 4–18 Years*—Vol. 1: Report of the Diet and Nutrition Survey. London: The Stationery Office.

Grenby, T. H. 1991. Advances in noncaloric sweeteners with dental health advantages over sugars. *Proc. Finn. Dent. Soc.* 87 (4): 489–499.

Guthrie, J. F. and J. F. Morton. 2000. Food sources of added sweeteners in the diets of Americans. *J. Am. Diet. Assoc.* 100: 43–48.

Hammer, H. F., K. D. Fine, A. C. A. Santa, J. L. Porter, L. R. Schiller, and J. S. Fordtran. 1989. Carbohydrate malabsorption: Its measurement and its contribution to diarrhoea. *J. Clin. Invest.* 84: 1056–1062.

Hargreaves, M., D. L. Costill, A. Coggan, W. J. Fink, and I. Nishibata. 1984. Effect of carbohydrate feedings on muscle glycogen utilization and exercise performance. *Med. Sci. Sports Exercise* 16: 219–222.

Hargreaves, M., D. L. Costill, A. Katz, and W. J. Fink. 1985. Effect of fructose ingestion on muscle glycogen usage during exercise. *Med. Sci. Sports Exercise* 17: 360–363.

He, J., D. Heber, and W. Shi. 2006. Antibacterial compounds from *Glycyrrhiza uralensis*. *J. Nat. Prod.* 69: 121–124.

Hermansen, L., E. Hultman, and B. Saltin. 1967. Muscle glycogen during prolonged severe exercise. *Acta Physiol. Scand.* 71: 129–139.

Hildebrandt, G., L. Lee, and J. Hodges. 2010. Oral mutans streptococci levels following use of a xylitol mouth rinse: A double-blind, randomized, controlled clinical trial. *Spec. Care Dentist.* 30: 53–58.

Holub, I., A. Gostner, S. Theis, L. Nosek, T. Kudlich, R. Melcher, and W. Scheppach. 2010. Novel findings on the metabolic effects of the low glycaemic carbohydrate isomaltulose (Palatinose). *Br. J. Nutr.* 103: 1730–1737.

Horowitz, J. F., R. Mora-Rodriguez, L. O. Byerley, and E. F. Coyle. 1997. Lipolytic suppression following carbohydrate ingestion limits fat oxidation during exercise. *Am. J. Physiol.* 273: E768–E775.

Hu, C.-H., J. He, R. Eckert, X.-Y. Wu, L.-N. Li, Y. Tian, R. Lux, J. A. Shuffer, F. Gelmann, J. Mentes, S. Spackman, J. Bauer, M. H. Anderson, and W.-Y. Shi. 2011. Development and evaluation of a safe and effective sugar-free herbal lollipop that kills cavity-causing bacteria. *Int. J. Oral Sci.* 3: 13–20.

Ilback, N. G., M. Alzin, S. Jahrl, H. Enghardt-Barbieri, and L. Busk. 2003. Estimated intake of the artificial sweeteners acesulfame-K, aspartame, cyclamate and saccharin in a group of Swedish diabetics. *Food Addit. Contam.* 20 (2): 99–114.

Ischayek, J. I. and M. Kern. 2006. U.S. honeys varying in glucose and fructose content elicit similar glycemic indexes. *J. Am. Diet. Assoc.* 106: 1260–1262.

Jain, N. K., V. P., Patel, and C. S. Pitchumoni. 1987. Sorbitol intolerance in adults: Prevalence and pathogenesis on two continents. *J. Clin. Gastroenterol.* 9 (3): 317–319.

Jentjens, R. and A. E. Jeukendrup. 2005. High exogenous carbohydrate oxidation rates from a mixture of glucose and fructose ingested during prolonged cycling exercise. *Br. J. Nutr.* 93: 485–492.

Jentjens, R., J. Achten, and A. E. Jeukendrup. 2004. High rates of exogenous carbohydrate oxidation from multiple transportable carbohydrates ingested during prolonged exercise. *Med. Sci. Sports Exercise* 36: 1551–1558.

Jentjens, R., C. Shaw, T. Birtles, R. H. Waring, L. K. Harding, and A. E. Jeukendrup. 2005. Oxidation of combined ingestion of glucose and sucrose during exercise. *Metabolism* 54: 610–618.

Jeukendrup, A. and M. Gleeson. 2010. *Sport Nutrition*, 2nd edition. Champaign, IL: Human Kinetics.

Jeukendrup, A., F. Brouns, A. J. Wagenmakers, and W. H. Sarris. 1997. Carbohydrate-electrolyte feedings improve 1-h time trial cycling performance. *Int. J. Sports Med.* 18: 125–129.

Johnson, L., A. P. Mander, L. R. Jones, P. M. Emmett, and S. A. Jebb. 2007. Is sugar-sweetened beverage consumption associated with increased fatness in children? *Nutrition* 23 (7–8): 557–563.

Kerksick, C., T. Harvey, J. Stout, B. Campbell, C. Wilborn, R. Kreider, D. Kalman, T. Ziegenfuss, H. Lopez, J. Landis, J. L. Ivy, and J. Antonio. 2008. International Society of Sports Nutrition position stand: Nutrient timing. *J. Int. Soc. Sports Nutr.* 5: 17–23.

Kim, Y. S., C. S. Park, and D. K. Oh. 2006. Lactulose production from lactose and fructose by a thermostable β-galactosidase from *Sulfolobus solfataricus*. *Enzyme Microb. Technol.* 39: 903–908.

King, N. A., K. Appleton, P. J. Rogers, and J. E. Blundell. 1999. Effects of sweetness and energy in drinks on food intake following exercise. *Physiol. Behav.* 66: 375–379.

Knopp, R. H., K. Brandt, and R. A. Arky. 1976. Effects of aspartame in young persons during weight reduction. *J. Toxicol. Environ. Health* 2 (2): 417–428.

Koizumi, N., M. Fujii, R. Nonomija, Y. Inoue, T. Kagawa, and T. Tsukamoto. 1983. Studies on transitory laxative effects of sorbitol and maltitol: Differences in laxative effects among various foods containing the sweetening agents. *Chemosphere* 12: 105–116.

Kolida, S. and G. R. Gibson. 2007. Prebiotic capacity of inulin-type fructans. *J. Nutr.* 137: S2503–S2506.

Kolida, S., K. M. Tuohy, and G. R. Gibson. 2002. Prebiotic effects of inulin and oligofructose. *Br. J. Nutr.* 87: S193–S197.

Kontiokari, T., M. Uhari, and M. Koskela. 1998. Antiadhesive effects of xylitol on otopathogenic bacteria. *J. Antimicrob. Chemother.* 41: 563–565.

Koutsou, G. A., A. Lee, A. Zumbe, B. Flourié, and M. Storey. 1996. Dose-related gastrointestinal response to the ingestion of either isomalt, lactitol or maltitol in milk chocolate. *Eur. J. Clin. Nutr.* 50: 17–21.

Kral, T. V., A. J. Stunkard, R. I. Berkowitz, V. A. Stallings, R. H. Moore, and M. S. Faith. 2008. Beverage consumption patterns of children born at different risk of obesity. *Obesity (Silver Spring)* 16 (8): 1802–1808.

Krall, E., C. Hayes, and P. Garcia. 1998. How dentition status and masticatory function affect nutrient intake. *J. Am. Dent. Assoc.* 129: 1261–1269.

Kreider, R., C. Rasmussen, J. Lundberg, P. Cowan, M. Greenwood, C. Earnest, and A. Almada. 2000. Effects of ingesting carbohydrate gels on glucose, insulin and perception of hypoglycemia. *FASEB J.* 14: A490–A497.

Kruger, D., R. Grossklaus, T. Ziese, and S. Koch-Gensencke. 1991. Caloric availability of Palatinit (isomalt) in the small intestine of rats: Implications of dose dependency on the energy value. *Nutr. Res.* 11: 669–678.

Kurotobi, T., K. Fukuhara, H. Inage, and S. Kimura. 2010. Glycemic index and postprandial blood glucose response to Japanese strawberry jam in normal adults. *J. Nutr. Sci. Vitaminol.* 56: 198–202.

Langkilde, A. M., H. Anderson, T. F. Schweizer, and P. Wursch. 1994. Digestion and absorption of sorbitol, maltitol and isomalt from the small bowel: A study in ileostomy subjects. *Eur. J. Clin. Nutr.* 48: 768–775.

Launiala, K. 1968. The mechanism of diarrhoea in congenital disaccharide malabsorption. *Acta Paediatr. Scand.* 57: 425–432.

Lavin, J. H., S. J. French, and N. W. Read. 1997. The effect of sucrose- and aspartame-sweetened drinks on energy intake, hunger and food choice of female, moderately restrained eaters. *Int. J. Obes. Relat. Metab. Disord.* 21: 37–42.

Leroy, P. 1982. Report on the tolerance test of lycasin 80/55 administered in the form of sweets to adult human volunteers: Unpublished report from Roquette Freres, Lestrem, France. Quoted according to JECFA: Hydrogenated glucose syrups. Toxicological evaluation of certain food additives and contaminants. 29th Meeting of the Joint FAO/WHO Expert Committee on Food Additives. Cambridge University Press, 1987.

Lian-Loh, R., G. G. Birch, and M. E. Coates. 1982. The metabolism of maltitol in the rat. *Br. J. Nutr.* 48: 477–481.

Lif Holgerson, P., C. Stecksen-Blicks, I. Sjostrom, M. Oberg, and S. Twetman. 2006. Xylitol concentration in saliva and dental plaque after use of various xylitol-containing products. *Caries Res.* 40: 393–397.

Lillemoe, K. D., J. L. Romolo, S. R. Hamilton, L. R. Pennington, J. F. Burdick, and G. M. Williams. 1987. Intestinal necrosis due to sodium polystyrene (kayexalate) in sorbitol enemas: Clinical and experimental support for the hypothesis. *Surgery* 101 (3): 267–272.

Livesey, G. 1990. The impact of the concentration and dose of Palatinit in foods and diets on energy value. *Food Sci. Nutr.* 42F: 223–243.

Livesey, G. 2003. Health potential of polyols as sugar replacers, with emphasis on low glycaemic properties. *Nutr. Res. Rev.* 16: 163–166.

Livesey, G. 2005. Low-glycaemic diets and health—implications for obesity. *Proceedings of the Nutrition Society*, 161, 105–113.

Livesey, G. 2006. Glycaemic responses and toleration. In: *Sweeteners and Sugar Alternatives in Food Technology*, edited by H. Mitchell. Blackwell Publishing.

Ludwig, D. S., K. E. Peterson, and S. L. Gortmaker. 2001. Relation between consumption of sugar sweetened drinks and childhood obesity: A prospective, observational analysis. *Lancet* 357: 505–508.

Lutsey, P. L., L. M. Steffen, and J. Stevens. 2008. Dietary intake and the development of the metabolic syndrome: The atherosclerosis risk in communities study. *Circulation* 117 (6): 754–761.

Macdonald, I. and C. A. Williams. 1988. Effects of ingesting glucose and some of its polymers on serum glucose and insulin levels in men and women. *Ann. Nutr. Metab.* 32: 23–27.

Maffeis, C. 2009. The use of low-calorie sweeteners: Can they help in weight control and diabetes mellitus? 10th Panhellenic Nutrition Conference. November 2009.

Makinen, K., C. Bennett, P. Hujoel, P. Isokangas, K. Isotupa, and H. J. Pape. 1995. Xylitol chewing gums and caries rates: A 40-month cohort study. *J. Dent. Res.* 74: 1904–1913.

Makinen, K. K., K. P. Isotupa, T. Kivilompolo, P. L. Makinen, J. Toivanen, and E. Soderling. 2001. Comparison of erythritol and xylitol saliva stimulants in the control of dental plaque and mutans *streptococci*. *Caries Res.* 35: 129–135.

Mattes, R. 1990. Effects of aspartame and sucrose on hunger and energy intake in humans. *Physiol. Behav.* 47: 1037–1044.

Mattes, R. D. and B. M. Popkin. 2009. Nonnutritive sweetener consumption in humans: Effects on appetite and food intake and their putative mechanisms. *Am. J. Clin. Nutr.* 89: 1–14.

Maughan, R. J. 2010. Fasting and sport: An introduction. *Br. J. Sports Med.* 44: 473–475.

Maughan, R. J., C. E. Fenn, and J. B. Leiper. 1989. Effects of fluid, electrolyte and substrate ingestion on endurance capacity. *Eur. J. Appl. Physiol.* 58: 481–486.

McConell, G. K., S. Fabris, J. Proietto, and M. Hargreaves. 1994. Effect of carbohydrate ingestion on glucose kinetics during exercise. *J. Appl. Physiol.* 77: 1537–1541.

McConell, G., K. Kloot, and M. Hargreaves. 1996. Effect of timing carbohydrate ingestion on endurance exercise performance. *Med. Sci. Sports Exercise* 28: 1300–1304.

Mela, D. J. 1997. Impact of macronutrient-substituted foods on food choice and dietary intake. *Ann. N. Y. Acad. Sci.* 819: 96–107.

Melanson, K. J., T. J. Angelopoulos, V. Nguyen, L. Zukley, J. Lowndes, and J. M. Rippe. 2008. High-fructose corn syrup, energy intake, and appetite regulation. *Am. J. Clin. Nutr.* 88 (Suppl.): 1738S–1744S.

Melanson, K. J., L. Zuckley, J. Lowndes, T. J. Angelopoulos, and J. M. Rippe. 2007. Effects of high-fructose corn syrup and sucrose consumption on circulating glucose, leptin, ghrelin, and on appetite in lean women. *Nutrition* 23: 103–112.

Mezitis, N. H., C. A. Maggio, P. Koch, A. Quddoos, D. B. Allison, and F. X. Pi-Sunyer. 1996. Glycemic effect of a single high oral dose of the novel sweetener sucralose in patients with diabetes. *Diabetes Care* 19: 1004–1005.

Michael, C. G., N. S. Javid, F. A. Collaizzi, and C. H. Gibbs. 1990. Biting strength and chewing forces in complete denture wearers. *J. Prosthet. Dent.* 63: 549–553.

Milgrom, P., K. A. Ly, M. C. Roberts, M. Rothen, G. Mueller, and D. K. Yamaguchi. 2006. Mutans *Streptococci* dose response to xylitol chewing gum. *J. Dent. Res.* 85: 177–181.

Millard-Stafford, M., L. B. Rosskopf, T. K. Snow, and B. T. Hinson. 1997. Water versus carbohydrate–electrolyte ingestion before and during a 15-km run in the heat. *Int. J. Sport Nutr.* 7: 26–38.

Miller, J. C. and I. Lobbezoo. 1994. Replacing starch with sucrose in a high glycemic index breakfast cereal lowers glycemic and insulin responses. *Eur. J. Clin. Nutr.* 48: 749–752.

Miller-Jones, J. 2007. Glycemic response definitions. *Cereal Foods World* 52: 54–55.

Mitchell, H. L. 2008. The glycemic index concept in action. *Am. J. Clin. Nutr.* 87 (Suppl.): 244S–246S.

Monro, J. A. and M. Shaw. 2008. Glycemic impact, glycemic glucose equivalents, glycemic index, and glycemic load: Definitions, distinctions, and implications. *Am. J. Clin. Nutr.* 87 (Suppl.): 237S–243S.

Monsivais, P., M. Perrigue, and A. Drewnowski. 2007. Sugars and satiety: Does the type of sweetener make a difference? *Am. J. Clin. Nutr.* 86: 116–123.

Morgan, K. J., V. J. Stults, and M. E. Zabik. 1982. Amount and dietary sources of caffeine and saccharin intake by individuals aged 5 to 18 years. *Regul. Toxicol. Pharmacol.* 2 (4): 296–307.

Moynihan, P. J. 2005a. The role of diet and nutrition in the etiology and prevention of oral diseases. *Bull. W. H. O.* 83: 694–699.

Moynihan, P. 2005b. The interrelationship between diet and oral health. *Proc. Nutr. Soc.* 64: 571–580.

Moynihan, P. J. and P. E. Petersen. 2004. Diet, nutrition and the prevention of dental diseases. *Public Health Nutr.* 7 (1A): 201–226.

Murray, R., G. L. Paul, J. G. Seifert, and D. E. Eddy. 1991. Responses to varying rates of carbohydrate ingestion during exercise. *Med. Sci Sports Exercise* 23: 713–718.

Murray, R., J. G. Seifert, D. E. Eddy, G. L. Paul, and G. A. Halaby. 1989. Carbohydrate feeding and exercise: Effect of beverage carbohydrate content. *Eur. J. Appl. Physiol.* 59: 152–158.

Navia, J. 1994. Dietary carbohydrates and dental health. *Am. J. Clin. Nutr.* 59 (Suppl.): 719S–727S.

Nemoseck, T. M., E. G. Carmody, A. Furchner-Evanson, M. Gleason, A. Li, H. Potter, L. M. Rezende, K. J. Lane, and M. Kern. 2011. Honey promotes lower weight gain, adiposity, and triglycerides than sucrose in rats. *Nutr. Res.* 31: 55–60.

Nettleton, J. A., J. F. Polak, R. Tracy, G. L. Burke, and D. R. Jacobs Jr. 2009. Dietary patterns and incident cardiovascular disease in the multiethnic study of atherosclerosis. *Am. J. Clin. Nutr.* 90 (3): 647–654.

O'Connor, T. M., S. J. Yang, and T. A. Nicklas. 2006. Beverage intake among preschool children and its effect on weight status. *Pediatrics* 118 (4): e1010–e1018.

Okada, H., E. Fukushi, A. Yamamori, N. Kawazoe, S. Onodera, J. Kawabata, and N. Shiomi. 2010. Novel fructopyranose oligosaccharides isolated from fermented beverage of plant extract. *Carbohydr. Res.* 345: 414–418.

O'Reilly, J., S. H. S. Wong, and Y. Chen. 2010. Glycaemic index, glycaemic load and exercise performance. *Sports Med.* 40: 27–39.

Osterberg, T. and B. Steen 1982. Relationship between dental state and dietary intake in 70-year-old males and females in Goteborg Sweden: A population study. *J. Oral Rehabil.* 9: 509–512.

Patel, S. and A. Goyal. 2011. Functional oligosaccharides: Production, properties and applications. *World J. Microbiol. Biotechnol.* 27: 1119–1128.

Perman, J. A., S. Modler, and A. C. Olson. 1981. Role of pH in production of hydrogen from carbohydrate by colonic bacterial flora. *J. Clin. Invest.* 67: 643–650.

Persano Oddo, L. and R. Piro. 2004. Main European unifloral honeys: Descriptive sheets. *Apidologie* 35: S38–S81.

Phelan, S., W. Lang, D. Jordan, and R. R. Wing. 2009. Use of artificial sweeteners and fat-modified foods in weight loss maintainers and always-normal weight individuals. *Int. J. Obes. (London).* 33 (10): 1183–1190.

Pierce, W. D., C. D. Heth, J. C. Owczarczyk, and S. D. Russell Proctor. 2007. Overeating by young obesity-prone and lean rats caused by tastes associated with low energy foods. *Obesity (Silver Spring)* 15: 1969–1979.

Pitts, N. B., J. Boyles, Z. J. Nugent, N. Thomas, and C. M. Pine. 2004a. The dental caries experience of 14-year-old children in England and Wales: BASCD Survey Report. Available at http://www.bascd.org/annual_survey_results.php.

Pitts, N. B., J. Boyles, Z. J. Nugent, N. Thomas, and C. M. Pine. 2004b. The dental caries experience of 5-year-old children in England and Wales: BASCD Survey Report. Available at http://www.bascd.org/annual_survey_results.php.

Porikos, K. P. and H. S. Koopmans. 1988. The effect of nonnutritive sweeteners on body weight in rats. *Appetite* 11: (Suppl. 1) 12–15.

Porikos, K. P., G. Booth, and T. B. Van Itallie. 1977. Effect of covert nutritive dilution on the spontaneous food intake of obese individuals: A pilot study. *Am. J. Clin. Nutr.* 30: 1638–1644.

Probert, H. M., J. H. A. Apajalahti, and N. Rautonen, J. Stowell, and G. R. Gibson. 2004. Polydextrose, lactitol and fructo-oligosaccharide fermentation by colonic bacteria in a three stage continuous culture system. *Appl. Environ. Microbiol.* 70: 4505–4511.

Qiang, X., C. YongLie, and W. QianBing. 2009. Health benefit application of functional oligosaccharides. *Carbohydr. Polym.* 77: 35–441.

Renwick, A. G. and S. M. Molinary. 2010. Sweet-taste receptors, low-energy sweeteners, glucose absorption and insulin release. *Br. J. Nutr.* 104: 1415–1420.

Rodearmel, S. J., H. R. Wyatt, N. Stroebele, S. M. Smith, L. G. Ogden, and J. O. Hill. 2007. Small changes in dietary sugar and physical activity as an approach to preventing excessive weight gain: The America on the Move family study. *Pediatrics* 120: e869–e879.

Rodin, J. 1990. Comparative effects of fructose, aspartame, glucose, and water preloads on calorie and macronutrient intake. *Am. J. Clin. Nutr.* 51: 428–435.

Rogers, P. J., J. A. Carlyle, A. J. Hill, and J. E. Blundell. 1988. Uncoupling sweet taste and calories: Comparison of the effects of glucose and three intense sweeteners on hunger and food intake. *Physiol. Behav.* 43: 547–552.

Rollo, I. and C. Williams. 2011. Effect of mouth-rinsing carbohydrate solutions on endurance performance. *Sports Med.* 41: 449–461.

Rolls, B. J. 1991. Effects of intense sweeteners on hunger, food intake, and body weight: A review. *Am. J. Clin. Nutr.* 53 (4): 872–888.

Roy, S. L., T. L. Davidson, and S. E. Swithers. 2007. The effects of high-intensity sweeteners (saccharin and acesulfame potassium) on food intake and body weight regulation in rats. *Appetite* 49: 272–341.

Rugg-Gunn, A. J., E. S. Fletcher, J. N. S. Matthews, A. F. Hackett, P. J. Moynihan, S. A. M. Kelly, J. C. Mathers, and A. J. Adamson. 2007. Changes in consumption of sugars by English adolescents over 20 years. *Public Health Nutr.* 10: 354–363.

Rusen, J., M. Krondl, and A. Csima. 1993. Perceived chewing satisfaction and food use of older adults. *J. Can. Diet. Assoc.* 54: 88–92.

Ruskoné-Fourmestraux, A., A. Attar, D. Chassard, B. Coffin, F. Bornet, and Y. Bouhnik. 2003. A digestive tolerance study of maltitol after occasional and regular consumption in healthy humans. *Eur. J. Clin. Nutr.* 57: 26–30.

Ruxton, C. H. S., F. J. S. Garceau, and R. C. Cottrell. 1999. Guidelines for sugar consumption in Europe: Is a qualitative approach justified? *Eur. J. Clin. Nutr.* 53: 503–513.

Saltin, B. and J. Karlsson. 1971. Muscle glycogen utilization during work of different intensities. In: *Muscle Metabolism during Exercise*, edited by B. Pernow and B. Saltin, pp. 289–299. New York: Plenum Press.

Samanta, A., A. C. Burden, and G. R. Jones. 1985. Plasma glucose responses to glucose, sucrose and honey in patients with diabetes mellitus: An analysis of glycaemic and peak incremental indices. *Diabetic Med.* 2: 371–373.

Sanchez, O., F. Guio, D. Garcia, E. Silva, and L. Caicedo. 2008. Fructooligosaccharides production by *Aspergillus* sp. N74 in a mechanically agitated airlift reactor. *Food Bioprod. Process.* 86: 109–115.

Sangwan, V., S. K. Tomar, R. R. B. Singh, A. K. Singh, and B. Ali. 2011. Galactooligosaccharides: Novel components of designer foods. *J. Food Sci.* 76 (4): R103–R111.

Sano, H., S. Nakashima, Y. Songpaisan, and P. Phantumvanit. 2007. Effect of a xylitol and fluoride containing toothpaste on the remineralization of human enamel *in vitro*. *J. Oral Sci.* 49: 67–73.

Sato, K., H. Arai, A. Mizuno, M. Fukaya, T. Sato, M. Koganei, H. Sasaki, H. Yamamoto, Y. Taketani, T. Doi, and E. Takeda. 2007. Dietary palatinose and oleic acid ameliorate disorders of glucose and lipid metabolism in Zucker fatty rats. *J. Nutr.* 137: 1908–1915.

Saunders, D. R. and H. S. Wiggins. 1981. Conservation of mannitol, lactulose, and raffinose by the human colon. *Am. J. Physiol.* 241: G397–G402.

Scheinin, A., K. K. Makinen, K. Ylitalo, and V. Turku. 1976. Final report on the effect of sucrose, fructose and xylitol diets on the caries incidence in man. *Acta Odontol. Scand.* 34: 179–198.

Schiweck, H. and S. C. Ziesenitz. 1986. Physiological properties of polyols in comparison with easily metabolisable saccharides. In: *Advances in Sweeteners*, edited by T. H. Grenby, pp. 68–75. London, U.K.: Blackie Academic and Professional.

Schulze, M. B., J. A. E. Manson, D. S. Ludwig, G. A. Colditz, M. J. Stampfer, W. C. Willett, and F. B. Hu. 2004. Sugar-sweetened beverages, weight gain, and incidence of type 2 diabetes in young and middle-aged women. *JAMA, J. Am. Med. Assoc.* 292 (8): 927–934.

Shambaugh, P., V. Worthington, and J. H. Herbert. 1990. Differential effects of honey, sucrose, and fructose on blood sugar levels. *J. Manipulative Physiol. Ther.* 13: 322–325.

Sherman, W. M., D. L. Costill, W. J. Fink, and J. M. Miller. 1981. Effect of exercise–diet manipulation on muscle glycogen and its subsequent utilization during performance. *Int. J. Sports Med.* 2: 114–118.

Spieth, L. E., J. D. Harnish, C. M. Lenders, L. Raezer, M. A. Pereira, S. J. Hangen, and D. S. Ludwig. 2000. A low-glycemic index diet in the treatment of pediatric obesity. *Arch. Pediatr. Adolesc. Med.* 154: 947–951.

Sreebny, L. M. 1982. Sugar availability, sugar consumption and dental caries. *Community Dent. Oral Epidemiol.* 10: 1–7.

Stanhope, K. L. and P. J. Havel. 2008. Endocrine and metabolic effects of consuming beverages sweetened with fructose, glucose, sucrose, or high-fructose corn syrup. *Am. J. Clin. Nutr.* 88 (Suppl. 6): 1733S–1737S.

Stanhope, K. L., S. C. Griffen, B. R. Bair, M. M. Swarbrick, N. L. Keim, and P. J. Havel. 2008. Twenty-four-hour endocrine and metabolic profiles following consumption of high-fructose corn syrup-, sucrose-, fructose-, and glucose-sweetened beverages with meals. *Am. J. Clin. Nutr.* 87: 1194–1203.

Stellman, S. D. and L. Garfinkel. 1986. Artificial sweetener use and one-year weight change among women. *Prev. Med.* 15: 195–202.

Stellman, S. D. and L. Garfinkel. 1988. Patterns of artificial sweetener use and weight: Change in an American Cancer Society Prospective Study. *Appetite* 11, (Suppl.1): 85–91.

Striegel-Moore, R. H., D. Thompson, S. G. Affenito, D. L. Franko, E. Obarzanek, B. A. Barton, G. B. Schreiber, S. R. Daniels, M. Schmidt, and P. B. Crawford. 2006. Correlates of beverage intake in adolescent girls: The National Heart, Lung, and Blood Institute Growth and Health Study. *J. Pediatr.* 148 (2): 183–187.

Swarbrick, M. M., K. L. Stanhope, and S. S. Elliott, J. L. Graham, R. M. Krauss, M. P. Christiansen, S. C. Griffen, N. L. Keim, and P. J. Havel. 2008. Consumption of fructose-sweetened beverages for 10 weeks increases postprandial triacylglycerol and apolipoprotein-B concentrations in overweight and obese women. *Br. J. Nutr.* 100: 947–952.

Swithers, S. E. and T. L. Davidson. 2008. A role for sweet taste: Calorie predictive relations in energy regulation by rats. *Behav. Neurosci.* 122: 161–173.

Swithers, S. E., A. A. Martin, and T. L. Davidson. 2010. High-intensity sweeteners and energy balance. *Physiol. Behav.* 100: 55–62.

Szöke, J., J. Bánóczy, and H. M. Proskin. 2001. Effect of after-meal sucrose-free gum-chewing on clinical caries. *J. Dent. Res.* 80 (8): 1725–1729.

Tapiainen, T., T. Kontiokari, L. Sammalkivi, I. Ikaheimo, M. Koskela, and M. Uhari. 2001. Effect of xylitol on growth of *Streptococcus pneumoniae* in the presence of fructose and sorbitol. *Antimicrob. Agents Chemother.* 45: 166–169.

Teff, K. L., S. S. Elliott, M. Tschop, T. J. Kieffer, D. Rader, M. Heiman, R. R. Townsend, N. L. Keim, D. D'Alessio, P. J. Havel. 2004. Dietary fructose reduces circulating insulin and leptin, attenuates postprandial suppression of ghrelin, and increases triglycerides in women. *J. Clin. Endocrinol. Metab.* 89: 2963–2972.

Tseveenjav, B., A. L. Suominen, H. Hausen, and M. M. Vehkalahti. 2011. The role of sugar, xylitol, toothbrushing frequency, and use of fluoride toothpaste in maintenance of adults' dental health: Findings from the Finnish National Health 2000 Survey. *Eur. J. Oral Sci.* 119: 40–47.

Tsintzas, K., R. Liu, C. Williams, I. Campbell, and G. Gaitanos. 1993. The effect of carbohydrate ingestion on performance during a 30-km race. *Int. J. Sport Nutr.* 3: 127–139.

Tsintzas, K. O., C. Williams, L. Boobis, and P. Greenhaff. 1995. Carbohydrate ingestion and glycogen utilization in different muscle fibre types in man. *J. Physiol.* 489: 242–250.

Tsintzas, K. O., C. Williams, W. Wilson, and J. Burrin. 1996. Influence of carbohydrate supplementation early in exercise on endurance running capacity. *Med. Sci. Sports Exercise* 28: 1371–1379.

Tsukamura, M., H. Goto, T. Arisawa, T. Hayakawa, N. Nakai, T. Murakami, N. Fujitsuka, and Y. Shimomura. 1998. Dietary maltitol decreases the incidence of 1,2-dimethylhydrazine induced cecum and proximal colon tumors in rats. *J. Nutr.* 128: 536–540.

Uhari, M., T. Kontiokari, M. Koskela, and M. Niemela. 1996. Xylitol chewing gum in prevention of acute otitis media: Double blind randomized trial. *BMJ [Br. Med. J.]* 313: 1180–1184.

van Can, J. G., T. H. Ijzerman, L. J. van Loon, F. Brouns, and E. E. Blaak. 2009a. Reduced glycaemic and insulinaemic responses following trehalose ingestion: Implications for postprandial substrate use. *Br. J. Nutr.* 102: 1395–1399.

van Can, J. G, T. H. Ijzerman, L. J. van Loon, F. Brouns, and E. E. Blaak. 2009b. Reduced glycaemic and insulinaemic responses following isomaltulose ingestion: Implications for postprandial substrate use. *Br. J. Nutr.* 102: 1408–1413.

Vermunt, S. H. F., W. J. Pasman, G. Schaafsma, and A. F. M. Kardinaal. 2003. Effects of sugar intake on body weight: A review. The International Association for the Study of Obesity. *Obes. Rev.* 4: 91–99.

Wallis, G. A., D. S. Rowlands, C. Shaw, R. Jentjens, and A. E. Jeukendrup. 2005. Oxidation of combined ingestion of maltodextrins and fructose during exercise. *Med. Sci. Sports Exercise* 37: 426–432.

Wayner, H. and H. H. Chauncey. 1983. Impact of complete dentures and impaired natural dentition on masticatory performance and food choice in healthy aging men. *J. Prosthet. Dent.* 49: 427–433.

Whelton, H., E. Crowley, D. O'Mullane, M. Cronin, and V. Kelleher. 2003. Children's Oral Health in Ireland 2002: A North–South Survey Coordinated by the Oral Health Services Research Centre University College Cork. Dublin, Republic of Ireland: Department of Health and Children.

Whitaker, R. C., M. S. Pepe, J. A. Wright, K. D. Seidel, and W. H. Dietz. 1998. Early adiposity rebound and the risk of adult obesity. *Pediatrics* 101 (3). Available at www.pediatrics.org/cgi/content/full/101/3/e5.

Williams, C. L., B. A. Strobino, and J. Brotanek. 2007. Weight control among obese adolescents: A pilot study. *Int. J. Food Sci. Nutr.* 58 (3): 217–230.

Wolever, T. M. 2003. Carbohydrate and the regulation of blood glucose and metabolism. *Nutr. Rev.* 61: S40–S48.

Woodward, M. and A. R. P. Walker. 1994. Sugar and dental caries: The evidence from 90 countries. *Br. Dent. J.* 176: 297–302.

World Health Organization Foreign Direct Investment (WHO/FDI). 1982. Global goals for oral health in the year 2000. *Int. Dent. J.* 23: 74–77.

World Health Organization. 1996. *Monitoring Caries in Adults Aged 35–44 Years*. Geneva: World Health Organization.

World Health Organization. 2003. Diet, Nutrition and the Prevention of Chronic Diseases: Report of a Joint WHO/FAO Expert Consultation. WHO Technical Report Series, 916. Geneva: World Health Organization.

Wright, D. A., W. M. Sherman, and A. R. Dernbach. 1991. Carbohydrate feedings before, during, or in combination improve cycling endurance performance. *J. Appl. Physiol.* 71: 1082–1088.

Yang, Q. 2010. Gain weight by "going diet?" Artificial sweeteners and the neurobiology of sugar cravings. Neuroscience. *Yale J. Biol. Med.* 83: 101–108.

Zero, D. T. 2004. Sugars—The arch criminal? *Caries Res.* 38: 277–285.

Ziesenitz, S. C. and G. Siebert. 1987. The metabolism and utilisation of polyols and other bulk sweeteners compared with sugar. In: *Developments in Sweeteners*, Vol. 3, edited by T. H. Grenby, pp. 109–149. Barking, U.K.: Elsevier Applied Science.

Zuckley, L., J. Lowndes, V. Nguyen et al. 2007. Consumption of beverages sweetened with high fructose corn syrup and sucrose produce similar levels of glucose, leptin, insulin, and ghrelin in obese females. *Exp. Biol.* 538.9 (abstr).

CHAPTER 12

Genetically Modified Herbicide-Tolerant Crops and Sugar Beet—Environmental and Health Concerns

The Case of Genetically Modified High-Fructose Corn Syrup

Theodoros Varzakas and Vasiliki Pletsa

CONTENTS

12.1 INTRODUCTION

Genetically modified organisms (GMOs) are defined as organisms in which the genetic material [deoxyribonucleic acid (DNA)] has been altered in a way that does not occur naturally. This is accomplished through the "recombinant DNA" or "genetic engineering" technology, sometimes also called "modern biotechnology" or "gene technology." It allows selected individual genes to be transferred from one organism into another, also between nonrelated species. Therefore, GMOs are understood to be plants, microorganisms, or animals into which foreign DNA coding one or more new genes has been integrated. However, so far, the term usually refers to genetically modified (GM) plants, which are then used to grow GM food crops. Foundation lines/hybrids are the conventional or unmodified parental or isogenic line/hybrids used in transformation events, and the resulting GMO line/hybrids are referred to as the transgenic line/hybrids. Both the products of GMOs and the GMOs themselves are potentially available for human and/or animal nutrition (WHO 2010).

Feed additives produced from GM microorganisms are already of considerable importance in animal nutrition. They are added to feedstuffs not merely to provide domesticated animals with essential nutrients to meet their needs (e.g., amino acids and vitamins) and are therefore of primary importance for animal health, performance, and the effective conversion of feed constituents into food of animal origin, but products of GMO are also used in animal nutrition as nonessential feed additives (e.g., enzymes; Flachowsky et al. 2005).

A working distinction is often made between the first and second generations of GM plants. This distinction is purely pragmatic or historical and does not reflect any particular scientific principle or technological development. The first generation is generally considered to be those crops carrying simple input traits such as increased resistance to pests or tolerance of herbicides. The proteins produced, which confer these benefits, occur in very low concentrations in the modified crops and therefore do not significantly change either the composition or feed value compared to the foundation lines (isogenic lines). In contrast, the second generation of GM plants, exemplified by the Golden Rice (Stein et al. 2006; Tang et al. 2009), includes crops in which the nutrient composition or availability has been deliberately changed by genetic engineering (Harlander 2002; ILSI 2004; Varzakas et al. 2007). Consequently, effects on the nutritional value of the food/feed are to be expected.

First-generation GM crop adoption has experienced an unprecedented rate of growth over the past 10 years. Worldwide, in 2008, there were 125 million ha of land under GM crops, with nearly 25 countries adopting this new technology (James 2008). The early adopters, namely, the top eight countries, growing more than 1 million ha of land are the United States, Argentina, Brazil, India, Canada, China, Paraguay, and South Africa. Together, they represent 98% of the 125 million ha of land under GMOs, out of which 57% is located in North America, 32% in Latin America, 6% in India, and 3% and 1.5% in China and South Africa, respectively. GM maize has been the major crop adopted by most of the countries (Scandizzo and Savastano 2010).

The rapid global spread of GM crops has been accompanied by an intense public debate. Supporters see great potential in the technology to raise agricultural productivity and reduce seasonal variations in food supply due to biotic and abiotic stresses. Against the background of increasing demand for agricultural products, natural resource scarcities, and additional challenges posed by climate change, productivity increases are a necessary precondition for achieving long-term food security. Second-generation GM crops, such as crops with higher micronutrient contents, could also help reduce specific nutritional deficiencies among the poor. Furthermore, GM crops could contribute to rural income increases, which is particularly relevant for poverty reduction in developing countries. In a recent analysis of the effects of insect-resistant Bt cotton, which has been adopted by millions of small-scale farmers, in India, China, and South Africa, farmers in developing countries seem to benefit from insecticide savings, higher effective yields, and sizeable income gains. Insights from India, in particular, suggest that Bt cotton is employment generating and poverty reducing. As an example of a second-generation technology, the likely impacts of β-carotene-rich

Golden Rice are analyzed from an *ex ante* perspective. Vitamin A deficiency is a serious nutritional problem, causing multiple adverse health outcomes. Simulations for India show that Golden Rice could reduce related health problems significantly, preventing up to 40,000 child deaths every year (Qaim 2010). Although these examples clearly demonstrate that GM crops can contribute to poverty reduction and food security in developing countries, their potential impacts on income, poverty, and nutrition in developing countries continue to be the subject of public controversy, because the realization of such social benefits on a larger scale requires more public support for research targeted to the poor, as well as more efficient regulatory and technology delivery systems.

Finally, supporters argue that reductions in the use of chemical pesticides through GM crops could alleviate environmental and health problems associated with intensive agricultural production systems. In an effort to assess the impact of this technology on global agriculture from both economic and environmental perspectives, Brookes and Barfoot (2008) examined specific global economic impacts on farm income and environmental impacts associated with pesticide usage and greenhouse gas (GHG) emissions for each of the countries where GM crops have been grown within the period 1996–2008. According to their analysis, there have been substantial net economic benefits at the farm level amounting to $6.94 billion in 2006 and $33.8 billion for the 11-year period (in nominal terms). The technology has reduced pesticide spraying by 286 million kg and, as a result, decreased the environmental impact associated with herbicide and insecticide use on these crops by 15.4%. GM technology has also significantly reduced the release of GHG emissions from this cropping area, which, in 2006, was equivalent to removing 6.56 million cars from the roads (Brookes and Barfoot 2008).

By contrast, biotechnology opponents emphasize the environmental and health risks associated with GM crops. Regarding the environmental aspect issues, as poor knowledge is available on the interactions among the different components of agroecosystems and natural ecosystems and on potential hazards posed by unintended modifications occurring during genetic manipulation, the majority of the scientific concerns involve the risks incurred when GM plants are grown in uncontrolled environments, such as agroecosystems. The increasing amount of reports on the ecological risks and benefits of GM plants stresses the need for experimental works aimed at evaluating the environmental impact of GM crops, taking also into account the fact that little is known of the fate of transgenes after their field release, except that, in nature, "everything goes everywhere" and genes can flow from one organism to another (Giovannetti et al. 2005). Therefore, pesticide-resistant crops could be toxic to nontarget organisms, and their growth may kill or harm nontarget organisms by changing soil chemistry, while herbicide-tolerant plants facilitate the use of certain herbicides by farmers. All of these environmental hazards are compounded by genetic pollution. Many crop plants disperse pollen, which may be carried by wind or insect pollinators; therefore, genetically engineered plants may cross-pollinate nonengineered plants, introducing the new genes into wild plant populations and into the ecosystem, affecting the food chain with several levels of consumers.

As mentioned above, doubts have also been raised with respect to the socioeconomic implications in developing countries. Some consider high-tech applications *per se* as inappropriate for smallholder farmers and disruptive for traditional cultivation systems. In addition, it is feared that the dominance of multinational companies in biotechnology and the international proliferation of intellectual property rights would lead to the exploitation of agricultural producers. In this view, GM crops are considered rather counterproductive for food security and development.

As for health issues, concerns regarding GM foods for human use mainly include the potential allergenicity as well as toxicity of novel foods and the promotion of resistance to antibiotics. A particular concern is raised in the U.S., where traditionally regulatory agencies have been strong in the protection of human health and the environment, but GM foods have entered the market almost unregulated. In 1992, the U.S. Food and Drug Administration (FDA) determined that GM foods were usually "the same as or substantially similar to substances commonly found in food" and are therefore not required to undergo specific safety testing before entering the market. The concept of "substantial equivalence," developed by the Organisation for Economic Cooperation and Development (OECD) in 1991, maintains that a novel food (e.g., GM foods) should be considered the same as and as safe as a conventional food if it demonstrates the same characteristics and composition. Substantial equivalence is important from a regulatory point of view. If a novel food is substantially equivalent to its conventional counterpart, then it could be covered by the same regulatory framework as a conventional food (OECD 1993). Substantial equivalence is the starting point of the nutritional and safety assessment of a GM material and can be described as a comparative approach to the assessment of safety (EFSA 2004). Due to the above-mentioned concerns and risks, currently, most national authorities consider that a specific risk assessment is necessary for GM foods. Therefore, specific systems have been set up for the rigorous evaluation of GM organisms and GM foods relative to both environment and human health.

This chapter comments on environmental and health issues arising from the growth and use of GM crops, outlines the main principles of the standard risk assessment methodology and traceability applied in the context of the European Union (EU) regulation, and finally comments on the particular case of the genetically modified herbicide-tolerant (GMHT) sugar beet and sugar cane.

12.2 ENVIRONMENTAL AND HEALTH CONCERNS

12.2.1 Environment

The question of whether GM plants are a threat to the environment remains of major concern and the debate continues. One of the main controversies around GM foods is the potential of these products to affect biodiversity. This is somewhat of a confusing area in the sense that it is difficult to assess and the effects tend to be long term rather than consequences that can be observed and measured in the short term. As such, it is challenging, because a GM crop may be approved but problems relating to biodiversity will not be evident for some time afterward. By that time, restoring biodiversity is a difficult, if not impossible, task.

The concept of biodiversity refers to a wide variety of life forms. A geographic area that is said to have high biodiversity will have many different species that call the land their abode. The variety that exists within even a single species is also an important contributor to biodiversity. In an ecosystem that has a high level of biodiversity, there will be numerous interactions between the different species. These interactions tend to keep the biodiversity high and the range of species similarly broad. When the environment changes or it is disturbed to some extent and if there is a high level of biodiversity, the geographic area can usually withstand some degree of change.

In this sense, an area of high biodiversity is generally resilient. Many experts will look to the biodiversity of an area to get a sense of the health of that particular ecosystem.

In the agricultural practice, the act of growing crops typically means that wild plants in that farm will be removed. If weeds grow around the crops, it means that the crops will be competing for important and scarce resources such as light, water, and the necessary nutrition from the soil. In turn, the farm will be less productive if weeds are left to grow. However, some level of wild plants can be helpful to the farm. If certain wild plants are left to grow, animals may consume them. These animals can then deter other pests on the farm. If a farm is carefully planned out and good growing practices are used, wild plants can exist alongside crops that grow well.

With GM crops, however, they may have "built-in" pest control, or they may negatively affect the health of an animal that consumes the crop. A concern is whether agriculture using GM crops can respect and maintain this important balance that supports a high level of biodiversity. There are a number of ways in which the biodiversity of an ecosystem could be compromised by GM crops. First, outcrossing can occur. This refers to a GM crop that passes on new traits to relatives in the wild. However, those relatives might be altered such that they now have a completely new ecological role, causing them to "outcompete" other species in the ecosystem. Another way in which GM crops might harm biodiversity relates to the use of only a small number of crop varieties. New characteristics that occur from genetic modification could be so advantageous that only a few crop varieties are used. In turn, there could be a greater number of outbreaks of disease or pests. In addition, any crop that is engineered to be resistant to pests or herbicides can influence biodiversity, because it may not only affect the target insect but could also be harmful to insects that are not targets but still consume the crop.

Thus far, there is some evidence to suggest that GM crops can harm biodiversity. When monarch butterflies feed on leaves that are covered in pollen from GM corn, their growth was slower, and they were more likely to die. Another study that investigated pink bollworm that fed on GM cotton found similar results. However, follow-up experiments on monarch butterflies found that biodiversity was not harmed, indicating that more research needs to be performed (Sears et al. 2001). Another study involved aphids that consumed GM potatoes. According to the study, GM nematode resistance was more compatible with aphid biological control than a systemic nematicide treatment; however, the food was found to have a detrimental effect on ladybirds that fed on aphids. Therefore, it was suggested that the consequences of commercial cultivation of GM nematode-resistant plants for aphid natural enemies and food web structure needs to be determined (Cowgill et al. 2004).

It is no easy task to encourage biotechnological development and progress while still respecting biodiversity and the environment. It is, however, a necessary task that must be done if GM foods are to have any long-term support worldwide. One important step is to obtain a better understanding of the ecology of a specific area prior to planting a GM crop. It is also difficult to create experiments that examine the impact of GM crops on biodiversity, but this is a challenge that must be met if biodiversity is to be respected.

Up to now, there is no evidence that the GM plants have caused environmental problems (Raven 2010; Phipps and Park 2002); however, each new GM variety needs to be closely examined as to its environmental impact. Therefore, issues of concern to be taken into account include the capability of the GMO to escape and potentially introduce the engineered genes into wild populations, the persistence of the gene after the GMO has been harvested, the susceptibility of nontarget organisms (e.g., insects that are not pests) to the gene product, the stability of the gene, the gene transfer to microorganisms, the reduction in the spectrum of other plants, including the loss of biodiversity, and the increased use of chemicals in agriculture. In addition, the environmental safety aspects of GM crops vary considerably according to local conditions.

In line with the above-mentioned concerns, current investigations focus on the potentially detrimental effect on beneficial insects or a faster induction of resistant insects (O'Callaghan et al. 2005), the potential generation of new plant pathogens (Dunfield and Germida 2004; Giovannetti et al. 2005), the potential detrimental consequences for plant biodiversity and wildlife and a decreased use of the important practice of crop rotation in certain local situations (Lövei et al. 2010), and the movement of herbicide-resistant genes to other plants (Frisvold et al. 2009; Duke and Powles 2009).

As mentioned above, environmental issues are difficult to handle, related hazards are difficult to predict, and despite the overall positive experience so far, a case-by-case approach should be followed concerning the safety issues. Furthermore, as the environmental safety aspects of GM crops vary considerably according to local conditions, continuous follow-up should be applied on this basis.

The example of the development of herbicide-tolerant weeds provides arguments for the adoption of this practice (Frisvold et al. 2009; Duke and Powles 2009): weeds, along with insect pests and plant diseases, are sources of biotic stress in crop systems that reduce yields, raise production costs, and contribute to income risk to farmers. Transgenic, herbicide-resistant (HR) crop varieties, first introduced in 1996, offer the promise of more effective weed control. By 2008, more than 79 million ha worldwide was planted HR varieties of soybean, maize, canola, cotton, alfalfa, and sugar beets (James 2008). Approximately 80% of the total area devoted to GM crops has been so far planted with HR crops, virtually all being glyphosate-resistant (GR) crops. Thus, a single genetic trait—glyphosate [*N*-(phosphonomethyl) glycine] resistance—accounts for most of the success of transgenic crops at this time. The widespread adoption of GR crops and glyphosate has had significant economic effects in agriculture, from the replacement of previous herbicide markets to cost savings for farmers in weed management, and has generally reduced the adverse environmental and health impacts of weed management. However, the overuse of this single weed management technology is jeopardizing this safe, highly effective, and economical tool due to the emergence of new weed species that are only poorly controlled by glyphosate (Owen 2008) and the evolution of GR weeds. The number and range of GR weeds has been increasing in the U.S. since the commercialization of Roundup Ready® (RR) crops (Heap 2009). Thus, benefits of HR crops are multifaceted, are difficult to quantify, and may be threatened by the evolution of weeds that are resistant to herbicides used in the process of their cultivation. The sustainable use of HR crops, therefore, requires strategies to delay the evolution of HR weeds. There is an increasing need to identify production practices to reduce the risk of weed resistance to glyphosate and other herbicides to maintain the benefits of GR and other herbicide-tolerant crops. Proposed practices include the incorporation of a residual herbicide into the GR weed management program and the rotation of herbicides from one growing season to the next. Using a residual herbicide with glyphosate helps mitigate or reduce the risk of resistance, because residual herbicides control weeds that escape glyphosate control and prevent them from setting seed. Rotating between herbicides with different modes of action helps reduce the risk of weed resistance to herbicides by reducing the selection pressure for resistance to any one herbicide (Hurley et al. 2009).

One of the least understood areas of the environmental risk assessment of GM crops is their impact on soil- and plant-associated microbial communities. Can microorganisms take up genetic material from plants and integrate it into their genomes? Could this be a way for genes from genetically engineered plants to spread in the environment? Safety research has been looking into these questions for almost 20 years (www.gmo-compass.org), as minor alterations in the diversity of the microbial soil community could affect soil health and ecosystem functioning. The impact that plant variety may have on the dynamics of the rhizosphere microbial population and, in turn, plant growth and health and ecosystem sustainability requires special attention.

The movement of genetic material between unrelated species is called horizontal gene transfer, a phenomenon known to occur in bacteria. One way by which this can happen is the direct incorporation of free DNA by bacterial cells. This direct form of gene transfer, for instance, in the soil or in the digestive tract of animals, is the most commonly predicted scenario for the transfer of genetic material from transgenic plants to microorganisms. So far, horizontal gene transfer can only be demonstrated under optimized laboratory conditions. Still, it occurs extremely infrequently and, thus, difficult to detect. Ongoing safety research is studying transgenic plants to see if horizontal gene transfer to microorganisms is possible or common or if it would cause any considerable consequences (Dunfield and Germida 2004; Giovannetti et al. 2005). One example for such a concern is with antibiotic resistance genes used as marker genes in the development of transgenic plants helping scientists to find out which cells successfully incorporated the gene of interest. If antibiotic-resistant genes from GM foods are taken up by bacteria in the gut during digestion, it would be very difficult to treat bacterial infections. Until now, according to numerous scientific studies in the last 15 years, there is no real proof that the antibiotic resistance genes in GM plants pose a threat. Regardless, as a precautionary measure, some still believe that antibiotic resistance markers should not be used at all, because if GM plants with antibiotic-resistant genes are planted over a very large area, the rare event of horizontal gene transfer could become significant. Accordingly, in the EU, GM plants with certain antibiotic-resistant markers have been given only limited authorization for release into the environment. Several criteria are considered when evaluating these genes (EFSA GMO panel 2009; Sparrow 2010), the most important ones being the medical importance of the antibiotic concerned, the distribution of microorganisms already possessing resistance genes in soil and water, as well as their presence in the digestive tracts of humans or other mammals. Therefore, transformation technologies not resulting in clinically relevant antibiotic resistance marker genes and/or leading to marker gene removal have been recently developed to construct GM crops (high-lysine maize compositions and event LY038 maize plants, Patent 7157281, issued on 2 January 2007). It is expected that GMOs without marker genes will simplify safety assessments.

12.2.2 Health

The three main issues debated in relation to the impact of GM crops on animal and human health are gene transfer, outcrossing, and tendencies to provoke allergic reaction (allergenicity).

Gene transfer from GM foods to cells of the body or to bacteria in the gastrointestinal tract would cause concern if the transferred genetic material adversely affects human health. This would be particularly relevant if antibiotic resistance genes, used in creating GMOs, were to be transferred. In the technique of transplanting genes, there is a need for a marker to identify which cells have taken up the foreign gene. One way is to attach a gene for antibiotic resistance. Following the attempt to insert the new genes, antibiotics can be applied to determine which cells survive and are therefore carriers of the implanted DNA. Once these resistant genes are into the food chain, more disease-causing bacteria may become antibiotic resistant, increasing the problems of public health, as it would be difficult to treat human disease. Already, in a 1999 report, the British Medical Association urged an end to antibiotic-resistant genes used as markers in GM crops. As mentioned above, the probability of horizontal gene transfer of GM plant genes into the human body or bacteria in the human body is very low. Still, it is unprecedented and could have possible adverse affects on human health; therefore, with many people and many GM crops, an expert panel from the Food and Agriculture Organization of the United Nations/World Health Organization (FAO/WHO) has recommended the use of technology without antibiotic resistance genes (WHO 2010).

The movement of genes from GM plants into conventional crops or related species in the wild, referred to as "outcrossing," as well as the mixing of crops derived from conventional seeds with those grown using GM crops, may have an indirect effect on food safety and food security. This risk, as shown in the case of StarLink Maize in the U.S., is real; hence, several countries have adopted strategies to reduce mixing, including a clear separation of the fields within which GM and conventional crops are grown. Feasibility and methods for postmarketing monitoring of GM food products, for the continued surveillance of the safety of GM food products, are discussed below.

Although traditionally U.S. regulatory agencies have been strong in the protection of human health and the environment, GM foods have entered the market almost unregulated. In 1992, the U.S. FDA determined that GM foods were usually "the same as or substantially similar to substances commonly found in food" and are therefore not required to undergo specific safety testing before entering the market. Then, in 1998, there was disclosure of a potato experiment by A. Pusztai in Scotland, who publicly announced that the results of his research showed that feeding GM potatoes to rats had negative effects on their stomach lining and immune system (Ewen and Pusztai 1999). The resulting controversy became known as the "Pusztai affair" and raised questions both in the U.K. and the U.S. about the safety of these crops. In 2000, it was discovered that corn being used for human food had been contaminated with GM StarLink™ Corn, which had the approval of the U.S. Environmental Protection Agency only as animal feed. U.S. regulatory authorities permitted the commercial sale of StarLink seed with the stipulation that crops produced must not be used for human consumption. This restriction was based on the possibility that a small number of people might develop an allergic reaction to the Bt protein used in StarLink, which is less rapidly digested than the version used in other Bt varieties. The StarLink corn controversy resulted in some 300 products recalled, mass litigation within the agriculture community, and drops in exports to foreign markets, including Japan. In the years since, farmers have had to worry about liability, markets, cross-pollination, and contamination in grain elevators (CDC 2001; FIFRA 2001). The U.S. corn supply has been monitored for the presence of the StarLink Bt proteins since 2001. No positive samples have been found since 2004, showing that, most probably, it was possible to withdraw this GM crop without leaving traces in the environment once it had been used in the field (North American Millers' Association 2008).

An additional concern has recently been raised by the alteration of endogenous gene expression, which can be an alternative method of producing useful phenotypes, especially regarding second-generation GM plants, including crops in which the nutrient composition or availability has been changed to add to the nutritional value of the food/feed (Harlander 2002; ILSI 2004; Rotthues et al. 2008). For example, ribonucleic acid (RNA)-associated mechanisms can be used to switch off genes, while the up-regulation of specific transcription factors can be used to enhance expression and thereby modify a plant's growth or response to stress. Since neither of these two mechanisms necessarily depends on the expression of a new heterologous protein(s), it is reasonable to ask if the safety assessment paradigm developed for and applied to transgenic plants that express novel proteins is appropriate for genetically engineered plants, in which gene expression has been altered. In a recent paper that examines the suitability of the currently used comparative safety paradigm to crops in which gene expression has been altered, the authors conclude that the safety assessment paradigm used to assure the safety of genetically engineered crops appears to be more than sufficiently robust for application to gene-modulated crops (Parrot et al. 2010).

It is certain that nowadays, especially in Europe, GM crops undergo rigorous safety assessment before being allowed to enter the market. One aspect of GM foods that has drawn much public attention because it could directly affect public health is the assessment of their potential allergenicity. Allergy is an abnormal immune reaction to naturally occurring protein substances, which are then called allergens. Protecting people with food allergies against accidental exposure to allergens has become an important focus for food manufacturers and regulators responsible for all food safety. A significant focus of the food industry is to keep food products that are not intended to contain a major allergen (e.g., peanut, milk, eggs, or wheat) from being contaminated with one of the major allergens. Likewise, the primary focus of the safety assessment for GM crops, as defined by the Codex Alimentarius Commission in 2003, is to prevent the transfer of a gene encoding a major allergenic protein (from any source) into a food crop that did not previously contain that protein. Producers of GM crops and regulatory authorities focus on preventing avoidable increases in the risk of allergy in producing and accepting new GM crops. It should, however, be recognized that the absolute avoidance of all risks is not achievable. Thus, the assessment that has been developed focuses on avoiding risks that are predictable and likely to cause common allergic reactions.

While traditionally developed foods are not generally tested for allergenicity, protocols for tests for GM foods have been evaluated by the FAO and WHO, and the allerginicity assessment of newly expressed proteins is an important component in the safety evaluation of GM plants (digestibility, homology, and mammalian tests are all performed as part of allerginicity assessment). The existence of multiple documents with diverging recommendations coming from different organizations has resulted in confusion and, sometimes, arbitrary inclusion of tests upon request from regulatory authorities. In some cases, regulators continued to base their judgment on nonvalidated (e.g., animal models) or even rejected (short-peptide matches) tests. In particular, regarding animal models that seemed to be quite promising when first included in the safety evaluation of GM plants, it is recognized that no current animal model is predictive of allergenicity in humans although they are certainly useful for mechanistic studies. To date, there is no documented proof that any approved, commercially grown GM crop has caused allergic reactions due to a transgenically introduced allergenic protein or that the generation of a GM crop has caused a biologically significant increase in the endogenous allergenicity of a crop, despite extensive biosafety tests in several countries, including the U.S. (Goodman et al. 2008).

12.3 RISK ASSESSMENT OF GENETICALLY MODIFIED CROPS

GM risk assessments play an important role in the decision-making process surrounding the regulation notification and permission to handle GMOs. The ultimate goal is to ensure the safe handling and containment of the GMO and assess any potential impacts on the environment and human health (Sparrow 2010; EC 2002).

The Codex Alimentarius Commission, under the FAO and the WHO, adopted guidelines in 2003 to harmonize the premarket risk assessment process for GM plants in the global market (Goodman et al. 2008). The guidelines were approved by the Codex Alimentarius Commission and are intended to guide countries in adopting consistent rules that provide a strong food safety evaluation process while avoiding trade barriers. Each new GM crop requires a premarket safety assessment to evaluate intended and unintended changes that might have adverse human health consequences caused by the transfer of the DNA. The goal is to identify hazards and, if found, to require risk assessment and, where appropriate, develop a risk management strategy (e.g., do not approve, approve with labeling and/or monitoring, or approve without restriction).

The process is based on scientific evidence and requires the use of methods and criteria that are demonstrated to be predictive. New methods should be validated and demonstrated to enhance the safety assessment.

The framework for guiding the evaluation of potential safety issues requires detailed characteristics of (Varzakas et al. 2006)

1. The GM plant and its use as food
2. The source of the gene
3. The inserted DNA and flanking DNA at the insertion site
4. The expressed substances (e.g., proteins and any new metabolites that result from the new gene product)
5. The potential toxicity and antinutritional properties of new proteins or metabolites
6. The introduced protein compared with those known to cause celiac disease if the DNA is from wheat, barley, rye, oats, or related grains
7. The introduced protein for potential allergenicity
8. Key endogenous nutrients and antinutrients, including toxins and allergens for potential increases for specific host plants (DNA recipients)

Certain steps in the assessment require the scientific assessment of existing information; others require experiments in which case assay validation, sensitivity, and auditable documentation are required.

The deliberate release of GM plants, within Europe, is governed by the 2001/2018 European Commission (EC) Directive (Sparrow 2010; EC 2002). This directive covers both experimental and commercial release for placing on the market. For every authorized release, the national authority provides the EC with a summary of the key information in the application [summary notification information format (SNIF)]. The SNIF document is then made public (http://gmoinfo.jrc.ec.europa.eu) for comment (European Commission Joint Research Center 2003). The key difference is that for research and development releases, decisions are made by individual member states, whereas for placing GMO products on the market, decisions are made by all member states, which often necessitates a voting procedure to address differences in opinion on risk. The European Food Standard Agency (EFSA) also oversees at this stage. EFSA oversees all food and feed applications (under Directive 1829/2003/EC; EC 2003a) as well as nonfood and nonfeed cases where agreement has not been met by all member states.

12.3.1 Coexistence and Traceability

In the United States and Canada, the labeling of GM foods is not required. However, in certain other regions, such as the European Union, Japan, Malaysia, and Australia, governments have required labeling so that consumers may have the choice among GM, conventional, or organic food (Miraglia et al. 2004; Beckmann et al. 2006). This legislation requires a labeling system as well as a system of traceability, ensuring the reliable separation of GM and non-GM organisms at the production level and throughout the whole processing chain (Miraglia et al. 2004). The traceability of GMOs is founded on two needs. First, consumers in many countries are reluctant to buy GM foods and are skeptical of the use of GM crops for animal feed. Consequently, the concept of coexistence has been developed to separate GM and non-GM supply chains and is only possible if all purchasers along the production chain know what they are buying. Second, although every GMO that is approved for commercialization must have passed a safety assessment, it may be necessary to withdraw a certain GMO from the market, for example, if new scientific evidence raises doubts about its safety. For these purposes, after 3 years of debate, the OECD countries came up with an identity code for GMOs in 2002. Initially, some member countries (e.g., the U.S.A., but also Canada and Australia) were opposed to the concept. The final decision requires the assignment of a "unique identifier" to each GMO event that is authorized in one or more OECD countries. The unique identifier is a code consisting of nine letters and/or numbers. The first two or three characters indicate the company submitting the application, while the following five or six characters specify the respective transformation event. The last digit serves as a verifier. All the crop varieties derived from one transformation event will share the same unique identifier.

The unique identifier has been integrated in the Cartagena Protocol on Biosafety and in the European Union legislation on the labeling and traceability of GMOs [Regulation (EC) No. 1830/2003; EC 2003b]. Detailing the unique identifier, the regulation demands the forwarding of the written documentation of the identity of a GMO at every stage of the production process. This allows a GMO to be traced even if, for example, due to intensive processing, it can no longer be detected. Using this unique identifier, information on all approved transgenic GMOs is accessible through the Biosafety Clearing-House, the information exchange platform of the Cartagena Protocol.

The cultivation of GMOs has implications for the organization of agricultural production. Conventional products—those produced without genetic modification—can unintentionally be contaminated by GM material during seed production, cultivation, harvesting, storage, transport, or processing. However, according to EC guidelines for the development of national strategies and best practices to ensure the coexistence of GM crops with conventional and organic farming, farmers should be able to cultivate the types of agricultural crops that they choose, be they GM, conventional, or organic. None of these forms of agriculture should be excluded in the EU. Coexistence refers to the ability of farmers to make a practical choice between conventional, organic, and GM crop production, in compliance with the legal obligations for labeling and purity standards (Jank et al. 2006).

In the last decade, a great number of EU-funded research projects such as GM and Non-GM Supply Chains: Their Coexistence and Traceability (Co-Extra), Sustainable Introduction of GMOs into European Agriculture (SIGMEA), and Developing Efficient and Stable Biological Containment Systems for Genetically Modified Plants (Transcontainer) aim at investigating improved methods for ensuring coexistence and providing stakeholders the tools required for the implementation of coexistence and traceability.

12.4 DETECTION

Testing on GMOs in food and feed is routinely done using molecular techniques like DNA microarrays or quantitative real-time polymerase chain reaction (qPCR). These tests can be based on screening genetic elements [such as p35S, *Agrobacterium tumefaciens* nopaline synthase terminator (tNOS), pat, or bar] or event-specific markers for the official GMOs (such as Mon810, Bt11, or GT73). The array-based method combines multiplex PCR and array technology to screen samples for different potential GMOs, combining different approaches (screening elements, plant-specific markers, and event-specific markers).

The qPCR is used to detect specific GMO events by the usage of specific primers for screening elements or event-specific markers. Controls are necessary to avoid false-positive or false-negative results. For example, a test for the cauliflower mosaic virus (CaMV) is used to avoid a false-positive in the event of a virus-contaminated sample.

Mbongolo Mbella et al. (2011) reported SYBR® Green qPCR methods for the detection of endogenous reference genes in commodity crops such as soybean, maize, oilseed rape, rice, cotton, sugar beet, and potato. Each qPCR method is shown to meet the performance criteria (specificity, limit of detection, and PCR efficiency) set by the European Network of GMO Laboratories (ENGL). When combined with the equivalent qPCR methods targeting GMO elements, these crop-specific SYBR Green qPCR methods can aid the development of an efficient tool for determining GMO presence in food and feed products.

An overview of the different techniques applied is presented in the work of Querci et al. (2010). Many of these approaches use two steps: (1) screening for the presence of crops and GM material using common genetic elements present in authorized GMOs (CaMV 35S promoter and tNOS) and (2) a GMO identification step where event-specific methods are used to identify which GMOs are present (Marmiroli et al. 2008).

Leimanis et al. (2008) used a microarray platform for a combination of crop, trait-, GM element-, construct-, and event-specific GMO screening targets. Morisset et al. (2008) used the NASBA implemented microarray analysis (NAIMA), a nucleic acid sequence–based amplification approach adapted on an array platform to develop a multitarget GMO screening tool.

Real-time quantification of the GMOs involves the choice of adequate primers, probes, and evaluation model, the appropriate standard material, and the DNA isolation method. The DNA extraction method should be cost and time effective, especially in the case of high amounts of samples. Previous examinations of different commercially available kits based on silica gel, magnetic beads, and precipitation were compared to lysis and precipitation with CTAB, showing that the conventional CTAB purification gave the highest yield with sufficient PCR amplification protocols (Holden et al. 2003). Besides the real-time experiments, Kunert et al. (2006) have also sequenced the 162-bp-long amplicon of the CaMV 35S promoter and found 100% homology for BT11, NK603, and MON810, giving evidence for the quantification of real samples with these primers. Additionally, they compared the amplicons of the invertase exon from three of their breedings with the Genbank entry (GI 1122438). In the cases of NK603 and MON810, they found some sequence divergence leading to different PCR efficiencies that resulted in problems with the interpretation of quantification (Arvanitoyannis and Varzakas 2006).

Pan et al. (2006) developed an event-specific detection method based on the flanking sequence of an exogenous integrant in the transgenic maize MON863 that contains the cry3Bb1 gene expressing a *Bacillus thuringiensis* Cry3Bb1 protein that is selectively toxic to a maize root worm pathogen. The 3′-integration junction between the host plant DNA and the integrated DNA of transgenic MON863 maize was isolated using thermal asymmetric interlaced–PCR. The event-specific primers and TaqMan probe were designed based on the isolated 30-integration junction sequence, and qualitative and quantitative PCR systems employing these designed primers and probe were established. In this system, the limit of detection of the qualitative PCR assay was estimated to be 40 initial haploid copies. The limit of quantitation of the quantitative PCR assay in authentic MON863 maize seeds was estimated to be approximately 80 haploid copies. GM MON863 contents were also quantified relative to the endogenous maize starch synthase IIb (zSSIIb) gene DNA, and the results were expressed as the percentage of the GM MON863 maize DNA relative to the total content of the maize DNA.

Qualitative and quantitative analytical methods were developed for the new event of GM maize MON863 by Lee et al. (2006).

Biotechnological advances have paved their way for very effective and reliable authenticity testing. The latter can be focused either on variety and geographical origin determination or traceability testing of a wide range of food products—produces of agricultural and animal origin and package counterfeiting. Apart from the widely employed DNA methods, other instrumental methods include site-specific natural isotope fractionation–nuclear magnetic resonance (SNIF-NMR) and more updated technology microsatellite markers, restriction fragment length polymorphism (RFLP), and single-strand conformation polymorphism (Varzakas et al. 2008).

Currently, the two most important approaches are immunological assays using antibodies that bind to the novel proteins and PCR-based methods using primers that recognize DNA sequences unique to the GM crop. The two most common immunological assays are enzyme-linked immunosorbent assay (ELISA) and immunochromatographic assay (lateral-flow strip tests). ELISA can produce qualitative, semiquantitative, and quantitative results in 1–4 h of laboratory time.

Nested PCR confirms the PCR product, allowing discrimination between specific and nonspecific amplification signals. Hence, the PCR product is reamplified using another primer pair, located in the inner region of the original target sequence (Anklam et al. 2002). It increases PCR sensitivity, allowing low levels of GMO to be detected (Zimmermann et al. 1998b). In order to detect the presence of RR soybean, a nested PCR method was applied to commercially available soy flour, infant formula containing soy protein isolate, and soymilk powder samples. Greiner et al. (2005) analyzed soy flour, infant foods, and soy protein isolates.

Electrochemiluminescence (ECL), where light-emitting species are produced by reactions between electrogenerated intermediates, has become an important and powerful analytical tool in recent years. An ECL reaction using tripropylamine (TPA) and *tris*(2,2′-bipyridyl)ruthenium(II) (TBR) has been demonstrated to be a highly sensitive detection method for quantifying amplified DNA. TPA and TBR are oxidized at approximately the same voltage on the anode surface. After deprotonation, TPA chemically reacts with TBR and results in an electron transfer. The resulting TBR molecule relaxes to its ground state by emitting a photon. TPA decomposes to dipropyl amine, is therefore consumed in this reaction and, on the other hand, is recycled. Because both reactants are produced at the anode, luminescence occurs there. ECL has the advantages that no radioisotopes are used, detection limits are extremely low, the extension of the dynamic range for quantification is over six orders of magnitude, labels are extremely stable compared to those of most other chemiluminescence systems, and simple and rapid measurement requires only a few seconds compared with other detection techniques—the ECL method is a chemiluminescent reaction of species generated electrochemically on an electrode surface. It is a highly efficient and accurate detection method. Liu et al. (2005) applied ECL PCR combined with two types of nucleic acid probes hybridization to detect GMOs. Whether the organisms contain GM components was discriminated by detecting the CaMV35S promoter and nopaline synthase terminator. The experimental results showed that the detection limit is 100 fmol of PCR products. The promoter and the terminator can be clearly detected in GMOs. The method may provide a new means for the detection of GMOs due to its simplicity and high efficiency. The instrument used was composed of an electrochemical reaction cell, a potentiostat, an ultra high sensitivity single photon counting module, a multifunction acquisition card, a computer, and the LabVIEW software. The electrochemical reaction cell contains a working electrode (platinum), a counter electrode (platinum), and a reference electrode ($Ag/AgCl_2$).

There are also in-process methods aimed at comparing the primal and GM plants divided into targeted and nontargeted approaches. Targeted approaches monitor directly the consequence of novel gene product presence on the GM plant phenotype. Moreover, changes in chemical composition are being detected. Nontargeted approaches consist of three basic levels: functional genomics, proteomics, and metabolomics. Functional genomics contain methods such as messenger RNA (mRNA) fingerprinting and DNA microarray. Considering proteomics, the protein composition of original and GM plants is being compared using methods such as two-dimensional electrophoresis (2-D ELPO; Gorg et al. 1999) and its modification or 2-D ELPO in connection with matrix-assisted laser desorption–ionization time-of-light mass spectrometry analysis (Andersen and Mann 2000). Difference gel electrophoresis is being employed for the testing of two samples on the same gel (Unlu 1999). The metabolomics level of analysis identifies and quantifies the maximum amount of particular components. This involves separating methods like gas chromatography, liquid chromatography, and high-performance liquid chromatography combined with various detection methods such as NMR, Fourier transform–impaired spectroscopy, mass spectroscopy, and flame ionization detector (Celec et al. 2005).

New methods and techniques have been developed within the framework of the European Research Project "Development of Methods to Identify Foods Produced by Means of Genetic Engineering" (Project No. SMT4-CT96-2072). In the scope of this project, DNA extraction methods have been compared (Zimmermann et al. 1998a), new primers and probes have been defined, ring tests with tomato, processed maize, and soya are being performed, and a database recording detailed information about GMO-containing food on the market, sequences, primers, and detection methods has been set up (Schreiber 1997).

12.4.1 Methods for Genome Analysis

12.4.1.1 Restriction Fragment Length Polymorphism

RFLP is a technique in which organisms may be differentiated by the analysis of patterns derived from the cleavage of their DNA. Organisms can differ in the distance between sites of the cleavage of a particular restriction endonuclease, and the length of the fragments produced will differ when the DNA is digested with a restriction enzyme. The restriction fragments are separated according to length by agarose gel electrophoresis. The resulting gel may further be analyzed by Southern blotting using specific probes. The similarity of the patterns generated can be used to differentiate even strains of the same organism. RFLP has been recently combined with microsynteny analysis comparing tomato to the *Arabidopsis* genome, and new microsyntenic expressed sequence tag markers were rapidly identified (Argyropoulos et al. 2008; Oh et al. 2002; Figure 12.1).

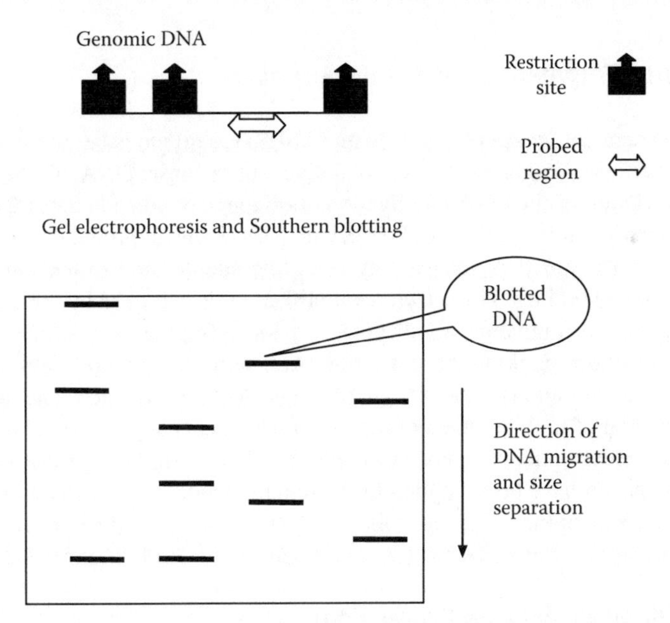

Figure 12.1 RFLPs are molecular markers used to create genetic maps of chromosomes. Combinations of enzymes and probes with digested DNA produce different profiles of DNA fragments, unique for each organism. The direction of DNA migration and size separation is shown by the arrow. Gel electrophoresis and Southern blotting show blotted DNA. (Adapted from Varzakas, T. et al., in *Tomatoes and Tomato Products, Nutritional, Medicinal and Therapeutic Properties*, edited by V.R. Preedy and R.R. Watson, pp. 515–536, Science Publishers, Enfield, NH, 2008. With permission.)

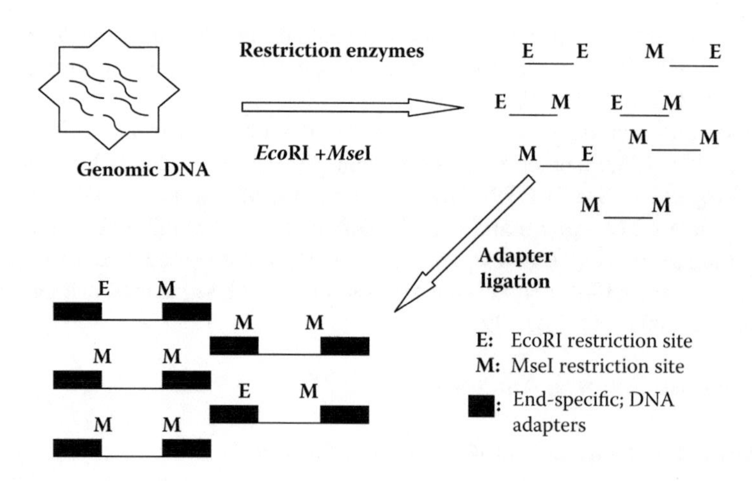

Figure 12.2 AFLP analysis employs the digestion of genomic DNA with *Mse*I and *Eco*RI enzymes. The resulting fragments are ligated to end-specific adapter molecules and are used in a preselective PCR with primers complementary to each of the two adapter sequences having an additional base at the 3′ end. Amplification of only 1/16 of *Eco*RI–*Mse*I fragments occurs. (Adapted from Varzakas, T. et al., in *Tomatoes and Tomato Products, Nutritional, Medicinal and Therapeutic Properties*, edited by V.R. Preedy and R.R. Watson, pp. 515–536, Science Publishers, Enfield, NH, 2008. With permission.)

12.4.1.2 Amplified Fragment Length Polymorphism

The amplified fragment length polymorphism (AFLP) technique is based on the selective PCR amplification of restriction fragments from a total digest of genomic DNA. The technique involves three steps: (1) restriction of the DNA and ligation of oligonucleotide adapters; (2) selective amplification of sets of restriction fragments; and (3) gel analysis of the amplified fragments. PCR amplification of restriction fragments is achieved by using the adapter and restriction site sequence as target sites for primer annealing. The selective amplification is achieved by the use of primers that extend into the restriction fragments, amplifying only those fragments in which the primer extensions match the nucleotides flanking the restriction sites. With this method, sets of restriction fragments are determined without prior knowledge of the nucleotide sequence. The method allows the specific coamplification of high numbers of restriction fragments and is usually accompanied with automated capillary electrophoresis (Vos et al. 1995). AFLP, simple sequence repeat, and single nucleotide polymorphism have been applied to the tomato genome for the assessment of polymorphism and for genome mapping (Suliman-Pollatschek et al. 2002), while new AFLP sequences are constantly added at the GenBank database (Argyropoulos et al. 2008; Figure 12.2).

12.4.1.3 Serial Analysis of Gene Expression

Serial analysis of gene expression (SAGE) was developed as an elegant means of analyzing mRNA populations by the large-scale sequence determination of short identifying stretches of individual messengers (Velculescu et al. 1995); however, the required depth of sequencing under different conditions makes it also labor intensive and time consuming (Figure 12.3).

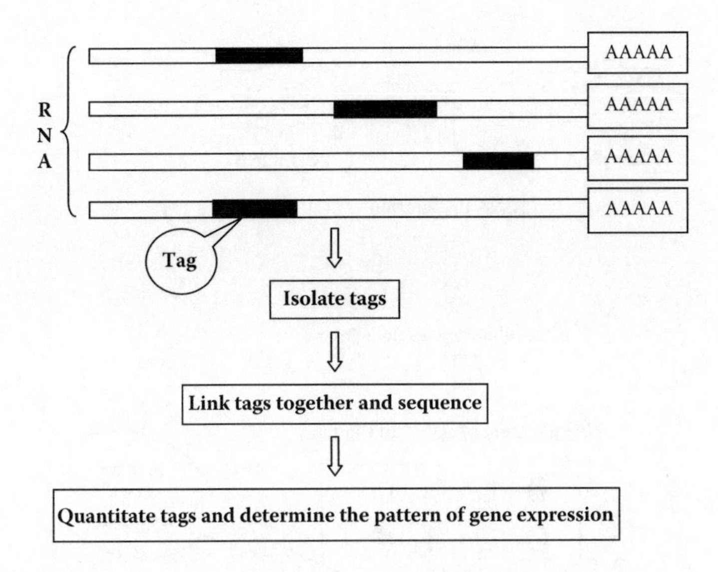

Figure 12.3 In the SAGE methodology, short sequence tags (10–14 bp) obtained from a unique position within each transcript uniquely identify transcripts. The expression level of the corresponding transcript is shown by the number of times a particular tag is observed. (Adapted from Varzakas, T. et al., in *Tomatoes and Tomato Products, Nutritional, Medicinal and Therapeutic Properties*, edited by V.R. Preedy and R.R. Watson, pp. 515–536, Science Publishers, Enfield, NH, 2008. With permission.)

12.4.1.4 Microarrays

More recent developments involve the use of DNA microarray technology in altered gene expression with a more efficient and informative way (Lockhart and Winzeler 2000; Panda et al. 2003; Mockler et al. 2005). Using DNA microarrays, the expression of a large number of genes can be analyzed simultaneously and in a semiquantitative manner. This allows for the analysis of different metabolic pathways in interaction and facilitates the identification of key responsive genes. For a limited number of species, microarrays that represent all identified metabolic routes and genes active therein have been constructed. These are the so-called whole genome arrays, oligo-arrays where all expressed gene sequences are represented by one or more short DNA sequences, usually up to 100 nucleotides (Mockler et al. 2005). Microarray analysis combined with suppression subtractive hybridization has been used in the identification of early salt stress response genes in tomato root (Ouyang et al. 2007; Figure 12.4).

An oligonucleotide microarray is a glass chip to the surface of which an array of oligonucleotides was fixed as spots, each containing numerous copies of a sequence-specific probe that is complementary to a gene of interest. To detect the presence of certain genes of interest in a sample, genomic DNA is isolated from a sample, amplified by PCR, and hybridized to the array. Hybridization of the sequences with their probes results on the microarray and can be detected by a fluorescence imaging system. The resulting patterns and relative intensities on the microarray will show whether the tested samples are carrying these certain genes.

There is a total of 20 probes for GMO detection in a DNA microarray system, which can be classified into three categories. The first category involves the screening of GMOs from nontransgenic plants based on general elements such as promoter, reporter, and terminator genes, the second category is based on target gene sequences such as herbicide resistance, or insect-resistant genes, and the third category screens for species-specific genes, that is, unique sequences for different plant

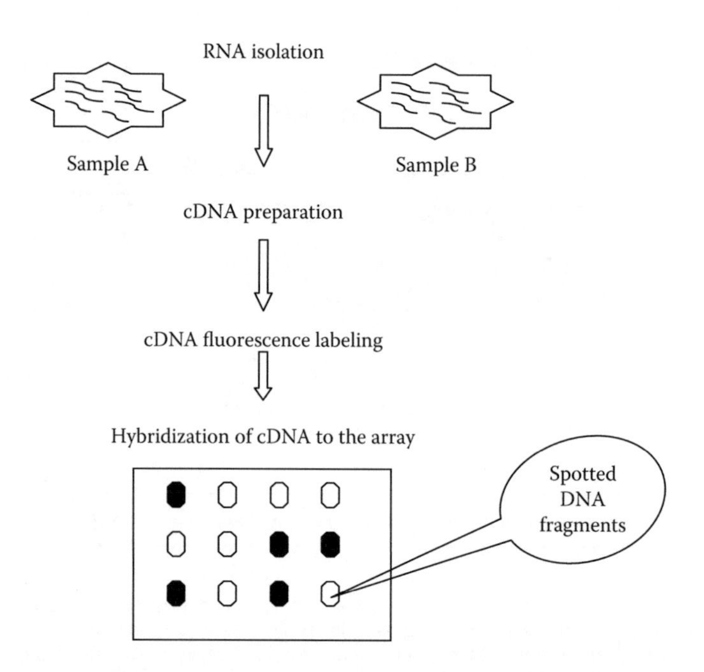

Figure 12.4 Microarrays used to study gene expression. mRNA is converted to complementary DNA (cDNA) and tagged with a fluorescent label following sample extraction. cDNA fluorescent samples are then applied to a microarray that contains DNA fragments corresponding to thousands of genes. Fluorescence intensity in each spot is used to estimate levels of gene expression. (Adapted from Varzakas, T. et al., in *Tomatoes and Tomato Products, Nutritional, Medicinal and Therapeutic Properties*, edited by V.R. Preedy and R.R. Watson, pp. 515–536, Science Publishers, Enfield, NH, 2008. With permission.)

species. To ensure the reliability of this method, different kinds of positive and negative controls are used in the DNA microarray.

12.4.1.5 Random Amplification of Polymorphic DNA

Random amplification of polymorphic DNA (RAPD) is a type of PCR reaction with DNA sequences amplified randomly. RAPD is applied by using several arbitrary, short primers (8–12 nucleotides) to the DNA based on the fact that some parts of the DNA will amplify. By resolving the resulting patterns by agarose gel electrophoresis, it is possible to differentiate strains of the same organism. The method has been used to assess populations of mRNA with differential display (Liang and Paedee 1992) or arbitrarily primed PCR (Welsh et al. 1992), amplification of random mRNA subsets, and subsequent analysis of the resulting fragment pool by gel electrophoresis (Argyropoulos et al. 2008).

12.4.1.6 Real-Time Polymerase Chain Reaction

In real-time PCR, the quantification of mRNA sequences is accomplished by absolute or relative analysis methods. Increasingly, the relative method of analysis is being used, as trends in gene expression can be better explained, but results depend on reference genes necessary to normalize sample variations (McMaugh and Lyon 2003; Weihong and Saint 2002).

A common technique in relative quantification is the choice of an endogenous control to normalize experimental variations caused by differences in the amount of the RNA added in the reverse transcription (RT) PCR reactions. Specifications of reliable endogenous controls (i.e., housekeeping genes) are that they need to be abundant, remain constant in proportion to the total RNA, and be unaffected by the experimental treatments. The best choices proposed to be used as normalizers of isolated mRNA quantities are mainly RNAs produced from glyceraldehyde-3-phosphate dehydrogenase (GAPDH; Bhatia et al. 1994), β-actin (Kreuzer et al. 1999), tubulin (Brunner et al. 2004), or rRNA (Bhatia et al. 1994). However, in general, depending on the developmental stage or environmental stimuli, the expression of certain reference genes is either up- or down-regulated.

The use of GAPDH mRNA as a normalizer is recommended with caution, as it has been shown that its expression may be up-regulated in proliferating cells (Zhu et al. 2001). The usage of 18S RNA as a normalizer is not always appropriate, as it does not have a polyA tail and thus prohibits the synthesis of cDNA with oligo-dT. Additionally, 18S RNA, being of ribosomal origin, may not always be representative of the entire cellular mRNA population and is in such overwhelming quantities relative to rare message that competimers must frequently be employed to obtain relevant normalization, making the results more complex. The expression of actin or tubulin often depends on the plant developmental stage (Diaz-Camino et al. 2005; Czechowski et al. 2005) and is affected upon environmental stresses (Jin et al. 1999), making their use as normalizers inappropriate.

Argyropoulos et al. (2006) proposed an alternative method for an internal control that would be applicable to different organisms without prior knowledge of genomic sequences, assuming that one wants to normalize against the total mRNA. In the presented methodology, synthetic DNA molecules (adapters) of known sequence tail cDNA during RT and are used instead of internal reference genes.

12.4.1.7 Nuclear Magnetic Resonance

Sobolev et al. (2010) used the ^{1}H NMR methodology in the study of GM foods. Transgenic lettuce (*Lactuca sativa* cv Luxor) overexpressing the *Arabidopsis* KNAT1 gene was presented as a case study. Twenty-two water-soluble metabolites present in leaves of conventional and GM lettuce were monitored by NMR and quantified at two developmental stages. NMR spectra did not reveal any difference in metabolite composition between the GM lettuce and its wild-type counterpart. Statistical analysis of metabolite variables highlighted metabolism variation as a function of leaf development as well as the transgene. A main effect of the transgene was the alteration of sugar metabolism.

12.4.2 Metabolic Engineering and Biotransformation

Metabolic engineering of plant cells has been employed for the biotransformation of hesperidin extracted from orange peels, by-products of the orange juice industry, into neohesperidin, a substrate for the production of the low-calorie sweetener and flavor enhancer neohesperidin dihydrochalcone (Frydman et al. 2005). Three steps were used: (1) extraction of hesperidin from orange peels; (2) hydrolysis of sugar moieties; and (3) biotransformation of hesperidin hydrolysis products into neohesperidin using metabolically engineered plant cell cultures (transgenic tobacco and carrot) expressing a recombinant flavanone-7-*O*-glucoside-2-*O*-rhamnosyltransferase.

Fong Chong et al. (2007) engineered sugarcane to synthesize sorbitol by introducing the gene from *Malus domestica* encoding the enzyme sorbitol-6-phosphate dehydrogenase (which catalyzes the reduction of glucose-6-phosphate to sorbitol-6-phosphate) and the gene from *Zymomonas mobilis* encoding the enzyme glucokinase. Motivated by the atypical development of the leaves in some sorbitolcane, the polar metabolite profiles in the leaves of those plants were compared to a group of control sugarcane plants. Eighty-six polar metabolites were detected in the leaf extracts. Principal component analysis of the metabolites indicated that three compounds were strongly associated with sorbitolcane. Two were identified as sorbitol and gentiobiose, and the third was unknown. Gentiobiose and the unknown compound were positively correlated with sorbitol accumulation. The unknown compound was a sorbitol–glucose conjugate (Fong Chong et al. 2010).

12.5 GENETICALLY MODIFIED HERBICIDE-RESISTANT SUGAR BEET

Sugar beet (*Beta vulgaris*) is a common crop in Europe, where it has been cultivated for sugar production since the late eighteenth century. It is a biennial plant. The large, succulent roots of sugar beet used for food and feed production are harvested at the end of the first year of growth. If left to grow, sugar beets will flower and produce seeds during the second year. Sugar beets are only allowed to flower for seed production, which mainly takes place in France and northern Italy.

Wild relatives of sugar beet originated in Asia Minor. Sugar beet's wild relative, sea beet (*B. vulgaris* ssp. *maritime*), is an annual in southern Europe and a biennial or perennial in northern latitudes (e.g., Scandinavia and Ireland).

In the case of sugar beets, outcrossing and hybridization are limited due to their harvesting before flowering. Only bolters—sugar beets that flower during the first growing season—can hybridize with each other or with wild relatives when present. However, sugar beets are cross compatible with other species in the genus *Beta* such as *B. vulgaris* ssp. *maritima*, *B. marcocarpa*, and *B. atriplicifolia*. There is no evidence that *B. vulgaris* can intercross with other genera in the Chenopodiaceae family.

Beets predominantly reproduce by seed, although plants can sometimes grow back from portions of roots left in the field after harvest. Volunteer sugar beets are rarely observed growing among other crops, in ditches, or on roadsides. If volunteer sugar beets were to occur in subsequent crops, they could be controlled by agricultural practices (e.g., herbicides or tillage during seedbed preparation). Most seeds left in the upper 5 cm of soil will germinate. Seeds that are ploughed deeper may remain dormant until conditions favor germination. Beet seeds can remain dormant for over 10 years. Problems with volunteer beets sometimes occur when planted on the same field for several consecutive years. Emerging annual weed beet from the seed bank can only be controlled by mechanical means and only to a certain degree. Remaining volunteers can reproduce and could potentially cross with bolting. Sugar beet is sensitive to frost and is poorly competitive in natural or agricultural habitats. Beet seeds are dispersed only over short distances. Pollen grains are dispersed mainly by wind, but insects can play a small role. Beet pollen, however, is quite sensitive and is viable for no more than 24 h under field conditions. In general, sugar beet does not survive in the environment. Only hybrids with wild beets can withstand natural competition and low winter temperatures (www.gmo-compass.org) Dewar et al. (2000).

As sugar beet is sensitive to weed competition, a relatively complex weed control program using five active ingredients has been used to control weeds (Buckmann et al. 2000), together with mechanical weeding in 50% of the crop. Popular postemergent herbicidal spray active ingredients include phenmedipham, metamitron, ethofumesate, desmedipham, triflusulfuron methyl, lenacil, and chloridazon.

GMHT sugar beet (RR sugar beet) has been genetically modified to be tolerant to the herbicide glyphosate, a broad-spectrum, nonselective herbicide that kills plants by binding to an enzyme (EPSPS), which prevents the production of essential amino acids in the plants. Thus, weed control within the GM crop can be achieved by applying glyphosate in place of the suite of herbicides referred to above. RR sugar beet, in general, would bring major agronomic and economic benefits to farmers (May 2003; Liu et al. 2008; Lisson et al. 2000; May 2001).

In the context of the ongoing debate concerning the possible environmental and human health impacts of growing GM crops, Bennett et al. (2004) reported the results of a life cycle assessment comparing the environmental and human health impacts of conventional sugar beet–growing regimes in the U.K. and Germany with those that might be expected if GMHT (to glyphosate) sugar beet is commercialized. The results presented for a number of environmental and human health impact categories suggest that growing the GMHT crop would be less harmful to the environment and human health than growing the conventional crop, largely due to lower emissions from herbicide manufacture, transport, and field operations. Emissions contributing to negative environmental impacts, such as global warming, ozone depletion, ecotoxicity of water, and acidification and nutrification of soil and water, were much lower for the herbicide-tolerant crop than for the conventional crop. Emissions contributing to summer smog, toxic particulate matter, and carcinogenicity, which have negative human health impacts, were also substantially lower for the herbicide-tolerant crop. Thus, in addition to herbicide use, RR sugar beet cultivation may reduce energy use and ecotoxicity.

Pidgeon et al. (2007) applied, for the first time, the "bow-tie" risk management approach in agriculture, for the assessment of land use changes, in a case study regarding the possible introduction of GMHT sugar beet into the U.K. This is an important issue not only due to the agronomic and economic benefits that it could bring to farmers but also because the EU sugar regime reform is reducing the profitability of sugar beet production seriously. The study concluded that, although there are certainly agronomic and economic benefits, indirect environmental harm from increased weed control poses a hazard.

In the U.S., between 2009 and 2011, the United States District Court for the Northern District of California considered the case involving the planting of GM sugar beets (McGinnis et al. 2010). This case involves Monsanto's breed of pesticide-resistant sugar beets. Earlier in 2010, Judge Jeffrey S. White allowed the planting of GM sugar beets to continue, but he also warned that this may be blocked in the future while an environmental review was taking place. On 13 August 2010, Judge White ordered a halt to the planting of the GM sugar beets in the U.S. He indicated that, "The Agriculture Department had not adequately assessed the environmental consequences before approving them for commercial cultivation." The decision was the result of a lawsuit organized by the Center for Food Safety, a U.S. nongovernmental organization that is a critic of biotech crops. On 25 February 2011, a federal appeals court for the Northern District of California in San Francisco overturned a previous ruling by Judge Jeffrey S. White to destroy juvenile GM sugar beets, ruling in favor of Monsanto, the Department of Agriculture Animal and Plant Health Inspection Service (APHIS), and four seed companies. The court concluded that, "The Plaintiffs have failed to show a likelihood of irreparable injury. Biology, geography, field experience, and permit restrictions make irreparable injury unlikely." In February 2011, The USDA allowed the commercial planting of GM sugar beet under closely controlled conditions. Analysis of potential targets for sugar beet improvement has included modification of gibberellin signaling (metabolism and signal transduction) as reported by Mutasa-Gottgens et al. (2009).

12.6 GENETICALLY MODIFIED CORN SYRUP

12.6.1 How High-Fructose Corn Syrup Is Made

High-fructose corn syrup (HFCS) comprises any of a group of corn syrup that has undergone enzymatic processing to convert some of its glucose into fructose to produce a desired sweetness. In the United States, consumer foods and products typically use HFCS as a sweetener. It has become very common in processed foods and beverages in the U.S., including breads, cereals, breakfast bars, lunch meats, yogurts, soups, and condiments (Wallinga et al. 2009).

The most widely used varieties of HFCS are HFCS 55 (mostly used in soft drinks; approximately 55% fructose and 42% glucose) and HFCS 42 (used in many foods and baked goods; approximately 42% fructose and 53% glucose). HFCS 90 (approximately 90% fructose and 10% glucose) is used in small quantities for specialty applications but is primarily used to blend with HFCS 42 to make HFCS 55 (White 2008).

HFCS is made from corn kernels. Actual syrup production necessitates a whole string of industrial processes, including high-velocity spinning and the introduction of three different enzymes to incite molecular rearrangements. The enzymes turn most of the glucose molecules in corn into fructose, which makes the substance sweeter. This 90% fructose syrup mixture is then combined with regular 100% glucose corn syrup to get the desired balance of glucose and fructose, somewhere between equal quantities of both to a ratio of 80%–20%. The final product is a clear, thick liquid sweeter than sugar.

HFCS is produced by milling corn to produce corn starch, processing that starch to yield corn syrup, which is almost entirely glucose, and then adding enzymes that change some of the glucose into fructose. The resulting syrup (after enzyme conversion) contains approximately 42% fructose (HFCS 42). Some of the 42% fructose is then purified to 90% fructose (HFCS 90). To make HFCS 55, HFCS 90 is mixed with HFCS 42 in appropriate ratios to form the desired HFCS 55. The enzyme process that changes the 100% glucose corn syrup into HFCS 42 is as follows:

1. Cornstarch is treated with α-amylase to produce oligosaccharides, that is, shorter chains of sugars.
2. Glucoamylase, which is produced by the *Aspergillus* mold in a fermentation vat, breaks down the sugar chains even further to yield the simple sugar glucose.
3. Xylose isomerase (glucose isomerase) converts glucose to a mixture of about 42% fructose and 50%–52% glucose with some other sugars mixed in.

While inexpensive α-amylase and glucoamylase are added directly to the slurry and used only once, the more costly xylose–isomerase is packed into columns, and the sugar mixture is then passed over it, allowing it to be used repeatedly until it loses its activity. This 42%–43% fructose–glucose mixture is then subjected to a liquid chromatography step, where the fructose is enriched to about 90%. The 90% fructose is then back-blended with 42% fructose to achieve a 55% fructose final product. Most manufacturers use carbon adsorption for impurity removal. Numerous filtration, ion exchange, and evaporation steps are also part of the overall process.

12.6.2 Pros and Cons of High-Fructose Corn Syrup

HFCS is not only sweeter and easier to blend into beverages than table sugar but is also a great preservative, so it can be used in processed foods to extend their shelf life. HFCS is easier to transport and more economical in countries where the price of sugar is twice the global price, such as the United States and Canada. The syrup can be 20%–70% cheaper than sugar.

HFCS is genetically modified. In 1982, when the artificial sweetener was introduced into the American food supply, children for the first time began getting type II diabetes, and obesity rates soared. In at least one study, the syrup has been linked to both. The syrup has also been shown to interfere with people's metabolism so that a person feels hungrier than he/she really is. This is because HFCS also limits the secretion of leptin into the body's system. Leptin is a hormone that signals to the brain when you are full, and without it, the amount of food you consume is not controlled. In parallel to this, the manufactured sweetener also encourages the production of ghrelin, a hormone responsible for controlling the appetite, sending the appetite into overdrive. The *American Journal of Clinical Nutrition* in 2004 published a study noting that the rise in HFCS consumption paralleled the rise in obesity rates in the U.S. and hypothesized that the way fructose is metabolized could be uniquely fattening. The authors later said that their study was meant to inspire further study, not to be a definitive declaration. Because there are no enzymes to digest HFCS, it is metabolized by the liver. The pancreas releases insulin the way it normally does for sugar, so fructose converts to fat more readily than any other sugar. An overworked liver produces significantly more uric acid, multiplying the risk for heart disease. Although a number of associations have claimed that HFCS is not unhealthy when consumed in moderation, it is hard to gauge just how much one consumes, because just like sugar, it is contained in so many foods one does not know about (http://www.highfructosecornsyrup.org/2009/02/guess-whats-lurking-in-your-food.html).

A study in 2005, published by Dufault et al. (2009), found trace amounts of mercury within nine of the 20 samples involved, having likely leached into the solutions during the process of their creation (Dufault et al. 2009). Concerns regarding the safety of HFCS are being continually studied.

The Corn Refiners Association disputes these claims and maintains that HFCS is comparable to table sugar (http://www.sweetsurprise.com/).

Studies by the American Medical Association suggest, "It appears unlikely that HFCS contributes more to obesity or other conditions than sucrose," but welcome further independent research on the subject (AMA 2008). HFCS has been classified as generally recognized as safe by the U.S. FDA since 1976 (FDA 2011).

The U.S. FDA, for example, notes that adult males and postmenopausal women who consume high amounts of HFCS have increased levels of blood lipids, which are related to heart disease. The amount of HFCS in a normal diet, however, does not appear to pose any risk to heart health according to the FDA. In a 2010 study at Princeton University, however, rats that consumed HFCS gained significantly more weight than rats that drank sugar–water or water sweetened with sucrose. The same researchers discovered that rats that consumed HFCS over a 6-month period displayed signs of metabolic syndrome, a serious condition that often leads to heart disease and stroke (Bauer 2011).

12.6.3 High-Fructose Corn Syrup—Genetically Modified

Two of the enzymes used to make the syrup—α-amylase and glucose isomerase—are genetically modified to make them more stable. Through genetic modification, specific amino acids are changed or replaced so that the enzyme's "backbone" would not break down or unfold. This allows the industry to use the enzymes at higher temperatures without them becoming unstable. Consumers trying to avoid GM foods should therefore avoid HFCS. It is almost certainly made from GM corn and processed with GM enzymes. Finally, a team of investigators at the United States Department of Agriculture discovered that a fructose diet might lead to many more health problems than a glucose diet (http://www.highfructosecornsyrup.org/2009/02/guess-whats-lurking-in-your-food.html).

While the corn used to produce HFCS may or may not have been produced using genetically enhanced corn, existing scientific literature and current testing results indicate that corn DNA cannot be detected in measurable amounts in HFCS (High Fructose Corn Syrup Health and Diet Facts; http://www.sweetsurprise.com/myths-and-facts/top-hfcs-myths/gmo-corn-hfcs).

HFCS—called "isoglucose" in Europe—is not banned in other countries but is subject to production quotas in Europe. These quotas are based on economic considerations, not health concerns. Because HFCS is a substitute for sugar in some commercial processes, it falls under European Union sugar regulations. European Council Regulation 1234/2007 of 22 October 2007 sets limits on the amount of HFCS that each member state may produce. By 2011, however, the demand for sugar and isoglucose in Europe exceeded production by 1 million tons. As a result, the EC decided to relax surplus levies on these two commodities to allow for greater internal production and increased imports.

Some corn grown in the United States comes from GM seed stock. Exports of HFCS made from these strains of corn are thus subject to bans in some countries on GM foods. While the European Union technically does not ban GM foods, such foodstuffs require approvals, and as of 2011, only two strains of GM crops have approval for import into Europe. Because some EU member states have serious concerns about such modified products, early 2011 draft EU legislation gives member states the right to ban GM foods on moral or religious grounds or due to public opposition for imports of such foods. These reasons, however, do not comply with World Trade Organization (WTO) regulations and may leave countries that impose such bans vulnerable to lawsuits according to Thijs Etty, assistant professor of law at the University of Amsterdam.

Now, almost all of the enzymes used to break down starch, in starch saccharification, are produced with the help of GM microorganisms. Some of these enzymes are economically impossible to produce without biotechnological methods. Certain procedures use "immobilized" enzymes, which are bound to a reaction surface. Rather than mixing freely, they remain fixed to a surface and are not present in the final product.

Labeling. It is impossible to tell by examining starch-derived sugar products if the source material was genetically modified or if the enzymes used were produced with the aid of GM microorganisms. Nonetheless, such products require labeling if they contain sugar products derived from the starch of GM plants.

Enzymes do not need to be declared or listed, regardless of the way they were produced (http://www.gmo-compass.org/eng/grocery_shopping/ingredients_additives/37.products_starch_corn_syrup_fructose_glucose.html).

12.6.4 Genetic Engineering—Starch and Enzymes

Genetic engineering can be associated with starch-derived sugars that are used in foods and beverages in two ways: the plant starch source can be genetically modified, and the enzymatic "tools" used for breaking down the starch can be made by GM microorganisms.

Plant used as a starch source. When maize is used as a source of starch, a certain portion of the raw material may be genetically modified, as GM maize is common in the U.S. and in other countries. When GM maize in Europe becomes more widespread, so will the proportion of GM content in starch processing. For potatoes, the second most important source of starch, GM cultivars with optimized starch content are getting closer to commercial cultivation.

12.6.5 Use as a Replacement for Sugar

HFCS replaces sugar in various processed foods in the United States. The main reasons for this switch are as follows:

1. Per relative sweetness, HFCS 55 is comparable to table sugar (sucrose), a disaccharide of fructose and glucose.
2. HFCS 90 is sweeter than sucrose; HFCS 42 is less sweet than sucrose.
3. HFCS is cheaper in the United States as a result of a combination of corn subsidies and sugar tariffs and quotas (Pollan 2003). Since the mid-1990s, the United States federal government has subsidized corn growers by $40 billion (Smith 2006; Engber 2009).
4. HFCS is easier to blend and transport, because it is a liquid (Hanover and White 1993).

In the United States, HFCS has become a sucrose replacement for honey bees. In 2009, a study by Leblanc et al. found that at temperatures above 45°C, HFCS rapidly forms hydroxymethylfurfural, which is toxic to the honey bees being fed HFCS.

12.7 CONCLUSION

The following conclusions could be drawn in the form of a displaying list.

1. The technology of genetic modification of plants remains a controversial issue that raises public debate. Even a part of the scientific community is skeptical regarding the benefits as well as the hazards of the GM technology. However, it is true that no food product is 100% safe.
2. The possible hazards mainly lie in allergies to be caused by novel proteins included in GM foods and in unpredictable risks arising from accidental gene insertion in the total genome or environmental effects.
3. There is already a plethora of processed foods containing ingredients deriving from GM crops, consumed by millions of people worldwide, and health effects have not been observed or registered. However, as this has been happening for almost 20 years, which is considered a short period to come up with concrete conclusions, wide, long-term epidemiological studies are necessary.
4. A case-by-case approach should be adopted in assessing GM crop technology and its use in agriculture, because scientific uncertainties exist, especially regarding the environment. Long-term

follow-up studies are needed to assess the environmental impact of GM varieties in certain ecosystems and, hence, biodiversity.

5. A case study of GM corn syrup has been described and followed.

GM HFCS has been discussed with regard to its advantages and disadvantages, an explanation about how it is made is given and finally its uses as sugar replacement.

REFERENCES

American Medical Association. 2008. The health effects of high fructose syrup: Report 3 of the Council on Science and Public Health (A-08).

Andersen, J.S. and M. Mann. 2000. Functional genomics by mass spectrometry. *FEBS Lett.* 480: 25–31.

Anklam, E., F. Gadani, P. Heinze, H. Pijnenburg, and G. Van den Eede. 2002. Analytical methods for detection and determination of genetically modified organisms in agricultural crops and plant-derived food products. *Eur. Food Res. Technol.* 214: 3–26.

Argyropoulos, D., C. Psallida, and C.G. Spyropoulos. 2006. Generic normalization method for real-time PCR application for the analysis of the mannanase gene expressed in germinating tomato seed. *FEBS J.* 273: 770–777.

Argyropoulos, D., T. Varzakas, C. Psallida, and I. Arvanitoyannis. 2008. Chapter 28: Methods for PCR and gene expression studies in tomato plants. In: *Tomatoes and Tomato Products, Nutritional, Medicinal and Therapeutic Properties*, edited by V.R. Preedy and R.R. Watson, 585–615. Enfield, NH: Science Publishers.

Arvanitoyannis, I. and T. Varzakas. 2006. Plant genetic engineering: General applications, legislations and issues. In: *Transgenic Horticultural Plants: Prospects and Controversies*, Vol. II, edited by R. Ray and O.P. Ward, pp. 1–84. Enfield, NH: Science Publishers, USA.

Bauer, M. 2011. Why is high-fructose corn syrup banned from other countries? Available at http://www.livestrong.com/article/458653-why-is-high-fructose-corn-syrup-banned-from-other-countries/ (accessed July, 2011).

Beckmann, V., C. Soregaroli, and J. Wesseler. 2006. Coexistence rules and regulations in the European Union. *Am. J. Agric. Econ.* 88: 1193–1199.

Bennett, R., R. Phipps, A. Strange, and P. Grey. 2004. Environmental and human health impacts of growing genetically modified herbicide-tolerant sugar beet: A life-cycle assessment. *Plant Biotechnol. J.* 2 (4): 273–278.

Bhatia, P., W.R. Taylor, A.H. Greenberg, and J.A. Wright. 1994. Comparison of glyceraldehyde-3-phosphate dehydrogenase and 28S-ribosomal RNA gene expression as RNA loading controls for Northern blot analysis of cell lines of varying malignant potential. *Anal. Biochem.* 216: 223–226.

Brookes, G. and P. Barfoot. 2008. Global impact of biotech crops: Socioeconomic and environmental effects, 1996–2006. *AgBioForum* 11 (1): 21–38.

Brunner, A.M., I.A. Yakovlev, and S.H. Strauss. 2004. Validating internal controls for quantitative plant gene expression studies. *BMC Plant Biol.* 4: 14–19.

Bückmann, H., J. Petersen, G. Schlinker, and B. Märländer. 2000. Weed control in genetically modified sugar beet—Two-year experiences of a field trial series in Germany. *Z. PflKrankh. PflSchutz*, Sonderheft XVII: 353–362.

Celec, P., M. Kukučková, V. Renczésová, S. Natarajan, R. Pálffy, R. Gardlík, J. Hodosy, M. Behuliak, B. Vlková, G. Minárik, T. Szemes, S. Stuchlík, and J. Turňa. 2005. Biological and biomedical aspects of genetically modified food. *Biomed. Pharmacother.* 59: 531–540.

Centers for Disease Control and Prevention (CDC). 2001. Investigation of human health effects associated with potential exposure to genetically modified corn: A report to the U.S. Food and Drug Administration from the Centers for Disease Control and Prevention. Atlanta, GA: Centers for Disease Control and Prevention, National Center for Environmental Health.

Codex Alimentarius Commission. 2003. Appendix III: Guideline for the conduct of food safety assessment of foods derived from recombinant-DNA plants; Appendix IV: Annex on the assessment of possible allergenicity. Alinorm 03/34: Joint FAO/WHO Food Standard Programme, Codex Alimentarius Commission, Twenty-Fifth Session, Rome, 30 June–5 July 2003, pp. 47–60.

Corn Refiners Association. 2010. Get the Facts About High Fructose Corn Syrup, A Sugar Made From Corn. Available at http://www.sweetsurprise.com (accessed October 2011).

Cowgill, S.E., C. Danks, and H.J. Atkinson. 2004. Multitrophic interactions involving genetically modified potatoes, nontarget aphids, natural enemies and hyperparasitoids. *Mol. Ecol.* 13: 639–647.

Czechowski, T., M. Stitt, T. Altmann, M.K. Udvardi, and W.R. Scheible. 2005. Genome-wide identification and testing of superior reference genes for transcript normalization in *Arabidopsis*. *Plant Physiol.* 139: 5–17.

FDA. 2011. Database of Select Committee on GRAS Substances (SCOGS) Reviews. Available at http://www.accessdata.fda.gov/scripts/fcn/fcnDetailNavigation.cfm?rpt=scogsListing&id=95. Retrieved 2011-11-06. (accessed 31 July 2011).

Dewar, A.M., L.A. Haylock, K.M. Bean, and M.J. May. 2000. Delayed control of weeds in glyphosate-tolerant sugar beet and the consequences on aphid infestation and yield. *Pest Manage. Sci.* 56: 345–350.

Diaz-Camino, C., R. Conde, N. Ovsenek, and M.A. Villanueva. 2005. Actin expression is induced and three isoforms are differentially expressed during germination of *Zea mays*. *J. Exp. Bot.* 56: 557–565.

Dufault, R., B. LeBlanc, R. Schnoll, C. Cornett, L. Schweitzer, D. Wallinga, J. Hightower, L. Patrick and W. J. Lukiw. 2009. Mercury from chlor-alkali plants: measured concentrations in food product sugar. Environmental Health 8: PubMed Central. http://ehjournal.net/content/8/1/2. Retrieved August 9, 2009.

Duke, S.O. and S.B. Powles. 2009. Glyphosate-resistant crops and weeds: Now and in the future. *AgBioForum* 12: 1–12.

Dunfield, K.E. and J.J. Germida. 2004. Impact of genetically modified crops on soil- and plant-associated microbial communities. *J. Environ. Qual.* 33: 806–815.

Engber, D. 2009. The decline and fall of high-fructose corn syrup. *Slate Magazine*. Available at http://www.slate.com/id/2216796 (accessed 06 November 2010).

European Commission (EC). 2002. Commission Decision (2002/623/EC) of 24 July 2002 establishing guidance notes supplementing Annex II to Directive 2001/18/EC of the European Parliament and of the Council on the deliberate release into the environment of genetically modified organisms and repealing Council Directive 90/220/EEC. *Off. J. Eur. Communities: Legis.* 200: 22–33. Available at http://europa.eu.int/eurlex/pri/en/oj/dat/2002/l_200/l_20020020730en00220033.pdf.

European Commission (EC). 2003a. Regulation (EC) 1829/2003 of the European Parliament and of the Council of 22 September 2003 on genetically modified food and feed. *Off. J. Eur. Communities: Legis.* 268: 1–23.

European Commission (EC). 2003b. Regulation (EC) 1830/2003 of the European Parliament and of the Council of 22 September 2003 concerning the traceability and labeling of genetically modified organisms and the traceability of food and feed products produced from genetically modified organisms and amending Directive 2001/18/EC. *Off. J. Eur. Communities: Legis.* 268: 24–28. Available at http://europa.eu.int/eur-lex/pri/en/oj/dat/2003/l_268/l_26820031018en00240028.pdf.

European Commission Joint Research Center. 2003. Review of GMOs under Research and Development and in the pipeline in Europe. Institute for Propsective Technological Studies. European Science and Technology Observatory. Available at http://gmoinfo.jrc.ec.europa.eu (accessed July 2011).

European Food Safety Authority (EFSA). 2004. Opinion of the Scientific Panel on genetically modified organisms on a request from the Commission related to the safety of foods and food ingredients derived from insect-protected genetically modified maize MON 863 and MON 863 × MON 810, for which a request for placing on the market was submitted under Article 4 of the Novel Food Regulation (EC No. 258/97) by Monsanto (Question No. EFSA-Q-2003-121). *The EFSA Journal* 50: 1–25.

European Food Safety Authority (EFSA) GMO Panel. 2009. EFSA evaluates antibiotic resistance marker genes in GM plants. July 2009. Science Publishers, USA.

European Union (EU) 2005. EC-sponsored research on safety of GMOs: GMO research in perspective. Report of a Workshop held by External Advisory Groups of the Quality of Life and Management of Living Resources Program (Fifth Framework).

Evaluating Safety: A Major Undertaking. Available at http://www.gmo-compass.org/eng/safety/human_health/41.evaluation_safety_gm_food_major_undertaking.html (accessed December 2011).

Ewen, S.W. and A. Pusztai. 1999. Effect of diets containing genetically modified potatoes expressing *Galanthus nivalis* lectin on rat small intestine. *Lancet* 354: 1353–1354.

Federal Insecticide, Fungicide, and Rodenticide Act (FIFRA). 2001. Scientific Advisory Panel Report No. 2001-09, July 2001. Available at http://www.epa.gov/scipoly/SAP/meetings/2001/july/julyfinal.pdf (accessed July 2011).

Flachowsky, G., A. Chesson, and K. Aulrich. 2005. Animal nutrition with feeds from genetically modified plants. *Arch. Anim. Nutr.* 59 (1): 1–40.

Fong Chong, B., W.P.P. Abeydeera, D. Glassop, G.D. Bonnett, M.G. O'Shea, and S.M. Brumbley. 2010. Coordinated synthesis of gentiobiitol and sorbitol, evidence of sorbitol glycosylation in transgenic sugarcane. *Phytochemistry* 71: 736–741.

Fong Chong, B., G.D. Bonnett, D. Glassop, M.G. O'Shea, and S.M. Brumbley. 2007. Growth and metabolism in sugarcane are altered by the creation of a new hexose-phosphate sink. *Plant Biotechnol. J.* 5: 240–223.

Frisvold, G.B., A. Boor, and J.M. Reeves. 2009. Simultaneous diffusion of herbicide resistant cotton and conservation tillage. *AgBioForum* 12: 249–257.

Frydman, A., O. Weisshaus, D.V. Huhman, L.W. Sumner, M. Bar-Peled, E. Lewinsohn, R. Fluhr, J. Gressel, and Y. Eyal. 2005. Metabolic engineering of plant cells for biotransformation of hesperidin into neohesperidin, a substrate for production of the low-calorie sweetener and flavor enhancer NHDC. *J. Agric. Food Chem.* 53: 9708–9712.

Giovannetti, M., C. Sbrana, and A. Turrini. 2005. The impact of genetically modified crops on soil microbial communities. *Biotech. Forum, Riv. Biol.* 98: 393–418.

Goodman, R.E., S. Vieths, H.A Sampson, D. Hill, M. Ebisawa, S.L. Taylor and R. van Ree. 2008. Allergenicity assessment of genetically modified crops—What makes sense? *Nat. Biotechnol.* 26: 73–81.

Gorg, A., C. Obermaier, G. Boguth, and W. Weiss. 1999. Recent developments in two-dimensional gel electrophoresis with immobilized pH gradients: Wide pH gradients up to pH 12, longer separation distances and simplified procedures. *Electrophoresis* 20: 712–717.

Greiner, R., U. Konietzny and A.L. Villavicencio. 2005. Qualitative and quantitative detection of genetically modified maize and soy in processed foods sold commercially in Brazil by PCR-based methods. *Food Control*, 16: 753–759.

Guess what's lurking in your food. 2009. Available at http://www.highfructosecornsyrup.org/2009/02/guess-whats-lurking-in-your-food.html. (accessed July 2011).

Guess what's lurking in your food. 2009. Available at http://www.highfructosecornsyrup.org/2009/02/guess-whats-lurking-in-your-food.html. (accessed October 2011).

Hanover, L.M. and J.S. White. 1993. Manufacturing, composition, and applications of fructose. *Am. J. Clin. Nutr.* 58 (Suppl. 5): 724S–732S.

Harlander, S.K. 2002. The evolution of modern agriculture and its future with biotechnology. *J. Am. Coll. Nutr.* 21: 161S–165S.

Heap, I.M. 2009. The international survey of herbicide resistant weeds [dataset]. Available at http://www.weedscience.org/. (accessed Julky 2011).

Holden, M.J., J.R. Blasic, L. Bussjaeger, C. Kao, L.A. Shokere, D.C. Kendall, L. Freese, and G.R. Jenkins. 2003. Evaluation of extraction methodologies for corn kernel (*Zea mays*) DNA for detection of trace amounts of biotechnology-derived DNA. *J. Agric. Food Chem.* 51: 2468–2474.

Hurley, T.M., P.D. Mitchell, and G.B. Frisvold. 2009. Effects of weed resistance concerns and resistance management practices on the value of Roundup Ready® crops. *AgBioForum* 12: 291–302.

International Life Sciences Institute (ILSI). 2004. Nutritional and safety assessments of foods and feeds nutritionally improved through biotechnology. *Compr. Rev. Food Sci. Food Safety* 3: 36–104.

James, C. 2008. Global status of commercialized biotech/GM crops. ISAAA Brief 39. Ithaca, NY: International Service for the Acquisition of Agri-Biotech Applications (ISAAA).

Jank, B., J. Rath, and H. Gaugitsch. 2006. Coexistence of agricultural production systems. *Trends Biotechnol.* 24 (5): 198–200.

Jin, S.M., R.L. Xu, Y.D. Wei, and P.H. Goodwin. 1999. Increased expression of a plant actin gene during a biotrophic interaction between leaved mallow, *Maiva pusillia*, and *Colletotrichum gloeosporioides* f. sp malvae. *Planta* 209: 487–494.

Kreuzer, K.A., U. Lass, O. Landt, A. Nitsche, J. Laser, H. Ellebrok, G. Pauli, D. Huhn, and C.A. Schmidt. 1999. Highly sensitive and specific fluorescence reverse transcription-PCR assay for the pseudogene-free detection of beta-actin transcripts as quantitative reference. *Clin. Chem.* 45: 297–300.

Kunert, R., J.S. Gach, K. Vorauer-Uhl, and H. Katinger. 2006. Validated method for quantification of genetically modified organisms in samples of maize flour. *J. Agric. Food Chem.* 54: 678–681.

Leblanc, W., G. Eggleston, D. Sammataro, C. Cornett, R. Dufault, T. Deeby, and E. St Cyr. 2009. Formation of hydroxymethylfurfural in domestic high-fructose corn syrup and its toxicity to the honey bee (*Apis mellifera*). *J. Agric. Food Chem.* 57 (16): 7369. doi:10.1021/jf9014526. ISSN 0021-8561. PMID 19645504.

Lee, H.S., D.M. Min, and J.K. Kim. 2006. Qualitative and quantitative polymerase chain reaction analysis for genetically modified maize MON863. *J. Agric. Food Chem.* 54: 1124–1129.

Leimanis, S., H. Hamels, F. Naze, G. Mbongolo Mbella, M. Sneyers, R. Hochegger et al. 2008. Validation of the performance of a GMO multiplex screening assay based on microarray detection. *Eur. Food Res. Technol.* 227: 1621–1632.

Liang, P. and A.B. Paedee. 1992. Differential display of eukaryotic messenger RNA by means of the polymerase chain reaction. *Science* 257: 967–971.

Lisson, R., J. Hellert, M. Ringleb, F. Machens, J. Kraus, and R. Hehl. 2010. Alternative splicing of the maize Ac transposase transcript in transgenic sugar beet (*Beta vulgaris* L.). *Plant Mol. Biol.* 74: 19–32.

Liu, J., D. Da Xing, X. Shen, and D. Zhu. 2005. Electrochemiluminescence polymerase chain reaction detection of genetically modified organisms. *Anal. Chim. Acta* 537: 119–123.

Liu, H., Q. Wang, M. Yu, Y. Zhang, Y. Wu, and H. Zhang. 2008. Transgenic salt-tolerant sugar beet (*Beta vulgaris* L.) constitutively expressing an *Arabidopsis thaliana* vacuolar Na/H antiporter gene, AtNHX3, accumulates more soluble sugar but less salt in storage roots. *Plant, Cell Environ.* 31: 1325–1334.

Lockhart, D.J. and E.A. Winzeler. 2000. Genomics, gene expression and DNA arrays. *Nature* 405: 827–836.

Lövei, G.L, T. Bøhn, and A. Hilbeck. 2010. *Biodiversity, Ecosystem Services and Genetically Modified Organisms*. Penang, Malaysia: Third World Network.

Marmiroli, N., E. Maestri, M. Gulli, A. Malcevschi, C. Peano, R. Bordoni, and G. De Bellis. 2008. Methods for detection of GMOs in food and feed. *Annal. Bioanal. Chem.* 392: 369–384.

May, M. 2001. Crop protection in sugar beet. *Pestic. Outlook* 12: 188–191.

May, M. 2003. Economic consequences for UK farmers of growing GM herbicide tolerant sugar beet. *Ann. Appl. Biol.* 142: 41–48.

Mbongolo Mbella, E.G., A. Lievens, E. Barbau-Piednoir, M. Sneyers, A. Leunda-Casi, N. Roosens, and M. Van den Bulcke. 2011. SYBR Green qPCR methods for detection of endogenous reference genes in commodity crops: A step ahead in combinatory screening of genetically modified crops in food and feed products. *Eur. Food Res. Technol.* 232(3): 485–96, doi: 10.1007/s00217-010-1408-2.

McGinnis, E.E., M.H. Meyer, and A.G. Smith. 2010. Sweet and sour: A scientific and legal look at herbicide-tolerant sugar beet. *The Plant Cell* 22: 1653–1657.

McMaugh, S.J. and B.R. Lyon. 2003. Real-time quantitative RT-PCR assay of gene expression in plant roots during fungal pathogenesis. *BioTechniques* 34: 982–986.

Miraglia, M., K.G. Berdal, C. Brera, P. Corbisier, A. Holst-Jensen, E. Kok, H. Marvin, H. Schimmel, J. Rentsch, J. Van Rie and J. Zagon. 2004. Detection and traceability of genetically modified organisms in the food production chain. *Food Chem. Toxicol.* 42: 1157–1180.

Mockler, T.C., S. Chan, A. Sundaresan, H. Chen, S.E. Jacobsen, and J.R. Ecker. 2005. Applications of DNA tiling arrays for whole genome analysis. *Genomics* 85 (1): 1–15.

Morisset, D., D. Dobnik, S. Hamels, J. Zel, and K. Gruden. 2008. NAIMA: Target amplification strategy allowing quantitative on-chip detection of GMOs. *Nucleic Acids Res.* 36: E118. doi: 10.1093/nar/gkn524.

Mutasa-Gottgens, E., A. Qi, A. Mathews, S. Thomas, A. Phillips, and P. Hedden. 2009. Modification of gibberellin signaling (metabolism and signal transduction) in sugar beet: Analysis of potential targets for crop improvement. *Transgenic Res.* 18: 301–308.

North American Millers' Association Press Release. 2008. Available at http://www.namamillers.org/PR_StarLink_04_28_08.html.

O'Callaghan, M., T.R. Glare, E.P.J. Burgess, and L.A. Malone. 2005. Effects of plants genetically modified for insect resistance on nontarget organisms. *Ann. Rev. Entomol.* 50: 271–292.

Oh, K.C., K. Hardeman, M.G. Ivanchenko, M.E. Ivey, A. Nebenführ, T.J. White, and T.L. Lomax. 2002. Fine mapping in tomato using microsynteny with the *Arabidopsis* genome: The *Diageotropica* (*Dgt*) locus. *Genome Biol.* 3: research 0049.1-0049.11.

Organisation for Economic Cooperation and Development (OECD). 1993. *Safety Evaluation of Foods Derived by Modern Biotechnology: Concepts and Principles*. Paris: Organisation for Economic Cooperation and Development.

Ouyang, B., T. Yang, H. Li, L. Zhang, Y. Zhang, J. Zhang, Z. Fei, and Z.J. Ye. 2007. Identification of early salt stress response genes in tomato root by suppression subtractive hybridization and microarray analysis. *Exp. Bot.* 58 (3): 507–520.

Owen, M.D.K. 2008. Weed species shifts in glyphosate-resistant crops. *Pest Manage. Sci.* 64: 377–387.

Pan, A., L.Yang, S. Xu, C. Yin, K. Zhang, Z. Wang, and D. Zhang. 2006. Event-specific qualitative and quantitative PCR detection of MON863 maize based upon the 30-transgene integration sequence *J. Cereal Sci.* 43: 250–257.

Panda, S., T.K. Sato, G.M. Hampton, and J.B. Hogenesch. 2003. An array of insights: Application of DNA chip technology in the study of cell biology. *Trends Cell Biol.* 3: 151–156.

Parrott W., B. Chassy, J. Ligon, L. Meyer, J. Petrick, J. Zhou, R. Herman, B. Delaney, and M. Levine. 2010. Application of food and feed safety assessment principles to evaluate transgenic approaches to gene modulation in crops. *Food Chem. Toxicol.* 48: 1773–1790.

Petersen, J., S. Koch, and K. Hurle. 2002. Weiterentwicklung von Mais- und Zuckerübenmulchsaatsystemen mit Hilfe der herbizidresistenten Sorten. Z. *Pflkrankh. Pflschutz*, Sonderheft XVIII, pp. 561–571.

Phipps, R.H. and J.R. Park. 2002. Environmental benefits of genetically modified crops: Global and European perspectives on their ability to reduce pesticide use. *J. Anim. Feed Sci.* 11: 1–18.

Pidgeon, J.D., M.J. May, J.N. Perry et al. 2007. Mitigation of indirect environmental effects of GM crops. *Proc. R. Soc. B* 274: 1475–1479.

Pollan, M. 2003. The Way We Live Now. The (Agri)Cultural Contradictions Of Obesity, *The New York Times, 12 October 2003*. http://www.nytimes.com/2003/10/12/magazine/the-way-we-live-now-10-12-03-the-agri-cultural-contradictions-of-obesity.html?pagewanted=all&src=pm.

Qaim, M. 2010. Benefits of genetically modified crops for the poor: Household income, nutrition, and health. *New Biotechnol.* 27: 552–557.

Querci, M., M. Van den Bulcke, J. Zel, G. Van den Eede, and H. Broll. 2010. New approaches in GMO detection. *Anal. Bioanal. Chem.* 396: 1991–2002.

Raven, P.H. 2010. Does the use of transgenic plants diminish or promote biodiversity? *N. Biotechnol.* 27: 528–533.

Rotthues, A., J. Kappler, A. Lichtfuss, D.U. Kloos, D.J. Stahl, and R. Hehl. 2008. Post-harvest regulated gene expression and splicing efficiency in storage roots of sugar beet (*Beta vulgaris* L.). *Planta* 227: 1321–1332.

Scandizzo, P. and S. Savastano. 2010. The adoption and diffusion of GM crops in the United States: A real option approach. *AgBioForum* 13 (2): 142–157.

Schreiber, G.A. 1997. The European Commission Research Project: Development of methods to identify foods produced by means of genetic engineering. In: *Food Produced by Means of Genetic Engineering*, Second status report, edited by G.A. Schreiber and K.W. Bogl. Berlin: Bundesinstitut fur gesundheitlichen Verbraucherschutz und Veterinarmedizin, BgVV. 01/1997.

Sears, M.K., R.L. Hellmich, D.E. Stanley-Horn, K.S. Oberhauser, J.M. Pleasants, H.R. Mattila, B.D. Siegfried, and G.P. Dively. 2001. Impact of Bt corn pollen on monarch butterfly populations: A risk assessment. *Proc. Natl. Acad. Sci. U.S.A.* 98: 11937–11942.

Smith, A.F. 2006. *Encyclopedia of Junk Food and Fast Food*, p. 258. Greenwood Publishing Group, Science Publishers, USA.

Sobolev, A.P., D. Capitani, D. Giannino, C. Nicolodi, G. Testone, F. Sanoro, G. et al. 2010. NMR-metabolic methodology in the study of GM foods. *Nutrients* 2: 1–15.

Sparrow, P.A.C. 2010. GM risk assessment. *Mol. Biotechnol.* 44: 267–275.

Stein, A.J. et al. 2006. Potential impact and cost-effectiveness of Golden Rice. *Nat. Biotechnol.* 24: 1200–1201.

Suliman-Pollatschek, S., K. Kashkush, H. Shats, J. Hillel, and U. Lavi. 2002. Generation and mapping of AFLP, SSRs and SNPs in *Lycopersicon esculentum*. *Cell Mol. Biol. Lett.* 7 (2A): 583–597.

Tang, G.W. et al. 2009. Golden Rice is an effective source of vitamin A. *Am. J. Clin. Nutr.* 89: 1776–1783.

Unlu, M. 1999. Difference gel electrophoresis. *Biochem. Soc. Trans.* 27: 547–549.

Varzakas, T., D. Argyropoulos, and I. Arvanitoyannis. 2008. Chapter 26: Gene transfer in tomato and detection of transgenic tomato products. In: *Tomatoes and Tomato Products, Nutritional, Medicinal and Therapeutic Properties*, edited by V.R. Preedy and R.R. Watson, pp. 515–536. Enfield, NH: Science Publishers, USA.

Varzakas, T., I.S. Arvanitoyannis, and H. Baltas. 2007. The politics and science behind GMO acceptance. *Crit. Rev. Food Sci. Nutr.* 47 (04): 335–361.

Varzakas, T., G. Chryssochoidis, and D. Argyropoulos. 2006. Approaches in the risk assessment of genetically modified foods by the Hellenic Food Safety Authority. *Food Chem. Toxicol.* 45 (4): 530–542.

Velculescu, V.E., L. Zhang, B. Vogelstein, and K.W. Kinzler. 1995. Serial analysis of gene expression. *Science* 270: 484–487.

Vos, P., R. Hogers, and M. Bleeker. 1995. AFLP: A new technique for DNA fingerprinting. *Nucleic Acids Res.* 23 (21): 4407–4414.

Wallinga, D., J. Sorensen, P. Mottl, and B. Yablon. 2009. *Not So Sweet: Missing Mercury and High Fructose Corn Syrup*. Institute for Agriculture and Trade Policy. Available at http://www.globe-expert.eu/quixplorer/filestorage/Interfocus/5-Climat_Environnement/58-Agriculture/58-SRCNL-IATP/200901/Jan._26_2009_Not_so_Sweet_Missing_Mercury_and_High_Fructose_Corn_Syrup_report_by_By_IATPDavid_Wallinga_M.D._Janelle_Sorensen_Pooja_Mottl_Brian_Yablon_M.D.pdf. Retrieved 2011-09-11.

Weihong, L. and D.A. Saint. 2002. Validation of a quantitative method for real-time PCR kinetics. *Biochem. Biophys. Res. Commun.* 294: 347–353.

Welsh, J., K. Chada, S.S. Dalal, D. Ralph, L. Cheng, and M. McClelland. 1992. Arbitrarily primed PCR fingerprinting of RNA. *Nucleic Acids Res.* 20: 4965–4970.

White, J.S. 2008. HFCS: How Sweet It Is? *Food Product Design*. Available at http://www.foodproductdesign.com/articles/2008/12/hfcs-how-sweet-it-is.aspx (accessed 06 September 2011).

White, J.S. 2011. Genetically modified corn. Available at http://www.sweetsurprise.com/myths-and-facts/top-hfcs-myths/gmo-corn-hfcs.

White, J.S. 2011. High Fructose Corn Syrup Health and Diet Facts. Available at http://www.sweetsurprise.com/ (accessed 06 November 2010).

World Health Organization (WHO). 2010. 20 Questions on Genetically Modified (GM) Foods. Available at http://www.who.int/foodsafety/publications/biotech/20questions/en/index.html (accessed 17 April 2010).

Zhu, G., Y. Chang, J. Zuo, X. Dong, M. Zhang, G. Hu, and F. Fang. 2001. Fuderine, a C-terminal truncated rat homologue of mouse prominin, is blood glucose regulated and can up-regulate the expression of GAPDH. *Biochem. Biophys. Res. Commun.* 281: 951–956.

Zimmermann, A., J. Luthy, and U. Pauli. 1998a. Quantitative and qualitative evaluation of nine different extraction methods for nucleic acids on soybean food samples. *Z. Lebensm.-Unters. Forsch. A* 207: 81–90.

Zimmermann, A., W. Hemmer, M. Liniger, J. Luthy, and U. Pauli. 1998b. A sensitive detection method for genetically modified MaisGardTM corn using a nested PCR-system. *LWT—Food Sci. Technol.* 31: 664–667.

CHAPTER 13

Bulking and Fat-Replacing Agents

Todor Vasiljevic and Theodoros Varzakas

CONTENTS

13.1 INTRODUCTION

Obesity-associated risk factors that contribute to the development of metabolic syndrome continue to present a challenge not only for the medical community but also for the food industry. The two dominant risk factors in this regard are central obesity and insulin resistance, which contribute to cardiovascular disease and type 2 diabetes. According to the Centers for Disease Control and Prevention (CDC), obesity is rapidly becoming the number one preventable cause of death in the United States (Mokdad et al. 2004). The alarming rise in obesity began in the 1970s and has been growing even more rapidly in recent years. In 2010, no state had a prevalence of obesity less than 20%. Thirty-six states had a prevalence of 25% or more (CDC 2010). A number of studies pointed to the increased consumption of calories from food as a primary cause (Wright et al. 2004). Weight-loss strategies to reduce these comorbidities include behavior modification and physical activity, as well as various forms of energy restriction with respect to dietary fat or carbohydrates. Therefore, low-calorie products are in demand and, consequently, so are the ingredients that make the production of these products possible. The use of low-calorie sugar-free products tripled in the final two decades of the twentieth century (Nabors 2002). A recent survey conducted by the Calorie Control Council (CCC, Atlanta, GA) found that 78% of adults reported consuming low-fat, reduced-fat, or fat-free foods and beverages as a method of weight loss (CCC 2011).

In general terms, low-calorie sweeteners can be categorized into two groups. The first group of sweeteners consists of substances with a very intense sweet taste and are used in small amounts to compensate for the sweetness loss due to the replacement of a much larger amount of sugar. One of the examples is the use of stevia/steviol glycosides as a sucrose replacer in the manufacture of chocolate (Shah et al. 2010). These sweeteners lack in bulking properties, with some of them either not globally approved due to the lack of the safety data or having a differing degree of aftertaste. Although these compounds are commonly applied in weight management strategies, an emerging body of evidence contradicts these intentions by suggesting that these substances provide little or no advantages in the weight control, and, in some instances, may even contribute to weight gain (Hampton 2008). The second group of sweeteners provides for not only sweetness but also physical bulk substation. This group includes the sugar alcohols (polyols) such as sorbitol, mannitol, xylitol, isomalt, erythritol, lactitol, maltitol, and hydrogenated starch hydrolysates, and hydrogenated glucose syrups, which are often termed as "sugar replacers" or "bulk sweeteners."

It has long been recognized that the ideal sweetener does not exist. The golden standard, sucrose, is not perfect either and has limited use in some applications. The industry has recognized these limitations and has been on a search for alternative sweeteners. Any sweetener should satisfy a number of criteria, including relative sweetness comparable to that of sucrose. Ideally, it should be odorless with a pleasant taste, with immediate onset and no lingering, colorless and noncariogenic, highly water soluble, and stable over a range of different processing and storage conditions (temperature, pH, and water activity). It must comply with rigorous safety criteria such as nontoxicity and normal metabolization and excretion. Therefore, it appears that the search for sweeteners alternative to sucrose must not only match the properties of this benchmark sugar but also comply to additional criteria recognized by the market and industry trends. It is known now that alternative sweeteners should provide specific or expand existing food and beverage choices by controlling the intake of a specific carbohydrate or energy density. As such, these novel ingredients contribute, thus, to the weight management, prevention, and treatment of the metabolic syndrome and assist in the prevention and control of dental caries. In addition to price and availability, the main factor in any application is the relative sweetness.

The sweetness of any sweetener is very subjective and affected by a number of intrinsic and extrinsic factors. The intrinsic factors include the chemical and physical composition and properties of the system in which the sweetener is dissolved. These include the pH and presence of taste enhancers or depressors and the concentration of the sweetener. An important extrinsic factor is temperature. The intensity of the sweetness of any substance is made on a weight basis relative to sucrose. The relative sweetness of some common alternative sweeteners is provided in Table 13.1. In many instances, the use of an individual sweetener may not be effective; thus, a multiple sweetener approach may be required. A variety of approved sweeteners is available, mainly due to the limitations of each individual compounds. Each sweetener, thus, can be used in the application where it is suited the best, or these sweeteners can be used in the mixtures. It has been recognized that the use of more than one sweetener provides a tool for overcoming mainly taste limitations. Very early, it was recognized that the sweetness intensity of many commercial sweeteners was reversely related to their concentration, that is, it decreased with increased concentration. However, by combining several sweeteners, this limitation may be overcome, mainly due to the compounding effect of relative sweetness resulting in the use of lower sweetener concentrations. The advantages of sweetener blends are many, and some of them include the formulation of products that closely resemble sucrose-sweetened products, flavor enhancement using a combination of different sweeteners, and cost effectiveness.

The first commercial sweetener blend was saccharin and cyclamate. The primary advantage of this blend was that saccharin (300 times the sweetness of sucrose) enhanced the sweetening power of cyclamate (30 times sweeter than sucrose). At the same time, cyclamate masked the lingering aftertaste associated with saccharin. Cyclamate was the major factor in launching the diet segment

Table 13.1 Relative Sweetness of Some Alternative Sweeteners (Compiled from Different Sources)

Sweetener	Sweetness Relative to Sucrose	Sweetener	Sweetness Relative to Sucrose
Lactitol	0.4	Fructose	1.5
Hydrogenated starch hydrolysates	0.4–0.9	Cyclamate (banned)	30
Trehalose	0.45	Aspartame	180
Isomalt	0.45–0.65	Saccharin	300
Isomaltulose	0.48	Stevioside	300
Sorbitol	0.6	Sucralose	600
Mannitol	0.7	Monellin	2000
Maltitol	0.9	Alitame	2000
Tagatose	0.9	Thaumatin	2500
Xylitol	1.0	Neotame	8000
High-fructose corn syrup	1.0		

of the carbonated beverage industry. By the time it was banned in the United States in 1970, the products and trademarks had been well established. Such a large market for diet beverages provided a tremendous incentive for the development of new sweeteners. After cyclamate was banned and removed from the market, saccharin was the only available low-energy alternative to sucrose. However, with more recent developments and the availability of a variety of sweeteners, including acesulfame potassium, aspartame, sucralose, and saccharin, the multiple-sweetener approach is an unavoidable reality in the food and beverage industry. Today, blends are frequently used. For example, in the United States, diet fountain soft drinks are generally sweetened with a combination of aspartame and saccharin, whereas in other parts of the world, soft drinks may contain as many as four sweeteners. Sugar-free gums and candies contain combinations such as saccharin/sorbitol and acesulfame potassium/isomalt.

Similarly, polyols are important adjuncts to sugar-free product development. These sweeteners, generally less sweet than sucrose (Table 13.1), provide the bulk of sugar. The main advantage of polyols is their compatibility with other low-energy sweeteners. This synergistic effect almost, in general, results in highly acceptable, good-tasting, low-energy products similar to their high–energy dense counterparts. In combination with a number of available carbohydrate-based fat replacers and low-energy bulking agents, such as polydextrose (PD) and some other soluble fibers, a novel, multiple ingredient approach in the development of novel foods can be applied to reduce the overall energy intake. In addition to the fact that consumers have an innate desire for sweet goods (Inglett 1970, 1974), the research has also indicated that the obese individuals and those who were once obese may have a greater preference for fatty liquids mixed with sugar than the others (nonobese persons; Beck 1974). Therefore, the fat and sugar replacements are important in the development of products that may aid in weight management and reduced energy intake.

Fat and oil consumption has been associated with a number of detrimental health effects, causing a shift in consumer preference toward low-fat products. However, although these concerns should be addressed, the approach needs to be more balanced, because the nutritional requirements with regard to fat intake should be met (Giese 1996). In fact, fats are important energy sources and, in some instances, conjoint building blocks in the human tissue, especially during early development. They provide approximately 37 kJ/g, whereas proteins and carbohydrates provide less than half of this amount. Some fatty acids such as linoleic and linolenic acids are regarded essential and play an important role in many metabolic pathways, aiding in the absorption of vital nutrients, regulation of smooth muscle contraction, regulation of blood pressure, and growth of healthy cells. On the other hand, most regulatory recommendations have stated that the consumption of high levels of fat is associated with obesity, certain cancers, and, possibly, gallbladder disease (Wilborn et al. 2005). Furthermore, strong evidence exists for a relationship between saturated fat intake, high blood cholesterol, and coronary disease. The majority of nutritionists, however, advocate balanced fat intake as a proportion of the daily energy requirements with concomitant limitation of levels of saturated fats in the diet. Current guidelines in many countries state that the total daily fat intake should be no more than 30% of the total energy, with saturated fat constituting less than 10% of calories. Intensive efforts from ingredient suppliers and product developers have reduced the use of ingredients rich in saturated fats such as lard, beef tallow, butterfat, coconut oil, and palm oil and increased the use of vegetable oils with higher percentages of monosaturated and polyunsaturated fats.

The majority of the consumed fat comes from salad and cooking oils, as well as frying fats and bakery shortenings. Furthermore, meat, poultry, fish, and dairy products (cheese, butter, and margarine) also contribute substantially to fat consumption. The fat plays a unique functional role in each food product; for example, in fried foods, oil serves as a heat conductor and, at the same time, is incorporated into the food. Due to this dual role, the oil must meet several important requirements: good thermal and oxidative stability, in addition to good flavor, good shelf life, and acceptable cost. Fats and oils provide important textural qualities to certain foods. The physical properties of food fats are influenced primarily by three factors: (1) polymorphism (structural, solidification, and transformation behavior); (2) the phase behavior of fat mixtures; and (3) the rheological and textural properties exhibited by fat crystal networks. In fat crystals, three main polymorphic forms, namely, α, β′, and β, are defined in accordance with subcell structure. Transformation from polymorphic form β′ to polymorph β frequently results in the deterioration of the end product, mainly due to changes in crystal morphology and network. Naturally occurring fats and lipids are mixtures of different types of triacylglycerol (TG). The complex behavior of fats with regard to melting, crystallization and transformation, and crystal morphology and aggregation is, in part, a result of the physical properties of the component TGs and, more importantly, a result of the phase behavior of the mixture. One of the most important macroscopic physical properties of food fats is rheology, which governs a number of important attributes, including the spreadability of spreads, the brittleness of chocolate, and the smoothness, mouthfeel, and stability of bulk fats and emulsion products (deMan 1999). The rheological properties of food fats are governed by numerous factors, which can be grouped as intrinsic, including the molecular compositions of fats (TGs, ingredients, and additives), the polymorphism of crystals of the constituent TGs, and the microstructure of fat crystals (morphology, crystal size distribution, and crystal network formation), and extrinsic, including processing conditions, such as temperature, shear, and flow velocity. Therefore, in order to replace fat, all these properties must be considered, which ultimately means that fat replacement carries many additional requirements than the fat itself. For example, additional ingredients are required to address mouthfeel, texture, aeration or structure, color, flavor, handling characteristics, and shelf stability. These are additionally accompanied by other factors such as cost, safety, and regulatory concerns.

The traditional approach to fat substitution was to replace it with either water or air, for example, using skim milk instead of whole milk in frozen dessert or alternative processing such as baking instead of frying. Fat replacers indicate food ingredients that can take the place of all or some of the fat in foods and yet give similar organoleptic properties to the foods (Akoh 1994, 1998). Depending on their chemical structure, fat replacers are classified into lipid-, protein-, and carbohydrate-based fat replacers. Another well-accepted classification is (1) fat substitutes and compounds, which are mainly lipid based, physically and chemically resemble TGs, and are often chemically or enzymatically synthesized, and (2) fat mimetics, constituting carbohydrate- and protein-based ingredients that imitate the organoleptic or physical properties of TGs. Most carbohydrate-based fat mimetics are hydrocolloids, which are mainly high-molecular-weight, hydrophilic biopolymers with a great water-holding capacity. Due to this property, their unique physical functionalities such as thickening, gelling, and emulsifying properties allow them to mimic the mouthfeel and flow properties that resemble those of fat in aqueous systems. The increased viscosity of dispersions containing carbohydrate-based fat mimetics allows for the incorporation of air bubbles and the structural stabilization of a food matrix (Lee et al. 2005), which may compensate for the volume and textural loss of reduced-fat foods. In addition, most fiber-based fat replacers in solutions exhibit shear-thinning flow behaviors that are known to provide a light and nonslimy mouthfeel to food products (Burkus and Temelli 2000). This chapter describes some carbohydrates, also considered dietary fibers (DFs), commonly used as either bulking agents or fat mimetics.

13.2 DIETARY FIBERS

The substantial scientific importance of DFs was realized over 20 years ago and has been growing steadily. Foods rich in DFs such as nuts, whole-grain flour, fruits, and vegetables have been associated with decreased blood cholesterol, lower insulin release, increased stool bulk, improved laxation, and weight management. Epidemiological studies have correlated high consumption of DFs with lower incidence of certain diseases such as cardiovascular disease and cancer of the colon and rectum (Park et al. 2005; Mann and Cummings 2009). Hipsley (1953) was the first to define DF as a nondigestible constituent of the plant cell wall. The most frequent definition has come from Trowell et al. (1985), who stated that, "Dietary fibres consist of remnants of plant cells resistant to hydrolysis (digestion) by the alimentary enzymes of man," whose components are hemicellulose, cellulose, lignin, oligosaccharides, pectins, gums, and waxes. The American Association of Cereal Chemists (AACC 2001) defined DF as the edible parts of a plant or analogous carbohydrates that are resistant to digestion and absorption in the human small intestine with complete or partial fermentation in the large intestine. A panel on the definition of DF (Institute of Medicine 2002) defined the DF complex to include DFs consisting of nondigestible carbohydrates and lignin, which are intrinsic and intact in plants, functional fibers consisting of isolated, nondigestible carbohydrates, which have beneficial physiological effects in humans, and total fiber as the sum of DF and functional fiber (Rodriguez et al. 2006).

DF is usually divided into four main categories: (1) total nonstarch polysaccharides (NSPs), which can further be categorized into insoluble and soluble NSPs; (2) resistant starch; (3) fructooligosaccharides (FOS); and (4) inulin and lignin. The energy released by fermentation varies with the source, with the most frequently cited value being 8.2 kJ/g (Wisker and Feldheim 1990, 1992). These carbohydrates can be classified into a number of ways: based on the structure (linear or branched), based on the source (plant, animal, or synthetic), and solubility (soluble or insoluble; BeMiller 2001). The chemical nature of DFs is complex, because they constitute a mixture of chemical entities, as shown by their classification into nondigestible carbohydrates, lignin, and other associated substances of plant origin, fibers of animal origin, and modified or synthetic nondigestible carbohydrate polymers (Table 13.2). Nondigestible carbohydrates constitute the following: polysaccharides—cellulose, β-glucan, hemicelluloses, gums, mucilage, pectin, inulin, and

Table 13.2 Classification of Dietary Fibers and Some Main Representatives

Class of Dietary Fiber	Representatives
Nonstarch polysaccharides and resistant oligosaccharides	Cellulose, hemicelluloses, arabinoxylans, arabinogalactans, polyfructose, inulin, oligofructans, galactooligosaccharides, mucilages, and pectins
Carbohydrate analogues	Indigestible dextrins, resistant maltodextrins, synthesizes carbohydrate compounds, polydextrose, methyl cellulose, and resistant starch
Lignin complexes	Waxes, phytate, cutin, saponins, suberin, and tannins

Source: DeVries, J. W., *Proc. Nutr. Soc.* 62, 37–43, 2004. With permission.

resistant starch; oligosaccharides—FOS, oligofructose (OF), PD, and galacto-oligosaccharides; and soybean oligosaccharides—raffinose and stachyose.

Chitosan is an example of fiber of animal origin, derived from the chitin contained in the exoskeletons of crustaceans and squid pens, and structurally similar to cellulose (Borderías et al. 2005). Cereals are the principal source of cellulose, lignin, and hemicelluloses, whereas fruits and vegetables are the primary sources of pectin, gums, and mucilage (Slavin et al. 1997; Slavin 2004). Each polysaccharide is characterized by its sugar residues and by the nature of the bond between them. The complexity of fibers—their chemical nature, the degree of polymerization (DP), and the presence of oligosaccharide and polysaccharide—governs their functionality in food systems.

DFs are classified as soluble or insoluble based on their ability to create stable colloidal dispersion when mixed with water (soluble) or not (insoluble). Soluble DFs include pectic substances, gums, mucilage, and some hemicelluloses, whereas cellulose, other types of hemicelluloses, and lignin are included in the insoluble fraction (BeMiller 2001). Solubility is related to the structure of polysaccharides; they can be set regularly (insoluble) or irregularly (soluble) on the backbone or as side chains. The presence of a substitution group such as COOH or SO_4^{2-} increases solubility, which is also dependent on the temperature and ionic strength of the system. The solubility of DF is one of the detriments of their physical and physiological functionalities. Soluble fibers are characterized by their capacity to increase viscosity in the gut and, thus, reduce the glycemic response and plasma cholesterol (Abdul-Hamid and Luan 2000). Insoluble fibers are, on the other hand, characterized by porosity, low density, and ability to increase fecal bulk and decrease intestinal transit (Lunn and Buttriss 2007). In food processing, the soluble fiber has a greater capacity to enhance viscosity, form gels, act as a surface agent, has no detrimental effect on the organoleptic properties, and, overall, is easier to handle and incorporate into foods and beverages. Tables 13.3 and 13.4 depict common DFs based on their solubility and functionality.

Table 13.3 Solubility-Based Classification of Dietary Fibers

Property	Representative
Insoluble	Cellulose
Hot-water soluble	Agars, aligns (+Ca^{2+}), amylose, κ-carrageenan (+K^+ or Ca^{2+}), gellan, konjac mannan, locust bean gum, low methoxyl-pectins (+Ca^{2+}), and granular starch
Soluble in water at room temperature but not at high temperature	Curldan, hydroxypropylcelluloses, hydroxylpropyl-methylcelluloses, and methylcellulose
Soluble	Alginate, amylopectin, carboxymethylcellulose, carrageenans, dextrins, furcellarans, guar gum, gum arabic, gum tragacanth, high methoxyl pectin, polydextrose, and xanthan

Source: BeMiller, J. N., *Carbohydrate Chemistry for Food Scientists*, AACC International, St. Paul, MN, 2007. With permission.

Table 13.4 Classification of Dietary Fibers Based on Their Physical Functionality

Property	Representative
Gelling	Agars, alginate, carrageenan, curdlan, gellan, gum arabic, konjac mannan, methylcellulose, pectins, starches, locus bean gum with κ-carrageenan, and xanthan
Thickening and stabilizing	Carboxymethylcellulose, gum arabic, gum tragacanth, hydroxypropylmethylcellulose, modified starch, and xanthan
Thickening primarily	Carboxymethylcellulose and guar gum
Fat mimetics	β-Glucan, cellulose, inulin, polydextrose, psyllium seed gums, and resistant starch

Source: With kind permission from Springer Science+Business Media: *Food Analysis*, edited by S. S. Nielsen, 2010, BeMiller, J. N.

13.3 RESISTANT STARCH

Starch is the major source of carbohydrates in the human diet (Ratnayake and Jackson 2008). It occurs in many plant tissues as granules, usually between 1 and 100 μm in diameter, depending on the botanical source. Chemically, starches are polysaccharides composed of α-D-glucopyranosyl units linked together with α-D-(1–4) and/or α-D-(1–6) linkages. Native starch granules have a crystallinity varying from 15% to 45% (Aprianita et al. 2010); thus, most native starch granules exhibit a Maltese cross when observed under polarized light. It is clear from the level of starch crystallinity that most starch polymers in the granule are in an amorphous state (Aprianita et al. 2010). The two major macromolecular components of starch are amylose and amylopectin (Figure 13.1). Amylose is the predominantly linear α-(1–4)-linked α-glucan with a DP as high as 600. Amylopectin is the major component of the granule (30%–99%) and is α-(1–4)-linked α-glucan with α-(1–6) branch points. Amylopectin contains about 5% of branch points, which impart profound differences in its physical and biological properties compared to amylose. Amylopectin possesses several populations of polymer chains that can be classified into short chains ($12 < DP < 20$), long chains ($30 < DP < 45$), and very long chains having an average $DP > 60$. The chains are further classified into A-, B-, and C-chains, where A-chains do not carry any other chains, B-chains carry one or more chains, and the C-chain is the original chain carrying the sole reducing end (Sharma and Yadav 2008; Haralampu 2000).

Two crystalline structures of starch have been identified (an "A" and a "B" type), which contain differing proportions of amylopectin. A-type starches are present in cereals, whereas B-type starches are found in tubers and amylose-rich starches. The third type, called "C-type," is a mixture of both A and B forms and is mainly present in legumes. In general, digestible starches are hydrolyzed by an enzymatic cascade in the small intestine to yield free glucose, which is subsequently absorbed (Nugent 2005). However, not all starch in the diet is digested and absorbed in the small intestine (Ratnayake and Jackson 2008). This recognition that some starch is neither completely digested nor absorbed in the small intestine has brought scientific and commercial interest in nondigestible starch fractions and their potential to perform functions similar to DF in the large intestine (Fuentes-Zaragoza et al. 2010). Resistant starch is not digested by pancreatic amylases

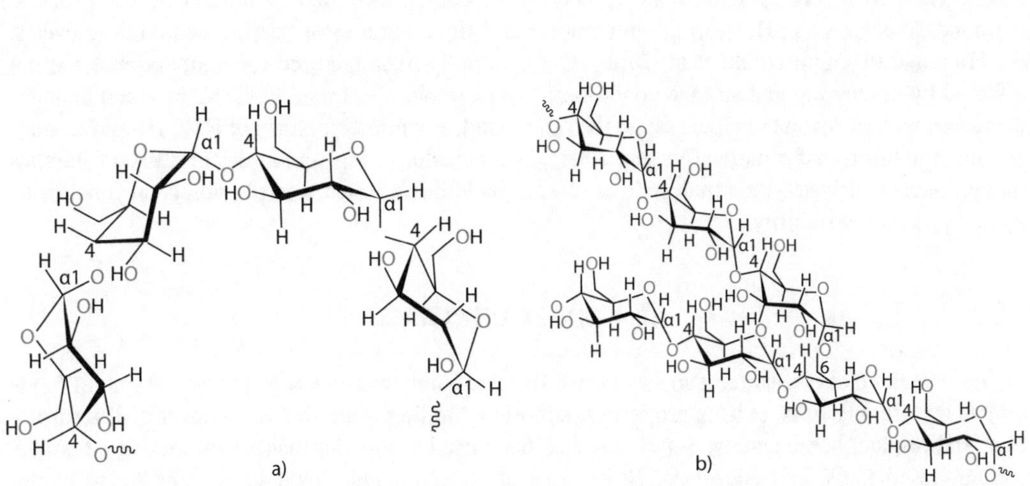

Figure 13.1 Schematic diagram of (a) amylose and (b) amylopectin. The Merck Index, Eleventh edition 1989; S. Budavari, Merck Co. Rahway, NJ.

in the small intestine but reaches the colon, where it provides additional benefits, including the growth of probiotics (Vasiljevic and Shah 2008). The fermentation of resistant carbohydrates by anaerobic bacteria results in the release of short-chain fatty acids (SCFAs), including acetic, propionic, and butyric acids, which are the respiratory fuel of cells lining the colon (Cummings and Englyst 1995).

Resistant starch has been classified into four general types, termed R1–R4, mainly based on the level of digestibility and physical properties: (1) RS1—physically inaccessible starch, which is entrapped within whole or partly milled grains or seeds; (2) RS2—some types of raw starch granules and high-amylose starches; (3) RS3—retrograded starch as a result of a food process; and (4) RS4—chemically modified starches to improve their resistance to enzymatic digestion (Baixauli et al. 2008; Sanz et al. 2009). The RS1 classification is assigned to resistant starch when it is physically inaccessible to digestion (Hernández et al. 2008). This type of resistant starch is heat stable under a variety of conventional cooking operations, which has broadened their application (Sajilata et al. 2006). The structural and conformational properties of RS2 starches prevent digestion (Hernández et al. 2008). In the diet, this starch is consumed in foods such as banana (Sajilata et al. 2006). The RS3 form is generally formed during the retrogradation of starch granules (Wepner et al. 1999) by heating starchy foods and then keeping them at low or room temperature (Hernández et al. 2008). This particular type is heat stable, which expands its applicability in a wide variety of conventional foods (Haralampu 2000). The application of moist heat during food processing may degrade the RS1 and RS2 types but may actually produce the RS3 form (Faraj et al. 2004). In addition to the three main types of RS, chemically modified starch has been assigned as RS4 type, similar to resistant oligosaccharides (ROs) and PDs (Wepner et al. 1999). These starches have been chemically modified by conversion, substitution, or cross linking in such a manner as to decrease their digestibility.

Resistant starches have desirable physical properties, including swelling, thickening, gelling, and water-binding capacity, which make them a useful ingredient in a variety of foods. These starches are characterized by a small particle size, white appearance, and bland flavor, all of which contribute to improved handling in processing and appropriate texture in the final product (Sajilata et al. 2006). For these properties, most resistant starches can replace flour on a one-for-one basis without detrimentally affecting dough rheology. As mentioned above, the processing conditions can affect the content of the resistant starch by influencing its gelatinization and retrogradation (Thompson 2000). Theoretically, it is achievable to increase its content in foods by modifying the processing conditions such as pH, heating temperature and time, number of heating and cooling cycles, freezing, and drying (Sajilata et al. 2006). Commercially manufactured resistant starches are not affected by processing and storage conditions. For example, the amount of RS2 in green bananas decreases with the extent of ripeness; on the other hand, a commercial form of RS2, Hi-maize, does not undergo this transformation (Nugent 2005). The inclusion of high quantities of resistant starches into products is driven with a multitude of effects, including physical and physiological functionality and processing stability.

13.4 MALTODEXTRINS

As described above, starch consists of two fractions: amylose and amylopectin. Starch in foods affects texture, viscosity, gelling properties, adhesion, binding, water-holding capacity, film formation, and product homogeneity. Starch has also been used in the pharmaceutical industry, textiles, alcohol-based fuels, and adhesives. Novel uses of starch include low-calorie substitutes, biodegradable packaging materials, thin films, and thermoplastic materials with improved thermal and mechanical properties (Biliaderis 1998). Starch is frequently modified to overcome shortcomings of native starches and, thus, diversify the utilization of starch in industrial applications. Starch modification decelerates retrogradation, improves gelling properties and prevents syneresis, and improves paste clarity and sheen, paste and gel texture, film formation, and adhesion (BeMiller 1997). Over the last few decades, starch has been modified by various methods to achieve functionalities suitable for various industrial applications. Basically, there are four broad-based kinds of modifications: chemical, physical, enzymatical, and genetical.

Maltodextrins depict the products of starch hydrolysis with dextrose-equivalent (DE) values of less than 20. The hydrolysis of starch is brought about by the application of heat and acid or a specific enzymatic treatment (Marchal et al. 1999). These treatments result in a variety of depolymerized oligomers, and the resulting mixture mainly consist of glucose, maltose, and a number of oligosaccharides and polysaccharides. The extent of hydrolysis is relative to glucose and is expressed as DE, with 100 depicting the complete hydrolysis to glucose monomers. The acid hydrolysis involves the treatment of starch with a strong acid at high temperatures, where the extent of hydrolysis and, thus, DE value is achieved by temperature and time control (Marchal et al. 1999). Enzymatic hydrolysis is mainly performed by α-amylase (1,4-α-D-glucan glucanohydrolase; EC 3.2.1.1) extracted from *Bacillus* sp. and pullulanase (pullalan 6-gluconhydrolase; EC 3.2.1.41). The α-amylases are distinguished by two features: the formation of products that have the α-configuration at the reducing end anomeric carbon of the newly formed products and an endomechanism of cleaving off glucose residues in the interior parts of the starch chain. As a result, maltodextrins produced by α-amylases are the results of the extensive hydrolysis of amylose and only partially of amylopectin. Their temperature optimum is in the range between 60°C and 90°C, and the pH optimum is approximately 6–7. Pullulanases are capable of cleaving starch at branch points, specifically α-(1,6) linkages; they have a lower pH (5) and temperature (60°C) optimum. Some endo α-amylases are capable of releasing small linear oligosaccharides such as maltotriose, maltotetraose, and maltohexaose during starch hydrolysis (Marchal et al. 1999).

Depending on the type of hydrolysis and the origin, the maltodextrins produced contain linear and branched oligomers and polymers. While DE is one of the important properties related to the number of glucose units, maltodextrins with the same DE value may have very different characteristics. The origin of starch (i.e., wheat, maize, oats, and cassava) is also an important factor governing the properties of maltodextrins. This is mainly due to varying ratios between amylose and amylopectin, which depend on the starch source (Aprianita et al. 2010). Considering this complexity, these molecules in the sol state are hydrated and expanded, with the extended helical regions interrupted by short disordered regions (Marchal et al. 1999). At concentrations above the critical, these molecules aggregate, forming crystalline domains—a characteristic of thermally reversible gels. Gels are characterized by low elasticity, small mechanical stability, high rigidity, and turbidity. Furthermore, due to varying DE values, maltodextrins have varying physicochemical properties; for example, hygroscopicity, solubility, osmolality, and ability to reduce water activity and depress the freezing point are directly correlated with the DE value, and as it increases, these properties will increase as well.

With DE values below 20, maltodextrins are highly digestible, easily blended with other ingredients, with fast dissolution and low viscosity in solution (Roller 1996). Low-DE maltodextrin gels have low energy density of approximately 4.2 kJ/g. Due to their ability to form weak (soft) thermoreversible gels that melt upon heating and reset upon cooling, their physical behavior is comparable to that of fat. Therefore, maltodextrins are capable of producing a fat-like mouthfeel that can be applied in a variety of products as fat mimetics (Roller 1996). Due to structural organization and particle size, maltodextrins are positioned in layers resembling the shape and size of fat crystals. In addition, they may act as bulking agents but retrograde as the native starches. For this reason, many stable starches used as fat mimetics are chemically modified to be resistant to temperature, shear, and low pH levels encountered in processing (Fuentes-Zaragoza et al. 2010). However, as noted above, some starches and low-DE maltodextrin derivatives would behave differently, depending on the source; for example, potato starch retrogrades at a slower rate than other starches due to presumably longer amylose molecules, reducing the tendency for turbidity and structural deformation (Roller 1996). This starch has additional properties that include the creation of thermoreversible gels that are pH stable (pH 3–7) with a plastic and shortening-like texture, fairly resistant to shearing and heating (Fuentes-Zaragoza et al. 2010). As such, maltodextrins can be used in a variety of products. The resulting texture of potato maltodextrins suits them ideally with applications in baking goods such as cakes, muffins, and cookies, where they increase dough viscosity and improve aeration. Furthermore, low-DE maltodextrins enhance the emulsification capacity, thus stabilizing water-in-oil emulsion, which makes them suitable in such applications as low-fat cheeses, spreads, soups, and salad dressings (Marchal et al. 1999).

13.5 POLYDEXTROSE

Market trends for low-energy dense foods have led to the discovery and development of several high-intensity sweeteners, as depicted in Table 13.1. By the rule of thumb, these products can replace the sweetness of sucrose; however, due to their sweetness intensity, their use is restricted to low concentrations, which ultimately means that the bulk of sugar requires additional replacement. Therefore, the need for bulking agents with physical properties equal to sucrose but substantially reduced energy content arose. Early bulking agents were usually insoluble fibers such as cellulose powders or high-viscosity food hydrocolloids, which often detrimentally affected the sensory perception by promoting a gritty texture or a gummy mouthfeel. By design, PD was developed to be functionally superior to these products. It is highly soluble in water, with low solution viscosity, very low taste intensity, and low digestibility, with almost absent or minimal adverse gastrointestinal effects at high levels of use (Auerbach et al. 2007). Therefore, PD is an ingredient designed to be the ultimate companion ingredient to high-intensity sweeteners. The key characteristic of PD, with regard to its bulking performance, is its energy value of 4.1 kJ/g. This attribute, in combination with the excellent solubility in aqueous systems, makes it almost irreplaceable in the development and processing of low-energy foods. As a sucrose or fat replacer, PD contributes only 25% or 11% of the energy content of the sucrose or fat, respectively (Shah et al. 2010).

The invention of PD was not accidental, but the result of a targeted research by scientists at Pfizer in 1970, in an attempt to develop a low-energy bulking agent. The objective was to develop an ingredient that could be used in conjunction with intense sweeteners when replacing fully digestible and, thus, energy dense carbohydrates in processed foods, and its production has been patented. PD is a highly branched, low-molecular-weight, randomly bonded polysaccharide composed of glucose units. An average DP is approximately 12 glucose units (Auerbach et al. 2007). As described in the patent, it is prepared commercially by the vacuum bulk polycondensation of a molten mixture of food-grade ingredients in an appropriate proportion. These ingredients are glucose, sorbitol as a plasticizer, and either citric or phosphoric acid as a catalyst, mixed in approximately 89:10:1 proportion. The resulting product is a mixture of different polymers with various types of glucosidic bonds in the structure (α-1,6 bonds predominate), which are weakly acidic and water soluble. The mixture usually contains minor amounts of sorbitol and citric or phosphoric acid. The theoretical chemical structure of PD is illustrated in Figure 13.1. The R group may be hydrogen, glucose, or a continuation of the PD polymer. As evidenced by this representative structure, PD is a very complex molecule, more highly branched than other carbohydrates, such as amylopectin, which contains mainly α-1,4 linkages, with about 4%–5% of α-1,6 linkages as branch points. The structural compactness and complexity of PD prevents digestive enzymes from readily hydrolyzing the molecule, resulting in reduced energy content (Figure 13.2).

PDs are often regarded as ROs, which escape the hydrolytic activity of human digestive enzymes and reach the lower intestine. It has been accepted in several countries that ROs are actually fermented in the large intestine. On the other hand, resistant polysaccharides (RPs) are also resistant

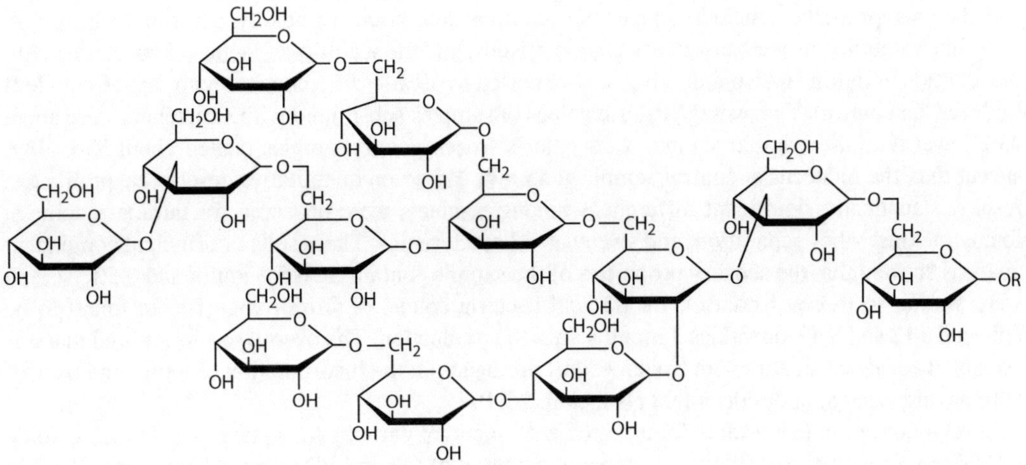

Figure 13.2 Chemical structure of polydextrose.

to hydrolysis but are only partially fermented in the colon, thus appearing in the feces. Considering that PD has an average DP of 12 and an average molecular weight of 2000, it should be classified as an RO; however, based on digestibility and fermentability, it may also be classified as an RP. PD is a white- to cream-colored powder with a clean taste, high water solubility, and good thickening capacity. Due to its involvement in the Maillard reaction, PD can be used in products where caramel and toffee flavors are required. Commercially available PDs are marketed under two brands: improved PD (Litesse) and superimproved PD (Litesse II) by Danisco. PD can act as a humectant, thus reducing the water activity of a product with simultaneous retention of moisture. This property is useful in applications where the control of water activity is important for the provision of extended shelf life. Due to its low sweetness, PD is preferred over sucrose in certain applications. Another colligative property of PD is its ability to depress the freezing point, thus aiding in achieving the correct consistency and mouthfeel of the finished product. Because it is not sweet, it is usually accompanied with a potent sweetener (Shah et al. 2010).

PD has been accepted as a food additive by the U.S. Food and Drug Administration (FDA) (Anon. 1981), covering application in eight food categories: baked goods and mixes; chewing gum; confections and frostings; dressings; frozen dairy desserts; gelatins, puddings, and fillings; hard candy; and soft candy. More recently, additional categories have been approved, including peanut spreads, syrups and toppings, sweet sauces, and fruit spreads. In these applications, PD may be used to address several functions: bulking—to provide a bulk to a food; formulation aid—to provide a desired physical property to a food; humectants—to retain moisture and regulate water activity; and, texturizer—to control the structure and mouthfeel of the food (Craig et al. 1998, 1999). PD has also been approved as a food additive internationally in 46 countries. The Joint FAO/WHO Expert Committee on Food Additives had reviewed the safety of PD, and they had no concerns about daily intake levels. In Japan, PD has also been approved under the Japanese Foods for Specified Health Use (FOSHU) law as an ingredient for which a claim stating that PD provides improved intestinal function may be made. Several other countries (Japan, Korea, Taiwan, PR China, Argentina, Egypt, and Poland) allowed the labeling of PD as a fiber.

Isik et al. (2011) produced a frozen yogurt containing low fat and no added sugar. Samples containing 5% PD, 0.065% aspartame and acesulfame potassium mixture, and different levels of inulin and isomalt (5.0%, 6.5%, and 8.0%) were produced at the pilot scale and analyzed for their physical and chemical properties, including proximate composition, viscosity, acidity, overrun, melting rate, heat shock stability, as well as sensory characteristics, and the viability of lactic acid bacteria. With the addition of inulin and isomalt, viscosity increased by 19%–52% compared with that of sample B (reduced-fat control). The average calorie values of samples substituted with sweeteners were about 43% lower than the original sample. Low-calorie frozen yogurt samples melted about 33%–48% slower than the reduced-fat control sample at 45 min. Based on quantitative descriptive profile test results, statistically significant differences among products were observed for hardness, iciness, foamy melting, whey separation, and sweetness characteristics. The results of principal component analysis showed that the sensory properties of the sample containing 6.5% inulin and 6.5% isomalt were similar to those of control. Lactic acid bacteria counts of frozen yogurt were found to be between 8.12 and 8.49 log values 3 months after the production. The overall results showed that it is possible to produce an attractive frozen yogurt product with the incorporation of inulin and isomalt with no added sugar and reduced fat (Isik et al. 2011).

Prebiotics are nondigestable food ingredients, made of carbohydrates targeting human colonic microflora. In the present study, the prebiotic potential of OF and PD in mixed sweetener (MS) in cake was investigated in healthy male volunteers. MS included PD (40.9%) and OF (20%). The aim of this study was to investigate the tolerable amount of MS and to evaluate the prebiotic effects of MS ingestion. This study was conducted in two steps. In the first step of the study, gastrointestinal system symptoms of the volunteers were examined during 4 weeks, and the tolerable amount of MS was detected. In the second step, the prebiotic effects of the tolerable dosage of MS (12-g/day) ingestion were investigated. At the end of the placebo and test periods, fecal samples were analyzed. Flatus was more frequent and intense in volunteers consuming MS48 than the other groups, and MS ingestion affected fecal weight in all groups. MS ingestion increased the amount of bifidobacteria, *Lactobacillus*, and total anaerobes (except for *Clostridium*) and decreased all aerobes. However, these changes were not statistically significant ($p > 0.05$). MS consumption decreased the amount of all aerobes, but only the reduction in the number of *Staphylococcus* was statistically significant compared to placebo period ($p < 0.01$). As a result, the 12-g/day consumption of MS generated prebiotic effects in the colon of healthy volunteers (Demircioğlu et al. 2008).

A study was made of various bulking agents as sucrose substitutes in the formulation of chocolate, aiming at obtaining a diet product in terms of sucrose and a light product in terms of calories (25% fewer calories than standard formulations containing sucrose) with good sensorial acceptance. The bulking agents used in this study were PD, inulin, FOS, lactitol, and maltitol. Sucralose was used as a high-intensity sweetener. The light chocolates were analyzed for moisture content (Karl Fischer), particle size (digital micrometer), and rheological properties (Casson plastic viscosity η_{ca} and yield strength τ_{ca}). The moisture content of the light chocolate varied from 1.23% to 2.12%, whereas the particle size varied from 19 to 24 μm, η_{ca} from 6.60 to 11.00 Pa·s, and τ_{ca} from 0.05 to 1.10 Pa. The formulations containing PD, PD and lactitol, and PD and maltitol were selected for a sensory analysis due to their good technological performance and adequate machinability of the chocolate mass in the different stages of the process. The sensory analysis revealed no statistically significant difference ($p > 0.05$) in the three evaluated formulations in terms of aroma, hardness, melting in the mouth, and flavor. There was no statistically significant difference ($p > 0.05$) in the intention to purchase the three chocolate formulations, although a preference was shown for the formulation containing PD (32.60%) and maltitol (15.57%; Gomes et al. 2007).

Fat mimetics, namely, Raftiline, Simplesse, C*deLight, and PD, diluted in water to give a gel with 200 g kg^{-1} of concentration, were used for partial fat replacement and polyols, namely, lactitol, sorbitol, and maltitol, for sugar replacement in low-fat, sugar-free cookies. Raftiline, Simplesse, or C*deLight combined with lactitol or sorbitol in 35% fat-reduced, sugar-free cookies resulted in products with hardness and brittleness comparable to those of the control. PD as a fat mimetic and maltitol as a sugar substitute resulted in very hard and brittle products. Further fat replacement to 50% was achieved using Raftiline, Simplesse, or C*deLight combined with a blend of lactitol and sorbitol; however, the final products were hard and brittle and did not expand properly after baking. Cookies prepared with Simplesse had the least acceptable flavor, whereas cookies prepared with C*deLight were rated as the most acceptable by a sensory panel. The textural properties were improved by either decreasing the amount of alternative sweetener or increasing the concentration of fat mimetic in the gel that was added to the cookies. All fat-reduced, sugar-free cookies prepared in this study had higher values of moisture content and water activity than the control, but these values were below the upper limit, which affects cookie shelf life (Zoulias et al. 2002).

13.6 FRUCTOOLIGOSACCHARIDES

Whole-grain finger millet and sorghum successively replaced commercial soft-type wheat flour in the formulation of multigrain cookies (MGCs) at 10%–30% levels each. MGCs were supplemented with FOS at levels of 40%, 60%, and 80% sugar replacement basis. The quality attributes of cookies were evaluated in terms of spread ratio, hardness, and nutritional characteristics. The spread ratio of control cookies (CCs) was 4.400 and that of MGCs with FOS ranged between 4.769 and 7.100. The initial hardness of CCs was 70.0 ± 1.6 N, and that of MGCs with FOS ranged from 69.7 ± 0.7 to 48.0 ± 1.2 N. MGCs with FOS were significantly ($p < 0.05$) less hard than CCs. Sensory data indicated moderate acceptability of MGCs with FOS at 60% sugar replacement level, 20% finger millet, and 30% sorghum. The total fiber including FOS (per 100 g) was estimated to be 17.4 and 1.3 g for MGCs with FOS and CCs, respectively. The caloric content of MGCs with FOS was 11.7% lower than the CCs. Acceptable cookies could be prepared with 50% whole-MG incorporation and up to 60% sugar replacement (Handa et al. 2011).

13.7 PHYSIOLOGICAL BENEFITS OF POLYDEXTROSE

Due to its low digestibility by pancreatic enzymes, PD acts like a soluble fiber. As opposed to insoluble fibers that contribute to fecal bulk and extended transit time, soluble fibers are highly fermentable in the large intestine and are associated with carbohydrate and lipid metabolism. The physiological effects of DFs are governed by their chemical and physical characteristics, including degradability, molecular weight, viscosity, particle size, organic acid absorption, and water-holding capacity (Eastwood and Morris 1992). Degradability enables the utilization of fiber by colonic bacteria during the fermentation in the large intestine, leading to subsequent pH decline, increase in the bacterial biomass, and the production of SCFAs (Vasiljevic and Shah 2008). Studies in animal and human models have shown that PD is partially fermented in the large intestine (Craig et al. 1999). Many DFs are metabolized by bacteria in the large intestine, generating hydrogen, methane, carbon dioxide, and SCFA. The SCFAs are rapidly absorbed from the gastrointestinal tract through the hepatic portal vein and contribute to the energy balance of the body. These contributions include the inhibition of hepatic cholesterol synthesis by propionate and the apoptosis of cancer cells by butyrate. It has been reported that a molar ratio of acetate/propionate/butyrate was 61:25:14 during the fermentation of PD, which was higher than many other DFs tested (Stowell 2009a,b). Selective fermentation by desirable colonic microflora results in a pH decline and overcrowding, which, on the other hand, suppresses the growth of harmful bacteria (Vasiljevic and Shah 2008). While this may have potential immunological effects, through growth suppression, the release of potentially harmful and carcinogenic compounds via bacterial metabolism can be reduced (Telang et al. 2005). PD may act as an insoluble fiber as well, for example, by increasing fecal volume, decreased transit time, and increased moisture of fecal samples (Oku et al. 1991). Reports on lowering plasma cholesterol levels in animals and humans by PD are contradictory. PD supplementation (10 g/day) significantly lowered the plasma total and low-density lipoprotein cholesterol after 18 days (Liu and

Tsai 1995), in contrast to another study with a greater supplementation but no effect observed on cholesterol levels over a 2-month period (Liu and Tsai 1995). However, the observed effects were contributed to a soluble fiber action similar to pectins or β-glucan through the production of SCFAs. These acids have a beneficial influence on the gut mucosa and may inhibit cholesterol synthesis in the liver. Other reports have shown that the hypoglycemic effect of certain fibers reduces elevated blood glucose levels (Zeng and Tan 2010).

13.8 β-GLUCAN

The health-promoting properties of β-glucan have been extensively reviewed (Murphy et al. 2010; Sirtori et al. 2009; Anderson et al. 2009). The major boost for β-glucan application in foods was the U.S. FDA's approval of the oat health claim (FDA 1997) for food products containing β-glucan from an oat source (bran, groats, or rolled oats), which guarantees ≥0.75 g of β-glucan per serving. This claim was approved for oat products rather than for individual components. Oats also contain very potent antioxidants such as avenanthramides and oat saponins, which, in minute amounts, have a hypocholesterolemic effect that equals that of β-glucan (Collins 1998). The oat health claim was filed by Quaker Oats (Chicago, IL). Besides the widely elaborated cholesterol-lowering and blood glucose–regulating effects, a number of studies examined the immunostimulating effects of β-glucan. Oat β-glucan showed the immunostimulating effect *in vitro* and *in vivo* in mice after intraperitoneal administration and increased the survival of mice infected with *Staphylococcus aureus* (Estrada et al. 1997). Causey et al. (1998) showed *in vitro* that even hydrolyzed barley β-glucan had an immunostimulating effect on human macrophages. When applied at a concentration of 100 μg/mL, β-glucan increased the production of white blood cells sixfold. In addition, the anticarcinogenic effect of barley bran (13% total dietary fiber [TDF]) on the incidence and development of certain tumor types was described by McIntosh et al. (1996).

The molecular structure of β-glucan has been determined to be 90% composed of cellotriosyl and cellotetraosyl regions connected with β-(1→3) bonds (Wood 1984; MacGregor and Fincher 1993), with the rest being longer cellulosic regions. The configuration of the molecule in three-dimensional space is not yet known with certainty. Some intrinsic viscosity measurements suggest that it is probably a partially stiffened worm-like cylinder (Gomez et al. 1997a,c). Straight (1 → 4) cellulosic regions are extremely rigid sequences resistant to mechanical and chemical action. In addition, they may align to form microcrystalline cellulosic regions with strong hydrogen bonds, which provide additional resistance to mechanical and chemical degradation. The β-(1 → 3) linkage is responsible for a kink in the β-glucan structure (Woodward et al. 1983). When present in a consecutive sequence, this bond creates a regularly shaped helical structure that can form aggregates and gel (Zhang et al. 1997).

Due to the length of their cellulosic regions, β-glucan molecules are able to associate with each other and create micelles (Varum et al. 1992; Grim et al. 1995) or gel (Burkus and Temelli 1999). β-Glucan from *Poria cocos* (a kind of mushroom used in traditional Chinese herbal medicine), which is entirely composed of β-(1→3)-d-glucan, was also able to create molecular aggregates that dissolved completely in cadoxen (saturated CdO solution in 29% ethylenediamine; Zhang et al. 1997). The mechanism of aggregation was probably the formation of hydrogen bonds between regularly shaped random coil regions of β-glucan chains. When heated above 60°C, micelles dissociate, and individual molecules require greater amounts of water for solvation, providing an explanation of the increased viscosity of β-glucan at low shear rates and elevated temperatures (70°C; Gomez et al. 1997b). Another important property of all polysaccharides to be considered is their compatibility with other biopolymers. Incompatibility is typical for biopolymers because of the large size of macromolecules and very low entropy that accompanies the mixing of biopolymers. Even when the corresponding monomer sugars are cosoluble in aqueous media in all proportions, polysaccharides assembled from these sugars are usually incompatible when their polymer chains differ in structure and/or composition. Incompatibility between barley β-glucan and milk proteins or starch was first described by Burkus (1996). Further experiments by Bansema (2000) attempted to establish the concentration–stability relationship between barley β-glucan and whey protein isolate (WPI). He found that a mixture was stable after 2 days of holding at refrigeration temperature below critical concentrations—β-glucan concentration was 0.25%, and that of WPI was ≤5%. A higher concentration of either ingredient resulted in phase separation. However, thermal treatment may cause instability, even at 0.25% β-glucan concentration. The incorporation of β-glucans into low-fat dairy products may improve their mouthfeel and sensory properties to resemble those of full-fat products. For example, β-glucan incorporation into low-fat cheese curds has beneficial effects on their gelation and rheological characteristics (Tudorica et al. 2004). The addition of β-glucan solutions to milk modifies curd formation, including reducing curd cutting time and increasing curd yields (Tudorica et al. 2004). However, when β-glucans are incorporated into a manufactured cheese system or low-fat yogurt, the texture can be altered deleteriously (Vasiljevic et al. 2007; Vithanage et al. 2008). It appears that the components known to be precipitated by β-glucan have a globular structure. An important parameter in the evaluation of incompatibility is the time factor, but it has not been studied. An increase in viscosity delays the sedimentation of a precipitated compound. Samples stored at refrigerator temperature have increased viscosity, and consequently, the suspension of the incompatible compound is enhanced. The decrease in protein solubility and solvation does not have to be complete. Observation of partial insolubility of protein may require times longer than those employed in quick laboratory experiments. The longer time frame associated with shelf life studies would provide sufficient time for proper incompatibility studies. However, care must be taken to exclude potential effects of all other factors, such as pH, temperature, and bacterial contamination of samples. Incompatibility with other food components may be a major obstacle for food and beverage applications of β-glucan, but this problem can be offset by the suspending ability of β-glucan, adjusting component concentrations and proper choice of ingredients (Figure 13.3).

Barley and oat β-glucans, with other NSPs, occur in the walls of endosperm cells, which encapsulate starch, matrix protein, and lipid reserves of the grain, creating a few obstacles for a successful extraction. The study of the physicochemical properties of isolated β-glucan fractions requires extraction procedures that optimize yield, purity, and retained integrity of the β-glucan molecule.

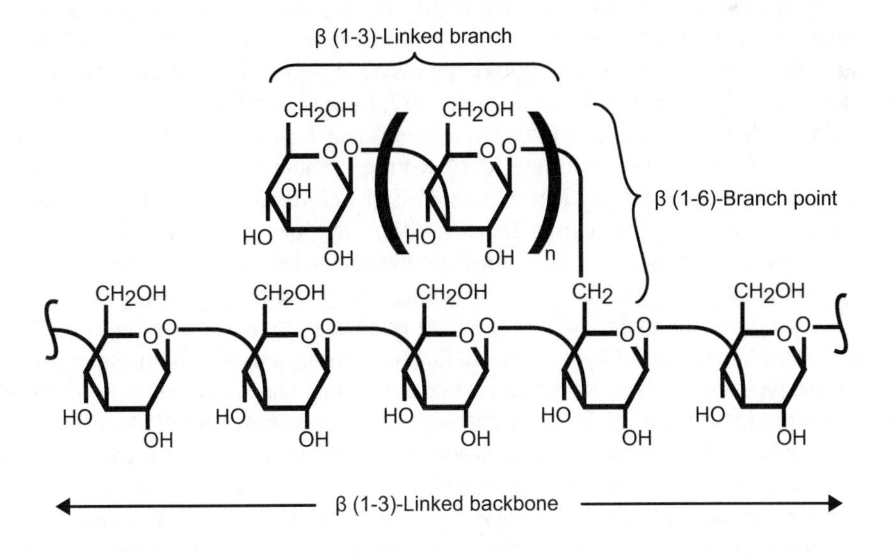

Figure 13.3 Structural organization of β-glucan.

At the same time, the process should be simple enough to ensure reproducibility. The cost effectiveness of the product, so that it can compete with other food hydrocolloids, is obviously a major concern. The extraction of β-glucans from cereal grains generally involves three basic steps: (1) inactivation of endogenous enzymes; (2) extraction; and (3) precipitation of the β-glucans. A high yield of β-glucan can be achieved with one-step aqueous alkali extraction (Wood et al. 1977, 1978). The yield and hydrolytic stability of β-glucan may also be influenced by a pretreatment of barley grains or flour, which is usually applied to stabilize the β-glucan extract by inactivating β-glucanase enzymes. Beer et al. (1996) and Burkus and Temelli (1998) found that the ethanol refluxing of flour lowered the yield of β-glucan, respectively, and did not deactivate β-glucanase. A high-purity product is always a goal of every extraction process. Further purification following extraction is always costly. The amount of impurities is primarily influenced by the extraction process parameters (Burkus 1996). Protein is considered to be the main impurity imparting a beige to brown color to β-glucan powders. Protein easily becomes incompatible with a hydrocolloid in the same solution and undergoes phase separation (Tolstoguzov 1991, 1997). When the gum is dissolved in water, proteins cause opalescence, which limits gum applications in certain products such as clear drinks. However, the lack of solution transparency may not be a problem in cloudy fruit beverages.

If β-glucan is to be competitive in the hydrocolloid market, it should be of the high-viscosity type. At least, that is the current thinking based on demonstrated health benefits, which are mostly linked to viscosity. Dawkins and Nnanna (1995) found that from the standpoint of viscosity, oat β-glucan is comparable to locust bean gum and better than guar gum whereas xanthan had much higher viscosity. Highly viscous β-glucan can be obtained from both oats and barley, although oats routinely outperforms barley (Wood 1991; Wood et al. 1991; Beer et al. 1997). The stability of β-glucan, in this case, is defined as constant viscosity over time, which is a desirable trait for a typical hydrocolloid. The main reason for the loss of viscosity is enzyme activity on the β-glucan chain, although viscosity loss due to agglomeration and precipitation was observed. The latter type of instability may be dealt with by reheating the β-glucan solution. Viscosity instability due to enzyme activity is a much harder problem for several reasons. Historically, barley has been selected for high enzymatic activity, and as such, it contains a great deal of β-glucanase enzymes to germinate the kernel quickly and transform it into malt or, actually, a new plant. These native enzymes work in the pH range of 5–7, with optimal temperatures of <40°C. Native β-glucanases are thermolabile, and a thermal process above 60°C degrades them relatively easily (Ballance and Meredith 1976).

Being a hydrocolloid, β-glucan may form highly viscous pseudoplastic solutions with a flow behavior index $n \ll 1$ if it is of the high-viscosity type (Autio et al. 1987; Autio 1996; Bhatty 1995; Burkus and Temelli 1998; Temelli et al. 2004), but even high-viscosity β-glucan at concentrations ≤0.25% behaves as a Newtonian fluid (Autio et al. 1987; Burkus and Temelli 1998). In the presence of other solutes, such as salt and sugar, β-glucan solutions may have increased viscosity (Autio et al. 1987; Dawkins and Nnanna 1995; Bansema 2000), but the increase is concentration dependent, and the order of hydration may affect the final solution viscosity (Burkus 1996). Gelation of low-viscosity β-glucan was described by Burkus and Temelli (1999) and Morgan and Ofman (1998). The firmness of the network and the rate of network formation were concentration and viscosity dependent.

Although extracted on a commercial scale, neither barley nor oat β-glucan gum is used as an ingredient in food products, mainly due to the high price of high-purity preparations (Wood and Beer 1998). Oatrim®, an extract of hydrolyzed oat flour containing 1%–10% β-glucan, is used as a fat replacer in some types of low-fat milk products (Pszczola 1996). Oatrim is used in several ConAgra Healthy Choice products: hot dogs, bologna, cheese, and 96% fat-free ground beef. Smaller companies are using the fat replacer in baked products such as muffins and cookies and in chocolate candy (Inglet and Grisamore 1991). Morgan and Ofman (1998) developed a hot water extraction procedure with the recovery of the β-glucan by freezing and thawing of the extract. The resulting product (Glucagel™) contained between 89% and 94% β-glucan, depending on the duration of the initial extraction, and is one β-glucan preparation commercially available as a food ingredient.

13.9 INULIN

Inulin is present in a wide range of plants, including common vegetables, fruits, and cereals. It is a polydisperse mixture of oligomers and polymers of β-(2-1)-fructose from the fructan family (Figure 13.4; De Leenheer and Hoebregs 1994). Native chicory inulin is extracted industrially from the root of the chicory plant (*Cichorium intybus*) by diffusion in hot water. In general, its structure can be represented by GF*n*, in which G stands for the glucosyl unit, F presents the fructosyl unit, and *n* is the number of linked fructosyl units linked by β-(2-1) linkages. The DP or number of fructose units of native chicory inulin varies between 3 and 60, with an average DP of 10. Inulin can be partially hydrolyzed by an endoinulinase, resulting in OFs, which the DP varies from 2 to 8 (average DP = 4). OF is composed of the same fructose monomers as inulin and is a mixture of both GF*n* (the inulin fraction containing the sucrose part) and F*n* molecules (Niness 1999).

The extensive use in the food industry is based on the nutritional and technological properties of inulin. The technological use of inulin is based on its properties as a sugar replacer (especially in combination with high-intensity sweeteners), a fat replacer, and a texture modifier. Inulin has very minimal influence on the sensory properties of a product due to its neutral taste and absence of color. A neutral taste is colorless. The solubility of inulin increases significantly with temperature, reaching 34% (w/v) at 90°C (Kim et al. 2001). The functionality of inulin depends on its effect on water solutions at various solid levels. At lower concentrations, it can be used as a rheology modifier, because it causes significant increase in viscosity while at a concentration of 40%–45%. An inulin gel or crème that is firm but with a fatty creamy feel is formed. In this form, inulin is stable in acidic conditions and at high temperatures (Murphy 2001). Upon complete dissolution in water, inulin forms a particle gel characterized with a white creamy structure and short-range spreadability. For these reasons, inulin is frequently used to almost completely replace fat. The created gel is composed of a tridimensional network of insoluble submicron particles of crystalline inulin in water. The gel is very stable due to the inulin's ability to immobilize large amounts of water (Franck 2002). As a fat mimetic in low-fat dairy products, inulin appears to contribute to an improved mouthfeel (Guggisberg et al. 2009; Paseephol et al. 2008). The physicochemical properties of inulin appeared to be linked to the DP. OF is much more soluble and sweeter than native and long-chain inulin and can contribute to improve mouthfeel, because its properties are closely related to other sugars. Long-chain inulin is less soluble, enhances viscosity, and, as such, can act as a texture modifier (Coussement 1999). Additionally, the DP affects melting (Blecker et al. 2003) and glass transition temperature (Schaller-Povolny et al. 2000, 2001), gel formation and gel strength (Meyer

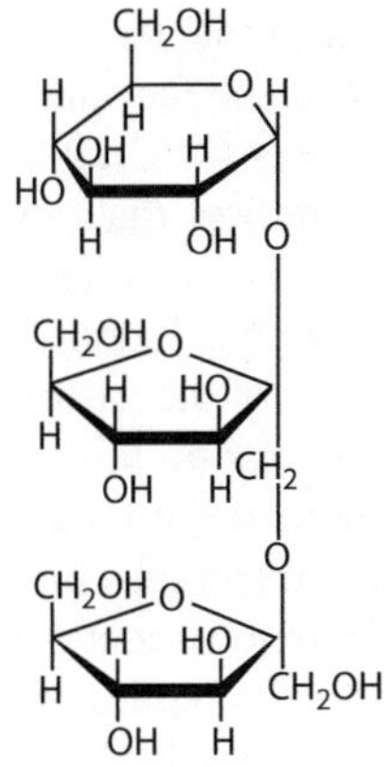

Figure 13.4 Structure of inulin.

and Blaauwhoed 2009), and compatibility with other components of the system such as starch and other hydrocolloids (Giannouli et al. 2004). As previously stated, long-chain inulin appears to act as a fat mimetic due to its ability to form microcrystals, which eventually may grow into a gel network (Hébette et al. 1998). Only longer chains with DP > 10 participate in the gel formation, with smaller chains remaining in the solution. The rate of crystallization, ultimate crystal size, and gel firmness are governed by inulin concentration, DP, applied shear, and temperature (Bot et al. 2004; Meyer and Blaauwhoed 2009). Due to lower solubility, longer chains will crystalyze at a greater rate than the shorter ones (Zimeri and Kokini 2003). In the presence of other food ingredients, inulin crystallization may be affected; they all compete for water and interact with each other through weak molecular interactions. Thus, it is expected that the final structure of the product will also be affected. Inulin has been assessed in a number of different models. For example, Bishay (1998) reported a synergistic effect between inulin and calcium alginate and a negative interaction between inulin and starch. Zimeri and Kokini (2003) similarly reported that inulin interfered with the network formation of starch acting as a diluent. There is less fundamental information available with regard to the rheological behavior of inulin in real food systems, but reports show that the properties of the system are influenced by the DP of inulin. For example, Shah et al. (2010) used different types of inulin as bulking agents in the production of sucrose-free chocolate. One of the observations was that the rate of chocolate tempering may be adjusted due to different viscoelastic behaviors. The addition of inulin produced a substantially different trend of phase shift in comparison to control. Sucrose-free chocolate samples displayed two distinct peaks: the first peak coincides with the rewarming of sucrose-free chocolate, whereas the second appeared around 20°C during the cooling phase. The increase of phase shift during the second step of tempering may indicate the structural rearrangement of inulin chains due to molecular relaxation and greater viscous behavior (Chiavaro et al. 2007). The second peak may be due to the repositioning of molten fat on the surface of solidified fat, which decreased friction between fat crystals and again improved viscous property (Kloek et al. 2005).

Inulin is used either as a supplement to foods or as a macronutrient substitute in foods. As a supplement to foods, it is added mainly for its nutritional properties. Such additions are usually in the range of 3–6 g per portion, not exceeding 10 g. As a macronutrient substitute, it is used mainly as a fat replacer. Typically, 1 g of fat is replaced by 0.25 g of inulin, which will lead to inulin concentrations of ~6 g per portion. Because inulin has been a natural component of many foods consumed safely by humans over millennia, it is therefore generally recognized as safe (Coussement 1999). Inulin-type fructans are classified as functional food ingredients. They are believed to target gastrointestinal functions, and, also, most likely via their effects on the gut and the gut microflora, systemic functions that are known to be related to health and well-being (Roberfroid 2005, 2007). The potential nutritional and health benefits of this ingredient are summarized in Table 13.5. Inulin and OF comply with most definitions of DFs, and they are labeled as such due to their fermentation pattern and selective stimulation of bacterial growth, inulin, and OF prebiotics.

A prebiotic is a nondigestible food ingredient that beneficially affects the host by selectively stimulating the growth, activity, or both of one or a limited number of bacterial species already resident in the colon (Ziemer and Gibson 1998). To exhibit such effects, a prebiotic must neither be hydrolyzed nor absorbed in the upper part of the gastrointestinal tract and must be selective for one or a limited number of potentially beneficial bacteria residing in the colon (Collins and Gibson 1999). The number of probiotics in the human gut tends to decrease with age (Mitsuoka 1992). Two major strategies have been proposed to maintain a high level of probiotics to sustain beneficial health effects: (1) continuous ingestion of probiotic-containing foods or (2) supplementation of food with prebiotics (Gomes and Malcata 1999). Probiotics are defined as "live microorganisms that, when administered in adequate amounts, confer a health benefit on the host" (Vasiljevic and Shah 2006, 2008). Over the years, many species of microorganisms have been identified, scientifically investigated, and commercially applied. They consist of not only lactic acid–producing bacteria

Table 13.5 Nutritional Effects and Potential Health Benefits of Inulin-Type Fructans

Enhanced colonic functions	Composition and activities of the gut microflora
	Stool production
	Absorption of minerals
	Production of gastrointestinal endocrine peptides
	Immunity and resistance to infections
	Digestion of high protein diets
Enhanced systemic functions	Lipid homeostasis
Reduction of disease risks	Intestinal infections
	Irritable bowel disease
	Colon cancer
	Osteoporosis
	Obesity
	Diabetes

(lactobacilli, streptococci, enterococci, lactococci, and bifidobacteria) but also *Bacillus* sp. and fungi such as *Saccharomyces* and *Aspergillus*. Postulated health advantages associated with probiotics intake are the improvement of lactose intolerance, increase in humoral immune responses, biotransformation of isoflavone phytoestrogen to improve postmenopausal symptoms, bioconversion of bioactive peptides for antihypertension, improvement of serum lipid profiles, increase in natural resistance to infectious disease in the gastrointestinal tract, suppression of cancer, reduction in serum cholesterol level, and improved digestion (Vasiljevic and Shah 2008).

REFERENCES

Abdul-Hamid, A., and Luan, Y. S. 2000. Functional properties of dietary fibre prepared from defatted rice bran. *Food Chem.* 68 (1): 15–19.

American Association of Cereal Chemists. 2001. The Definition of Dietary Fiber. AACC Report. *Cereal Foods World* 46: 112–126.

Akoh, C.C. 1998. Fat replacers. *Food Tech.* 52: 47–52.

Akoh, C.C. 1994. Synthesis of carbohydrate fatty acid polyesters. In: *Carbohydrate Polyesters as Fat Substitutes*, edited by C.C. Akoh and B.G. Swanson, pp. 9–35. New York: Marcel Dekker, Inc.

Anderson, J.W., Baird, P., Davis Jr., R.H., Ferreri, S., Knudtson, M.A., Koraym, A., Waters,V., and Williams, C.L. 2009. Health benefits of dietary fiber. *Nutr. Rev.* 67: 188–205.

Anon. 1981. Food and Drug Administration. Food Additive Petition 9A3441: Polydextrose. *Federal Register* 46: 112–126.

Aprianita, A., Purwandari, U., Watson, B., and Vasiljevic, T. 2010. Physico-chemical properties of flours and starches from selected commercial tubers available on the Australian market. *Int. Food Res. J.* 16: 507–520.

Auerbach, M.H., Craig, S.A.S., Howlett, J.F., and Hayes, K.C. 2007. Caloric availability of polydextrose. *Nutr. Rev.* 65 (12): 544–549.

Autio, K. 1996. Functional aspects of cell wall polysaccharides. In: *Carbohydrates in Food*, edited by A.-C. Eliasson, pp. 227–264. New York: Marcel Dekker.

Autio, K., Myllymaki, O., and Malkki, Y. 1987. Flow properties of solutions of oat β-glucans. *J. Food Sci.* 52: 1364–1366.

Baixauli, R., Salvador, A., Martinez-Cervera, S., and Fiszman, S.M. 2008. Distinctive sensory features introduced by resistant starch in baked products. *Food Sci. Tech.* 41: 1927–1933.

Ballance, G. M., and Meredith, W. O. S. 1976. Purification and partial characterization of an endo-β-1,3-glucanase from green malt. *J. Inst. Brew.* 82: 64–67.

Bansema, C. 2000. Development of a barley β-glucan beverage with and without whey protein isolate. M. Sc. Thesis. University of Alberta, Edmonton, AB.

Beck, C.I. 1974. Sweetness, character, and applications of aspartic acid-based sweeteners. In: *ACS Sweetener Symposium*, edited by G. E. Inglett, pp. 164–181. Westport, CT: AVI Publishing.

Beer, M.U., Arrigoni, E., and Amado, R. 1996. Extraction of oat gum from oat bran: Effects of process on yield, molecular weight distribution, viscosity and (1→3)(1→4)-β-D-glucan content of the gum. *Cereal Chem.* 73: 58–62.

Beer, M.U., Wood, P.J., Weisz, J., Fillion, N. 1997. Effect of cooking and storage on the amount and molecular weight of (1/3)(1/4)-D-β-glucan extracted from oat products by an in vitro digestion system. *Cereal Chem.* 74: 705–709.

BeMiller, J.N. 1997. Starch modification: Challenges and prospects. *Starch/Stärke* 49: 127–131.

BeMiller, J.N. 2001. Classification, structure and chemistry of polysaccharides of foods. In: *Handbook of Dietary Fiber*. edited by S.S. Cho and M.L. Dreher, pp. 603–611. New York: Marcel Dekker, Inc.

BeMiller, J.N. 2007. *Carbohydrate Chemistry for Food Scientists*. St. Paul, MN: AACC International.

BeMiller, J.N. 2010. Chapter 10: Carbohydrate analysis. In: *Food Analysis*, edited by S.S. Nielsen. New York: Springer Science and Business Media, LLC.

Bhatty, R.S. 1995. Laboratory and pilot plant extraction and purification of β-glucans from hull-less barley and oat bran. *J. Cereal Sci.* 22: 163–170.

Biliaderis, C.G. 1998. Structures and phase transitions of starch polymers. In: *Polysaccharide Association Structures in Foods*, edited by R.H. Walter pp. 57–168. New York: Marcel Dekker.

Bishay, I. E. 1998. Rheological characterization of inulin. In: *Gums and Stabilizers for the Food Industry*, edited by P.A. Williams and G.O. Phillips, vol. 11, pp. 403–408. Cambridge, UK: Royal Society of Chemistry.

Blecker, C., Chevalier, J.-P., Fougnies, C., Van Herck, J.-C., Deroanne, C., and Paquot, M. 2003. Characterization of different inulin samples by DSC: Influence of polymerisation degree on melting temperature. *Journal of Thermal Analysis and Calorimetry* 71 (1): 215–224.

Borderías, A.J., Sánchez-Alonso, I., and Pérez-Mateos, M. 2005. New application of fibre in foods: Addition to fishery products. *Trends in Food Science and Technology* 16: 458–465.

Bot, A., Erle, U., Vreeker, R. and Agterof, W.G.M. 2004. Influence of crystallization conditions on the large deformation rheology of inulin gels. *Food Hydrocolloids* 18: 547–556.

Burkus, Z. 1996. Barley β-glucan: Extraction, functional properties, and interaction with food components. M.Sc. thesis, University of Alberta, Edmonton, AB.

Burkus, Z., and Temelli, F. 1998. Effect of extraction conditions on yield, composition, and viscosity stability of barley β-glucan gum. *Cereal Chem.* 75, 805–809.

Burkus, Z., and Temelli, F. 1999. Gelation of barley β-glucan concentrate. *J. Food Sci.* 64: 198–201.

Burkus, Z., and Temelli, F. 2000. Stabilization of emulsions and foams using barley β-glucan. *Food Res. Int.* 33: 27–33.

Causey, J.L., McKeehen, J.D., Slavin, J.L. and Fulcher, R.G. 1998: Differential modulation of human macrophages by cereal cell wall β-glucans and phenolic acids. *Cereal Foods World* 43: 511–512.

Centers for Disease Control and Prevention. 2010. Chronic disease prevention. Fact Sheet: Overweight and obesity. http://www.cdc.gov/ncbddd/disabilityandhealth/documents/obesityFactsheet2010.pdf. Accessed September 2010.

CCC. 2010. Fat reduction in foods. 111 pp, Calorie Control Council, Atlanta, GA.

CCC. 2011. Survey: Most Americans are weight conscious. Calorie Control Council. Atlanta, GA. Available at http://www.caloriecontrol.org/pressrelease/national-survey-most-americans-now-worried-about-their-weight.

Chiavaro, E., Vittadini, E., and Corradini, C. 2007. Physicochemical characterization and stability of inulin gels. *European Food Research Technology* 225: 85–94.

Collins, B. 1998. Phytochemicals from oats. Where are we now and where are we going? Cereal Science Symposium–Updates and Inovations II. *The Oat Story. Strategies for Successful, Inovative Functional Food Developments*. Alberta Section of CIFST, Edmonton, AB, April 27.

Collins, M.D., and Gibson, G.R. 1999. Probiotics, prebiotics, and synbiotics: Approaches for modulating the microbial ecology of the gut. *Am. J. Clin. Nutr.* 69 (Suppl 1): 1042S–1057S.

Coussement, P.A.A. 1999. Inulin and oligofructose: safe intakes and legal status. *The Journal of Nutrition* 129 (Suppl 1): 1412S–1417S.

Craig, S.A.S., Holden, J.F., Frier, H., Troup, J.P., and Auerbach, A.H. Chapter 18. Polydextrose as soluble fiber and complex carbohydrate. In: *Complex Carbohydrates in Foods*, edited by L. Prosky, S.S. Cho, and M. Dreher. CRC Press 1999.

Craig, S.A.S., Holden, J.F., Troup, J.P., Auerbach, M.H. and Frier, H.I. 1998. Polydextrose as soluble fiber: Physiological and analytical aspects. *Cereal Foods World* 43: 370–376.

Cummings, J.H., and Englyst, H.N. 1995. Gastrointestinal effects of food carbohydrates. *Am. J. Clin. Nutr.* 61: 938S–945S.

Dawkins, N.L., and Nnanna, I.A. 1995. Studies on oat gum [(1→3, 1→4)-β-D-glucan]: Composition, molecular weight estimation and rheological properties. *Food Hydrocol.* 9: 1–7.

De Leenheer, L., and Hoebregs, H. 1994. Progress in the elucidation of the composition of chicory inulin. Starch 46: 193–196.

deMan, J.M. 1999. Relationship among chemical, physical, and textural properties of fats. In: *Physical Properties of Fats, Oils, and Emulsions*, edited by N. Widlak, pp. 79–95. Champaign, IL: AOCS Press.

Demírcíoğlu, Y., Başoğlu, S., Özkan, S., Şimşek, I., and Abbasoğlu, U. 2008. The prebiotic effects of mixed sweetener containing polydextrose and oligofructose substituted sugar in diet. *Turk. J. Pharm. Sci.* 5 (2): 95–106.

DeVries, J.W. 2004. On defining dietary fiber. *Proc. Nutr. Soc.* 62: 37–43.

Eastwood, M.A., and Morris, E.R. 1992. Physical properties of dietary fiber that influence physiological function: a model for polymers along the gastrointestinal tract. *Am. J. Clin. Nutr.* 55: 436–42.

Estrada, A., Yun, C.H., Van Kessel, A., Li, B., Hauta, S., and Laarveld, B. 1997. Immunomodulatory activities of oat beta-glucan in vitro and in vivo. *Microbiology and Immunology* 41: 991–998.

Faraj, A., Vasanthan, T., and Hoover, R. 2004. The effect of extrusion cooking on resistant starch formation in waxy and regular barley flours. *Food Res. Int.* 37: 517–525.

FDA. 1997. Food Labeling; Health Claims; Oats and Coronary Heart Disease; Final Rule Federal Register Doc. 97-1598, filed 1-22-97.

Food and Drug Admninistration, Dept. of Health and Human Services. 1997. Federal Register 62 FR 3583, January 23, 1997 – Food Labeling: Health Claims; Oats and Coronary Heart Disease; Final Rule. [Federal Register: January 23, 1997 (Volume 62, Number 15)], [Page 3583-3601]. http://www.fda.gov/Food/LabelingNutrition/LabelClaims/HealthClaimsMeetingSignificantScientificAgreementSSA/ucm074719.htm. Accessed December, 2011.

Franck, A. 2002. Technological functionality of inulin and oligofructose. *Br. J. Nutr.* 87: S287–S291.

Fuentes-Zaragoza, E., Riquelme-Navarrete, M.J., Sanchez-Zapata, E., and Perez-Alvarez, J.A. 2010. Resistant starch as functional ingredient: A review. *Food Res. Int.* 43: 931–942.

Giannouli, P., Richardson, R.K., and Morris, E.R. 2004. Effect of polymeric cosolutes on calcium pectinate gelation. Part 2. Dextrans and inulin. *Carbohydrate Polymers* 55 (4): 357–365.

Giese, J. 1996. Fats, oils, and fat replacers. *Food Technol.* 50 (4): 78–84.

Gomes, M.P, and Malcata, F.X. 1999. *Bifidobacterium* spp. and *Lactobacillus acidophilus*: Biological, biochemical, technological and therapeutical properties relevant for use as probiotics. *Trends Food Sci. Technol.* 10: 139–157.

Gomes, C.R., Vissotto, F.Z., Fadini, A.L., De Faria, E.V., and Luiz, A.M. 2007. Influence of different bulk agents in the rheological and sensory characteristics of diet and light chocolate. [Influência de diferentes agentes de corpo nas características reológicas e sensoriais de chocolates diet em sacarose e light em calorias]. *Cienc. Tecnol. Aliment. (Campinas, Braz.)* 27 (3): 614–623.

Gomez, C., Navarro, A., Manzanares, P., Horta, A., and Carbonell, J.V. 1997a. Physical and structural properties of barley (1 → 3),(1 → 4)-β-glucan. Part I. Determination of molecular weight and macromolecular radius by light scattering. *Carbohydrate Polymers* 32: 7–15.

Gomez, C., Navarro, A., Manzanares, P., Horta, A., and Carbonell, J.V. 1997b. Physical and structural properties of barley (1 → 3),(1 → 4)-β-glucan. Part II. Viscosity, chain stiffness and macromolecular dimensions. *Carbohydrate Polymers* 32: 17–22.

Gomez, C., Navarro, A., Garrier, C., Horta, A., and Carbonell, J.V. 1997c. Physical and structural properties of barley (1 → 3), (1 → 4)-β-glucan. Part III. Formation of aggregates analyzed through its viscoelastic and flow behavior. Carbohydrate Polymers 34: 141–148.

Grimm, A., Kruger, E. and Burchard, W. 1995. Solution properties of β-D-(1, 3)(1, 4)-glucan isolated from beer. *Carbohydrate Polymers* 27: 205–214.

Guggisberg, D., Cuthbert-Steven, J., Piccinali, P., Bótikofer, U., and Eberhard, P. 2009. Rheological, microstructural and sensory characterization of low-fat and whole milk set yogurt as influenced by inulin addition. *Int. Dairy Journal* 19: 107–115.

Hampton, T. 2008. Sugar substitutes linked to weight gain. *JAMA* 299: 2137–2138.

Handa, C., Goomer, S., and Siddhu, A. 2011. Effects of whole-multigrain and fructooligosaccharide incorporation on the quality and sensory attributes of cookies. *Food Sci. Technol. Res.* 17 (1): 45–54.

Haralampu, S.G. 2000. Resistant starch—A review of the physical properties and biological impact of RS3. *Carbohydr. Polym.* 41: 285–292.

Harkema, J. 1996. Starch-derived fat memetics from potato. In: *Handbook of Fat Replacers*, edited by S. Roller and S.A. Jones, pp. 119–129. Boca Raton, FL: CRC Press.

Hébette, C.L.M., Delcour, J.A., and Koch, M.H.J., Booten, K., Kleppinger, R., Mischenkod, N. et al. 1998. Complex melting of semi-crystalline chicory (*Cichorium intybus* L.) root inulin. *Carbohydrate Research* 310 (1–2): 65–75.

Hernández, O., Emaldi, U., and Tovar, J. 2008. In vitro digestibility of edible films from various starch sources. *Carbohydrate Polymers* 71: 648–655.

Hipsley E.H. 1953. Dietary "fiber" and pregnancy toxaemia. *British Medical Journal* 2: 420–422.

Inglett, G.E. 1970. Natural and synthetic sweeteners. *Hortscience* 5: 139–141.

Inglett, G.E. 1974. Symposium: Sweeteners. Westport, Connecticut: The Avi Publishing Company, Inc.

Inglett, G.E., and Grisamore, S.B. 1991. Maltodextrin fat substitute lowers cholesterol. *Food Technology* 45: 104.

Institute of Medicine. 2002. Dietary reference intakes. Proposed Definition of Dietary Fiber. Washington, DC: National Academies Press.

Isik, U., Boyacioglu, D., Capanoglu, E., and Nilufer Erdil, D. 2011. Frozen yogurt with added inulin and isomalt. *J. Dairy Sci.* 94 (4): 1647–1656.

Kim, Y., Faqih, M.N., and Wang, S.S. 2001. Factors affecting gel formation of inulin. *Carbohydrate Polymers* 46: 135–145.

Kloek, W., Van Vliet, T., and Walstra, P. 2005. Large deformation behavior of fat crystal networks. *Journal of Texture Studies* 36 (5–6): 516–543.

Lee, S., Warner, K., and Inglet, G.E. 2005. Rheological properties and baking performance of new oat β-glucan-rich hydrocolloids. J. Agric. Food Chem. 53: 9805–9809.

Liu, S. and Tsai, C.E. 1995. Effect of biotechnically synthesized oligosaccharides and polydextrose on serum lipids in the human. *J. Chin. Nutr. Soc.* 20: 1–12.

Lunn, J., and Buttriss, J.L. 2007. Carbohydrates and dietary fiber. *Nutrition Bulletin* 32: 21–64.

MacGregor, A.W., and Fincher, G.B. 1993. Carbohydrates of the barley grain. In: *Barley: Chemistry and Technology*, edited by A.W. MacGregor and R.S. Bhalty, pp. 73–130. St. Paul, MN: Am. Assoc. Cereal Chem. Inc.

Mann, J.I., and Cummings, J.H. 2009. Possible implications for health of the different definitions of dietary fibre. *Nutr Metab Cardiovasc Dis.* 19: 226–229.

Marchal, L.M., Beeftink, H.H., and Tramper, J. 1999. Towards a rational design of commercial maltodextrins. *Trends in Food Science and Technology* 10: 345–355.

McIntosh, G.H., Le-Leu, R.K., Royle, P.J., and Young, P.G. 1996. A comparative study of the influence of differing barley brans on DMH-induced tumors in male Sprague-Dawley rats. *Journal of Gastroenterology and Hepatology* 11: 113–119.

Meyer, D., and Blaauwhoed, J.-P. 2009. Inulin. In: Handbook of Hydrocolloids, edited by G.O. Phillips and P.A. Williams, 2nd ed., pp. 829–848. Cambridge, UK: Woodhead Publishing and CRC Press.

Mitsuoka, T. 1992. Intestinal flora and aging. *Nutr. Rev.* 50: 438–446.

Mokdad, A.H., Marks, J.S., Stroup, D.F., and Gerberding, J.L. 2004. Actual causes of death in the United States, 2000. *JAMA* 291(10): 1238–1245.

Morgan, K.R., and Ofman, D.J. 1998. Glucagel, a gelling β-glucan from barley. *Cereal Chemistry* 75: 879–881.

Murphy, E.A., Davis, J.M., and Carmichael, M.D. 2010. Immune modulating effects of β-glucan. *Curr. Opin. Clin. Nutr. Metab. Care* 13: 656–661.

Murphy, O. 2001. Non-polyol low-digestible carbohydrates: Food applications and functional benefits. *British Journal of Nutrition* 85(Suppl 1), S47–S53.

Nabors, L.O. 2002. Sweet choices: sugar replacements for foods and beverages. *Food Technology* 56: 28–32, 45.

Niness, K.R. 1999. Inulin and oligofructose: What are they? *J. Nutr.* 129: S1402–S1406.

Nugent, A.P. 2005. Health properties of resistant starch. *British Nutrition Foundation, Nutrition Bulletin,* 30: 27–54.

Oku, T., Fujii, Y., and Okamatsu, H. 1991. Polydextrose as dietary fiber: Hydrolysis by digestive enzymes and its effects on gastrointestinal transit time in rats. *J Clin Biochem Nutr* 11: 31–40.

Park, Y., Hunter, D.J., Spiegelman, D., Bergkvist, L., Berrino, F., van den Brandt, P.A., Buring, J.E., Colditz, G.A., Freudenheim, J.L., Fuchs, C.S., Giovannucci, E., Goldbohm, G.A., Graham, S., Harnack, L., Hartman, A.M., Jacobs, D.R., Kato, K., Krogh, V., Leitzmann, M.F., McCullough, M.L., Miller, A.B., Pietinen, P., Rohan, T.E., Schatzkin, A., Willett, W.C., Wolk, A., Zeleniuch-Jacquotte, A., Zhang, S.M., and Smith-Warner, S.A. 2005. Dietary fiber intake and risk of colorectal cancer a pooled analysis of prospective cohort studies. *JAMA* 294 (2): 2849–2857.

Paseephol, T., Small, D.M., and Sherkat, F. 2008. Rheology and texture of set yoghurt as affected by inulin addition. *Journal of Texture Studies* 39 (6): 617–634.

Pszczola, D.E. 1996. Oatrim finds application in fat-free, cholesterol-free milk. *Food Technol.* 50 (9): 80–81.

Ratnayake, W.S., and Jackson, D.S. 2008. Thermal behavior of resistant starches RS 2, RS 3, and RS 4. *Journal of Food Science* 73 (5): 356–366.

Roberfroid, M.B. 2005. Introducing inulin-type fructans. *British Journal of Nutrition* 93 (suppl. 1): 13–25.

Roberfroid, M.B. 2007. Inulin-type fructans: Functional food ingredients. *The Journal of Nutrition* 137: 2493S–2502S.

Rodriguez, R., Jimenez, A., Bolanos, J.F., Guillen, R., and Heredia, A. 2006. Dietary fiber from vegetable products as source of functional ingredients. *Trends Food Sci. Technol.* 17: 3–15.

Roller, S. 1996. Starch-derived fat mimetics: Maltodextrins. In: *Handbook of Fat Replacers*, edited by S. Roller and S.A. Jones, pp. 98–118. Boca Raton: CRC Press.

Sajilata, M.G., Singhal, R.S., and Kulkarni, P.R. 2006. Resistant starch–A review. *Comprehensive Reviews in Food Science and Food Safety* 5: 1–17.

Sanz, T., Salvador, A., Baixauli, R., and Fiszman, S.M. 2009. Evaluation of four types of resistant starch in muffins. II. Effects in texture, color and consumer response. *Europ. Food Res. Technol.* 229: 197–204.

Schaller-Povolny, L.A., Smith, D.E., and Labuza, T.P. 2000. Effect of water content and molecular weight on the moisture isotherms and glass transition properties of inulin. *International Journal of Food Properties* 3: 173–192.

Schaller-Povolny, L.A., and Smith, D.E. 2001. Viscosity and freezing point of a reduced fat ice cream mix as related to inulin content. *Milchwissenschaft–Milk Science International* 56: 25–29.

Shah, A., Jones, G., and Vasiljevic, T. 2010. Sucrose-free chocolate sweetened with stevia and containing different bulking agents—Effects on physicochemical and sensory properties. *Int. J. Food Sci. Technol.* 45: 1426–1435.

Sharma, A., and Yadav, B.S. 2008. Resistant starch: Physiological roles and food applications. *Food Reviews International* 24: 193–234.

Sirtori, C.R., Galli, C., Anderson, J.W., Sirtori, E., and Arnoldi, A. 2009. Functional foods for dyslipidaemia and cardiovascular risk prevention. *Nutr. Res. Rev.* 22: 244–261.

Slavin, J. 2004. Whole grains and human health. *Nutrition Research Reviews*, 17, 99–110.

Slavin, J., Jacobs, D., and Marquart, L. 1997. Whole-grain consumption and chronic disease: Protective mechanisms. *Nutrition and Cancer* 27: 14–21.

Stowell, JD. 2009a. Prebiotic potential of polydextrose. In: Charalampopoulos D, Rastall RA, editors. Prebiotics and probiotics science and technology. Reading: Springer, pp. 337–352.

Stowell, J.D. 2009b. Polydextrose, In: *Fiber Ingredients–Food Applications and Health Benefits*, edited by S.S. Cho and P. Samuel, pp. 173–201. CRC Press: Boca Raton.

Telang, J., Shah, N.P., and Vasiljevic, T. 2005. Probiotics and prebiotics: A food-based approach in the prevention of colorectal cancer. *Milchwissenschaft* 60 (3): 241–245.

Temelli, F., Bansema, C., and Stobbe, K. 2004. Development of an orange-flavored barley β-glucan beverage. *Cereal Chem.* 81 (4): 499–503.

Thompson, D.B. 2000. On the non-random nature of amylopectin branching. *Carbohydrate Polymers* 43: 223–239.

Tolstoguzov, V.B. 1991. Functional properties of food proteins and role of protein–polysaccharide interaction. *Food Hydrocolloids* 4: 429–468.

Tolstoguzov, V.B. 1997. Protein–polysaccharides interactions. In: *Food Proteins and Their Applications*, edited by S. Damodaran and A. Paraf, pp.171–199. New York: Marcel Dekker.

Trowell. H. 1985. In: *Dietary Fiber, Fiber-Depleted Foods and Disease*, edited by H. Trowell, D. Burkitt, and K. Heatons, Chapter 1, pp. 1–20. Academic Press: London.

Tudorica, C.M., Jones, E., Kuri, V., and Brennan, C.S. 2004. The effects of refined barley β-glucan on the physico-structural properties of low-fat dairy products: Curd yield, microstructure, texture and rheology. *J Sci Food Agri* 84: 1159–1169.

Varum, K.M., Smidsrod, O., and Brant, D.A. 1992. Light scattering reveals micelle-like aggregation in the (1→3),(1→4)-β-D-glucans from oat aleurone. *Food Hydrocolloids* 5: 497–511.

Vasiljevic, T., Mishra, V.K., and Kealy, T. 2007. β-Glucan addition to a probiotic containing yogurt. *J. Food Sci.* 72: C402–C411.

Vasiljevic, T., and Shah, N.P. 2006. Fermented milk: health benefits beyond probiotic effect. In: *Handbook of Food Products Manufacturing*, edited by Y.H. Hui, doi:10.1002/9780470113554.ch51. Hoboken, NJ: John Wiley & Sons, Inc.

Vasiljevic, T., and Shah, N.P. 2008. Probiotics from Metchnikoff to bioactives. *Int. Dairy Journal* 18: 714–728.

Vithanage, C.J., Mishra, V.K., Vasiljevic, T., and Shah, N.P. 2008. Use of β-glucan in development of low-fat Mozzarella cheese. *Milchwissenschaft* 63: 420–423.

Wepner, E. Berghofer, E., and Miesenberger, K. 1999. Tiefenbacher. Citrate starch: Application as resistant starch in different food systems. *Starch* 51 (10): 354–361.

Wilborn, C., Beckham, J., Campbell, B., Harvey, T., Galbreath, M., La Bounty, P., Nassar, E., Wismann, J., and Kreider, R. 2005. Obesity: Prevalence, theories, medical consequences, management, and research directions. *J. Int. Soc. Sports Nutr.* 2: 4–31.

Wisker, E., and Feldheim, W. 1990. Metabolizable energy of diets low or high in dietary fiber from fruits and vegetables when consumed by humans. *Journal of Nutrition* 120: 1331–1337.

Wisker, E., and Feldheim, W. 1992. Fecal bulking and energy value of dietary fiber. In: *Dietarv Fiber—a Component in Foods and Nutritional Function in Health and Disease*, edited by T.F. Schweizer and C.A. Edwards, pp. 223–246. London: Springer-Verlag.

Wood, P.J. 1984. Physicochemical properties and technological and nutritional significance of cereal β-glucansin. In: *Cereal Polysaccharides in Technology and Nutrition*, edited by V.F. Rasper, pp. 35–78. St. Paul, MN: Am. Assoc. Cereal Chem. Inc.

Wood, P.J. 1991. Oat β-glucan-physicochemical properties and physiological effects. *Trends in Food Science and Technology* 2: 311–314.

Wood, P.J., and Beer, MU. 1998. Functional oat products. In: *Functional foods: Biochemical and Processing Aspects*, edited by G. Mazza, pp. 1–37. Lancaster, PA: Technomic Publishing.

Wood, P. J., Paton, D., and Siddiqui, I.R. 1977. Determination of β-glucan in oats and barley. *Cereal Chem.* 54: 524–533.

Wood, P.J., Siddiqui, I.R., and Paton, D. 1978. Extraction of high-viscosity gums from oats. *Cereal Chem.* 55: 1038–1049.

Wood, P. J., Weisz, J., and Mahn, W. 1991. Molecular characterization of cereal β-glucans II, size exclusion chromatography for comparison of molecular weight. *Cereal Chem.* 68: 530–536.

Woodward, J.R., Fincher, G.B., and Stone, B.A. 1983. Water soluble (1→3)(1→4) -β-D-glucans from barley (*Hordeum vulgare*) endosperm. II. Fine structure. *Carbohydrate Polymers* 3: 207–225.

Wright J.D., Kennedy-Stephenson J., Wang C.Y., McDowell M.A., and Johnson C.L. 2004. Trends in intake of energy and macronutrients–United States, 1971–2000. (Reprinted from MMWR, vol 53, pg 80–82, 2004). *Journal of the American Medical Association* 291: 1193–1194.

Zeng, J., and Tan, Z. 2010. Metabolic homeostasis and colonic health: The critical role of short chain fatty acids. *Curr. Nutr. Food Sci.* 6: 209–222.

Zhang L., Ding, Q., Zhang, P., Zhu, R., and Zhou, Y. 1997. Molecular weight and aggregation behavior in solution of β-D-glucan from *Poria cocos* sclerotium. *Carbohydr. Res.* 303: 193–197.

Ziemer, C.J., and Gibson, G.R. 1998. An overview of probiotics, prebiotics and synbiotics in the functional food concept: Perspectives and future strategies. *Int. Dairy Journal* 8: 473–479.

Zimeri, J.E., and Kokini, J.L. 2003. Morphological characterization of the phase behavior of inulin-waxy maize starch systems in high moisture environments. *Carbohydrate Polymers* 52: 225–236.

Zoulias, E.I., Oreopoulou, V., and Kounalaki, E. 2002. Effect of fat and sugar replacement on cookie properties. *J. Sci. Food Agric.* 82 (14): 1637–1644.

CHAPTER 14

Risk Assessment of Sweeteners Used as Food Additives

Alicja Mortensen and John Christian Larsen

CONTENTS

14.1 BACKGROUND FOR THE SAFETY ASSESSMENT OF SWEETENERS USED AS FOOD ADDITIVES

A number of sweeteners are authorized as food additives in the European Union (EU; Table 14.1). They are divided into two classes based on their relative sweetness compared to sucrose: intense sweeteners, which are several times sweeter than sucrose, and bulk sweeteners, whose sweetness is a little less or comparable to that of sucrose.

Intense sweeteners are potentially high-consumption food additives, because they are used in products consumed in high volume such as soft drinks and tabletop sweeteners.

Bulk sweeteners are hydrogenated carbohydrates, also referred to as sugar alcohols or polyols. They offer certain functional advantages over sucrose in food preparation (e.g., lowering of the freezing point of an ice cream mix and reducing caramelization at heating) or certain dietetic advantages (e.g., being noncariogenic or not creating an immediate demand for insulin in contrast to sucrose).

Safety in the use of the sweeteners authorized as food additives in the EU has been evaluated in accordance with internationally agreed principles for the safety evaluation of food additives and

Table 14.1 Acceptable Daily Intakes of Sweeteners Evaluated by the SCF/EFSA or the JECFA

Evaluating Body		SCF/EFSA[a]		JECFA	
Class/E Number/Name		ADI (mg/kg bw/day)	Year of Last Evaluation	ADI (mg/kg bw/day)	Year of Last Evaluation
		Intense Sweeteners			
E 950	Acesulfame potassium	0–9	2000	0–15	1990
E 951	Aspartame	0–40	2002	0–40	1981
E 952	Cyclamate: cyclamic acid and its Na and Ca salts	0–7	2000	0–11	1982
E 959	Neohesperidin DC	0–5	1988	–	–
E 961	Neotame	0–2	2010	0–2	2003
E 954	Saccharin and its Na, K, and Ca salts	0–5	1995	0–5	1993
E960	Steviol glycosides	4	2010	0–4	2009
E 955	Sucralose	0–15	2000	0–15	1990
E 957	Thaumatin	Acceptable	1988	Not specified	1985
		Bulk Sweeteners			
E 968	Erythritol	Acceptable[b]	2003	Not specified[c]	1999
E 953	Isomalt	Acceptable[b]	1988	Not specified[c]	1985
E 966	Lactitol	Acceptable[b]	1988	Not specified[c]	1983
E 965	Maltitol and maltitol syrup	Acceptable[b]	1999	Not specified[c]	1997
E 421	Mannitol	Acceptable[b]	1999	Not specified[c]	1986
E 420	Sorbitol	Acceptable[b]	1984	Not specified[c]	1982
E 967	Xylitol	Acceptable[b]	1984	Not specified[c]	1983

[a] AFC or ANS Panels.

[b] Provided that their laxative effect should be borne in mind. (From Scientific Committee for Food (SCF). Sweeteners—Opinion expressed on 14 September 1984. Reports of the Scientific Committee for Food (16th series). EUR 10210 EN. Luxembourg: Commission of the European Communities.1985. With permission.)

[c] The laxative effect in man and animals at high doses—a common feature of all polyols—should be taken into account when considering appropriate levels of use of polyols alone and in combination (From Scientific Committee for Food (SCF). Sweeteners—Opinion expressed on 14 September 1984. Reports of the Scientific Committee for Food (16th series), EUR 10210 EN, Commission of the European Communities, Luxembourg, 1985. With permission. Joint FAO/WHO Expert Committee on Food Additives (JECFA), Evaluation of certain food additives—Twenty-seventh report of the Joint FAO/WHO Expert Committee on Food Additives, WHO Technical Report Series 696, 1983, and Corrigendum. [1983, TRS 696-JECFA 27], 1983a. With permission. Joint FAO/WHO Expert Committee on Food Additives (JECFA), Toxicological evaluation of certain food additives. WHO Food Additive Series No. 18. [1983, FAS 18-JECFA 27], 1983b. With permission.)

other chemicals in commercial use in the food production. These principles were agreed upon by the World Trade Organization (WTO) at the Uruguay Round and are laid down in the Agreement on the Application of Sanitary and Phytosanitary Measures (SPS Agreement). This agreement requires health and safety measures to be based on sound scientific risk assessment.

WTO recognizes the Food and Agriculture Organization of the United Nations/World Health Organization (FAO/WHO) Codex Alimentarius Commission (CAC or "Codex") standards as a reference point for the safety of foodstuffs traded internationally (WTO 1994). These standards are established by the Codex Committee on Food Additives (CCFA). The CCFA uses the Joint FAO/WHO Expert Committee on Food Additives (JECFA) as an advisory committee with regard to the safety evaluation of food additives and contaminants.

In the EU, the European Commission Scientific Committee for Food (SCF) was the scientific guarantor for the safety of food additives in use until March 2003. Since then, the safety assessment of food additives has been taken over by the European Food Safety Authority (EFSA). Within the EFSA, food additives were evaluated by the Scientific Panel on Food Additives, Flavorings, Processing Aids, and Materials in Contact with Food (AFC Panel) until July 2008, when this task was taken over by the Scientific Panel on Food Additives and Nutrient Sources Added to Food (ANS Panel).

14.2 PRINCIPLES OF THE SAFETY EVALUATION OF SWEETENERS USED AS FOOD ADDITIVES

JECFA issued principles for the safety assessment of food additives in 1987 (WHO 1987). These principles have recently been updated (WHO 2009). The SCF issued its first guidance for the safety assessment of food additives in 1980 (SCF 1980). In 2001, the SCF adopted its new guidance (SCF 2001). Within the EFSA and the AFC Panel and its followers since July 2008, the ANS Panel formally adopted the SCF's guidance from 2001. A new guidance document has been prepared by the ANS Panel and is expected to be adopted this year (2012). The guidance document advices which data are necessary for the evaluation of the safety of a chemical intended for use as a food additive.

14.3 DATABASE NECESSARY FOR THE SAFETY EVALUATION OF SWEETENERS AS FOOD ADDITIVES

The database to be used for the safety evaluation of a food additive should be provided by the party interested in obtaining an authorization for its use in food and/or beverages (a petitioner). In the EU, the database should be prepared in accordance with the guidance on submission for food additive evaluations, which are adopted by the EFSA at the time of submission. The submission, referred to as a "dossier," should provide administrative data, technical data (e.g., identity of the substance, purity, proposed specifications, stability and breakdown products, methods of analysis in foods, and manufacturing process), a presentation of technical need, proposed applications and levels of use in different food categories, the estimated exposure of consumers resulting from the proposed uses and use levels, and biological data. The latter includes results from studies on absorption, organ and tissue distribution, metabolism, and excretion (ADME) in animals and humans and from the toxicological studies *in vitro* (bacteria and cell cultures) and in laboratory animals (*in vivo* testing).

14.4 TOXICOLOGICAL TESTING

The aim of toxicological testing is to determine whether the compound, when used in the manner and quantities proposed, would pose any appreciable risk to the health of the average consumer and of those whose pattern of food consumption and physiological or health status may make them vulnerable, for example, "a high consumer," young age, pregnancy, and diabetes (SCF 2001). The toxicological testing of a chemical intended for food additive use includes a large number of studies such as ADME studies, short-term (3 months in rats) and long-term chronic (2 years in rats) toxicity, and short-term and *in vitro* testing for genotoxicity (mutagenicity or clastogenicity). Studies on carcinogenicity, often combined with long-term studies, as well as studies on reproductive and developmental toxicity, are also required. In addition, special *in vitro* and *in vivo* studies on effects on the nervous system, the immune system, and the endocrine system, as well as on the mechanism of action, may be needed. The relevance and scope of any additional studies to support the safety of use should be considered thoroughly by the petitioner, and the welfare and the reduction, refinement, replacement (3R) principle with respect to the use of laboratory animals (Gad 1990) should be kept in mind. Human tolerance studies may also be informative.

14.5 RISK ASSESSMENT OF FOOD ADDITIVES

The safety evaluation of a sweetener prior to the acceptance as a food additive is performed in the scientific process called risk assessment. Risk assessment is a part of a risk analysis, which includes risk assessment, risk evaluation, and risk communication.

The risk assessment of sweeteners as food additives is performed in the same manner as the risk assessment of any other chemical in food. It requires expertise in toxicology and exposure (intake) assessment. Risk assessment results in an upper limit for the exposure to the compound through the consumption of foods, at which level or below the dietary exposure to the compound is not expected to cause health-damaging (adverse) effects.

The procedure comprises four steps: hazard identification, hazard characterization, exposure assessment, and risk characterization (Renwick et al. 2003).

14.5.1 Hazard Identification

Hazard identification identifies all the adverse health effects that the compound may inherently cause. For this purpose, scientific data from experiences from human exposures, studies in laboratory animals, and *in vitro* assays are needed. As human data are seldom available, risk assessors have to rely on results from animal studies and *in vitro* assays.

14.5.2 Hazard Characterization

Hazard characterization (also referred to as dose–response assessment) estimates the relationship between the dose and the incidence and severity of an effect and leads to the selection of the pivotal data set in which the critical adverse effect is identified. Subsequently, this data set is used to describe and evaluate the dose–response relationship for the critical effect of the test compound. If the data demonstrate that the test compound is nongenotoxic, the identification of a point of departure for the establishment of a health based reference value, such as the acceptable daily intake (ADI) is performed. A point of departure can be a no-observed-adverse-effect level (NOAEL) or a lowest observed adverse effect level (LOAEL; IPCS 1994).

The NOAEL is defined as "the highest concentration or amount of a substance, found by experiment or observation, which causes no detectable adverse alteration of morphology, functional capacity, growth, development, or life span of the target organism under defined conditions of exposure."

The LOAEL is defined as "the lowest concentration or amount of a substance, found by experiment or observation, which causes an adverse alteration of morphology, functional capacity, growth, development, or life span of the target organism distinguishable from normal (control) organisms of the same species and strain under the same defined conditions of exposure."

As an alternative to the traditional NOAEL approach, the benchmark dose (BMD) concept has been proposed for use in the quantitative assessment of the dose–response relationship (Crump 1984; Barnes et al. 1995). The BMD is the dose of a substance that is expected to result in a prespecified level of effect (Setzer and Kimmel 2003). The BMD approach was not applied in the safety evaluation of sweeteners authorized for food use up to date. However, a recent opinion from the EFSA Scientific Committee now advocates the use of the BMD approach in future risk assessments of chemicals in food (EFSA 2009c).

For nongenotoxic chemicals, it is generally accepted that there is a threshold for toxicity and exposure below the threshold is safe. Based on the type of critical effect in a pivotal study, the dose–response relationship of a critical effect, and the overall quality of the scientific database, an uncertainty factor is selected. This factor takes into account species differences between the test animal and the human, and variation within humans (Renwick 1993; Renwick and Lazarus 1998; IPCS 1994; WHO 2009). When the database is considered adequate, an uncertainty factor of 100 is used by default, but it may be modified when adequate human data are available. The NOAEL identified for the test compound is then divided by the selected uncertainty factor to establish an acceptable daily intake (ADI) for humans.

The ADI is a measure for the innocuousness of a substance. It is the amount of a food additive, expressed on a milligram per kilogram body weight (bw) basis, that can be ingested daily over a lifetime without incurring any appreciable health risk (WHO 1987). All intense sweeteners permitted for the food use in the EU, except for thaumatin, have been allocated a numerical ADI (Table 14.1).

When the database is not optimal but there are no indications of any short-term health problems from the anticipated use of a food additive, a temporary ADI may be established using a larger uncertainty factor (200 by default), requesting additional studies within a given time frame.

When, on the basis of the available toxicological, biochemical, and clinical data, the total daily intake of the substance, arising from its natural occurrence and/or its present use(s) in food at the level necessary to achieve the desired technological effect, will not represent a hazard to health, the establishment of a numerical ADI is not considered necessary. Instead, terms such as "ADI acceptable" (SCF 1985) or "ADI not specified" are used (WHO 1987). Both terms do not mean, however, that any amount of the substance is toxicologically acceptable. Such an additive must be used according to good manufacturing practice; that is, it should be technologically efficacious, should be used at the lowest level necessary to achieve the technological effect, should not conceal inferior food quality or adulteration, and should not create nutritional imbalance. All bulk sweeteners authorized for food use in the EU have been allocated an ADI acceptable (Table 14.1).

The ADI does not apply to neonates and infants before the age of 12 weeks. The susceptibility of infants and children to food additives has been an important subject in the discussion of food chemical safety (Larsen and Pascal 1998). Food additives to be used in infant formulas require a special evaluation.

It is a basic rule that a substance that is genotoxic and causes cancer in animal studies by directly damaging the genes is not accepted as a food additive.

14.5.3 Exposure Assessment

Exposure assessment is based on information on the levels of a compound intended for use as a food additive in different food categories and data on the intake of the relevant food items in the country or region in question. Information on food consumption may be derived from food supply data, household surveys, individual dietary surveys, total diet studies, and/or the use of biomarkers. The estimates of exposure should include average, medium, and maximum intake figures from regular foods, special foods, and all foods (regular and special foods) and should concern the whole population, segments of the population, and individuals.

14.5.4 Risk Characterization

Risk characterization is the step of risk assessment that integrates information from exposure assessment and hazard characterization into advice suitable for risk management. Based on the actual exposure assessment, this step determines which fraction, if any, of the population would have intakes higher than the ADI and estimates the magnitude by which the exposures would exceed the ADI. The health impact, if any, of this excess is evaluated in a given population as well as the seriousness of any health risk. The conclusions of risk assessors can be that the expected exposure is safe according to the established ADI or that it will lead to exceeding of the ADI.

An occasional exceeding of the ADI, although undesirable from a toxicological point of view, does not pose an immediate health risk (Barlow et al. 1999; Larsen and Richold 1999). As a permanent exceeding of the ADI is associated with an appreciable health risk and, therefore, is not acceptable, reductions in exposure are needed to comply with the ADI. It is up to risk managers to establish provisions needed to ensure the compliance of exposure to the ADI. Furthermore, it must be controlled to ensure that the implemented provisions have the desired effect. In the EU, the risk management of sweeteners like other food additives is within the EU Commission.

14.6 EU REGULATIONS OF THE USE OF SWEETENERS IN FOOD PRODUCTS

The use of sweeteners in the EU has been regulated by a framework directive (Council Directive 89/107/EEC of 21 December 1988 on the approximation of the laws of the Member States concerning food additives authorized for use in foodstuffs intended for human consumption, as amended by Directive 94/34/EC) and a specific directive, European Parliament and Council Directive 94/35/EC of June 1994 on sweeteners for use in foodstuffs, amended by Directives 96/83/EC and 2003/115/EC. The annexes to the specific directives provided the information on which sweeteners were permitted in different foodstuffs or groups of foodstuffs together with the maximum permitted levels.

The specific directives have provisions for the periodic monitoring of the use of food additives. The EU monitoring system is based on recommendations given in the report of the Working Group on the Development of Methods for Monitoring Intake of Food Additives in the EU, Task 4.2 of the Scientific Cooperation on Questions Relating to Food (published on the Commission website, http://europa.eu.int/comm/food/fs/sfp/addit_flavor/flav15_en.pdf).

The regulation of food additives is now governed by Regulation (EC) No. 1333/2008 of 16 December 2008 on food additives. The authorization of a new food additive shall now follow Regulation (EC) No. 1331/2008 of 16 December 2008 establishing a common authorization procedure for food additives, food enzymes, and food flavorings.

The review of published data on the intake of intense sweeteners in the EU up to 1997 indicated that their average intakes were below the relevant ADI values. The intakes by the highest consumers of sweeteners other than cyclamate were also below their ADIs. The highest estimated intakes of cyclamate by diabetics and children were close to or slightly above the ADI (Renwick 1999).

Studies on the intake of intense sweeteners in different countries of the EU published since 1999 indicate that the average and 95th percentile intakes of acesulfame K, aspartame, cyclamate, and saccharin by adults are below the relevant ADIs (Renwick 2006; Leth et al. 2008).

Bulk sweeteners (polyols) permitted in the EU are not included in the EU monitoring system. Monitoring of the exposure to bulk sweeteners is not feasible, as no upper limits for their use has been specified. They are also authorized for purposes other than sweetening in all foods where additives may be used, except for beverages other than liqueurs.

14.7 RISK EVALUATIONS OF THE SWEETENERS PERMITTED FOR FOOD USE IN THE EU

The SCF opinions, the JECFA reports, and monographs for sweeteners are publicly available (http://europa.eu/comm./food/fs/sc/scf/outcome_en.html; http://www.who.int/foodsafety/chem/jecfa/publications/en/index.html). The EFSA opinion on sweeteners in which safety was evaluated by the AFC or ANS Panels are published in the *EFSA Journal* and can be found through the EFSA home page (www.efsa.europa.eu).

The summary of the safety evaluations in form of the ADIs of the sweeteners authorized for food use in the EU is presented in Table 14.1.

14.7.1 Intense Sweeteners

14.7.1.1 Acesulfame Potassium (E950)

The SCF expressed its opinion on acesulfame K for the first time in 1984 and established an ADI of 0–9 mg/kg bw/day (SCF 1985). The SCF considered the highest level of acesulfame K, tested in rats and dogs at 3% of the diet, as the NOAEL equivalent to 1500 mg/kg bw/day in rats and 900 mg/kg bw/day in dogs. The ADI was established based on data from the study, in which dogs were found to be the most sensitive species. Subsequently, the safety of acesulfame K was updated by the Committee in 1991 (SCF 1992) and in 2000 (SCF 2000a). On both occasions, the SCF was asked to consider whether the 2-year study in rats (instead of the 2-year study in dogs) could be considered a basis for the ADI. However, taking into account previously available and new toxicokinetic data in various species, including man, and based on limited evidence of toxicokinetics similarity between humans and dogs and the observation that, for the same total daily dose, the plasma peak concentration in dogs is several folds higher than for rats, the SCF considered that dogs remained the appropriate species on which to base the ADI and reaffirmed its previous ADI (SCF 1992, 2000a). Additionally in 2000, the SCF considered new mutagenicity studies and claims that the old long-term studies indicated that acesulfame K had a carcinogenic potential. The SCF found that such claims could not be substantiated on the basis of the available data and maintained the ADI of 0–9 mg/kg bw/day (SCF 2000a).

The JECFA evaluated acesulfame K for the first time in 1981 (JECFA 1981a,b). No ADI was allocated because of shortcomings in the long-term/carcinogenicity studies in mice and rats. In 1983, an ADI of 0–9 mg/kg bw/day was established based on the 2-year dog study (JECFA 1983a,b), which was also considered the pivotal study by the SCF (as mentioned above). In 1990, the JECFA changed the ADI to 0–15 mg/kg bw/day (JECFA 1991a,b). At the time of the evaluation, the JECFA considered that, because acesulfame K was not metabolized in any species tested, including man, and further studies in rats in which repeated doses were given did not reveal any induction of metabolism or change in pharmacokinetic behavior, rats appeared to be an appropriate model for humans. Consequently, the JECFA decided that the ADI should be based on the no-observed-effect-level (NOEL) in rats, which was 1500 mg/kg bw/day, since the 2-year study in rats represented a greater portion of the life span of the species than the 2-year study in dogs, and the rat study included exposure *in utero* (i.e., from conception and throughout gestation).

14.7.1.2 Aspartame (E951)

The SCF expressed its opinion on aspartame and established an ADI of 0–40 mg/kg bw/day in 1984 (SCF 1985). In 1988, the SCF evaluated new data on the effects of aspartame on blood and tissue levels of phenylalanine and the possibility of behavioral and other neurotoxic effects due to the consumption of aspartame. The ADI was maintained (SCF 1989). In 1997, the SCF examined a report alleging a connection between aspartame and increases in the incidence of brain tumors in the U.S. (Olney et al. 1996) and concluded that there was no new evidence to justify a reevaluation of aspartame (SCF 1997). In 2002, SCF carried out a further review of all the original and more recent data on aspartame and concluded that there was no need to revise the previously established ADI (SCF 2002). The EFSA AFC and ANS Panels have evaluated, on the request of the European Commission, two long-term carcinogenicity studies in rats on aspartame conducted by the European Rammazzini Foundation of Oncology and Environmental Sciences (ERF; Soffritti et al. 2005, 2006, 2007). In both cases, the Panels concluded that, on the basis of all the evidence available from the ERF studies, other new data emerged since 2002 on aspects other than carcinogenicity, as well as previous evaluations that there was no reason to revise the previously established ADI for aspartame (EFSA 2006b, 2009a,b). In 2010, a carcinogenicity study in mice on aspartame (Soffritti et al. 2010) and an epidemiological study on the association between intakes of artificially sweetened soft drinks and increased incidence of preterm delivery (Halldorsson et al. 2010) were published. Although the epidemiological study did not specifically address aspartame, as intense sweeteners other than aspartame are also used in soft drinks, the authors speculated that exposure to aspartame or its metabolite methanol could be a causative factor for premature deliveries. Subsequently, EFSA, in its statement from 2011, concluded that these studies gave no reason to reconsider previous evaluations of aspartame or of other sweeteners currently authorized for food use in the EU (EFSA 2011).

The JECFA evaluated aspartame in 1980 and allocated an ADI of 0–40 mg/kg bw/day based on the estimated level causing no toxicological effects in rats (JECFA 1980a,b). In 1981, the JECFA evaluated an additional long-term study in rats on aspartame and the diketopiperazine impurity and further biochemical studies of aspartame in humans. The ADI of 40 mg/kg bw/day was confirmed (JECFA 1981a,b).

14.7.1.3 Cyclamate (E952)

The mixture of 10 parts of cyclamate and one part of saccharine was widely used in foods and beverages during the 1960s. In 1969, however, cyclamate was prohibited in many countries, because bladder tumors were found in rats fed with the 10:1 cyclamate–saccharin mixture (Price et al. 1970). Since then, several additional toxicity and carcinogenicity studies were conducted with cyclamate, the cyclamate–saccharin mixture, and the cyclamate metabolite cyclohexylamine (CHA). These studies were considered negative with regard to the carcinogenic effect of both cyclamate and CHA (Bopp et al. 1986).

The SCF expressed its opinion on cyclamate and its two metabolites (CHA and dicyclohexylamine) in 1984 and established a temporary ADI of 0–11 mg/kg bw/day expressed as cyclamic acid, for cyclamic acid, and its calcium and sodium salts (SCF 1985). The ADI was based on a NOAEL of 100 mg/kg bw/day for CHA, a metabolite produced by the intestinal microflora, with respect to testicular toxicity in a 90-day toxicity study in rats. An uncertainty factor of 100 was used for CHA, and a conversion factor of 11 was applied to account for the amount of CHA formed from cyclamate. The ADI was made temporary because of uncertainties relating to the relevance for man of the testicular damage found in rats treated with CHA. Thereafter, the committee reviewed cyclamate on several occasions, when additional data became available (SCF 1989, 1992, 1997). Each time, the temporary ADI of 0–11 mg/kg bw/day was maintained. Finally, in 2000, the SCF removed the temporary status and established an ADI of 0–7 mg/kg bw/day, expressed as cyclamic acid, for cyclamic acid, and its calcium and sodium salts (SCF 2000b).

The JECFA established a temporary ADI for cyclamate of 0–4 mg/kg bw/day based on a no-effect level for CHA (free base) of 74 mg/kg bw/day with respect to testicular atrophy in 1977 (JECFA 1977). The JECFA used an uncertainty factor of 200 to establish a temporary ADI for CHA, a factor of 2 to convert the ADI for CHA to an ADI for cyclamic acid if all cyclamic acid was converted into

CHA, and, finally, a factor of 5.5 to adjust for the fact that only 18% of the ingested cyclamic acid would be converted to CHA by the gut microflora in humans. Following a review of new data in 1980, the JECFA maintained the temporary ADI of 0–4 mg/kg bw/day (JECFA 1980a). Finally, in 1982, an ADI of 0–11 mg/kg bw/day expressed as cyclamic acid, for cyclamic acid, and its calcium and sodium salts were allocated based on the new data (JECFA 1982a,b). This ADI was based on the NOAEL of 100 mg/kg bw/day for CHA and adjusted for the percentage of absorption of cyclamate by humans and the human conversion rate of cyclamate to CHA.

14.7.1.4 Neohesperidin Dihydrochalcone (E959)

The SCF expressed its opinion on neohesperidin dihydrochalcone (DC) for the first time in 1984. The committee concluded that the compound was not toxicologically acceptable due to the lack of data (SCF 1985). However, in 1988, the SCF established an ADI for neohesperidin DC of 0–5 mg/kg bw/day (SCF 1989).

The sweetener has not been evaluated by the JECFA.

14.7.1.5 Neotame (E961)

A safety evaluation of neotame was performed by the EFSA in 2007. The AFC Panel of EFSA established an ADI for neotame of 0–2 mg/kg bw/day based on the application of a safety factor (at present referred by the EFSA as an uncertainty factor) to a NOAEL of 200 mg/kg bw/day from a 52-week dog study for an increase in serum alkaline phosphatase, which was considered the critical end point by the AFC Panel (EFSA 2007).

Neotame was evaluated by the JECFA on its 61st meeting in June 2003. The JECFA established an ADI of 0–2 mg/kg bw/day for neotame on the basis of a NOEL of 200 mg/kg bw/day in a 1-year study in dogs and a 100-fold safety factor (JECFA 2003a,b). According to the JECFA, the only consistent treatment-related effect observed was an increase in serum alkaline phosphatase activity in the 13-week and 1-year studies in dogs fed neotame in the diet. While the increase in alkaline phosphatase was moderate, reversible, and not accompanied by other evidence of liver toxicity, the observed change was reproducible, of high statistical significance, and treatment related. The JECFA agreed that the data were insufficient to discount this effect and, therefore, accepted the dog as the most sensitive species with a NOEL for neotame of 200 mg/kg bw/day on the basis of the 1-year study in dogs.

14.7.1.6 Saccharin (E954)

There has been controversy over the safety of saccharin in the past. Rat feeding studies indicated that saccharin at high doses produced tumors in the bladder of male rats (SCF 1977). Since then, several animal studies have provided information on the mechanisms behind this carcinogenic response in the male rat and demonstrated no carcinogenic effect of saccharin in other animal species. Furthermore, extensive research on the human population has established no association between saccharin and cancer.

Saccharin and its sodium, potassium, and calcium salts were first evaluated by the SCF in 1977 when a temporary ADI of 0–2.5 mg/kg bw/day was allocated (SCF 1977). The SCF again reviewed saccharin in 1984 and decided to maintain the temporary ADI set in 1977 until the questions concerning the mechanism and relevance of the male rat bladder tumors would be clarified by new data (SCF 1985). Following the submission of new data and a request from the industry for a reevaluation of the temporary ADI, the SCF considered saccharin in 1988 (SCF 1989). The temporary ADI was not changed. First, in 1995, the SCF removed the temporary status and established an ADI for sodium saccharin of 0–5 mg/kg bw/day (which is 0–3.8 mg/kg bw/day when the ADI is expressed as the free acid; SCF 1995). At that time, the SCF considered both previous and new experimental information available and the extensive epidemiological data with no evidence of any relationship between saccharin intake and bladder cancer in humans. Saccharin is not genotoxic but is a urinary bladder carcinogen in male rats when given in very large doses. Although it is unlikely that this carcinogenic effect is relevant in humans, the ADI has been based on the NOEL of 500 mg/kg bw/day for male rats and a safety factor of 100.

The JECFA has reviewed the safety of saccharin on several occasions. In 1967 and 1974, the review of the safety of the sweetener resulted in an unconditional ADI of 0–5 mg/kg bw/day and a conditional ADI of 0–15 mg/kg bw/day for dietetic purposes only (JECFA 1967, 1974). In 1977, the ADI was changed to a temporary ADI of 0–2.5 mg/kg bw/day, and the conditional ADI was abolished (JECFA 1977). This action was based on the results of studies demonstrating that high doses of saccharin induced bladder tumors in male rats, but no such effect has been found in other animal species. Following the evaluation of new data in 1980 and 1982, the temporary ADI was extended (JECFA 1980a, 1982a,b). The temporary ADI was confirmed in 1984 based on a no-effect level of 1% in the diet (equivalent to 500 mg/kg bw/day in rats and a safety factor of 200; JECFA 1984a,b). In 1993, a review of new experimental data and epidemiological data that provided evidence of no relationship between saccharin intake and bladder cancer in humans resulted in an ADI of 0–5 mg/kg bw/day for saccharin and its calcium, potassium, and sodium salts based on the NOAEL of 500 mg/kg bw/day in a two-generation long-term feeding study in rats and a safety factor of 100 (JECFA 1993a,b).

14.7.1.7 Steviol Glycosides (E960)

Steviol glycosides, accepted in the EU as a food additive, are mixtures of steviol glycosides that comprise not less than 95% of stevioside and/or rebaudioside A, which can be extracted from the leaves of the *Stevia rebaudiana* Bertoni plant. Safety evaluation of steviol glycosides was performed by the EFSA in 2010. The ANS Panel of EFSA established an ADI for steviol glycosides, expressed as steviol equivalents, of 4 mg/kg bw/day based on the application of a 100-fold uncertainty factor to the NOAEL in a 2-year carcinogenicity study in rats (EFSA 2010a).

In the past, stevioside, one of the major glycosides in the leaves of the *S. rebaudiana* Bertoni plant, was evaluated as a sweetener by the SCF on several occasions (SCF 1985, 1989, 1999a). The SCF concluded that the use of stevioside was toxicologically not acceptable due to insufficient available data to assess its safety.

The JECFA evaluated toxicological data on steviol glycosides (stevioside and the aglycone steviol) in 1998 at its 51st meeting (JECFA 1998), and the need for further data was expressed. At the 63rd meeting, the JECFA reviewed new data and information and established tentative specifications for the material referred to as "steviol glycosides" (JECFA 2004a,b). The additional biochemical and toxicological data provided the basis for the allocation of a temporary ADI of 0–2 mg/kg bw/day for steviol glycosides expressed as steviol. The temporary ADI was based on the NOEL of 2.5% stevioside in the diet, equal to 970 or 383 mg/kg bw/day expressed as steviol, in a 2-year study in rats using an uncertainty factor of 200. On the same occasion, the JECFA specified the need for studies with repeated exposure of normotensive and hypotensive individuals and patients with insulin-dependent and non insulin-dependent diabetes. On the 68th meeting of the committee, a subsequent review of additional data did not raise concerns regarding the safety of steviol glycosides. However, the temporary ADI was maintained, pending submission of the results of ongoing clinical studies (JECFA 2007). The tentative specifications were revised (JECFA 2008a). Finally, the results of the finalized clinical studies were found by the JECFA sufficient to allow the removal of the temporary status, and a full ADI of 0–4 mg/kg bw/day, expressed as steviol, was established (JECFA 2008b,c).

14.7.1.8 Sucralose (E955)

The SCF expressed its first opinion on sucralose in 1989 (SCF 1989). At that time, the SCF considered sucralose to be toxicologically unacceptable, as several outstanding questions emerged from the evaluation of the available data. In 2000, the SCF considered further studies and established an ADI of 0–15 mg/kg bw (SCF 2000c).

The JECFA reviewed the safety of sucralose on two occasions. At its 33rd meeting, the committee allocated a temporary ADI of 0–3.5 mg/kg bw/day based on a NOEL of 750 mg/kg bw/day in a 1-year dog study and an uncertainty factor of 200 (JECFA 1988a,b). The JECFA requested further data on absorption and metabolism in humans after prolonged oral dosing, studies to show that sucralose produced no adverse effects on individuals with insulin-dependent and maturity-onset diabetes, studies on the elimination of sucralose from pregnant animals and the fetus to exclude bioaccumulation, and a short-term study on one of the potential intermediates in the metabolism of sucralose. Further studies on sucralose were evaluated by the JECFA in 1991 (JECFA 1991a,b), and an ADI of 0–15 mg/kg bw/day was allocated based on a NOAEL of 1500 mg/kg bw/day from a long-term study in rats, which included *in utero* exposure, and an uncertainty factor of 100.

14.7.1.9 Thaumatin (E957)

The SCF expressed its opinion on thaumatin for the first time in 1984, found the sweetener temporarily acceptable, and requested additional data on possible receptor binding and endocrine activity (SCF 1985). After considering the additional data, the SCF found the sweetener acceptable (SCF 1989).

The JECFA evaluated thaumatin in 1983 (JECFA 1983a), but no ADI was allocated at that time. In 1985, the JECFA allocated an ADI not specified (JECFA 1985a,b).

14.7.2 Bulk Sweeteners (Polyols)

All the polyols authorized for food use as sweeteners in the EU have been allocated an ADI acceptable by the SCF or the AFC Panel (Table 14.1).

The SCF expressed its opinion on most polyols that are used as food additives in the EU in 1984 (SCF 1985). The SCF considered additional information on isomalt (E953) and lactitol (E966) in 1988 (SCF 1989). The SCF evaluated maltitol syrup with a new specification in 1999 and found the continued use of the compound acceptable (SCF 1999b). In 2006, the AFC Panel expressed its opinion on a new production method for maltitol (EFSA 2006a). The SCF published an opinion on erythritol (E968) in 2003 (SCF 2003). In 2010, the ANS Panel provided a scientific opinion on the safety of erythritol in light of a pediatric study on the gastrointestinal (GI) tolerability of the compound. The ANS Panel concluded that there was a safety concern with respect to GI tolerability for the use of erythritol in beverages at a maximum use level of 2.5% for nonsweetening purposes (EFSA 2010b).

The JECFA has evaluated polyols on several occasions. At present, all the polyols used as sweeteners and evaluated by the JECFA have received an ADI not specified (Table 14.1).

Erythritol (E968) was evaluated by the JECFA at its 53rd meeting in 1999 (JECFA 1999a,b).

Isomalt (E953) was evaluated on two occasions. In 1981, isomalt was allocated a temporary ADI of 0–25 mg/kg bw/day based on a NOEL with respect to laxation (JECFA 1981a,b), but this ADI was later changed to "not specified" (JECFA 1985a,b).

Lactitol (E966) was evaluated in 1983 (JECFA 1983a,b).

Maltitol (E965) was evaluated several times. In 1980, the available information was found inadequate for evaluation (JECFA 1980a). In 1983, a temporary ADI of 0–25 mg/kg bw/day was allocated for hydrogenated glucose syrup, with maltitol being a main component of 50%–90% of the product. At the same time, the JECFA requested that data from a lifetime feeding study should be made available (JECFA 1983a). Then, in 1985, the ADI for maltitol was changed to "not specified," and the JECFA found that the previously requested lifetime study was not needed (JECFA 1985a,b). Finally, in 1988, the JECFA changed the name hydrogenated glucose syrup to maltitol syrup, and the ADI not specified was confirmed and also extended to cover maltitol as well (JECFA 1988a). In 1993, the JECFA reviewed a long-term carcinogenicity study in rats and confirmed the existing ADI (JECFA 1993a,b). Finally, in 1997, the ADI not specified was confirmed, and the JECFA concluded that this ADI could be applied to maltitol syrup with the revised specifications (JECFA 1997a,b).

Mannitol (E421) was evaluated by the JECFA on several occasions. A temporary ADI of 0–50 mg/kg bw/day was allocated in 1974 (JECFA 1974). This ADI was retained in 1976 (JECFA 1976) and extended until 1986 (JECFA 1985a), when finally an ADI not specified was allocated (JECFA 1986a,b).

Sorbitol (E420) was evaluated by the JECFA several times. The first evaluation led to the allocation of an ADI not specified (JECFA 1973a,b), which was changed to a temporary ADI not specified in 1978 and confirmed in 1980 (JECFA 1978a,b, 1980a,b). Finally, an ADI not specified was established in 1982 (JECFA 1982a).

Xylitol (E967), when evaluated in 1977 (JECFA 1977) and in 1978, was not allocated an ADI, as the data were insufficient (JECFA 1977). In 1983, an ADI not specified was established (JECFA 1983a,b).

Polyol bulk sweeteners have a laxative effect when consumed in excessive amounts due to the osmotic effects of unabsorbed polyol reaching the colon. Polyols differ in the potency to cause laxation. For instance, human tolerance studies demonstrated a laxative effect of mannitol at intake levels as low as 10–20 g/day (SCF 1985), of maltitol at 30–50 g/day (SCF 1985), of lactitol at about 50 g/day (SCF 1989), and of sorbitol and xylitol at intake levels above 50 g/day (SCF 1984). Erythritol causes a laxative effect at higher doses than other polyols (SCF 2003; JECFA 1999b). Consumption on the order of 20 g/person/day of polyols is unlikely to cause undesirable laxative symptoms. In accepting the continued use of polyols (ADI acceptable), the SCF emphasized that this should not be interpreted as the acceptance of unlimited use in all foods at any technological level but that the laxative effect should be borne in mind (SCF 1985, 1989). Similarly, the JECFA considered that the fact that polyols exert a laxative effect in man and animals at high doses should be taken into account when considering appropriate levels of the use of polyols alone and in combination (JECFA 1983a, 1985a, 1986a).

14.8 REEVALUATION OF THE SAFETY OF SWEETENERS

The risk assessment of sweeteners (and other food additives) is based on the knowledge and the data available at the time of assessment. When new toxicity data become available in the scientific literature, national experts and international expert committees consider them with caution, and the reevaluation of the safety may be performed. Depending on the outcome, three scenarios are possible: the ADI can be sustained, the ADI can be changed, or the use of a sweetener as a food additive can be found unacceptable.

Although there was no provision for periodic reviews of the safety of the permitted food additives in the past, some of the sweeteners were evaluated by the SCF/EFSA several times, when new data or new requests for reevaluation were acquired (i.e., cyclamate, saccharin, and aspartame).

In December 2008, a new regulatory package on food improvement agents was adopted by the European Parliament. Regulation (EC) No. 1333/2008 requires that food additives are subject to safety evaluation by the EFSA before they are permitted for use in the EU. In addition, it is foreseen that food additives must be kept under continuous observation and must be reevaluated by the

EFSA. For this purpose, a program for the reevaluation of food additives that were already permitted in the EU before 20 January 2009 has been set up under Regulation (EU) No. 257/2010. This regulation also foresees that food additives are reevaluated whenever necessary in light of changing conditions of use and new scientific information. In accordance with this legislation, EFSA has to reevaluate the sweeteners by 31 December 2020.

14.9 CONCLUDING REMARKS

All sweeteners permitted for use in foods have been subjected to risk assessment. The use of the sweeteners, authorized as food additives in food products, has been regulated on the basis of the ADI. This regulation has ensured that the amounts permitted in foods of a given food additive would not result in the daily consumption of amounts exceeding the ADI. The consumption of sweeteners in quantities within the ADI does not constitute a health hazard to consumers.

REFERENCES

Barlow, S., Pascal, G., Larsen, J.C., and Richold, M.R. 1999. ILSI Europe workshop on the significance of excursion of intake above the acceptable daily intake (ADI). *Regul. Toxicol. Pharmacol.* 30: S1–S121.

Barnes, D., Daston, G., Evans, J. et al. 1995. Benchmark dose workshop: Criteria of use of a benchmark dose to estimate a reference dose. *Regul. Toxicol. Pharmacol.* 21: 296–306.

Bopp, A.B., Sonders, R.C., and Kesterson, J.W. 1986. Toxicological aspects of cyclamate and cyclohexylamine. *Crit. Rev. Toxicol.* 16: 213–306.

Crump, K.S. 1984. A new method for determining allowable daily intakes. *Fundam. Appl. Toxicol.* 32: 133–153.

European Food Safety Authority (EFSA). 2006a. Opinion of the Scientific Panel on Food Additives, Flavourings, Processing Aids and Materials in Contact with Food (AFC) on the request from the Commission related to maltitol syrup E965 (ii) new production process. *EFSA J.* 354: 1–7.

European Food Safety Authority (EFSA). 2006b. Opinion of the Scientific Panel on Food Additives, Flavourings, Processing Aids and Materials in Contact with Food (AFC) on the request from the Commission related to a new long-term carcinogenicity study on aspartame. *EFSA J.* 356: 1–44.

European Food Safety Authority (EFSA). 2007. Scientific opinion of the Panel on Food Additives, Flavourings, Processing Aids and Materials in Contact with Food (AFC) on the request from the European Commission on neotame as a sweetener and flavour enhancer. *EFSA J.* 581: 1–43.

European Food Safety Authority (EFSA). 2009a. Scientific opinion of the Panel on Food Additives and Nutrient Sources Added to Food on a request from the European Commission related to the 2nd ERF carcinogenicity study on aspartame. *EFSA J.* 945: 1–18.

European Food Safety Authority (EFSA). 2009b. Updated scientific opinion of the Panel on Food Additives and Nutrient Sources Added to Food on a request from the European Commission related to the 2nd ERF carcinogenicity study on aspartame taking into consideration study data submitted by the Ramazzini Foundation in February 2009. *EFSA J.* 1015: 1–18.

European Food Safety Authority (EFSA). 2009c. Guidance of the Scientific Committee: Use of the benchmark dose approach in risk assessment. *EFSA J.* 1150: 1–72.

European Food Safety Authority (EFSA). 2010a. Scientific opinion of the Panel on Food Additives and Nutrient Sources (ANS) on the safety of steviol glycosides for the proposed uses as a food additive. *EFSA J.* 8: 1537 [85 pp.].

European Food Safety Authority (EFSA). 2010b. Statement of the Panel on Food Additives and Nutrient Sources (ANS) in relation to the safety of erythritol (E 968) in light of new data, including a new paediatric study on the gastrointestinal tolerability of erythritol. *EFSA J.* 8: 1650. [17 pp.].

European Food Safety Authority (EFSA). 2011. EFSA statement on the scientific evaluation of two studies related to the safety of artificial sweeteners. *EFSA J.* 9: 2089 [16 pp.].

Gad, S.C. 1990. Recent development in replacing, reducing, and refining animal use in toxicologic research and testing. *Fundam. Appl. Toxicol.* 15: 8–16.

Halldorsson, T.I., Strøm, M., Petersen, S.B., and Olsen, S.F. 2010. Intake of artificially sweetened soft drinks and risk of preterm delivery: A prospective cohort study in 59,334 Danish pregnant women. *Am. J. Clin. Nutr.* 92: 626–633.

Joint FAO/WHO Expert Committee on Food Additives (JECFA). 1967. Specifications for the identity and purity of food additives and their toxicological evaluation: Some flavouring substances and nonnutritive sweetening agents—Eleventh report of the Joint FAO/WHO Expert Committee on Food Additives FAO Nutrition Meetings Series No. 44, 1968; WHO Technical Report Series 383, 1968, and Corrigendum. [1968, NMRS 53/TRS 383-JECFA 11].

Joint FAO/WHO Expert Committee on Food Additives (JECFA). 1973a. Toxicological evaluation of certain food additives, with a review of general principles and of specifications—Seventeenth report of the Joint FAO/WHO Expert Committee on Food Additives. FAO Nutrition Meetings Series No. 53, 1974; WHO Technical Report Series 553, 1974, and Corrigendum. [1973, NMRS 53/TRS 539-JECFA 17].

Joint FAO/WHO Expert Committee on Food Additives (JECFA). 1973b. Toxicological evaluation of some food additives including anticaking agents, antimicrobials, antioxidants, emulsifiers, and thickening agents. FAO Nutrition Meeting Report Series No. 53A, 1974; WHO Food Additives Series No.5, 1974. [1973, FAS 5/NMRS 53A-JECFA 17].

Joint FAO/WHO Expert Committee on Food Additives (JECFA). 1974. Evaluation of certain food additives. Eighteenth Report of the Joint FAO/WHO Expert Committee on Food Additives. FAO Nutrition Meetings Series No. 54. WHO Technical Report Series 557, 1974, and Corrigendum. [1974, NMRS 54/TRS 557-JECFA 18].

Joint FAO/WHO Expert Committee on Food Additives (JECFA). 1976. Evaluation of certain food additives—Twentieth report of the Joint FAO/WHO Expert Committee on Food Additives. FAO Nutrition Meeting Series No. 54, 1974. WHO Technical Report Series 559, 1976. [1976, FNS 1/TRS 559-JECFA 20].

Joint FAO/WHO Expert Committee on Food Additives (JECFA). 1977. Evaluation of certain food additives—Twenty-first report of the Joint FAO/WHO Expert Committee on Food Additives. WHO Technical Report Series 617, 1978. [1977, TRS 617-JECFA 21].

Joint FAO/WHO Expert Committee on Food Additives (JECFA). 1978a. Evaluation of certain food additives and contaminants—Twenty-second report of the Joint FAO/WHO Expert Committee on Food Additives. WHO Technical Report Series 631, 1978. [1978, TRS 631-JECFA 22].

Joint FAO/WHO Expert Committee on Food Additives (JECFA). 1978b. Summary of toxicological data of certain food additives and contaminants. WHO Food Additives Series No. 13, 1978. [1978, FAS 13-JECFA 22].

Joint FAO/WHO Expert Committee on Food Additives (JECFA). 1980a. Evaluation of certain food additives—Twenty-fourth report of the Joint FAO/WHO Expert Committee on Food Additives. WHO Technical Report Series 653, 1980. [1980, TRS 653-JECFA 24].

Joint FAO/WHO Expert Committee on Food Additives (JECFA). 1980b. Toxicological evaluation of certain food additives. WHO Food Additives Series No. 15, 1980. [1980, FAS 15-JECFA 24].

Joint FAO/WHO Expert Committee on Food Additives (JECFA). 1981a. Evaluation of certain food additives—Twenty-fifth report of the Joint FAO/WHO Expert Committee on Food Additives. WHO Technical Report Series 669, 1981. [1981, TRS 669-JECFA 25].

Joint FAO/WHO Expert Committee on Food Additives (JECFA). 1981b. Toxicological evaluation of certain food additives. WHO Food Additive Series No. 16, 1981. [1981, FAS 16-JECFA 25].

Joint FAO/WHO Expert Committee on Food Additives (JECFA). 1982a. Evaluation of certain food additives—Twenty-sixth report of the Joint FAO/WHO Expert Committee on Food Additives. WHO Technical Report Series 683, 1982. [1982, TRS 683-JECFA 26].

Joint FAO/WHO Expert Committee on Food Additives (JECFA). 1982b. Toxicological evaluation of certain food additives. WHO Food Additive Series, 1982, No. 17. [1982, FAS 17-JECFA 26].

Joint FAO/WHO Expert Committee on Food Additives (JECFA). 1983a. Evaluation of certain food additives—Twenty-seventh report of the Joint FAO/WHO Expert Committee on Food Additives. WHO Technical Report Series 696, 1983, and Corrigendum. [1983, TRS 696-JECFA 27].

Joint FAO/WHO Expert Committee on Food Additives (JECFA). 1983b. Toxicological evaluation of certain food additives. WHO Food Additive Series No. 18. [1983, FAS 18-JECFA 27].

Joint FAO/WHO Expert Committee on Food Additives (JECFA). 1984a. Evaluation of certain food additives and contaminants—Twenty-eighth report of the Joint FAO/WHO Expert Committee on Food Additives. WHO Technical Report Series 710, 1984, and Corrigendum. [1984, TRS 710-JECFA 28].

Joint FAO/WHO Expert Committee on Food Additives (JECFA). 1984b. Toxicological evaluation of certain food additives. WHO Food Additive Series No. 19, 1984. [1984, FAS 19-JECFA 28].

Joint FAO/WHO Expert Committee on Food Additives (JECFA). 1985a. Evaluation of certain food additives and contaminants—Twenty-ninth report of the Joint FAO/WHO Expert Committee on Food Additives. WHO Technical Report Series 733, 1986, and Corrigendum. [1985, TRS 733-JECFA 29].

Joint FAO/WHO Expert Committee on Food Additives (JECFA). 1985b. Toxicological evaluation of certain food additives and contaminants. WHO Food Additive Series No. 20, 1987. [1985, FAS 20-JECFA 29].

Joint FAO/WHO Expert Committee on Food Additives (JECFA). 1986a. Evaluation of certain food additives and contaminants—Thirtieth report of the Joint FAO/WHO Expert Committee on Food Additives. WHO Technical Report Series 751, 1987, and Corrigendum. [1986, TRS 751-JECFA 30].

Joint FAO/WHO Expert Committee on Food Additives (JECFA). 1986b. Toxicological evaluation of certain food additives and contaminants. WHO Food Additive Series No. 21, 1987. [1986, FAS 21-JECFA 30].

Joint FAO/WHO Expert Committee on Food Additives (JECFA). 1988a. Toxicological evaluation of certain food additives and contaminants—Thirty-third report of the Joint FAO/WHO Expert Committee on Food Additives. WHO Technical Report Series 776, 1989, and Corrigendum. [1988, TRS 776-JECFA 33].

Joint FAO/WHO Expert Committee on Food Additives (JECFA). 1988b. Toxicological evaluation of certain food additives and contaminants. WHO Food Additive Series No. 24, 1989. [1989, FAS 24-JECFA 33].

Joint FAO/WHO Expert Committee on Food Additives (JECFA). 1991a. Evaluation of certain food additives—Thirty-seventh report of the Joint FAO/WHO Expert Committee on Food Additives. WHO Technical Report Series 806, 1991, and Corrigendum. [1991, TRS 806-JECFA 37].

Joint FAO/WHO Expert Committee on Food Additives (JECFA). 1991b. Toxicological evaluation of certain food additives. WHO Food Additive Series No. 28, 1991. [1991, FAS 28-JECFA 37].

Joint FAO/WHO Expert Committee on Food Additives (JECFA). 1993a. Evaluation of certain food additives—Forty-first report of the Joint FAO/WHO Expert Committee on Food Additives. WHO Technical Report Series 837, 1993. [1993, TRS 837-JECFA 41].

Joint FAO/WHO Expert Committee on Food Additives (JECFA). 1993b. Toxicological evaluation of certain food additives. WHO Food Additive Series No. 32, 1993. [1993, FAS 32-JECFA 41].

Joint FAO/WHO Expert Committee on Food Additives (JECFA). 1997a. Evaluation of certain food additives and contaminants—Forty-ninth report of the Joint FAO/WHO Expert Committee on Food Additives. WHO Technical Report Series 884, 1999. [1997, TRS 884-JECFA 49].

Joint FAO/WHO Expert Committee on Food Additives (JECFA). 1997b. Toxicological evaluation of certain food additives. WHO Food Additive Series No. 40, 1998 [1997, FAS 40-JECFA 49].

Joint FAO/WHO Expert Committee on Food Additives (JECFA). 1998. Evaluation of certain food additives—Fifty-first report of the Joint FAO/WHO Expert Committee on Food Additives. Geneva, Switzerland. WHO Technical Report Series No. 891, 2000. [1998, TRS 891-JECFA 51].

Joint FAO/WHO Expert Committee on Food Additives (JECFA). 1999a. Evaluation of certain food additives and contaminants—Fifty-third report of the Joint FAO/WHO Expert Committee on Food Additives. WHO Technical Report Series No. 896. Geneva, 2000. [1999, TRS 896-JECFA 53].

Joint FAO/WHO Expert Committee on Food Additives (JECFA). 1999b. Safety evaluation of certain food additives and contaminants. WHO Food Additives Series 44, Geneva, 2000. [1999, FAS 44-JECFA 53].

Joint FAO/WHO Expert Committee on Food Additives (JECFA). 2003a. Evaluation of certain food additives and contaminants—Sixty-first report of the Joint FAO/WHO Expert Committee on Food Additives. WHO Technical Report Series No. 922, 2004. [2003, TRS 922-JECFA61].

Joint FAO/WHO Expert Committee on Food Additives (JECFA). 2003b. Safety evaluation of certain food additives. WHO Food Additives Series No. 52, 2004. [2003, FAS 52-JECFA61].

Joint FAO/WHO Expert Committee on Food Additives (JECFA). 2004a. Evaluation of certain food additives—Sixty-third report of the Joint FAO/WHO Expert Committee on Food Additives. WHO Technical Report Series No. 928, 2005. Geneva, Switzerland. [2004, TRS 928-JECFA63].

Joint FAO/WHO Expert Committee on Food Additives (JECFA). 2004b. Safety evaluation of certain food additives. WHO Food Additives Series No. 54, 2005. [2004, FAS 54-JECFA63].

Joint FAO/WHO Expert Committee on Food Additives (JECFA). 2007. Evaluation of certain food additives—Sixty-eight report of the Joint FAO/WHO Expert Committee on Food Additives. WHO Technical Report Series 947, 2007. [2007, TRS 947-JECFA 68].

Joint FAO/WHO Expert Committee on Food Additives (JECFA). 2008a. Monograph 5: Steviol glycosides. Compendium on Food Additive Specifications. Available at http://www.fao.org/ag/agn/jecfa-additives/details.html?id=898.

Joint FAO/WHO Expert Committee on Food Additives (JECFA). 2008b. Evaluation of certain food additives—Sixty-ninth report of the Joint FAO/WHO Expert Committee on Food Additives. WHO Technical Report Series 952, 2009. [2008, TRS 952-JECFA 69].

Joint FAO/WHO Expert Committee on Food Additives (JECFA). 2008c. Safety evaluation of certain food additives. WHO Food Additives Series No. 60, 2009. [2008, FAS 60-JECFA 69].

Larsen, J.C. and Pascal, G. 1998. Workshop on applicability of the ADI to infant and children: Consensus summary. *Food Addit. Contam.* 15 (Suppl.): S2–S12.

Larsen, J.C. and Richold, M. 1999. Report of workshop on the significance of excursion of intake above the ADI. *Regul. Toxicol. Pharmacol.* 30: S2–S12.

Leth, T, Jensen, U., Fagt, U., and Andersen, R. 2008. Estimated intake of intense sweeteners from non-alcoholic beverages in Denmark, 2006. *Food Addit. Contam.* 25: 662–668.

Olney, J.W., Farber, N.B., Spitznagel, E., and Robins, L.N. 1996. Increasing brain tumor rates: Is there a link to aspartame? *J. Neuropathol. Exp. Neurol.* 55: 1115–1123.

Price, J.M., Biava, C.G., Oser, B.L., Vogin, E.E., Steinfeld, J., and Ley, H.L. 1970. Bladder tumors in rats fed cyclohehylamine or high doses of a mixture of cyclamate and saccharin. *Science* 167: 1131–1132.

Renwick, A.G. 1993. Data-derived safety factors for the evaluation of food additives and environmental contaminants. *Food Addit. Contam.* 10: 275–305.

Renwick, A.G. 1999. Intake of intense sweeteners. *World Rev. Nutr. Diet.* 85: 178–200.

Renwick, A.G. 2006. The intake of intense sweeteners—An update review. *Food Addit. Contam.* 23: 327–338.

Renwick, A.G. and Lazarus, N.R. 1998. Human variability and noncancer risk assessment—An analysis of the default uncertainty factor. *Regul. Toxicol. Pharmacol.* 27: 3–20.

Renwick, A.G., Barlow, S.M., Hertz-Picciotto, I. et al. 2003. Risk characterization of chemicals in food and diet. *Food Chem. Toxicol.* 41: 1211–1271.

Scientific Committee for Food (SCF). 1977. Saccharin—Opinion expressed on 24 June 1977. Reports from the Scientific Committee for Food (4th series). CB-AH-77-004 EN-C. Available at http://ec.europa.eu/comm/food/fs/sc/scf/reports/scf_reports_04.pdf.

Scientific Committee for Food (SCF). 1980. Guidance for the safety assessment of food additives—Opinion expressed on 22 February 1980. Reports of the Scientific Committee for Food (10th series). EUR 6892. Luxembourg: Commission of the European Communities. Available at http://ec.europa.eu.comm/food/fs/sc/scf/reports/scf_reoprts_10.pdf.

Scientific Committee for Food (SCF). 1985. Sweeteners—Opinion expressed on 14 September 1984. Reports of the Scientific Committee for Food (16th series). EUR 10210 EN. Luxembourg: Commission of the European Communities. Available at http://ec.europa.eu/comm/food/fs/sc/scf/reports/scf_reports_16.pdf.

Scientific Committee for Food (SCF). 1989. Sweeteners—Opinion expressed on 11 December 1987 and 10 November 1988. Reports of the Scientific Committee for Food (21st series). EUR 11617 EN. Luxembourg: Commission of the European Communities. Available at http://ec.europa.eu/comm/food/fs/sc/scf/reports/scf_reports_21.pdf.

Scientific Committee for Food (SCF). 1992. Recommendation on cyclamates—Opinion expressed on 21 June 1991. Reports from the Scientific Committee for Food (27th series). EUR 14181 EN. Luxembourg: Commission of the European Communities. Available at http://ec.europa.eu/comm/food/fs/sc/scf/reports/scf_reports_27.pdf.

Scientific Committee for Food (SCF). 1995. Opinion of the Scientific Committee on Food on Saccharin (adopted on 2 June 1995). Available at http://europa.eu.int/comm/food/fs/sc/scf/outcome_en.html.

Scientific Committee for Food (SCF). 1997. Minutes of the 107th Meeting of the Scientific Committee for Food held on 12–13 June 1997 in Brussels. Available at http://europa.eu.int/comm./food/fs/sc/oldcom7/out13_en.htm.

Scientific Committee for Food (SCF). 1999a. Opinion on steviol glycosides as a sweetener (adopted on 17 June 1999). Belgium: European Commission, Health and Consumer Protection Directorate-General, Scientific Committee for Food. [CS/ADD/EDUL/167 Final].

Scientific Committee for Food (SCF). 1999b. Opinion of the Scientific Committee on Food on a maltitol syrup not covered by current specifications (expressed on 2 December 1999). Available at http://europa.eu.int/comm/food/fs/sc/scf/out48_en.html.

Scientific Committee for Food (SCF). 2000a. Opinion on the reevaluation of acesulfame K with reference to the previous SCF opinion of 1991 (expressed on 9 March 2000). SCF 2000. Available at http://europa.eu.int/comm./food/fs/scf/out52_en.pdf.

Scientific Committee for Food (SCF). 2000b. Revised opinion of the Scientific Committee on Food on cyclamic acid and its sodium and calcium salts (expressed on 9 March 2000). Available at http://ec.europa.eu/comm/food/fs/sc/scf/out53_en.pdf.

Scientific Committee for Food (SCF). 2000c. Opinion of the Scientific Committee on Food on sucralose (adopted on 7 September 2000). 2000. Available at http://ec.europa.eu/comm/food/fs/sc/scf/out68_en.pdf.

Scientific Committee for Food (SCF). 2001. Guidance on submission for food additive evaluations—Opinion expressed on 11 July 2001. Available at http://ec.europa.eu/comm/foodfs/sc/scf/out98_en.pdf.

Scientific Committee for Food (SCF). 2002. Opinion of the Scientific Committee on Food: Update on the safety of aspartame (expressed on 4 December 2002). Available at http://europa.eu.int/comm/food/fs/sc/scf/outcome_en.html.

Scientific Committee for Food (SCF). 2003. Opinion of the Scientific Committee on Food on erythritol (expressed on 5 March 2003). Available at http://ec.europa.eu/comm/food/fs/sc/scf/out175_en.pdf.

Setzer Jr., R.W. and Kimmel, C.A. 2003. Topic 3.13: Use of NOAEL, benchmark dose, and other models for human risk assessment of hormonally active substances. *Pure Appl. Chem.* 75: 2151–2158.

Soffritti, M., Belpoggi, F., Esposti, D.D., and Lambertini, L. 2005. Aspartame induces lymphomas and leukaemias in rats. *Eur. J. Oncol.* 10: 107–116.

Soffritti, M., Belpoggi, F., Esposti, D.D., Lambertini, L., Tibaldi, E., and Rigano, A. 2006. First experimental demonstration of the multipotential carcinogenic effects of aspartame administered in the feed to Sprague–Dawley rats. *Environ. Health Perspect.* 114: 379–385.

Soffritti, M., Belpoggi, F., Tibaldi, E., Eposti, D.D., and Lauriola, M. 2007. Life-span exposure to low doses of aspartame beginning during prenatal life increases cancer effects in rats. *Environ. Health Perspect.* 115: 1293–1297.

Soffritti, M., Belpoggi, F., Manservigi, M., Tibaldi, E., Lauriola, M., Falcioni, L., and Bua, L. 2010. Aspartame administered in feed, beginning prenatally through life span, induces cancers of the liver and lung in male Swiss mice. *Am. J. Ind. Med.* 53: 1197–1206.

World Health Organization (WHO). 1987. Environmental Health Criteria 70: Principles for the safety assessment of food additives and contaminants in food. International Programme on Chemical Safety (IPCS). Geneva: World Health Organization.

World Health Organization (WHO). 1994. Environmental Health Criteria No. 170—Assessing human health risk of chemicals: Derivation of guidance values for health-based exposure limits. International Programme on Chemical Safety (IPCS). Geneva: World Health Organization.

World Health Organization (WHO). 2009. Environmental Health Criteria 240: Principles and methods for the risk assessment of chemicals in food—A joint publication of the Food and Agriculture Organization of the United Nations and the World Health Organization.

World Trade Organization (WTO). 1994. Agreement on the application of sanity and phytosanity measures. Committee on Sanity and Phytosanity Measures, World Trade Organization. LT/UR/A-1A/12. 15 April 1994.

Index

Page numbers followed by f and t indicate figures and tables, respectively.

B

C

D

G

H

I

J

K

L

M

N

V

W

X

Y

Z